Environmental Handbook

**Documentation on monitoring and
evaluating environmental impacts**

Environmental Handbook

Documentation on monitoring and evaluating environmental impacts

Volume II
Agriculture, Mining/Energy, Trade/Industry

German Federal Ministry for Economic Cooperation and Development

Bundesministerium für wirtschaftliche Zusammenarbeit und Entwicklung

(BMZ)

Die Deutsche Bibliothek – CIP-Einheitsaufnahme

Environmental handbook : documentation on monitoring and
evaluating environmental impacts / German Federal Ministry
for Economic Cooperation and Development (BMZ). [Transl.:
GTZ Language Services].
  Dt. Ausg. u.d.T.: Umwelt-Handbuch
NE: Deutschland / Bundesministerium für Wirtschaftliche
  Zusammenarbeit und Entwicklung

Vol. 2. Agriculture, mining/energy, trade/industry. – 1995
  ISBN 978-3-663-09950-5   ISBN 978-3-663-09948-2 (eBook)
  DOI 10.1007/978-3-663-09948-2

Publication support and distribution: Friedrich Vieweg & Sohn
Verlagsgesellschaft mbH, Braunschweig
Vieweg is a subsidiary company of the Bertelsmann Professional Information.
Translation: GTZ Language Services
Typesetting: GTZ Wordprocessing Services

ISBN 978-3-663-09950-5

VOLUME 1

VI   *Contents*

VOLUME 2

VOLUME 3 Compendium of Environmental Standards

Agriculture

Plant Production

27. Plant Production

Contents

1. Scope

The following **terms** recur frequently in this environmental brief and therefore require **definition:**

- **Single cropping** involves growing only **one crop on a particular area of land,** e.g. rice. The **sequence in which various single crops are grown one after the other** in a field is known as the **crop rotation.**

- **Intercropping** is a system in which **a number of different crops** grow together for the entire vegetation period or part of it, e.g. a combination of cassava, cowpeas and millet.

- **Annual crops** are generally herbaceous plants with a **one-year vegetation cycle** (e.g. cereals, legumes, various vegetables, tobacco).

- **Perennial crops** are **plants which are used over a number of years**; each plant is sowed or planted only once, e.g. fruit trees, tea, coffee and cocoa.

- **Monoculture** involves **growing a particular crop on the same area of land over a number of cultivation periods**, e.g. sugar cane.

Taking into account the production of wood, self-regenerating raw materials, animal fodder and crops used in the manufacture of semi-luxury goods, **plant production** represents - in terms of area - **man's major form of interference** with the Earth's natural balance.

Traditional farming systems are usually based on intercropping and tend to be **subsistence-oriented**. External inputs such as fertilisers and pesticides are uncommon and are used on only a small scale.

By contrast, **large-scale plantation farming** generally takes the form of **monoculture** (sugar cane, cotton) or **permanent cropping** (coffee, tea, cocoa). These forms of cultivation are market-oriented and dependent on external inputs.

Plant production involves **activities** in areas such as

- plant protection
- agricultural engineering and animal traction
- irrigation
- species and variety selection
- tillage and fertilising

-        crop tending and weed control, harvesting, post-harvest treatment, storage
-        erosion protection and control.

Crops are grown to **meet the needs of the producer or the market.** They also **play
a role in protecting** soil, air and water.

Plant production is carried out on **farms,** for the most part using **family labour,**
in order to **ensure subsistence** and **earn monetary income.**

2.   Environmental Impacts and Protective Measures

In **agroecosystems, man becomes** the dominant element in the ecosystem
(anthropogenically oriented ecosystems). Agroecosystems differ in particular from
natural ecosystems in that **natural regulation processes take second place to
control by man.**

In the natural environment, **plants form part of the ecosystem and play a key
role in preserving it.** Depending on the cropping method used, the nature, intensity
and interaction of cultivation measures give rise to specific **environmental impacts.**
These may cause a **reduction in the diversity of species, disruption of the soil
structure** and **pollution** of the soil, water and air (pesticides, salts resulting from
irrigation and fertilisation, nitrate etc.) **Natural ecosystems** with their wide variety
of functions are **replaced** by artificial land-use systems poor in species.

Growing use of industrially produced **inputs** (fertilisers, pesticides, machines,
energy) and inappropriate cultivation techniques lead to contamination of drinking
water by fertilisers and pesticides, as well as causing soil erosion, desertification and
genetic erosion.

2.1   Environmental impacts

2.1.1   Soil

Soil forms the **basis for plant production** and thus performs a vital function in
guaranteeing human survival.

Soil conservation is **essential** if man's living environment is to be maintained in
a healthy state and a sustained supply of high-quality foods is to be ensured.

Opportunities for changing the conditions prevailing on a particular site are
limited. Cultivation measures must therefore be geared to the **natural conditions**
under which the land is used.

Erosion - in other words the removal of soil by water and wind - is one of the most problematic **consequences of agriculture**, particularly in the tropics.

The actual extent of erosion depends on the type of crop and form of cropping. To **minimise** erosion, efforts should be made to ensure that there is **ground cover all year round**. In the case of **monoculture and single cropping**, the **risk of erosion** becomes greater the more slowly the young plants develop (e.g. maize, grain legumes), the lower the planting density is and the more comprehensive the weed control measures are. As **annual crops** such as cereals, tubers and grain legumes entail frequent tillage, they have an adverse effect on the soil structure and are thus **conducive to erosion**.

Perennial crops such as fruit trees generally **prevent soil erosion** once the stand is complete; they provide permanent **shade**, which has a **positive** effect on the soil structure.

A **soil's erodibility** depends among other things on its **physical properties**. Fine sand and abraded particles can be displaced most easily, whereas a high stone and clay content inhibits erosion. A high humus content stabilises the soil structure and increases the water storage capacity; both of these factors inhibit erosion.

The most important ways of **controlling erosion** are:

- adequate ground cover (intercropping, underseeding etc.);
- "storeyed" cultivation through integration of trees and shrubs;
- division of cropping areas into small units and creation of windbreaks at right angles to the direction of the prevailing wind;
- avoidance of overstocking and measures to prevent animals from grazing on newly sown areas (see environmental brief Livestock Farming).

Excessive **mechanisation** of tillage and harvesting can lead to **compaction, plough sole formation and puddling**, particularly in the case of tropical soils with a weak structure. This may have the **adverse** effect of reducing water infiltration and the air supply for the soil flora and fauna as well as for the crops. Mechanisation can also lead to changes in the division of labour between men and women.

Although **frequent tillage** generally has a **stimulating effect** on microbial activity and thus also on replenishment of the nutrient supply, it has **disadvantages in the tropics**:

- humus decomposition is excessively rapid on account of the high temperatures,
- the soil fauna are adversely affected and formation of new humus is thereby delayed.

Single cropping promotes the **spread of pests** on a large scale and tends to necessitate substantial **use of pesticides**. **Introduction** of pesticides into the soil has **adverse effects** on the soil fauna and flora.

Organic matter plays a major role in the dynamics of tropical soils. It stores water, provides a living environment for soil organisms, promotes structural stability, and both supplies and stores nutrients. It is above all in storing nutrients that organic matter performs an especially vital function, as tropical soils seldom contain high-quality nutrient-fixing clay minerals. Use of **mineral fertilisers** therefore depends on the proportion of organic matter in the soil. If the amount of fertiliser used is not in correct proportion to the organic matter, there is a danger of leaching and of the fertiliser passing into deeper soil layers. **Use of too much fertiliser** is thus **ecologically undesirable** and **economically disadvantageous**.

The risk of unbalanced **nutrient depletion** is greatest in the case of **monoculture** and **single cropping**, e.g. in the case of maize, cocoa, root crops and tubers. Where **a number of plant species** are grown in an intercropping or crop rotation system this risk becomes smaller, as **differing nutrient requirements** have to be met. As such forms of cropping incorporate plants with different root systems (shallow, deep) and nutrient requirements (high, low), **competition for nutrients, water and light is substantially reduced**.

2.1.2 Water

The erosion referred to above can lead to **eutrophication** of bodies of water through the introduction of nutrients, e.g. liquid manure and nitrate, and to contamination with **toxic pesticide residues**.

2.1.3 Air

The **climate in multi-storeyed stands growing in an intercropping system is better** i.e. more balanced, than that in stands of annual crops forming a monoculture or single cropping system. The **wind velocity** is **lower** and thus better for crops **susceptible** to the wind (e.g. bananas).

Air **pollution** caused as a result of plant production stems primarily from **chemical plant protection** measures. **Evaporation of ammonia** during application of solid or liquid manure has hitherto been of only **minor significance**. Under tropical conditions (high temperatures, low soil sorption capacity), up to 80% of the total nitrogen may evaporate.

Pollution of the air and atmosphere is caused by **waste gases** resulting from **use of machinery, slash-and-burn techniques** and **burning-off** of crop residues, as well as by **discharge of gases** such as methane and nitrous oxide by swamp rice and large herds of cattle. These factors play a part in the **greenhouse effect**.

2.1.4 Biosphere

The **risk** of both the **loss of species** and a **change in the balance of species** increases in proportion to the intensity of plant production activities. Controlled **shifting cultivation** - observing the necessary fallow periods - encroaches least on the natural environment in terms of area if only level areas are cleared on a selective basis. This helps not only to **preserve the forests, particularly the rainforests** and their resources, but also to **protect forest-dwellers**, who often possess know-how about things such as plants with potential pharmacological uses and the ecological interrelationships within their living environment.

Systematic cultivation of crops and the related mechanical and chemical forms of weed control cause **wild plants** to be largely **displaced**, leading to a reduction in the number of species.

In regions subject to periodic droughts, large-scale cultivation of certain woody plants in a monoculture system substantially increases the **fire risk**. In addition to nutrient and leaching losses, this can also result in unwanted destruction of grass and tree species not resistant to fire.

Displacement and destruction of plants leads to a **reduction in biological diversity**. Extensive use of rainforests also substantially reduces the variety of animal species, e.g. in the case of primates and birds.

Natural ecosystems are adversely affected not only by **land being required** for plant production but also by being **broken up** (e.g. by traffic routes), which can result in a **loss of stability**.

Use of land for plant production generally leads to the loss of forest, dry, wet and aquatic biotopes and causes the **landscape to take on a uniform nature**, e.g. as a consequence of land clearance, drainage, levelling and irrigation.

By comparison with the natural vegetation, **plant production destroys habitats** and **reduces regional diversity**. **Standardisation** of products for the market and **breeding** to obtain specific traits (e.g. yield, shape, colour) play a part in the loss of local varieties (**genetic erosion**).

2.2   Protective measures

2.2.1   General conditions

Plant production is influenced to a particularly large extent by **general conditions**; these may relate not only to **climate** but also to **national** (e.g. land ownership situation) or **international** (economic relations) factors.

Many climatic and vegetation zones are highly sensitive to **interference by man**, whose activities generally **destroy the vegetation**, as in the following cases:

-       clearance of the tropical rainforest in the Amazon basin for the purpose of obtaining high-grade timber
-       slash-and-burn land clearance by arable farmers in Nigeria's tree-studded savannah, where the transition to permanent cultivation no longer allows the land the opportunity to regenerate
-       overgrazing in the Sahel zone as a result of overstocking with large numbers of livestock which remove the already sparse vegetation.

The **consequences** are **disastrous**, not only in the humid tropics but also in places which receive less rainfall. As there are virtually no plants left to provide ground cover, the soil undergoes changes within the space of a few years; a key role is played here by the increased decomposition of organic matter in the soil and the fact that the introduction of new organic matter is reduced to a minimum.

Within the existing world economic order, the **terms of trade** for the countries concerned have steadily deteriorated. It is above all these countries which have been hit by the increased cost of energy and finished products. **International agricultural policy** likewise does nothing to ensure balanced promotion of plant production.

Rapid **population growth** means that **farms** are becoming increasingly small and the **land** is thus **being used more and more intensively**. Farms in Latin America today already have an average size of only 2.7 hectares; those in Africa on average cover 1.3 hectares, while the corresponding figure for Asia is less than one hectare. What is more, 10% of persons deriving their living from agriculture in Africa, 25% of those in the Middle East and 30% of those in Latin America own no land at all. Two thirds of those who do possess land own only a tiny area and cannot afford capital-intensive technical inputs such as pesticides, herbicides and mineral fertilisers.

As **land becomes increasingly scarce**, farming systems undergo a **transition** from shifting cultivation to semi-permanent and eventually permanent arable farming. This process has already been largely completed in Asia, while in much of Africa and Latin America it is still under way. The changeover to **permanent arable farming** means that there are no longer any fallow periods (forest, bush, pasture) which allow the soil to regenerate; soil fertility declines and eventually remains at a fairly low level permitting only substantially smaller yields. The shortage of land also necessitates **use of areas such as slopes at risk from erosion** and thus contributes to environmental degradation.

The relative importance of the **crops grown** also changes. In the humid and semi-humid tropics the cultivation of yams, sorghum, and maize declines in significance, while crops such as cassava and sweet potatoes become more important. The last-mentioned crops produce **relatively good yields** even on poor sites, but at the same time **cause the soil to become exhausted more quickly.**

In many countries, both **intensification of agriculture** and the **industrialisation process** are having increasingly **adverse impacts on the environment.** Waterlogging, salinisation and sedimentation cause the **irrigated cropping areas** - often created at considerable expense - to lose their fertility after only a few years, which gives a rise to a considerable **drop in yield**. Traces of **persistent pesticides** are being increasingly found in **bodies of surface water** and **groundwater reservoirs**. The past decade has seen a sharp rise in the number of people suffering pesticide poisoning, while at the same time there has been an enormous increase in the number of **pest species resistant** to the commonly used pesticides.

The **factors** described here are generally to be found wherever efforts are being made to **raise yields** through targeted, conventional **modernisation of agriculture**. However, such problems are not simply consequences of **large-scale agricultural projects**, but also arise as the **cumulative result of numerous activities on the part of smallholders**.

As the actual **environmental costs** have little or no impact from the farm management viewpoint, there is no incentive to take measures aimed at conserving natural resources or producing sustained improvements in efficiency. **Land law, taxation policy and subsidisation policy**, along with **ascertainment of the external costs** involved in production and consumption, are **areas** which the **state** must tackle in the interest of promoting environmentally oriented plant production.

There are certain **concepts**, such as that of ecodevelopment, which are based on the necessary **integrated approaches**. Tried and tested measures such as **integrated plant protection, ecofarming** and others point the way towards **sustainable development**.

2.2.2 Ecofarming

Ecofarming aims to achieve a high **sustained level of productivity** on the site in question under "low external input" conditions and at the same time to preserve or recreate a **balanced ecosystem**.

This applies in particular in densely populated regions with smallholder-based farming structures and under economic conditions which largely preclude **use of external inputs** (e.g. mineral fertilisers), for in many cases such inputs are economically non-viable, unaffordable or unavailable on account of supply shortages. **Intensification** of agriculture must therefore be based on **more productive use of scarce goods** (nutrients, water, energy) and **underutilised idle resources** (e.g. labour, individual initiative).

The **demand for stability and sustainability** stems from the obligation of each generation to pass on to future generations an environment that remains capable of guaranteeing the fundamentals of human existence. The **demand for productivity coupled with stability** is often seen as a **conflict of objectives** between irreconcilable short-term and long-term (and frequently also between microeconomic and macroeconomic) viewpoints; in most cases it is the short-term microeconomic considerations that prevail. **Ecofarming** must endeavour **to achieve both objectives to an equal extent**.

Ecofarming, or "site-appropriate agriculture" as it is also known, involves treating both **regions used for agriculture** and individual farms as **ecological systems**. However, the concept of "site" must not be restricted to natural conditions (soil, climate).

Consideration must also be given to **economic development** (price-cost ratios, incomes), farm-specific conditions (access to factors of production) and the internal forces influencing a farm's operations (self-sufficiency, risk minimisation, preservation of soil fertility). Last but not least, it is essential that man, together with his culture, needs, taboos and habits, be viewed as a **component of the ecological system** and **not as an outsider**.

This **integrated approach** requires a certain degree of **geographical differentiation**. Agriculture in many countries is affected by a **growing shortage of raw materials and energy** and by the accompanying **rise in prices**. This is particularly true of countries which are in debt and possess little foreign exchange. It is thus these countries above all which must develop **forms of agriculture** that permit a **high degree of self-sufficiency** (within a self-contained system) and **decentralisation** (as well as self-regulation) at national and regional level and within individual farms.

The major **elements of ecofarming** are as follows:

- creation of appropriate vegetation
    - inclusion of trees and shrubs in arable farming
    - creation of erosion-protection strips parallel to the incline on slopes and planting of hedges to divide a farm into numerous small fields
    - afforestation on the poorest and most degraded soils
- intercropping, alternating with intensive fallow
- organic manuring
- integrated livestock husbandry
- improved mechanisation
- supplementary use of mineral fertiliser
- integrated plant protection and selective weed control

The elements listed above are given **in order of precedence**. As it is impossible to introduce the entire package of measures immediately, this form of classification indicates which **measures must be given top priority** for the purpose of preserving, increasing and stabilising soil productivity.

The following key **areas of activity and options** in the plant production sector should be combined with one another according to the nature of the site:

- farm planning and organisation (information systems, economic thresholds, soil investigations, climatic data)
- design of cropping system (single cropping, intercropping etc.)
- variety and seed selection (resistance, quality, quantity)
- tillage
    - conventional
    - minimum tillage
    - direct drilling
- cultivation and land use (crop rotation, sustainable cropping capacity)
- plant nutrition (fertilising)
    - organic
    - mineral
- plant protection
    - mechanical
    - biological
    - chemical

To sum up, it can be said that **environmentally sound, site-appropriate agriculture aims**

-       to guarantee that plant production is geared to natural conditions, i.e. site-appropriate;
-       to preserve the soil structure, the biological processes taking place in the soil and the soil's fertility;
-       to prevent erosion damage;
-       to prevent contamination of groundwater and bodies of surface water;
-       to prevent adverse impacts on biotopes adjacent to agricultural land as a result of the introduction of substances or other consequences of cultivation measures;
-       to preserve typical landscape features;
-       to take account of the requirements of nature conservation and protection of species, particularly as regards preservation of ecologically valuable biotopes, within the scope of overall consideration of the environment;
-       to make livestock husbandry an integral component of environmentally sound agriculture.

3.   Notes on the Analysis and Evaluation of Environmental Impacts

In the plant production sector, the following assessment criteria lend themselves to **direct or indirect measurement**:

-       changes in the biotope (diversity of species of flora and fauna)
-       impacts on finite natural resources (minerals, ores, water, atmosphere)
-       impacts on global ecological relationships (net energy production: energy audit comparing energy fixed by a crop plant/harvested product and energy used in its production)
-       contamination levels (chemical products, salts, dusts, gases)

Limits varying from one country to another have been laid down for many substances occurring in agriculture. Although many countries have **maximum-quantity regulations** covering immissions in water, air and soil, these are generally concerned with the effect of pollutants on human health.

As the **properties and sensitivity** of tropical soils **vary** greatly, a **site survey** must always be conducted before project planning commences. Such a survey involves **mapping** the soil types with regard to their heat, water, air and nutrient balances as well as their **susceptibility to erosion**. The **soil type** can be **determined** in the field or by means of granulometric analysis in a laboratory; once this has been done it is possible to assess the risk of compaction. **Measurement of the infiltration rate** permits more accurate appraisal of the erosion risk. Tolerance limits for **humus decomposition** can be formulated only on the basis of the soil conditions and the land use situation. The **humus content** can be roughly **ascertained** in the field; precise determination can be carried out in the laboratory by means of ignition loss, wet incineration or gas chromatography.

Spade analysis can be used for simple **assessment of soil structure and biological activity**; rooting characteristics are of particular importance. The findings can be substantiated in the laboratory by means of wet screening (aggregate stability), analysis of the C/N ratio (nitrogen availability) etc. The presence of effective root symbionts (nitrogen-fixing organisms, mycorrhiza) can be detected only by way of infection tests.

The extent of the **leaching risk** (particularly for nitrate and pesticides) can be ascertained by determining the field capacity of the soil profile down to the effective rooting depth. This can be estimated in the field with the aid of a drilling stock; it is advisable to conduct a pore analysis of typical soil horizons in order to calibrate the response. However, excavation of a profile is essential in some cases, above all if waterlogging or crusting is suspected.

Deficiency or toxicity symptoms in crops may prompt determination of nutrient status or contamination level. Measurement of the pH value as a function of soil depth can often reduce the necessary scope of analysis and provides information about the **lime requirement**. Measurement of effective cation exchange capacity and base saturation yields pointers regarding **nutrient imbalances** and the **degree of salinisation**. In the case of trace elements and heavy metals, plant analysis is to be preferred. The results allow appropriate **recommendations** to be made regarding **fertilising or - where necessary - rehabilitation**.

A body of water can be characterised with relative ease by means of **quality classification**, which is carried out by determining the pH value, temperature, oxygen content and important indicator organisms. If such organisms are not present or are unknown, the water's ammonium and phosphate content can also yield information about the **trophic level**. Analysis of biochemical and chemical oxygen demand (BOD, COD) allows conclusions to be drawn regarding the **degree of pollution with degradable organic substances**. The requirements to be fulfilled in terms of water quality will vary depending on the planned use.

It is above all in semi-arid regions that hydrogeological investigations are necessary for **assessing the groundwater reserves**. Such investigations can yield information on subsoil conditions and the location of the catchment areas. Current annual evaporation and groundwater recharge rate can then be estimated on the basis of the land use and soil distribution determined in the course of the site survey. If the rate at which the groundwater is tapped (drinking water, irrigation) permanently exceeds the recharge rate, **lowering of the groundwater** may cause severe damage to land which is in a near-natural state or has undergone reforestation. In such cases the groundwater must also fulfil more stringent quality requirements, since its use as drinking water must not be restricted.

Areas used for plant production often serve to **neutralise or reduce emissions emanating from other areas.** Correctly designed intensive agroecosystems can in fact sometimes perform such functions more effectively than the potential natural vegetation, because it becomes profitable, from a certain yield level upwards, to neutralise immission-induced damage through appropriate **use of inputs** (e.g. liming to offset the introduction of acid). The same applies to climatic effects, which can be positively influenced if suitable land and the correct forms of cropping are selected.

Summarising assessments of energy flows and natural cycling systems, which also yield information about loading capacities, will be highly unreliable in the absence of adequate familiarity with the species involved and their interrelationships.

4.   Interaction with Other Sectors

Plant production always has impacts on the environment, either directly or through its links with other areas. By virtue of its objectives and impacts it has **particularly close links** with the following areas **featured in farming systems**:

- Plant protection
- Forestry
- Livestock farming
- Aquaculture
- Agricultural engineering
- Irrigation

The **objectives** pursued in these sectors (see relevant environmental briefs) may be **compatible** with those of plant production, have **no bearing on them or conflict with them.** In the same way, **impacts** of plant production may be **increased, reduced** or **offset** by measures in these areas. When assessments are being carried out, attention must be paid to the possibility that **impacts** generated by activities in different areas could have a **cumulative effect** and thereby increase the amount of damage done. Such processes can be **regulated** with the aid of **research and advisory work**, backed up by instruments in fields such as legislation, poverty alleviation, self-help and advancement of women.

If plant production is on a scale extending beyond subsistence level, it also has links with **agroindustry**. Sinking of wells as part of schemes to provide **rural water supplies** can accelerate the desertification process, which has disastrous consequences for plant production.

As many countries **require** an increasing amount of **land** for settlement, transport systems, trade and industry and sometimes have to meet this need by developing areas formerly used for plant production, conflicts inevitably arise (spatial and regional planning, location planning, transport and traffic, large-scale hydraulic engineering). Although **improvement of the transport system** facilitates access to inputs (fertilisers, workshops) and sale of produce, **land development within natural ecosystems** can **accelerate the destruction** of such systems. The need for **erosion control measures** generally arises as a result of erosion caused by forms of cropping inappropriate to the site concerned. The availability of **renewable energy sources** and **compostable domestic waste** can also be of importance for plant production.

5. Summary Assessment of Environmental Relevance

In order to **prevent** plant production from giving rise to **unintentional developments**, ascertainment of the initial situation and appraisal of potential consequences must be followed by **regular assessment** of forecast and actual changes in environmental conditions. The same applies to **social conditions**, as **there is a close interrelationship between cultural and economic factors** on the one hand and the **natural environment** on the other hand.

The **impacts of plant production** generally consist in **reduction of the diversity of species, adverse effects on the nutrient balance** as well as on the physical and **chemical properties of the soil**, and **contamination of the environment with pollutants**.

Appropriate planning **techniques** and technical **measures** have been **developed** and must be taken into consideration. It is essential to refute the opinion that plant production activities (including biological erosion protection measures) have little or no impact on the environment.

Resource-depleting impacts are generally **unwanted side-effects** which are not directly related to the production goals. It is precisely when **these side-effects are ignored** that the natural environment will suffer **damage** and adverse long-term consequences will arise in the economic and social spheres.

Careful planning and implementation will ensure that plant production **has minimal environmental impacts, has no undesirable social consequences** and is **economically efficient**.

6. References

ALKÄMPER, J. AND MOLL, W. (Eds.), 1983: Möglichkeiten und Probleme intensiver Bodennutzung in den Tropen und Subtropen. Gießener Beiträge zur Entwicklungsforschung, Reihe 1, Band 9, Gießen.

ARBEITSGRUPPE BODENKUNDE, 1982: Bodenkundliche Kartieranleitung. Hanover.

ASIAN DEVELOPMENT BANK, 1986: Environmental Planning and Management, Regional Symposium on environmental and natural planning, Manila, 19. - 21.02.1986.

BARBIER, E.B. and CONRAY, G.R., 1988: After the Green Revolution: Sustainable and equitable agricultural development. Futures, 651-670.

BAUMANN, W., BAYER, H., GEUPNER, P., KRAFT, H., LAUTERJUNG, E., LAUTERJUNG, H., MOLIEN, H., WOLKEWITZ, H. AND ZEUNER, G., 1984: Ökologische Auswirkungen von Staudammvorhaben, Erkenntnisse und Folgerungen für die entwicklungspolitische Zusammenarbeit. Forschungsberichte des BMZ, BAND 60, Weltforumverlag, Munich.

BEHRENS-EGGE, M., 1991: Möglichkeit und Grenzen der monetären Bewertung in der Umweltpolitik. Zeitschrift für Umweltpolitik 1, pp. 71-94.

BINSWANGER, H. C.; M. FABER AND R. MANSTETTEN, 1990; The dilemma of Modern Man and Nature: an Exploration of the Faustian Imperative. Ecological Economics 2, with case studies. In: BISWAS and GEPING, 1987, pp. 3-64.

BISWAS, A.K. AND GEPING, Q., 1987: Environmental impact assessment for Developing Countries. Tycooly Publishing, London.

BONUS, H., 1984: Marktwirtschaftliche Konzepte im Umweltschutz. In: (Eds.) Agrar- und Umweltforschung in Baden-Württemberg, Bd. 5, Stuttgart.

BRODBECK, U., FORTER, D., ISELIN, G. AND WYLER, M. (Eds.), 1987: Die Umsetzung der Umweltverträglichkeitsprüfung in der Praxis. Eine Herausforderung für die Wissenschaft. Verlag Paul Haupt, Berne.

CAESAR, K., 1986: Einführung in den tropischen und subtropischen Pflanzenbau. DLG-Verlag, Frankfurt/a.M.

CGIAR, 1990a: Sustainable Agricultural Production: Final Report of the committee on sustainable agriculture. Consultative Group Meeting, May 21-25, The Hague, The Netherlands.

CGIAR, 1990b: To Feed a Hungry World: The urgent role of development agencies. Committee on Agricultural Sustainability for Developing Countries.

CONWAY, G.R., 1985: Agroecosystem Analysis. Agricultural Administration 20, 31-55.

CONWAY, G.R., 1987: The properties of agroecosystems. Agricultural Systems 24, 95-117.

DEUTSCHER FORSTVEREIN, 1986: Erhaltung und nachhaltige Nutzung tropischer Regenwälder. Forschungsberichte des Bundesministeriums für wirtschaftliche Zusammenarbeit [BMZ - German Federal Ministry for Economic Cooperation and Development]. Weltforum-Verlag, Munich.

DEUTSCHES INSTITUT FÜR TROP. UND SUBTROP. LANDWIRTSCHAFT (DITSL), 1989: The extent of soil erosion - regional comparisons. Witzenhausen.

DIXON, J. A. AND M. M. HUFSCHMIDT, 1986: Economic Valuation Techniques for the Environment. A case study workbook. The John Hopkins University Press.

DIXON, J. A.; CARPENTER, R. A.; FALLON, L. A.; SHERMAN, P. B. AND MANIPOMOKE, S. 1988: Economic Evaluation of the Environmental Impacts of Development Projects. ADB.

DUMANSKI, J., 1987: Evaluating the Sustainability of Agricultural Systems. Africaland - Land Development and Management of Acid Soils in Africa II: IBSRAM Proceedings 2nd Regional Workshop, Lusaka and Kasama, Zambia, 9-16 April 1987, pp. 195-205.

DUMANSKI, J. et al. 1990: Guidelines for evaluating sustainability of land development projects. entwicklung + ländlicher raum, p. 3-6.

EHUI, S.K. and SPENCER, D. S. C., 1990: Indices for Measuring the Sustainability and Economic Viability of Farming Systems. Research Monograph No. 3. Resource and Crop Management Program, International Institute of Tropical Agriculture.

ENGELHARDT, T., 1989: Angewandte Projektökonomie, GTZ.

EWUSIE, J. Y., 1980: Elements of tropical ecology. Heinemann education books, London.

FOOD and Agriculture Organization (FAO), 1981: Agriculture: toward 2000. FAO, Rome.

FOOD and Agriculture Organization (FAO), 1984: Land, Food and People. FAO Economic and Social Development Series 30. Rome.

FOY, G. and DALY, H., 1989: Allocation, distribution and scale as determinants of environmental degradation: Case studies of Haiti, El Salvador and Costa Rica. WORLD BANK, Environmental Department Working Paper Nr. 19.

GÖLTENBOTH, F., 1989: Subsistence agriculture improvement - Manual for the humid tropics. Margraf-Verlag, Weikersheim.

HARTJE, V., 1990: Das UVP-Verfahren in der Entwicklungszusammenarbeit. Vortrag zum Symposium des Wissenschaftlichen Beirates beim BMZ "Umweltschutz in der Entwicklungszusammenarbeit", Bonn.

HARRISON, D. AND RUBINFELD, D. O., 1978: Hedonic Prices and the Demand for Clear Art. In: Journal of Environmental Economics and Management, Vol. 5, 81-102.

HAS, M., 1986: Großtechnologie - aber welche Alternative, in: STÜBEN, P.E. (Ed.) 1986: Nach uns die Sintflut. Staudämme, Entwicklungs"hilfe", Umweltzerstörung und Landraub. Focus Verlag, Gießen, pp. 194-213.

HESKE, H., (Ed.), 1987: Ernte-Dank? - Landwirtschaft zwischen Agro-Business, Gentechnik und traditionellem Landbau. Ökozid Nr. 3, Focus Verlag, Gießen.

HUFSCHMIDT, M. M.; JAMES, D. E.; MEISTER, A. D.; BLAIR, T. B. AND DIXON, J. A.: 1983: Environmental Natural Systems and Development. An Economic Evaluation Guide. The John Hopkins University Press.

INSTITUT FÜR UMWELTSCHUTZ DER UNIVERSITÄT DORTMUND, 1977: Umweltindikatoren als Planungsinstrumente (Beiträge zur Umweltgestaltung: B; H. 11) E. Schmidt-Verlag, Berlin.

INTERNATIONALE VEREINIGUNG BIOLOG. LANDBAUBEWEGUNGEN, (Ed.), 1989: Basisrichtlinien der IFOAM für den ökologischen Landbau. SÖL-Sonderausgabe Nr. 16, 7. Auflage, Kaiserslautern.

INTERNATIONAL LAND DEVELOPMENT CONSULTANTS (ILACO, B. V.), 1981: Agricultural compendium for rural development in the tropics and subtropics. Elsevier Scientific Publishing Company, Amsterdam (detailed standard works for site surveys).

INTERNATIONAL RICE RESEARCH INSTITUTE (IRRI), 1980: Soil-related constraints to food production in the tropics. Los Banos, Philippines.

KIRKBY, M. J. AND MORGAN, R. P. C., 1980: Soil erosion. Wiley and Sons, Chichester.

KOTSCHI, J. AND ADELHELM, R., 1984: Standortgerechte Landwirtschaft zur Entwicklung kleinbäuerlicher Betriebe in den Tropen und Subtropen. GTZ, Eschborn.

KOTSCHI, J.; WATERS-BAYER, A.; ADELHELM, R. AND HOESLE, U., 1989: Ecofarming in agricultural development. Margraf-Verlag, Weikersheim.

LAL, R.; ANDRUSSEL, E. W., 1981: Tropical agricultural hydrology. Wiley and Sons, Chichester.

LAUER, W. (Ed.), 1984: Natural environment and man in tropical mountain ecosystems. Steiner Verlag, Wiesbaden.

LYNAM, J. K. AND HERDT, R. W., 1989: Sense and Sustainability: Sustainability as an objective in international agricultural research. Agricultural Economics, Vol. 3, No. 4, pp. 381-398.

MÄCKEL, R. AND SICK, W. D., 1988: Natürliche Ressourcen und ländliche Entwicklungsprobleme der Tropen. Reihe Erdkundliches Wissen, Steiner Verlag, Wiesbaden.

MAGRATH, W. AND ARENS, P., 1989: The cost of soil erosion on Java: a natural resource accounting approach. WORLD BANK. Environment Department Working paper Nr. 18.

MARTEN, G. G., 1988: Productivity, Stability, Sustainability, Equitability and Autonomy as Properties for Agroecosystem Assessment. Agricultural Systems 26, 291-316.

MESSERLI, B.; BISAZ, A.; KIENHOLZ, A. AND WINIGER, M., 1987: Umweltprobleme und Entwicklungszusammenarbeit. Arbeitsgemeinschaft Geographica Bernesia, Berne.

MÜLLER-HOHENSTEIN, K.; GROSSER, L; AND RAPPENHÖHNER, D., 1988: Umweltverträglichkeitsstudie zum GTZ-Projekt "Zucker-rohrversuchsanbau", Marokko (PN 80.2116.4-01.100). Identifizierung und Gewichtung der Umweltbeeinflussungen, sowie Empfehlungen für zukünftige Maßnahmen und weitere Studien. GTZ, Eschborn.

MÜLLER-SÄMANN, K. M., 1986: Bodenfruchtbarkeit und standortgerechte Landbau-Maßnahmen und Methoden im tropischen Pflanzenbau. Schriftenreihe der GTZ Nr. 195, Eschborn.

NAIR, P. K. R., (Ed.) 1989: Agroforestry systems in the tropics. Kluwer Academic Publishers, Dordrecht.

PREUSCHEN, G., 1983: Die Kontrolle der Bodenfruchtbarkeit - eine Anleitung zur Spatendiagnose. IFOAM-Sonderausgabe Nr. 2, 2. Auflage, Kaiserslautern.

REHM, S. AND EPSIG, G., 1990: The cultivate plants of the tropics and subtropics. Margraf-Verlag, Weikersheim.

ROTTACH, P., (Ed.), 1984: Ökologischer Landbau in den Tropen. Alternative Konzepte, C. F. Müller Verlag, Karlsruhe.
1988: Ökologischer Landbau in den Tropen/Ecofarming in Theorie und Praxis. C. F. Müller Verlag, Karlsruhe.

RUTHENBERG, H., 1980: Farming systems in the tropics. Clarendon Press, Oxford.

SCHEFFER, F. AND SCHACHTSCHABEL, P., 1989: Lehrbuch der Bodenkunde. 12. Auflage, Enke-Verlag, Stuttgart.

SMIT, B. AND M. BRKLACICH, 1989: Sustainable development and the analysis of rural systems. Journal of Rural Studies, 5, (4), 405-414.

STOCKING, M., 1987: Measuring land degradation. In: P. BLAKIE and H. BROOKFIELD, (Eds.), Land degradation and society. London: Methuen, 49-63.

WHANTHONGTHAM, S., 1990: Economic and Environmental Implications of two alternative citrus production systems. MS thesis Asian Institute of Technology.

WEITSCHET, W., 1977: Die ökologische Benachteiligung der Tropen. Teubner Verlag, Stuttgart.

WÖHLCKE, M., 1987: Ökologische Aspekte der Unterentwicklung. Fakten, Tendenzen und Handlungsbedarf in Bezug auf den Umwelt- und Ressourcenschutz in der Dritten Welt. Stiftung Wissenschaft und Politik. Ebenhausen.

WÖHLKE, M., 1990: Umwelt und Ressourcenschutz in der Internationalen Entwicklungspolitik. AK. Materialien zur Internationalen Politik. Baden-Baden.

YOUNG, A., 1976: Tropical soils and soil survey. Cambridge University Press, Cambridge.

ZENTRALSTELLE FÜR ERNÄHRUNG UND LANDWIRTSCHAFT [German Food and Agriculture Development Centre], 1988: Agroforstwirtschaft in den Tropen und Subtropen - Aktualisierung und Orientierung der Forschungsaktivitäten in der Bundesrepublik Deutschland.

ZÖBISCH, M. A., (ed.), 1986: Probleme und Möglichkeiten der Landnutzung in den Tropen und Subtropen unter besonderer Berücksichtigung des Bodenschutzes. Der Tropenlandwirt, Beiheft Nr. 31, Witzenhausen.

Agriculture

Plant Protection

28. Plant Protection

Contents

1.   Scope

Plant protection measures are carried out to **limit performance and yield losses** in crop production during the growing season and afterwards (storage protection) as well as for **quarantine purposes**. They serve primarily to **safeguard yields**, although in combination with other cultivation measures they can also help to **raise yields**.

A wide variety of **individual measures** - with varying ecological, economic and socio-economic impacts - are available for **keeping harmful organisms** (diseases, pests, weeds) **below the economic threshold**. To **reduce the probability of damage, preventive measures** are taken in the areas listed below. Some of these can be regarded as belonging to the field of plant production (cf. environmental brief Plant Production), which reflects the close links between the two sectors:

- site design (hedges, border strips etc.)
- site and variety selection
- sowing, planting
- healthy seed and planting stock
- crop rotation, intercropping
- tillage, land improvement
- fertilising
- crop tending
- harvesting
- storage

Measures in these areas are backed up by the following **direct forms of plant protection:**

- physical methods
- chemical methods
- biotechnical methods
- biological methods
- integrated methods

Physical methods directly **destroy harmful organisms**, aim to **retard their development** or **prevent them from spreading**. They can be divided into **mechanical and thermal measures**. The former include **tillage to control weeds** and pests (hoeing, removal of affected parts of plants and intermediate hosts), **flooding of fields** to combat soil-borne harmful organisms (e.g. Fusarium oxysporum, which causes banana wilt), **laying of sticky belts** to trap flightless insect pests and other measures for **catching** pests or **keeping them away** from crops, such as fences, trenches (locust control), traps and picking-off of pests. **Thermal methods** utilise the harmful organisms' sensitivity to high or low temperatures. They include **hot-water treatment** of seed and planting stock (e.g. to combat viruses and bacteria in sugar cane cuttings), **solarisation** (covering the surface of the ground with plastic sheeting produces phytosanitary effects by virtue of the greenhouse effect resulting from insolation, e.g. for controlling parasitic seed plants, soil-borne harmful organisms etc.), **burning-over** to control weeds and **burning** of crop residues. **Low temperatures** inhibit the spread of certain storage pests.

Eradicative, protective and curative methods are used in **chemical plant protection** to **destroy** harmful organisms or **keep them away** from plants, to **protect** plants against attack and penetration by harmful organisms and to **cure** plants (or parts of plants) that have already become infested or diseased. Although chemical methods can be subdivided in this way on the basis of their effects, the **boundaries** between the individual categories are somewhat **fluid**, as many **pesticides have more than one type of effect**. Pesticides generally **kill** the harmful organism **by influencing** vital metabolic processes or disrupting the conduction system. **Selectivity can be varied** through appropriate **selection of the active ingredient, formulation, application technique and time of application**.

Biotechnical and biological methods of plant protection have **gained in significance**, among other things because the risks and limits of chemical measures are today assessed more realistically. Biotechnical methods utilise the **natural reactions** of the (almost exclusively motile) harmful organisms to physical and chemical stimuli in order to bring about changes in their behaviour for the purpose of plant protection (e.g. light and colour traps, chemical attractants, antibodies, pheromones, hormones, growth regulators). The emphasis is on measures which aim **not to directly kill the harmful organisms**, but rather to permit **population monitoring for the purpose of forecasting, defensive action and deterrence**. The **harmful organisms** can be **killed** by **combining** biotechnical methods **with chemical measures**.

Biological plant protection involves using organisms and their activity to **protect** plants and **enhance their resistance** to biotic (harmful organisms) and abiotic limiting factors. For the purpose of pest and disease control, **beneficial organisms** are specifically **preserved and fostered, released in large numbers** or **introduced** into habitats where they have not been found hitherto. Biological control of weeds has to date primarily involved introducing beneficial organisms into new habitats.

Another biological method is that of **inducing resistance to disease**. This can be done, for example, by **infecting plants with pathogens having low virulence**.

There are **close links** between **biological and integrated plant protection** in that both methods **attach major importance to regulation by means of biotic limiting factors**. If such methods are to prove effective, moreover, there must be **little or no use of preventive and broad-spectrum pesticides**. Biological methods can be **applied on only a limited scale in intensively used agrobiocoenoses which are poor in species**, but can play a **more important role in areas where extensive farming is practised and in coenoses comprising a greater variety of species**. Their limits are determined above all by the **efficiency of the beneficial organisms** and the latter's **dependence on environmental conditions**.

Integrated plant protection is a concept which involves coordinated use of all ecologically and economically justifiable methods in order to keep harmful organisms below the **economic threshold**. The emphasis is on **utilising natural limiting factors**. The **main aim** is to **preserve the balance of nature as far as possible**; this is to be achieved by reducing use of chemical plant protection methods and simultaneously employing a variety of measures from the other categories. It is here that the links with the plant production sector are particularly close. **Use of pesticides is to be reduced** to the essential minimum by **abandoning the practice of routine or calendar-based spraying, gearing pesticide dosage** to actual conditions, **refraining from the use of broad-spectrum persistent agents (liable to harm beneficial organisms)** and selecting the **time of application** such that **beneficial organisms suffer no adverse effects**.

Integrated plant protection methods generally prove **more successful in permanent crops** than in short-lived crops, since the biocoenoses of the former are more stable and can be more permanently influenced whereas those of the latter are inevitably subject to constant change. The **limits** and **risks** attaching to these methods become clear if the work is performed by untrained personnel. Use of integrated plant protection methods generally calls for **detailed knowledge of biological, ecological and economic factors**.

2.   Environmental Impacts and Protective Measures

2.1   Plant protection in general

• **Environmental impacts**

The **environmental impacts** of plant protection are caused by the **influence** of **substances** and/or **forms of energy on organisms and their functioning** as well as on **soil, water and air**. The extent to which a plant protection measure is **harmful**, and in particular the degree to which it is liable to cause **lasting** harm, is determined by its varied **influences on the functioning** of the ecosystem. Adverse environmental impacts are likely if plant protection measures fail to take adequate account of ecological considerations. **Repeated, one-sided application** of a particular active ingredient will cause the harmful organism to **develop resistance** to it. Although **non-specific control methods** curb the spread of a harmful organism, they also unintentionally **affect numerous beneficial organisms**. They thus **adversely influence the diversity of species and biological regulation mechanisms**, creating a risk that harmful organisms may multiply more rapidly and consequently necessitating additional plant protection measures. **Effects on the abiotic environment** are also likely (e.g. soil erosion caused by tillage carried out for the purpose of plant protection).

When combined with other plant production measures, **plant protection extends** the **ecophysiological cultivation limits** of numerous crops. Cultivation of potatoes or tomatoes in humid mountain regions necessitates increased plant protection measures for combating fungi. Plants whose underground storage organs constitute the harvested crop (e.g. potatoes, taro) jeopardise the **sustainability of land use**, particularly when grown on slopes, on account of the erosion risk and increased mobilisation of nutrients.

Chemical plant protection came to occupy its position of major importance by virtue of the fact that **pesticides are easy to use and fast-acting**. There is thus at the same time also a **risk of misuse**, e.g. uneconomical use of pesticides.

Socio-economic conditions can be **influenced** to a considerable extent by the introduction of - or changes in - plant protection methods, which at the same time constitute a key element of the production system. This is particularly true of countries whose economy is based primarily on agriculture. The transition from a cropping system incorporating fallow periods to permanent cultivation, for example, necessitates substantially **increased financial outlay** on weed control, giving rise to corresponding socio-economic effects. What is more, changes in the spectrum of field flora will also become apparent, with species that are more difficult to control gaining the upper hand.

The changeover from weed control by means of hoeing to use of herbicides can bring **disadvantages for the population groups** (children, women, men, ethnic groups) which previously performed the work. The introduction of new methods may also have an influence on **health, earning capacity and standard of living**. At the same time, **social goals** and **ethical and moral concepts** provide the framework within which plant protection must operate (e.g. bans on killing certain types of animal; assessment of water/air quality, freedom from residues, job safety, work preferences, leisure needs).

- **Protective measures**

The **aim** of environmental protection measures is to minimise **the long-term ecological damage caused by plant protection. To this end, macroeconomic goals** must be weighed against **microeconomic goals** and the **"polluter pays" principle** consistently applied. The **control threshold** should be determined on the basis of ecological and economic criteria, taking **long-term aspects** into account.

Efforts should be made to achieve this goal by making extensive **use of natural limiting factors** (cf. environmental protection measures described in the environmental brief Plant Production) and by reducing the **probability of damage** (see 1. above). The potential **consequences** of plant protection for the production system and ecosystem, e.g. resulting from expansion of cropping to include sites with a greater risk of pest infestation, must be taken into account along with possible impacts on economic and social conditions.

2.2   Specific plant protection methods

2.2.1   Physical methods

- **Environmental impacts**

Thermal methods often require the input of **sizeable amounts of energy** in order to kill harmful organisms through the effects of heat (burning-over, production of steam or hot water). The **environmental impacts of energy generation** must be borne in mind (cf. environmental briefs Overall Energy Planning and Renewable Sources of Energies). Although solarisation uses **solar energy, plastic sheeting -** generally made of polyethylene - has to be placed over the entire area concerned or between the crop rows in order to achieve the greenhouse effect and many countries have **still to find a satisfactory way** of disposing of this sheeting. The **effects** of thermal methods on the biocoenosis are in most cases non-selective, so that microflora and microfauna populations must then re-establish themselves and achieve equilibrium in a **biological vacuum** in soil which is generally pasteurised or sterilised. **Mechanical** weed control **methods** involving **tillage measures** will lead to **changes in the soil's susceptibility to erosion**, an effect which must be given particular consideration where **slopes** are concerned. There is also **a risk of damaging plant organs** and thereby creating **portals of entry** for mechanically transmitted viruses and secondary parasites. Both thermal and mechanical methods generally promote **mobilisation of nutrients** from organic matter. This **humus decomposition,** accompanied by the destruction of clay-humus complexes and a deterioration in the soil structure, leads to a **reduction in soil fertility**. There is also a danger that **nutrients** may be **leached out** or introduced into other ecosystems. **Flooding** to curb the spread of soil-borne harmful organisms has a major impact - albeit only in the short term - on biotic and abiotic soil factors, with the **soil structure and nutrient dynamics** being **adversely affected**. Physical plant protection methods generally require **a considerable amount of labour** and their **effectiveness** against harmful organisms is highly **limited** in terms of both duration and area. Use of such methods may be restricted on account of labour shortages and for economic reasons.

- **Protective measures**

In terms of timing, location and intensity, thermal and mechanical methods are to be employed such that they combine maximum effectiveness with minimum detriment to beneficial organisms. Where mechanical methods are used, the **role played by the vegetation in protecting** the soil structure and soil organisms must be borne in mind. **Covering the ground with pieces of vegetable matter** (mulch) is one way of **controlling weeds** and at the same time **preventing erosion**. Use of mechanical methods is promoted by the development of labour-saving and effectiveness-enhancing techniques which make it possible to avoid the damage caused by other techniques.

2.2.2   Chemical methods

- **Environmental impacts**

The **environmental impacts** of chemical plant protection essentially comprise **three overlapping areas:**

a)   acute and chronic toxic effects
b)   contamination of harvested crops, soil, water and air with pesticides and their conversion products, as well as accumulation of such substances in the system
c)   impacts at system level (biocoenosis)

a) **Classifying** chemical pesticides on the basis of target groups **gives the false impression that their toxic effect is in each case limited** to their target group (herbicides - plants, fungicides - fungi, insecticides - insects etc.). Most **agents are non-selective** and have a lethal or inhibiting effect on organisms, as they interfere with basic metabolic processes (photosynthesis, ATP (adenosine triphosphate) formation, membrane development and functioning etc.). The toxicity of pesticides gives rise to **significant impacts.** The World Health Organisation (WHO) estimates that 1.5 million people are **poisoned** by pesticides each year, 28,000 of them fatally (54). Apart from their **active ingredients**, pesticides also contain **additives** to ensure adhesion and wettability as well as to perform various other functions. Out of 1,200 additives tested by the US Environmental Protection Agency, 50 were classified as toxic (24).

    Particular risks emanate from **poor-quality products**, which are often to be found on the market in countries with **liberal registration requirements** (68). Recurrent problems include pesticides which have aged beyond the point where they can still be safely used, contamination, poor formulation and active-ingredient concentrations deviating from those declared.

    Pesticides can give rise to **environmental pollution** during **storage and transportation** (soil, water, air), primarily as a result of leaking containers and subsequent problems caused by sale of large quantities.

There is also a **risk of food contamination** if **pesticides and foods** are **not stored separately** or are **sold together,** which is frequently the case in some countries.

    As **pesticides** generally **deteriorate** within a short time (often less than two years), the **hitherto unsolved** problem of **proper disposal** arises. Dangerous "time bombs" exist in many countries, with sizeable quantities of pesticides sometimes concentrated in a storage area of a few square metres.

    If dealers and farmers lack **adequate information, knowledge and training,** pesticides are liable to be **incorrectly used** (mix-ups, incorrect dosage, failure to observe waiting periods, etc.).

- The **absence of adequate information** on the containers (pictogram, labelling in a foreign language) can also result in incorrect use. Local dealers often put pesticides in **food containers** (fruit juice bottles, bags), while pesticide containers are frequently **re-used** for household purposes.

- Depending on the application technique and weather conditions, the risk of poisoning exists for pesticide users, members of their family participating in the farm work (particularly children) and neighbours. **Protective clothing suitable for the tropics** is virtually **unavailable.** Pesticides **sprayed from aircraft** are particularly likely to **drift** onto houses, neighbouring crops, pastures, bodies of water etc.

-       **Correct use** of pesticides is based on purchase as and when needed, together with considerable outlay on **appropriate** storage methods and application techniques. It calls for **sizeable inputs of capital**.

b)   **Contamination of harvested crops**, food and animal fodder with pesticide active ingredients or their residues and accumulation of such substances, giving rise to **health risks** for both man and animals [particularly likely in the case of incorrect use (see above), e.g. wrong dosages, failure to observe waiting periods etc.]. Use of chlorinated hydrocarbons on root vegetables, for example, led to accumulation in the harvested crop and intake by babies through baby food, which resulted in a subsequent ban on use of chlorinated hydrocarbons for vegetables.

-       **Contamination of soil, water and air** with pesticide active ingredients and their conversion products: Over half of the pesticide applied is discharged directly into the **atmosphere** upon **atomisation** and is transported in **aerosol** form, sometimes over long distances, before **rainfall** washes it into the **soil and water**. Most of the remainder directly contaminates soil and water. The risk that active ingredients will undergo a change to the **gaseous phase** is particularly great in the tropics, which is why pesticides with a high vapour pressure are unsuitable for use in such regions. Failure to take ecological and toxicological aspects into account can lead to **cultivation problems at a later date and to restrictions on cropping on account of the site's toxic load** (use of cuprous agents on bananas). If the soil's **sorption capacity** (retention capacity) is low, as is the case with sandy soils, pesticides and residues can be **leached into the groundwater**. Their **persistence** may increase with soil depth, e.g. as a result of the decline in microbial activity.

c)   The **non-specific action** of most pesticides and their conversion products has a variety of **direct and indirect impacts on biotic and abiotic components of ecosystems**, even at a considerable distance from the application site. The indirect impacts in particular are generally impossible to forecast; unforeseeable **"cascade effects"** may occur within the functional structure of ecosystems. Pimentel (61) calculates that the **damage caused to the biocoenosis** in North America by the use of chemical pesticides corresponds annually to a figure of US $ 500 million. Well over half of these costs can be attributed to reductions in the number of beneficial organisms and development of resistance to pesticides.

Impacts of this type include **elimination of pollinating insects** and other **beneficial organisms** (natural limiting factors) as system regulation and control elements. Use of insecticides in (irrigated) swamp-rice systems endangers fish and entomofauna, which can be seen as an indication of the conflict between aquaculture and pesticide use. The biological activity of earthworms and nitrifying bacteria is adversely affected by the use of methyl bromide for soil disinfection.

Beneficial organisms can be indirectly affected if, for example, the population density of a pest which at the same time represents the specific basis of beneficial organisms' food supply is radically reduced by the use of pesticides. Decimation of a species can **weaken the pest's biocoenotic ties, leading to increased reproduction and multiplication on a large scale.** For example, use of broad-spectrum insecticides in fruit growing to combat the apple-leaf sucker led to the fruit-tree red spider mite becoming a problem, as pesticides had an inadequate effect on the latter and caused beneficial organisms to be destroyed.

Pesticides can influence a **crop plant's susceptibility** to a particular group of harmful organisms on which the pesticide applied has no effect (for example, where a high level of fertilising is practised use of herbicides containing triazine or urea derivates can cause cereals to become more susceptible to mildew).

Lasting changes within the biocoenosis: Certain species remain **unaffected** by the agents used or **develop resistance** to them (one-sided use of atrazine in maize promotes weed infestation in the form of millet, while exclusive use of hormone weedkillers in cereals promotes the growth of grasses). Insecticides can also **have an effect on pollinating insects.** For instance, use of carbaryl to combat the mango leafhopper endangered or killed (wild) honeybees, thereby reducing smallholders' yield of honey and wax (32).

Over 400 arthropod species - half of them crop pests - have been found to have developed **resistance** to one or more active ingredients (10) (e.g. resistance of the boll weevil to DDT and other chlorinated hydrocarbons).

- **Protective measures**

In countries like the Federal Republic of Germany with strict **legislation on the distribution and use of pesticides, agents must not be recommended and used unless** they have gone through the necessary **registration procedure.** This procedure yields information about a pesticide's **toxicological, carcinogenic, teratogenic and other properties** as well as its effects on, and risks for, the **balance of nature.** Active ingredients are accordingly assigned to **toxicity classes. Fields of application, suitable disposal methods, analysis techniques and the ways in which conversion products are broken down** are also indicated. The **FAO Code of Conduct,** adopted in 1985, contains **recommendations** on the registration, distribution and use of pesticides. In countries such as the USA where legislation is strict, numerous pesticides involving comparatively high risks have been taken off the market (i.e. **banned altogether**) and/or **restrictions imposed on their use in terms of time and place**.

The reasons why certain products should not be used generally apply in all countries (30). In particular, use of persistent, broad-spectrum agents is **internationally proscribed**. The **"dirty dozen"** comprise the following fifteen active ingredients which should be banned in view of the substantial risks attaching to them:

* **Insecticides**

Chlorinated hydrocarbons: aldrin, chlordane, DDT, dieldrin, endrin, HCH-mixed isomers, heptachlor, lindane, camphechlor
Carbamates: aldicarb (proprietary name: Temik)
Organophosphates: parathion (E 605)
Other insecticides: dibromochloropropane (DBCP), chlordimeform, penta-chlorophenol (PCP).

* **Herbicides**
  2,4,5-T (proprietary name: Weedone)

Pesticide containers must bear a **description of their content**, the necessary **safety precautions**, the **permissible form of use and suitable antidotes**. It must be ensured that the **information given** can be **understood** by population groups potentially at risk. The necessary **information should be given in English and at least one national language** and should be backed up by **pictograms on labels that cannot easily be removed**. The criteria for marketing of a pesticide are determined by the **users' degree of illiteracy and awareness of the potential risks**.

If **chemical methods** are used **to combat harmful organisms, accompanying protective measures** must be laid down and enforced. These **minimum requirements** relate above all to appropriate selection of the **product** to be used, the **safety and functioning of the application technique** and **environmentally sound disposal of leftover pesticide and empty packaging**.

National plant protection organisations must conduct **training programmes** in order to ensure that extension officers, users and everyone coming into contact with pesticides are **aware of the risks involved. Internationally valid regulations** governing the prerequisites for distribution and use of pesticides are to be **developed** and compliance with them **monitored** by high-level authorities.

Preference should be given to **pesticides with low toxicity, a selective action and low persistence**. Effects, possibilities of misuse, special regional factors, water conservation areas and ecological conservation zones must be taken into account as **criteria for registration and use** of pesticides. **Use of dressed seed as food or animal fodder** is to be prevented by means of **adequate labelling**. It must also be ensured that **pesticide containers are not re-used for household purposes**; this can be done by way of awareness-raising measures, appropriate container labelling and possible also special container design. Pesticides should be **sold** only in **small containers holding a specific amount. Development of resistance** on the part of harmful organisms can be counteracted by **changing the active ingredient used**.

Unauthorised production and distribution of pesticides by pirate firms is a particular problem in many countries. This underscores the importance of **stringent and effective legislation on pesticides (registration)** and of enforcing **strict import controls** (with clearance certificates required if necessary to confirm that products are pure and in perfect condition). In addition, **access** to pesticides can, for example, be made **contingent** upon **production of an official "prescription"**, proof of **adequate know-how** and **use of pesticides within the framework of integrated plant protection methods**.

Government subsidisation of pesticides - which is common in many countries - creates **special risks** as regards misuse and environmental hazards (42). It must be established whether assistance measures of this type actually reach the target group and to what extent environmentally sound use and disposal of pesticides are ensured.

2.2.3   Biotechnical methods

•   **Environmental impacts**

If harmful organisms are attracted by a stimulus or killed by combining such measures with use of a poison, other organisms can also be **affected at the same time** (see environmental impacts described in 2.2.2 above). **Light traps** attract most nocturnal winged insects. **Use of noise to frighten off bird pests** is non-specific and has an effect on other organisms, whose mode of life (nesting, mating, rearing of young) can be disturbed. Repeated use of growth regulators (hormones) has been shown to promote **development of resistance** on the part of the target organisms. There is also **a risk of adverse effects on beneficial organisms**; for example, bee larvae and other insects which consume contaminated pollen or the like may be prevented from moulting.

- **Protective measures**

Non-specific biotechnical methods are to be **avoided** (e.g. light traps **attracting all nocturnal insects**). **Use of noise** to combat bird pests is to be **restricted in terms of time and place** to the extent necessary for directly averting crop damage. The times at which **growth regulators** are used and the technique employed are to be chosen such that little or no harm is done to beneficial organisms. Where appropriate, use of attractants should be combined with application of insecticides. **Development of resistance** is to be **counteracted** through **appropriate choice of agents**.

2.2.4 Biological methods

- **Environmental impacts**

Although the relationship between beneficial organism and host is in many cases **highly specific** and is thus likely to have only **minor unwanted impacts**, biological methods too give rise to **environmental risks**. Use of predators, parasites, pathogens and genetically modified organisms involves a danger that **other beneficial organisms may be displaced or harmed**. Indeed, there is even a risk that the **biocoenosis will undergo extensive and uncontrollable changes** as a result of the inherent momentum of biological processes. For instance, biological control of the coffee berry beetle with the aid of the fungus Beauveria bassiana jeopardises silk production in the coffee-growing region, as the fungus also attacks the silkworm (Bombyx mori).

In another case, a non-indigenous species of toad was introduced to combat insect pests in sugar cane. However, these toads switched to a different source of food and themselves became an almost uncontrollable nuisance.

Where plants develop artificially-**induced resistance** to a pathogenic virus following initial infection with low-virulence strains of the same virus or a similar one, there is a **risk of virus mutation** or - if other viruses are also present - a danger of **synergistic effects.**

- **Protective measures**

To prevent adverse environmental impacts, **biological plant protection measures**, particularly those in the field of genetic engineering, must be subject to **statutory regulations** and **controls**.

The (further) development of **genetic engineering techniques** in connection with which the **risk of uncontrollable biological processes** can be predicted or discerned beforehand is to be **prevented** by way of effective legislation (cf. risks arising from biological agents, as described in the environmental brief Analysis, Diagnosis, Testing). **Biological pest control programmes** must be subject to effective **government control**. **Organisations** to investigate and record the import of predators and parasites are to be **set up** (quarantine).

2.2.5 Integrated methods

- **Environmental impacts**

Depending on the combination of measures chosen from the range of available options, the resultant environmental impacts will be similar to those described above for the individual types of method, albeit on a far **smaller scale**. **Economic-threshold concepts** are to be **further developed**, taking into account their practical applicability. Where pesticides with low active-ingredient dosages are used frequently, certain strategies may well **promote development of resistance** on the part of the harmful organisms. To permit repeated application of plant protection measures, **permanent vehicle access** to a site is often necessary and there is thus a risk of **damage to the soil structure**, e.g. compaction in wet weather. In many cases the only way of solving this problem is to use **lightweight vehicles,** which require a **sizeable input of capital**.

- **Protective measures**

The comments already made regarding the individual types of measures also apply to integrated methods involving a combination of individual measures from the four areas discussed above.

3. Notes on the Analysis and Evaluation of Environmental Impacts

Plant protection measures have a wide variety of impacts on the environment. As there are **no universally applicable concepts**, methods must be assessed by comparing their environmental impacts. In order to weigh up alternative plant protection methods, **assessment criteria** are needed. This calls for **indicators** which convey **qualitative and quantitative impacts** - including their **duration** - as accurately as possible so as to permit comparison (cf. environmental brief Plant Production). The active ingredients, additives and conversion products of pesticides are analysed to establish their **physical and chemical characteristics** (persistence, evaporability, adsorption, desorption etc.). Reproducible **measured values** incorporating safety factors are used to determine their **toxicity and residue properties** (acute 50-values), chronic toxicity (no-effect level, acceptable daily intake [ADI]), maximum-quantity regulation (permissible level). The values serve as **indicators or limits** and must be compared with the actual contamination levels in foods and animal fodder, flora and fauna, soil, water and air. Synergistic and additive effects resulting from use of pesticides can be identified only by studying the relationships between environmental impacts (e.g. decline in particularly sensitive species, use of indicator plants, diversity studies etc.). These **relationships** are **as yet known only in part** and are to some extent **obscured** by the effects of other measures; in many cases they thus cannot be ascribed to plant protection measures alone.

Findings which have emerged during implementation of plant protection measures (e.g. depletion of resources or adverse social consequences resulting from such measures) provide pointers for additional assessment criteria.

Where negative environmental impacts are likely, it must be considered whether these can be remedied without excessive outlay. **Risks of irreversible damage** must be ascertained separately and assessed accordingly. **Plant protection methods** have an influence on **employment structures** (e.g. division of labour between men and women, workload and capital requirements). Further assessment criteria can be developed on the basis of their impacts on farm structures and production.

4.   Interaction with Other Sectors

Plant protection is linked to other plant production measures and is thus **subordinate to the goals of plant production** (cf. environmental brief Plant Production). Measures in the field of plant production also have a **bearing** on the goals and environmental impacts of the **following sectors**:

-     Livestock farming (fodder, quality control)
-     Fisheries (prevention of water pollution)
-     Agro-industry (quality standards)
-     Health and nutrition, including drinking-water supplies (toxicology, residues)
-     Analysis, diagnosis, testing (quality control, development, analytical techniques)
-     Chemical industry (pesticide production)

Decisions on plant protection measures may therefore be influenced by measures in these areas. When assessments are being made, attention must be paid to the possibility that impacts generated by the various sectors could have a **cumulative** effect and thereby **increase the amount of damage done.**

5. Summary Assessment of Environmental Relevance

Plant protection measures must be assessed within the context of the overriding goals of plant production, taking into account site-specific conditions as well as economic and socio-economic factors. The **substances and forms of energy** used in plant protection may have **adverse impacts** on humans, flora, fauna, foods, animal fodder, soil, water and air. Measures to control harmful organisms affect the diversity of species as well as the population density of individual species and have impacts at system level (biocoenosis).

Numerous options are available in terms of **plant protection methods**. Analysis and evaluation of their environmental impacts should lead to selection of **methods** which are comparatively **environment-friendly**, thereby ensuring that **undesirable or unjustifiable impacts are avoided**.

Environmentally oriented plant protection strategies are characterised by **targeted fostering and use of ecosystem-specific natural limiting factors**, backed up by other measures from the wide range of physical, chemical, biotechnical and biological methods.

6. References

Basic literature/General

1.   BAIER, C., HURLE, K. AND KIRCHHOFF, J., 1985: Datensammlung zur Abschätzung des Gefahrenpotentials von Pflanzenschutzmittel-Wirkstoffen in Gewässern. Schriftenreihe des deutschen Verbands für Wasserwirtschaft und Kulturbau e.V., Heft 74.

2.   BICK, H., HANSMEYER, K.H., OLSCHOWY, G. AND SCHMOOCK, P. (Eds.), 1984: Angewandte Ökologie - Mensch und Umwelt. Band I: Einführung, räumliche Strukturen, Wasser, Lärm, Luft, Abfall. G. Fischer Verlag, Stuttgart.

3.   BÖRNER, H., 1989: Pflanzenkrankheiten und Pflanzenschutz. 6. Auflage, Ulmer Taschenbuchverlag, Stuttgart.

4.   BUNDESGESETZBLATT BGBL (Federal Law Gazette), 1986: Gesetz zum Schutz der Kulturpflanzen (PflSchG) vom 15.09.86 BGBL Teil I, Nr. 49, 1505 - 1519.

5.   BUNDESMINISTERIUM FÜR ERNÄHRUNG, LANDWIRTSCHAFT UND FORSTEN (German Federal Ministry of Food, Agriculture and Forestry), 1986: Biologischer Pflanzenschutz. Schriftenreihe des BMELF, Reihe A, Nr. 344, Landwirtschaftsverlag, Münster-Hiltrup.
      1988: Schonung und Förderung von Nützlingen. Schriftenreihe des BMELF, Reihe A, Nr. 365, Landwirtschaftsverlag, Münster-Hiltrup.

6.   EESA, N.M. AND CUTKOMP, L.K., 1984: A glossary of pesticide toxicology and related terms. Fresno: Thomson, 84 p.

7.   FIGGE, K., KLAHN, J. AND KOCH, J., 1985: Chemische Stoffe in Ökosystemen. Schriftenreihe Ver. Wasser-, Boden-, Lufthygiene 61: 1-234.

8.   FOOD AND AGRICULTURE ORGANIZATION, 1986: International code of conduct on the distribution and use of pesticides. Rome.

9.   HEINRICH, D., AND HERGT, M., 1990: DTV-Atlas zur Ökologie. Tafeln und Texte. DTC-Verlag, Munich.

10. HEITEFUSS, R., 1987: Pflanzenschutz. Grundlagen der praktischen Phytomedizin. 2. Auflage, Thieme-Verlag, Stuttgart.

11. HOLDEN, P.W., 1986: Pesticides and groundwater quality. National Academic Press, Washington.

12. INTERNATIONAL ORGANIZATION OF CONSUMERS UNIONS (IOCIJ), 1986: The Pesticide Handbook - profiles for action. Penang, Malaysia.

13. IVA (Industrieverband Agrar) (Ed.), 1990: Wirkstoffe in Pflanzenschutz- und Schädlingsbekämpfungsmitteln. Physikalisch-chemische und toxikologische Daten. BLV-Verlagsgesellschaft, Munich.

14. KORTE, F. et al., 1987: Lehrbuch der ökologischen Chemie, Thieme Verlag, Stuttgart.

15. KRANZ, J., SCHMUTTERER, H., AND KOCH, W., 1979: Krankheiten, Schädlinge und Unkräuter im tropischen Pflanzenbau, Parey Verlag, Hamburg.

16. FRANZ, J.M. AND KRIEG, A., 1976: Biologische Schädlings-bekämpfung. 2. Auflage, Pareys Studientexte 12, Verlag P. Parey, Hamburg.

17. MORIARTY, F., 1988: Ecotoxicology. The study of pollutants in ecosystems. 2nd ed., Acad. Press, London.

18. MÜLLER-SÄMANN, K.M., 1986: Bodenfruchtbarkeit und standortgerechte Landwirtschaft. Maßnahmen und Methoden im tropischen Pflanzenbau. Schriftenreihe der GTZ Nr. 195, Roßdorf.

19. NATIONAL RESEARCH COUNCIL (Ed.), 1986: Pesticide resistance - strategies and tactics for management. NAT. Acad. Press, Washington.

20. PERKOW, W., 1988: Wirksubstanzen der Pflanzenschutz- und Schädlingsbekämpfungsmittel. Parey, Hamburg (published in loose-leaf form).

21. SCHEFFER, F. AND SCHACHTSCHABEL, P., 1989: Lehrbuch der Bodenkunde. 12. Auflage, Enke Verlag, Stuttgart.

22. SCHMIDT, G.H., 1986: Pestizide und Umweltschutz. Vieweg Verlag, Braunschweig.

23. SCHUBERT, R., 1985: Bioindikatoren in terrestrischen Ökosystemen. G. Fischer Verlag, Stuttgart.

24. SCHWAB, A., 1989: Pestizideinsatz in Entwicklungsländern - Gefahren und Alternativen. Margraf Verlag, Weikersheim.

25. STOLL, G., 1988: Natural crop protection - based on local farm resources in the tropics and subtropics. 3rd edition, Margraf Verlag, Weikersheim.

26. UBA (UMWELTBUNDESAMT) (German Federal Environmental Agency (Ed.)), 1984: Chemikaliengesetz, Prüfung und Bewertung der Umweltgefährlichkeit von Stoffen. UBA Bewertungsstelle Chemikaliengesetz, Berlin.

27. VOGTMANN, H. (Ed.), 1988: Gentechnik und Landwirtschaft - Folgen für Umwelt und Lebensmittelerzeugung. Alternative Konzepte 64, C.F. Müller Verlag, Karlsruhe.

28. WARE, G., 1986: Fundamentals of pesticides. - Fresno: Thomson, 2nd edition.

29. WEISCHET, W., 1977: Die ökologische Benachteiligung der Tropen. Teubner Verlag, Stuttgart.

30. WITTE, I., 1988: Gefährdungen der Gesundheit durch Pestizide - Ein Handbuch über Kurz- und Langzeitwirkungen. Fischer Verlag, Frankfurt.

Further/supplementary literature

31. ANON., 1987: EPA's new policy on inerts; in: Farm Chemicals International Vol. 1, No. 4, Summer 1987, pp. 22-25.

32. ARBEITSGRUPPE TROPISCHE UND SUBTROPISCHE AGRARFORSCHUNG (ATSAF), 1987: Möglichkeiten, Grenzen und Alternativen des Pflanzenschutzmitteleinsatzes in Entwicklungsländern. Sachstandbericht zu Projekten der deutschen Agrarforschung 1980-1987, Bonn.

33. AREEKUL, S., 1985: Ecology and environmental considerations of pesticides; Department of Entomology, Kasetsart University, Bangkok (working paper).

34. BOLLER, E., BIGLER, F., BIERI, M., HÄNI, F. AND STÄUBLI, A., 1989: Nebenwirkungen von Pestiziden auf die Nützlingsfauna landwirtschaftlicher Kulturen. Schweiz. Landw. For. 28: 3-40.

35. BIJLL, D., 1982: A growing problem - pesticides and the Third World Poor. Oxford: Oxfam.

36. CAIRNS, J., 1986: The myth of the most sensitive species. BioScience 36: 670-672.

37. CARL, K.P., 1985: Erfolge der biologischen Bekämpfung in den Tropen. Giessener Beitr. Entwicklungsforsch. I, 12: 19-35.

38. CHIANG, H.L., 1982: Factors to be considered in refining a general model of economic threshold. Entomophaga 27 (special issue): 99-103.

39 DEUTSCHE GESELLSCHAFT FÜR TECHNISCHE ZUSAMMEN-ARBEIT (GTZ) GMBH,
  1978: Rückstandsprobleme im Pflanzenschutz in der Dritten Welt. GTZ-Schriftenreihe Nr. 63, Eschborn.
  1987: Nebenwirkungen bei der Anwendung chemischer Pflanzenschutz-mittel. Arbeitsunterlagen für Projekte im ländlichen Rahmen Nr. 8, Eschborn.
  1988: Technische Zusammenarbeit im ländlichen Raum - Pflanzen- und Vorratsschutz. Schriftenreihe der GTZ, Eschborn.
  1989: ZOPP-Unterlagen, Forstschutz, Marokko, Eschborn.
  1990: Kaffeerostbekämpfung in der Dominikanischen Republik, Gutachten, Eschborn.
  1990: Bericht über die Fortschrittskontrolle zum Projekt Ausbildung und Beratung im Pflanzenschutz, Eschborn.

40. DOMSCH, K.H., JAGNOW, G. AND ANDERSON, T.M., 1983: An ecological concept for the assessment of side effects of agrochemicals on soil micro-organisms. Residue Review 86: 65-105.

41. FOOD AND AGRICULTURE ORGANIZATION, UNITED NATIONS ENVIRONMENT PROGRAMME, 1989: Integrated pest control. Report of the 14th session of the FAO/UNEP panel of experts meeting.

42. GOODELL, G., 1984: Challenges to international pest management re-search and extension in the Third World: Do we really want IPM to work? in: Bulletin of the Entomological Society of America, Vol. 30, No. 3.

43.   HAQUE, A. AND PFLUGMACHER, J., 1985: Einflüsse von Pflanzen-
      schutzmitteln auf Regenwürmer. Ber. Landwirtschaft 198: 176-188.

44.   HASSAN, S.A., 1985: Standard methods to test the side effects of pesticides
      on natural enemies of insects and mites developed by the IOBC/WPRS
      Working Group "Pesticides and beneficial organisms". J. Appl. Ent. 105:
      321-329.

45.   HEINISCH, E. AND KLEIN, S., 1989: Einsatz chemischer Pflanzen-
      schutzmittel - ein Spannungsfeld von Ökonomie und Ökotoxikologie.
      Nachrichten Mensch + Umwelt, 17: 53-66.

46.   HUISMANS, J.W., 1980: The international register of potentially toxic
      chemicals (IRPTC). Ecotox. Environm. Safety: 276-283.

47.   IGLISCH, I., 1985: Bodenorganismen für die Bewertung von Chemikalien.
      Z. Angew. Zool. 72: 395-431.

48.   KNEITZ, G., 1983: Aussagefähigkeit und Problematik eines Indikator-
      konzepts. Verh. Deutsch. Zool. Ges. 1983: 117-119.

49.   KOCH, W., SAUERBORN, J., KUNISCH, M. AND PÜLSCHEN, L., 1990:
      Agrarökologie und Pflanzenschutz in den Tropen und Subtropen. PLITS
      1990/8(2), Verlag J. Margraf, Weikersheim.

50.   KOCH, R., 1989: Umweltchemie und Ökotoxikologie - Ziele und Aufgaben.
      in Umweltchemie Ökotox. 1: 41-43.

51.   KÖNIG, K., 1985: Nebenwirkungen von Pflanzenschutzmitteln auf die
      Fauna des Bodens. Nachrichtenbl. Deut. Pflschutz. 37: 8-12.

52.   KRANZ, J. (Ed.), 1985: Integrierter Pflanzenschutz in den Tropen. Gießener
      Beiträge zur Entwicklungsforschung. Reihe 1, Band 12, Gießen.

53.   LEVIN, S.A., HARWELL, M.A., KELLY, J.R. AND KIMBALL, K.D.
      (Ed.), 1989: Ecotoxicology: Problems and approaches. - Springer, New
      York.

54.   LEVINE, R. S., 1986: Assessment of mortality and morbidity due to unin-
      tentional pesticide poisonings. Geneva, WHO. Document, VBC, 86, 929.

55.   MALKOLMES, H.-P., 1985: Einflüsse von Pflanzenschutzmitteln auf
      Bodenmikroorganismen und ihre Leistungen. Ber. Landw. 198: 134-146.

56. MAY, R.M., 1985: Evolution of pesticide resistance. Nature 315: 12-13.

57. MÜLLER, P., 1987: Ecological side effects of Dieldrin, Endosulphan and Cypernlethrin application against the TseTse flies in Adamoua, Cameroon. Initiated by the GTZ and World Bank, Eschborn and Washington.
1988: Ökotoxikologische Wirkungen von chlorierten Kohlenwasser-stoffen, Phorsäureestern, Carbamaten und Pyrethroiden im nordöstlichen Sudan. Im Auftrag der GTZ, Eschborn.

58. OTTOW, J.C.G., 1982: Pestizide- Belastbarkeit, Selbstreinigungsvermögen und Fruchtbarkeit von Böden. Landwirtschaftliche Forschung 35, 238-256.

59. OWESEN, H.A., 1976: Artendiversität in der Ökologie. SFB 95, Rep. 16, Kiel.

60. PAN (PESTICIDE ACTION NETWORK), 1987: Monitoring and reporting the implementation of the international code of conduct on the use and distribution of pesticides. Final report. Nairobi, Kenya: Environm. Liaison Centre.

61. PIMENTEL, D. et al., 1980: Environmental and social costs of pesticides: A preliminary assessment; in: OIKOS 34, p. 126-140.

62. SCHMID, W., 1987: Art, Dynamik und Bedeutung der Segetalflora in mais-betonten Produktionsystemen Togos. PLITS 1987/3(2), Verlag J. Margraf, Weikersheim.

63. SCHMUTTERER, H., 1985: Versuche zur biologischen und integrierten Schädlingsbekämpfung in den Tropen. Giessener Beiträge Entwick-lungsforsch. I, 12: 143-149.

64. SCHOENBECK, F., KLINGAUF, F. AND KRAIJS, P., 1988: Situation, Aufgaben und Perspektiven des biologischen Pflanzenschutzes. Ges. Pfl. 40: 86-96.

65. STREIT, B., 1989: Zum Problem der Bioindikatoren aus zoologisch-ökolo-gischer Sicht. Geomethodica 14: 19-45.

66. SWIFT, M.J. et al., 1977: Persistent pesticides and tropical soil fertility. Meded. Fac. Landbouw. Rijksuniv. Gent 42: 845-852.

67. VÖGELE, J.M., KASKE, R. AND SCHMUTTERER, H., 1989: Biologische Schädlingsbekämpfung im Südpazifik. Gutachten im Auftrag der GTZ, Eschborn.

68. WAIBEL, H., 1987: Die Einstellung von Kleinbauern in Südostasien zum Pflanzenschutz, in: Möglichkeiten, Grenzen und Alternativen des Pflanzenschutzmitteleinsatzes in Entwicklungsländern. DSE/ZEL, Feldafing.

Agriculture

Forestry

29. Forestry

Contents

1. Scope

1.1 General

Forestry is generally defined as **utilisation of forests** to **satisfy human needs**. Characteristic features of forestry are the **extremely long production periods** extending over several decades and, in the case of timber production, the fact that the product is identical with the production input. **Production of commercial timber** and the use of **return on investment** as a **success criterion** for forest owners, generally state authorities (cf. REPETTO), are **unsuitable** for integrated problem solutions in view of the socio-economic conditions.

Probably the most important aspects of forestry activities today are **protection of species and biotopes** and **preservation of the human habitat**. In virtually no other sector are the "limits of growth" demonstrated as clearly as by the global destruction of the forests. The **impacts** of this process are **no longer confined to specific regions**, but are **interlinked on a global scale**, as forests - along with the oceans - represent the most important terrestrial **bioregulators** of global cycling systems and the Earth's climate.

The **tropical forest** is particularly **badly affected**, with around 20 million hectares being clear-cut or degraded each year (cf. ENQUETE). **Tropical moist forests** cover only around 6% of the land on Earth, yet they provide a **living environment** for over half of all species of fauna and flora and for millions of people.

Although the worldwide **destruction of the forests** has a wide variety of forms, causes and consequences, it can nevertheless be ascribed above all to **exploitation and conversion of forest resources** to satisfy **short-term** economic and personal interests. This form of activity eventually leads to the degradation and loss of human habitats.

This situation imposes **new demands** in terms of executing agency, project size and site. **Holistic approaches** which also include the sectors immediately related to forestry (CARILLO et al.) are therefore essential.

1.2 Subsectors

The forestry sector differs from every other sector of an economy by virtue of its **production period** and the fact that it **produces both tangible goods and intangible benefits**. It thus calls for special forms of biological production. Four **subsectors** are involved: planning, formation and utilisation of forest stands, and harvesting. Attention will be drawn where necessary to the special features of tropical forest management.

1.2.1   Biological production/Planning

Biological production in the forestry sector is controlled through the various methods of **forest management, agroforestry and product harvesting**. The purpose of these methods is to **achieve**, by controlling site production potential, the **management goals** laid down during planning; such goals may involve production, conservation or a combination of the two.

Planning, establishment of a stand and utilisation are classed as **subsectors** of biological production.

All forestry planning is based on **inventories** (e.g. ZOEHRER), which provide the framework for forest management over a period of ten years in most cases. Apart from **static** elements such as timber stock, **dynamic** and **structural elements** such as increment and horizontal and vertical species and diameter distribution must also be recorded, particularly in tropical moist forests, otherwise the **sustainability** of yield regulation cannot be guaranteed.

In addition to **measurement** of quantitative stand parameters, multifunctional forest management also calls for comprehensive **site mapping** (e.g. DENGLER, MAYER, WENGER) which ascertains **geoecological factors** for each stand individually.

The combination of **continuous forest inventories, site mapping and specification of planning targets** is known as **forest management planning**, the details of which are laid down in forest management plans. The management and recording unit is the compartment. The **division** of a forest into compartments should reflect the topographical and hydrological differences between stands. **Simple geometrical shapes** are appropriate only in **exceptional cases** for lowland plantations. In heterogeneous tropical moist forests, division can be extended to **water catchment areas** (sub-catchments).

The individual **management goals** are laid down separately for each compartment on the basis of forest management planning and economic analyses. Apart from the traditional economic indicators such as internal rate of return, cost-effectiveness analyses (cf. FÄHSER, WENGER) can be used to assess the relative advantages of alternative management goals.

The function and structure of forest resources vary depending on the intensity with which they are utilised and the geoecological zone in question (MUELLER-HOHENSTEIN). The **management concepts** must be geared to the characteristic features of each type of resource, e.g. high and low forests, savannahs, mangroves and agroforestry systems or resources for gathering. The site-specific **limiting factors** - namely forest area, water, nutrients and light - **restrict** the scope for optimising forestry operations from the management viewpoint.

The different **types of resource** are closely **interrelated** and can **complement** one another in functional terms. These interrelationships necessitate **integrated, interlinked planning**. Storeyed, **species-rich high forests** of primary or secondary origin are best able to fulfil the necessary protective and utility functions over time and area. **Commercial-timber plantations** can alleviate demographic pressure on natural forests if the local population are involved in plantation management.

1.2.2   Establishment of a stand

Forests are **regenerated** by artificial or natural means at the end of their rotation periods or if they become overmature.

- **Planting**

Artificial forms of the establishment of a stand comprise conventional **afforestation** on open sites (EVANS, GOOR), **planting under shelterwood** to fill gaps in naturally regenerated stands or **improvement measures** in over-used forests (LAMPRECHT). The form of **soil preparation** and the **planting technique** depend on species, soil fertility, water balance and land condition. **Brushwood and felled-area flora** can be placed in windrows or burned unless this appears inadvisable on account of nutrient losses and soil erosion.

Planting techniques range from **hole planting** in clearings and deep soils through **cuvette planting** ("micro catchments") on slight slopes in arid regions to **terracing** on steep slopes in high mountain regions. For crusted laterite soils, **complete turning** of the soil is usual.

Irrespective of the technique used, **artificial regeneration** produces **forests having a simple structure** and largely open cycling systems. The tree species selected are generally vigorously growing, competitive **pioneer species**. The number of species and the degree to which they are mixed remain low for reasons of manageability. The biotic and abiotic **risks** to such plantations become greater as aridity increases (shortage of water, fire) and as soil fertility decreases, particularly on geologically old substrates in the inner humid tropics. The relatively **small range of species** on **marginal sites** is a result of **natural** factors.

Measures to protect stands against fire, storms, water stress, nutrient deficiencies and disease are frequently necessary (see FRANZ, for example, with regard to biological pest control). Controlled **burning** of organic surface layers (known as "prescribed burning") is a special plantation-management practice employed in arid regions to remove highly inflammable organic layers (BROWN, GOLDAMMER). Use of this method is restricted by the demands of **soil and water conservation**.

The **planting stock** needed for artificial afforestation is produced in **nurseries**, which usually concentrate on generative raising of planting stock from seed. Vegetative forms of plant production, such as propagation by cuttings, tissue cultures or seed production in seed plantations, are often **capital-intensive** and usually have to be carried out **on a central basis** by **agencies possessing the necessary expertise** (e.g. KRUESSMANN). It must be ensured that the consequences of **genetic impoverishment** can be overcome by means of an appropriate silvicultural strategy. **Consumption and use** of the **inputs** commonly required by nurseries (land, seed, fertiliser, substrate, water, pesticides, means of transport) depend on the specifications of the forest management plans (management goal type planning) and the propagation method employed. The high **substrate consumption** involved in production of pot or container plants calls for **careful logistical planning**.

Natural-forest programmes could involve **small** or **temporary nurseries** in the vicinity of the stand, as the provenances and varieties necessary to achieve the related management goals can best be produced on a decentralised basis. **Regional afforestation programmes** can be supplied by large **central nurseries**.

• **Natural regeneration**

Growth dynamics, phenology and adherence to the given spatial arrangement constitute **criteria** for selection of **natural regeneration techniques** (ASSMANN, GOLLEY, DENGLER, LEIBUNGUT, MAYER, WEIDELT, WHITEMORE). A general distinction is made between **methods** for clear-felling high forests, involving shelterwood and strip techniques, and the selection method for storeyed, species-rich stands. **Special demands** in terms of species protection and soil conservation arise in connection with conversion systems in tropical **moist forests** (LAMPRECHT) and management of **high-altitude forests** (MAYER).

Characteristic of all methods are a specific sequence and number of **individual fellings** over a long period, with care taken to **preserve the dominated stand**. During the regeneration process, the soil usually **remains almost totally covered**. Working of the soil is generally confined to preparing the germination bed with implements such as cultivators or harrows. Prescribed burning may become necessary for regeneration of pyrophytes (cf. section on planting).

Irrespective of the method used, the outcome will ideally be **stands** which, both horizontally and vertically, are more or less **heterogeneous, species-rich and multi-storeyed**, and which have closed natural cycling systems. Such stands are **highly resistant** to injurious factors and there is thus little need for protective measures and artificial raising of plants.

1.2.3 Stand utilisation

In generalised terms, distinctions can be made between the following forms of utilisation:

- conventional **forest management** for production of timber and to realise protective functions (e.g. DENGLER, LAMPRECHT, MAYER)
- **agroforestry** for integrated production of agricultural and forestry products (e.g. ICRAF)
- **gathering,** generally in connection with non-timber products (e.g. DE BEER).

Irrespective of the form of utilisation, it is essential to **control** the relevant **economic** (e.g. McNEELY) and **demographic conditions** in order to prevent over-utilisation. A **key role** is played by **forest tending**, without which it is impossible to utilise the wide range of opportunities offered by multiple-use forestry.

- **Forest management**

Distinctions are generally made between **three types of forestry operation** (e.g. DENGLER, MAYER):

- high forest created by means of natural or artificial regeneration, incorporating clear-felling (age-class) forest and all-aged (selection) forest
- middle forest originating from coppices and planting
- low forest originating from coppices

Low and middle forests are generally clear-cut in short rotation cycles. They are highly suited to production of **fuelwood** and other small dimensions for supplying local markets, provided that care is taken to implement the necessary protective measures such as regulation of forest grazing, terracing or cultivation of site-appropriate provenances. **Management criteria** are based on economic and technical data such as diameter and timber stock.

In **high-forest management** a distinction is made between final felling during stand regeneration and intermediate felling during forest tending. Long rotation periods and periodic tending measures make high forests suitable for achieving multifunctional goals such as **production of high-grade timber** and ensuring of **welfare functions**.

The purpose of **forest tending** is to raise **stable stands** by controlling the stages of stand development (e.g. MAYER). The **management criteria** relate to aspects of **silviculture and forest yield science**, such as basal area, number of trees, species and diameter distribution, or target diameter. Irrespective of the stage concerned, control of the limiting factor represented by **light** is a characteristic element of forest tending (e.g. BAUMGARTNER in MAYER).

In generalised terms, a distinction can be made between young growth tending, or **pre-commercial thinning**, and **commercial thinning**. Both **chemical and mechanical methods** are used, the **latter** being performed **manually or with the aid of machines**. **Chemical methods** are feasible only if **non-persistent agents** can be applied on a targeted basis. Systematic methods such as row thinning are commonly employed on monotypic plantations; only in favourable locations are they unlikely to give rise to problems (soil erosion).

Selective methods, known as **selection thinning** (e.g. MAYER) or timber stand improvement (WEIDELT, WENGER, WÖLL), are **the most effective in terms of yield and from the ecological viewpoint**. A characteristic feature of such methods is **regulation of the growing space** of trees preselected for harvesting through removal of competing adjacent trees. In all-aged forests, growing-space regulation, intermediate felling, final felling and regeneration can be combined, whereas in artificially formed plantations this is possible only if at least two tree species have coordinated degrees of shade tolerance.

Tropical moist forests have **special requirements** where timber production is concerned (e.g. LAMPRECHT; WEIDELT; WHITEMORE 1984).**Selective methods** are in most cases particularly suitable for such forests in view of the diversity of species, the horizontal mosaic structure, the vertical layering, the nutrient balance (see Section 3) and the phase-controlled growth dynamics. In simplified terms, it is possible to distinguish four cyclically linked development stages (WHITEMORE, 1978): terminal stage - gap/group stage - build-up stage - maturity/climax stage. Forest regeneration and timber harvesting must be geared to these growth dynamics.

In principle, even **timber harvesting in primary forests** containing irregularly fruiting tree species with uneven diameter distribution is justifiable if only **dying individual trees** in the main stand are removed using **highly mobile harvesting methods** (see 1.2.4) after fruiting ("mortality pre-emption", SEYDACK 1990). In stands consisting of tree species with even diameter distribution, located on fertile and erosion-resistant sites, the trees can be **removed in groups**. Less selective methods are possible where **stand conditions are homogeneous**, as is the case for example in many natural tropical coniferous forests. However, harvesting of hitherto **unutilised tree species** (lesser-known species) is acceptable only if the nutrient balance (see Section 3) remains in equilibrium and the reproductive biology of the species concerned is known.

Management of **special forms** of tropical forest vegetation, such as gallery forests, savannahs and mangroves, **cannot be treated** within the scope of this environmental brief (cf. CHAPMAN, GOLLEY et al. 1978).

• **Agroforestry**

It is above all in the humid tropics that increasing population pressure is blurring the dividing line between agriculture and forestry. In **areas on the fringes** of intact forests, **combining** agricultural and forestry operations is often the only way of meeting the population's food and **wood requirements**. Agroforestry operations exhibit a **higher degree of ecological stability** than purely agricultural ones and in many places are the only means of permitting permanent cultivation (e.g. JORDAN).

Although there are no universally valid definitions, distinctions can be made for practical purposes between **agroforestry, silvopastoral** and **agrosilvopastoral** systems.

The degree to which agricultural and forestry elements are **integrated in terms of time and area** (e.g. ICRAF) depends on the available know-how, the availability of water, the soil fertility and the market. In **marginal areas remote from markets** it is generally possible to realise only **simple forms** of agroforestry, such as slash-and-burn agriculture (also known as shifting cultivation) or pasture farming in savannahs (PRATT).

• **Gathering**

In many geologically old, tropical moist forests, gathering of **non-timber products**, known as "minor produce", is often the only possible form of sustainable use. This is particularly true of Central Amazonian sites highly deficient in nutrients, where intensive forms of roundwood production result in a negative **nutrient balance**. In many parts of South-East Asia, for example, the **production value** of the minor produce far exceeds that of timber production (DE BEER). As tropical forests yield an immense number of non-timber products, it is impossible to cover all the region-specific aspects of this area within the scope of this environmental brief.

1.2.4   Harvesting techniques

Forests and trees yield numerous **products** important to man: commercial timber, pharmaceutical products, spices, resins, rattan, foods and tanning agents. Each of these products requires a **tailor-made harvesting method** (CAPREZ, STAAF, DE BEER).

• **Timber**

Of all activities in the forestry sector, **timber harvesting** requires the **greatest input of capital** and is **most likely to cause damage**. The strained nutrient balance means that timber harvesting is often impossible in tropical moist forests located on geologically old substrates. Planning and execution of timber harvesting must therefore be based on both **economic and ecological criteria**. The **paramount aim** of all timber harvesting measures is to **minimise damage** to the soil and stand. The following **criteria** must be taken into consideration in **selecting the method to be used**:

-       management goal (rights of use, protective forest, commercial forest)
-       stand density (number of trees, structure, nutrient dynamics)
-       type of felling (intermediate felling, final felling, timber assortments)
-       topography and soil (skidding distance, soil erosion)
-       infrastructure (accessibility, construction costs)

From the operational viewpoint, actual **felling** of the trees is considered separately from **hauling** of the felled trunks. While **mobile harvesting machines** are generally used in non-tropical regions, felling of trees in the tropics is performed **manually** with the aid of an axe, handsaw or power saw. The degree of success is largely determined by the training and remuneration of the forest workers and by the way in which work at the felling site is organised. **Resource-conserving methods of timber harvesting** include the following features:

-       marking of stand before felling (inventory)
-       directional felling
-       conversion of shortwood at the felling site before hauling

In **roundwood handling**, distinctions are made in **organisational** terms between skidding or hauling within the stand and (long-distance) timber transportation (road, rail, waterway), in **technical** terms between manual, animal-powered and mechanical techniques, and in **method-related** terms between whole-tree methods and roundwood methods. The **damage** done to the stand **increases** in proportion to **engine power, intensity of utilisation, slope inclination, degree of accessibility, trunk length** and the amount of **ground skidding** involved. Inappropriate hauling methods can **cause** soil compaction, rill erosion along wheel tracks, destruction of forest soil flora and the dominated stand, and butt and root damage to the rest of the stand. The most important **hauling methods** are given below in generalised terms, listed in order of the potential degree of damage that may result from them:

-       Ground-based methods
    •   skidding hoists for clear-cutting, final felling, moderate or long skidding distances, suitable anywhere from lowland to high mountain regions
    •   wheeled and tracked forest tractors for clear-cutting, final and intermediate felling, short to moderate skidding distances, hilly terrain

- • animals (horses, oxen, water buffalo etc.) for intermediate felling, smallwood, short distances, lowland regions

- Gravity methods
  - • manual floating for intermediate and final felling, short distances, in high mountain regions
  - • log chutes (wooden or earthen chutes), generally for final felling, long distances, high mountain regions

- Airborne methods
  - • travelling winches for intermediate and final felling in high mountain regions
  - • cable cranes for universal use
  - • helicopters for transportation of high-grade timber

In view of the mosaic structure of **tropical moist forests, timber harvesting** in such regions is liable to cause damage to resources unless the methods used take account of the conditions of single-tree harvesting in groups by way of mobility, off-ground hauling and a low-density road network (HODGSON). **Homogeneous forms of stand** in lowland regions allow less complicated methods to be used. **Full-tree or whole-tree methods** are suitable only for nutrient-rich lowland sites that are resistant to erosion.

After felling and hauling, **timber** is **stored** for a short time in the forest by the side of the road until it is removed by the purchaser. It is thus not usually necessary to protect timber stored in this way. In exceptional circumstances, for example after natural disasters, it may be essential to **store large quantities** of timber for lengthy periods in specially created log dumps. Steps must then be taken to **limit the amount of land required and the use of pesticides** and to **dispose of bark shavings**.

Appropriate **options are to be selected** on the basis of time studies, forest damage analyses and economic criteria. In addition to conventional economic **assessment tools** such as cost-benefit analyses, cost-effectiveness analyses (e.g. WENGER) should also be employed. Such analyses must relate to the entire rotation period (production period), rather than being confined to individual operations.

Timber harvesting can have **indirect effects** of environmental relevance in that it **opens up** forest areas in a manner permitting **their subsequent use**. Apart from **selection of environmentally sound timber harvesting methods**, an **efficient forest administration** capable of carrying out surveillance of forest use is **essential** to **minimise damage to the stands**.

- **Non-timber products**

As **non-timber products** encompass such a broad range, the effects of **harvesting** them cannot be described in detail here. It is essential to **draw upon available local know-how** in this connection.

A distinction must be made between products harvested for the harvester's **own use** and those harvested for **marketing**, as there is generally **no danger of over-use** where products are intended merely to meet **subsistence needs**. Special precautions must be taken in harvesting tree products such as resin, bark or climbing plants (e.g. rattan), as the function of the trees as means of production or support can be permanently impaired. Harvesting of "non-tree products" such as fruit or game requires less in the way of specific management if the products in question are not to be marketed.

2.   Environmental Impacts and Protective Measures

2.1   Sector-typical influences on the environment

In terms of area, forests constitute the Earth's **most important terrestrial eco-systems**. Since the "invention" of arable farming around 10,000 years ago, they have been continuously fragmented and degraded and today **cover** less than a third of the Earth's inhabitable land surface, extending over an area of around 42 million km^2 (STARKE). As forests can perform their **protective functions** only where they cover a **large area**, man's living environment in certain regions is in jeopardy. **Four protective functions** will be discussed here:

- **Climate regulation**

Together with the oceans, the Earth's forests constitute a **biological climate regulator**. By means of their high evapotranspiration, they generate a large proportion of the precipitation themselves in some places. Evaporation of this water absorbs up to three quarters of the radiation energy, particularly in the tropics, and thereby prevents excessive warming of the atmosphere. Large quantities of the greenhouse gas CO_2 are fixed as well. These two **climate-regulating functions** can be most efficiently controlled by means of **near-natural, long-lived types of forest containing abundant stocks and covering large areas**. By virtue of their more favourable assimilation/respiration ratio, many temperate forest formations, such as the coastal forests in the north-west of the USA, store up to three times as much CO_2 as tropical rainforests (STARKE, 1991).

•   Protection of genetic resources

Although tropical moist forests cover only a fraction of the Earth's surface (6%), they contain around 90% of all apes, at least 80% of all insects, at least two thirds of all plant species and roughly 40% of all species of birds of prey. As the majority of these species can exist only in **near-natural forms of forest** extending over large areas, monotypic artificial forests covering small areas are unsuitable for protecting species and genetic resources.

•   Soil conservation

Storeyed high forests are the **most efficient biological means of soil conservation**. **Soil erosion** and **soil formation** under such stands are **balanced** and in line with the geoecological norm. The simpler stand structures found in dry forests or grass savannahs mean that such regions differ less markedly from artificial forests. The same applies to alternative forms of forest in lowland regions. Under humid tropical conditions and in high mountain regions, the **erosion rates in artificial forests** may far exceed the natural soil loss rate (MORGAN).

•   Protection of human habitats

Rapid deforestation is **constricting** the human **habitat**, particularly in tropical moist forests, while at the same time destroying jobs. Tax concessions for **large-scale projects** (timber exploitation, mining, cattle rearing) can accelerate this process locally and displace the labour-intensive methods involved in traditional resource utilisation. It is thus above all in natural-forest and agroforestry projects that **training and upgrading** can play an important part in raising decision-makers' awareness of the relevant issues.

2.2   Sector-typical protective strategies

Forests perform vital **protective functions**, but at the same time **require protection themselves** in their function as biotopes housing a variety of plant and animal communities. However, effective protection of forests is possible only if the state, industry and the local population all have an **interest** in their long-term preservation. The ways in which forests are used must therefore ensure **protection of forest resources** and sustainable **generation of added value**, besides being **acceptable** to all interest groups involved. From the hygiene viewpoint, for example, the clearance of African savannahs infested with the tsetse fly is highly beneficial. For Iko bushmen and other game hunters, however, it means the destruction of their living environment, while for hydrologists it means flooding in low-lying areas and for nature conservationists it represents the destruction of biotopes.

Depending on site conditions, a **protective strategy** in the forestry sector will include components such as the **following**:

-       Political/economic instruments

    •   regulating forest utilisation by interlinking protective, buffer, ex-
            ploitation and settlement areas
    •   ensuring the generation of added value through utilisation of forests
            by means of diversification in the producer region and reinvestment
            of profits, e.g. in forest-tending programmes
    •   participative planning, implementation and monitoring of forest
            utilisation concepts
    •   moratorium on timber exploitation in primary forests located in
            tropical and temperate zones
    •   market-oriented incentives such as input and output taxes or subsi-
            dies for substitutes (e.g. use of cable cranes instead of bulldozers in
            timber harvesting)

-       Technical/ecological instruments

    •   reducing wood consumption through improvements in wood proc-
            essing
    •   function- and needs-oriented forest management by means of silvi-
            cultural planning on a single-stand basis
    •   simulation of natural growth dynamics and forest tending through
            long rest periods and natural-regeneration periods

An implementation-oriented discussion of the above elements can be found in the reference literature (cf. BMZ, ENQUETE).

3.   Notes on the Analysis and Evaluation of Environmental Impacts

All action in the forestry sector is **based** on the principle of **sustainability**. This requires a form of utilisation which is in line with the potential of the natural resources and which preserves both the steady state of the natural cycling systems and the **ecosystems' capacity for self-regulation** (VESTER). Sustainability thus does not imply a constant annual yield level - in timber production, for example - but rather the achievement of goals such as **ensuring species-rich natural re-generation** through the **use of resource-conserving timber-harvesting methods** (SEYDACK et al. 1990).

As intervention by man disrupts these cycles, it is essential to use not only sustainability indicators such as annual cut but also **ecological and socio-economic indicators**:

- **Nutrient balance**

Nutrient cycles are a **function** of stand density, soil exchange capacity, nutrient storage and allochthonous introduction of substances via the atmosphere. As it is virtually impossible to control exchange capacity, storage and introduction of substances, management concepts must aim to **minimise nutrient losses**. If use of mineral fertilisers is to be avoided (since they generally necessitate use of non-renewable energy sources), nutrient losses **can be offset** by means of **allochthonous introduction of substances** only where **small quantities of stem timber are removed over long production periods**. Nutrient-deficient sites severely restrict production of large timber and biomass (see GOLLEY, RUHIYAT 1989 in WEIDELT 1989, ULRICH). Relevant **indicators** are

- nutrient reserves in kg/ha, broken down according to ecological compartments such as soil, roots, stem timber, branches and foliage, and

- nutrient flows between the individual compartments in kg/ha/a, including introduction and removal of substances.

- **Water balance**

Water is a **limiting factor** in many habitats. Its availability varies according to hydrogeological and bioclimatic conditions. As these components of the natural environment cannot be changed, the **intensity of utilisation** must be **geared** to the dynamics of the water balance in individual catchment areas. Near-natural storeyed forests are most capable of controlling the water balance. The components of the **water balance** - i.e. interception, evapotranspiration, run-off and groundwater recharge - can be controlled by means of **forest tending** and **species selection** (cf. MITSCHERLICH, WENGER). Depending on purpose, individual components can be used as sustainability indicators, for example to quantify groundwater recharge in arid regions.

• Soil erosion

Soil erosion is essentially a function of **stocking, precipitation** and **relief intensity**. It forms part of the Earth's natural cycling system. The **smallest degree** of soil erosion occurs under **species-rich, storeyed high forests**. The **indicator** for soil erosion is

-    the site-specific geological norm (kg/ha/a), which can be ascertained on ecologically undisturbed sites by means of simple field trials (e.g. FAO) or, if this is impossible in totally degraded regions,

-    the tolerable soil loss threshold, which can be ascertained by empirical means with the aid of the general soil loss equation after WISCHMEIER (e.g. in MORGAN).

Both of these indicators provide a criterion for the **intensity of utilisation** and the technical and biological **protective measures** required.

• Forest area

The **minimum amount** of forestry land **required** is determined by the **population's requirements** in terms of forest products and the economically necessary **protective functions**. The amount of land required depends on site-specific factors and the habits of the local people. In addition to ecological criteria, wood requirements and wood consumption must be taken into account along with the degree of fragmentation (ELLENBERG, PIELOU) of formerly continuous forest areas. One indicator is the **forest area balance**, expressed (in hectares) as the difference between the existing forest area and that which is economically necessary.

In the event of **changes in the intensity of forest utilisation** it is essential to know the parameters serving as a kind of **early warning system** which make it possible to spot new problems as soon as they start to develop. Apart from the **ecological** indicators mentioned above, such indicators may also be **biological** (pioneer plants, particular types of animal as anthropophilous species) or **socioeconomic** (increased market supply of gathered products hitherto used only locally).

Economic assessment of forest resources involves various factors of uncertainty. Conventional monetary methods do not adequately cover the forest's indirect functions or the non-timber products generated "informally" to meet the population's own requirements. Cost-effectiveness and risk-analysis methods must therefore be used for evaluations in the forestry sector (BMZ, EWERS, KASBERGER-SANFTL).

4.   Interaction with Other Sectors

Against the background of **population growth** and steady **depletion of resources,** it becomes clear that the core problem in the forestry sector, namely destruction of the forests in pursuit of economic interests, cannot be solved by technical means alone. **Back-up measures** in related sectors play a crucial part in permitting **interdisciplinary** management of **general conditions** for the purpose of preserving human habitats.

4.1   Complementarity

Conflicts over use of resources can be **avoided** by ensuring that the individual **sector plans complement** one another. To achieve this, it is essential to raise decision-makers' awareness of the relevant issues. Implementation of comprehensive development approaches is restricted by politico-economic realities (national and international corruption, international trade agreements, function of timber exports as a source of foreign exchange for non-diversified economies). **Integrated approaches** employ **tools** such as the following:

- **population policy**, for limiting population growth and mobilising young people as a potential labour force
- **economic policy**, for conserving natural resources by limiting demand and reducing debt
- **regional planning**, e.g. for implementing large-scale afforestation programmes as a means of rehabilitating the environment and alleviating poverty
- **energy policy**, for conserving natural resources by enhancing efficiency and promoting the use of non-biological, renewable energy sources (solar power, water power, wind etc.)
- **agricultural policy**, for achieving food security through land reforms, raising of productivity and refrainment from large-scale resettlement programmes

The environmental briefs on related sectors can be consulted where necessary. Among those of particular relevance are the following:

For biological production

- Agriculture
  - Plant Production and Plant Protection
  - Livestock Farming

- Infrastructure
  - Spatial and Regional Planning
  - Overall Energy Planning
  - Water Framework Planning
  - Mining and Energy
  - Renewable Sources of Energy

For harvesting techniques

- Agriculture
  - Agricultural Engineering
- Infrastructure
  - Road Building and Maintenance
- Trade and Industry
  - Timber, Sawmills, Wood Processing and Wood Products

4.2   Social environment

Socio-cultural factors play a major role in determining the success of measures in the forestry sector. Apart from acceptance, the following **factors** are among the most important:

- traditional forest utilisation rights and obligations
- system of social controls regulating resource utilisation
- target group's income situation
- health and food supply
- training

The **complexity** of the social environment means that difficulties are liable to be encountered in recording sociological data. Techniques such as **rapid rural appraisals** (CHAMBERS) may prove useful for **small-scale projects** but are generally **inadequate for integrated approaches**.

5. Summary Assessment of Environmental Relevance

Characteristic features of the forestry sector are the **extremely long production periods** and the **large areas** needed to permit regulation of key global cycling systems. The impacts of management errors are thus difficult to limit in terms of both time and area, as the **consequences** of choosing the wrong species of tree may not become apparent until more than a **century** has passed.

To ensure the **success** of forestry measures, it is thus essential to **simulate** natural cycling processes. **Involving the local population** in the forestry production process plays an important role as a social management tool, particularly in marginal living environments threatened with destruction.

The **concepts for forest utilisation** must therefore be **multifunctional** and **needs-oriented**. Monotypic plantations may thus prove site-appropriate under certain conditions, for example to provide a fuelwood supply in arid regions. In general, however, **integrated management goals** can be achieved only in **near-natural mixed forests**. Negative impacts on the environment can be minimised by employing techniques which **refrain from measures along the lines of clear-cutting** and **contribute to creating and preserving heterogeneous stands**.

6.   References

ANDERSON, P., 1989: The myth of sustainable logging: The case for a ban on tropical timber imports, The Ecologist Vol. 19, No. 5.

ASSMANN, E., 1970: The Principles of Forest Yield Science, Clarendon Press, Oxford.

BIRN + JUNG 1990: Abfallbeseitigungsrecht für die betriebliche Praxis, WEKA Fachverlag, Kissing.

BROWN, A.A. + DAVIS, K.P., 1973: Forest Fire: Control and Use, McGraw Hill, New York.

BOSSEL, H., 1990: Umweltwissen, Forschungsgruppe Umweltanalyse Gesamthochschule Kassel.

BUCHWALD, K. et al., 1978: Handbuch für Planung, Gestaltung und Schutz der Umwelt, Munich, Berlin.

BUNDESMINISTER FÜR LANDWIRTSCHAFT (German Federal Ministry of Agriculture), 1990: Tropenwaldprogramm, Bonn.

BURGER, D., 1989: Perspektiven standortgerechter Landnutzung im Amazonasgebiet, in: Amazonien im Umbruch, Berlin.

BURSCHEL, P. and HUSS, J., 1987: Grundriß des Waldbaus, Verlag Paul Parey, Hamburg.

CAPREZ, G. + STEPHANI, P., 1984: Die Holzernte, Friedrich Reinhardt Verlag, Basel.

CARILLO, A. and FALKENBERG, C. M. + SANDVOSS, F., 1990: Kritische Wertung des TFAP-Standes, GTZ, Eschborn.

CAULFIELD, C., 1982: Tropical Moist Forests: The Resource, the People and the Threat, International Institute for Environment and Development, London.

CHANDLER, T. and SPURGEON, D., 1979: International Cooperation in Agroforestry, ICRAF, Nairobi.

CHAPMAN, V. J., 1977: Wet Coastal Ecosystems, Elsevier Scientific Publishing Company, Oxford.

DANIEL, T. W. et al., 1979: Principles of Silviculture, McGraw Hill Book Company (MGBK), New York.

DAVIS, K. P., 1966: Forest Management: Regulation and Valuation, MGBK, New York.

DE BEER, J. H. + McDERMOTT, M. J., 1989: The Economic Value of non-timber Forest Products in South-East Asia, IUCN.

DENGLER, A., 1971 and 1972: Waldbau auf ökologischer Grundlage, Band 1 and Band 2, 4. Auflage, Verlag Paul Parey, Hamburg.

DIETZ, P., KNIGGE, W. and LÖFFLER, H., 1984: Walderschließung, Verlag Paul Parey (VPP), Hamburg.

DURELL, L., 1985: GAIA - Der Ökoatlas unserer Erde, Fischer Taschenbuch Verlag (FTV), Frankfurt.
1987: GAIA - Die Zukunft der Arche, FTV, Frankfurt.

ELLENBERG, H. et al., 1988: PIRANG - Ecological Investigations in a Forest Island in the Gambia, Stiftung Walderhaltung in Afrika, Hamburg.

ENGELHARDT, W., FITTKAU, E. J., 1984: Tropische Regenwälder: Eine globale Herausforderung, Spixiana, Munich.

ENQUETE KOMMISSION SCHUTZ DER ERDATMOSPHÄRE, 1990: Schutz der tropischen Wälder, 2. Bericht, Dt. Bundestag, Bonn.

EVAN, J., 1982: Plantation Forestry in the Tropics, Clarendon Press, Oxford.

EWERS, H. J., 1986: Zur monetären Bewertung von Umweltschäden am Beispiel der Waldschäden, Berlin.

FÄHSER, L., 1987: Die ökologische Orientierung der Forstökonomie, Forstarchiv, Hanover.

FAO, 1986: Databook on Endangered Tree and Shrub Species and Provenances, FAO Forestry Paper No. 77, Rome.
1988: Environmental Guidelines for Resettlement Projects in the Humid Tropics, FAO Environment and Energy Paper No. 9, Rome.

FRANZ, J. M., KRIEG, A., 1985: Biologische Schädlingsbekämpfung, Verlag Paul Parey, Hamburg.

GIL, N., 1979: Watershed Development, FAO Soils Bulletin No. 44, Rome.

GOLDAMMER, J. G., 1988: Rural Land-use and Wildland Fires in the Tropics, Agroforestry Systems, Dortrecht.

GOODALL, D. W. (Ed.), 1983 and 1989: Tropical Rainforest Ecosystems Vol. 14a + 14b, Elsevier Scientific Publishing Company, Oxford.
1982: Savanna and Savanna Woodland, Vol. 13.
1978: Wet Coastal Ecosystems, Vol. 1.

GOOR, A. Y. and BARNEY, C. W., 1968: Forest Tree Planting in Arid Zones, The Ronald Press Company, New York.

GOSSOW, H., 1976: Wildökologie, BLV Verlagsgesellschaft, Munich.

GRAMMEL, R., 1987: Forstliche Arbeitslehre, Verlag Paul Parey, Hamburg.

GREY, G. W. and DENEKE, F. J., 1986: Urban Forestry, John Wiley & Sons, New York.

v.d. HEYDE, B. et al., 1988: Timber Stand Improvement Field Manual, Manila.
1991: Resource-compatible timber harvesting for commercial class II forests of Sabah, Forestry Department, Sandakan, Malaysia.

HODGSON, G. and DIXON, J. A., 1989: Logging versus fisheries in the Philippines, The Ecologist, Vol. 19, No. 4.

HUMMEL, F. C., 1984: Forest Policy, Martinus Nijhof Dr. W. Junk Publishers, Dortrecht, Boston.

JOST, D., 1990: Die neue TA-Luft, WEKA Fachverlag für technische Führungskräfte, Kissing.

JORDAN, C. F., 1987: Amazonian Rain Forests: Ecosystem Disturbance and Recovery, Springer Verlag, Berlin.

KELLEY, H. W., 1983: Keeping the Land Alive, FAO Soils Bulletin No. 50, Rome.

KASBERGER-SANFTL, G., 1988: Kriterien und Methoden zur Bewertung von forstlichen Projekten im Rahmen der Entwicklungshilfe, University of Munich.

KNIGGE, W. et al., 1966: Grundriß der Forstbenutzung, Verlag Paul Parey, Hamburg.

KRUESSMANN, G., 1981: Die Baumschule, Verlag Paul Parey, Hamburg.

LAMPRECHT, H., 1986: Waldbau in den Tropen: Die tropischen Waldökosysteme und ihre Baumarten - Möglichkeiten und Methoden zu ihrer nachhaltigen Nutzung, Hamburg.

LOVELOCK, J. E., 1984: GAIA: Unsere Erde wird überleben: GAIA, Eine optimistische Ökologie, Wilhelm Heyne Verlag, Munich.

MAYER, H., 1976: Gebirgswaldbau - Schutzwaldpflege, Gustav Fischer Verlag, Stuttgart.
1977: Waldbau auf soziologisch-ökologischer Grundlage, Gustav Fischer Verlag, Stuttgart.

McNEELY, J. A., 1988: Economics and Biological Diversity, IUCN.

MITSCHERLICH, H., 1970, 1971 and 1975: Wald, Wachstum und Umwelt, Band 1, Band 2 and Band 3, J. D. Sauerländer's Verlag, Frankfurt.

MORGAN, R. D. C., 1980: Soil Erosion, Longman House, London.

MUELLER-HOHENSTEIN, K., 1989: Die geoökologischen Zonen der Erde, GS 59, Bayreuth.

NAIR, P. K. R., 1989: Agro-forestry systems in the Tropics, Kluwer Academic Publishers, London.

PANCEL, L., 1982: Bewertung von Aufforstungsprojekten und Aufbau einer Datenbank für Holzgewächse: Beitrag zur Frage der Evaluierung von Projekten der Entwicklungszusammenarbeit, Hamburg.

PIELOU, E. C., 1974: Biogeography, Wiley Interscience Publication.

POORE, M. E. D. and FRIES, C., 1985: The Ecological Effects of Eucalyptus, FAO Forestry Paper No. 59, Rome.

PRATT, D. J. and GWYNNE, M. D., 1978: Rangeland Management and Ecology in East Africa, Hodder and Stoughton, London, Sydney.

PRICE, C., 1989: The Theory and Application of Forest Economics, Basil Blackwell, Oxford.

REPETTO, R., 1988: The Forest for the Trees? Government Policies and Misuse of Forest Resources, World Resources Institute, Washington.

SAMSET, I., 1985: Winch and Cable Yarding Systems, Martinus Nijhof Dr. W. Junk Publishers, Dortrecht, Boston.

SCHEFFER, F. and SCHACHTSCHABEL, P., 1989: Lehrbuch der Bodenkunde, 12. Auflage, F. Enke Verlag, Stuttgart.

v. SCHLABRENDORFF, F., 1987: The Legal Structure of Transnational Forest-based Investments in Developing Countries, ETH Zurich.

SCHMIDTHUESEN, F., 1986: Forest Legislation in selected African Countries, FAO Forestry Paper No. 65, Rome.

SEYDACK, A. H. W. et al., 1990: Yield regulation in the Knysna Forests, South African Forestry Journal No. 152.
1982: Indigenous forest management planning: A new approach, University Stellenbosch.

SIEBERT, M.: Agroforstwirtschaft als standortgerechtes Landnutzungssystem in Gebirgsregionen der Philippinen, University of Freiburg.

SPEIDEL, G., 1967: Forstliche Betriebswirtschaftslehre, Verlag Paul Parey, Hamburg.
1972: Planung im Forstbetrieb, Verlag Paul Parey, Hamburg.

STAAF, K. A. G. and WIKSTEN, N. A., 1985: Tree Harvesting Techniques, Martinus Nijhof Dr. W. Junk Publishers, Dortrecht.

STARKE, S. (Ed.), 1988, 1989, 1990, 1991: State of the World Report 1988, World Watch Institute, New York.

STORM, P. C. et al., 1988: Handbuch Umweltverträglichkeitsprüfung, Berlin.

SYNGE, H., 1979: The Biological Aspects of Rare Plant Conservation, John Wiley & Sons.

SYNNOT, T. J., 1979: A Manual of Permanent Plot Procedures for Tropical Rain Forests, Commonwealth Forestry Institute, Oxford.

ULRICH, B., 1981: Destabilisierung von Waldökosystemen durch Biomassenutzung, Forstarchiv Heft 6, Hanover.

VARESCHI, V., 1980: Vegetationsökologie der Tropen, Verlag Eugen Ulmer, Stuttgart.

VESTER, F., 1983: Ballungsgebiete in der Krise: Vom Verstehen und Planen menschlicher Lebensräume, Deutscher Taschenbuchverlag (DTV), Munich.
1984: Neuland des Denkens, DTV Munich.
1985: Ein Baum ist mehr als ein Baum, Kösel Verlag, Munich.

WEIDELT, H. J., 1986: Die Auswirkungen waldbaulicher Pflegemaßnahmen auf die Entwicklung exploitierter Dipterocarpaceen-Wälder, Heft, University of Göttingen.
1989: Die nachhaltige Bewirtschaftung des tropischen Feuchtwaldes, Forstarchiv Heft 3, Hanover.

WEISCHET, W., 1980: Die ökologische Benachteiligung der Tropen, B. G. Teubner Verlag, Stuttgart.

WENGER, K. F., 1984: Forestry Handbook, John Wiley & Sons, New York.

WHITEMORE, T. C., 1984: Tropical Rain Forests of the Far East, Clarendon Press, Oxford.

WÖLL, H. J., 1988: Struktur und Wachstum von kommerziell genutzten Dipterocarpaceen-Wäldern und die Auswirkung von waldbaulicher Behandlung auf deren Entwicklung, University of Hamburg.

YOUNG, A., 1989: Agro-forestry for Soil Conservation, ICRAF, BPCC Wheatons Ltd., Exeter.

ZOEHRER, F., 1980: Forstinventur, Verlag Paul Parey, Hamburg.

29

Glossary of selected terms

Basal area: Total of the trunk cross-sections of all trees in a stand exceeding a minimum diameter, given in square metres per hectare and serving as a measure of the stand density.

Biocybernetics: Subdiscipline of cybernetics (from the Greek "kybernetes", meaning "helmsman"), which describes the control and automatic regulation of interlinked, closed-loop processes with minimum energy input in biological systems

Biomass utilisation: In forestry, limited to timber utilisation in the form of full trees, i.e. stem timber including bark, leaves and branches, or whole trees, i.e. full trees plus root wood.

Biotope: The habitat occupied by an organism or community (biocoenosis) within an ecosystem, determined by physical and chemical factors

Compartment: Permanent physical unit of forest division, serving simultaneously as a unit for planning, execution and monitoring of measures .

Cost-effectiveness analysis: Comparison of operational alternatives in which the inputs are of a monetary nature but the outputs cannot be measured in monetary terms.

Ecosystem: A unit within the natural environment, consisting of a community and its habitat (biotope) and characterised by balanced cycling systems, i.e. dynamic steady states.

Management goal: Production goal for forestry operations setting out the range and order of precedence of all requirements on the part of the forest owner and/or the general public, both material (timber, non-timber products) and intangible (soil and water conservation, nature conservation, recreation). Distinctions are made between product goals, security goals and monetary goals, with times being set for their achievement.

Management goal type: Management goal for a stand or compartment.

Savannah: Form of vegetation found in the semi-humid tropics, generally between the inner humid tropics and the latitudes marking the Tropics of Cancer and Capricorn, consisting of grassland with scattered trees and shrubs.

Silviculture: Science of forestry production concerned with systematic creation and tending of forests so as to ensure that society's material and intangible needs can be permanently satisfied

Site: Complex of location-related - i.e. natural, economic and social - factors influencing forestry operations.

Stand: Group of trees which exhibit similar features, occupy an unbroken minimum area and all require similar silvicultural treatment.

Welfare function: Also referred to as the indirect effects or non-wood beneficial effects of a forestry operation, i.e. "production" of goods with economic relevance such as water, soil conservation and recreation.

Agriculture

Livestock Farming

30. Livestock Farming

Contents

1.   Scope

As a biological process, livestock farming influences, and is influenced by, the environment. With respect to the environment the aim is to change it in such a way that a maximum of food and raw materials can be obtained on a sustainable basis.

Environmental impacts vary depending on the **form of livestock husbandry and type of farm** involved. There are three basic forms of livestock husbandry:

- pasture usage
- pasture use with supplementary feeding
- confinement

Farming systems can be divided into the following types:

- ranches (cattle, sheep)
- traditional pastoralism (cattle, sheep, goats, camelids, equids, often mixed herds)
- smallholder livestock husbandry (cattle, buffalo, camelids, equids, sheep, goats, poultry, pigs, small animals such as guinea pigs, rabbits and bees; a farm often keeps a variety of different animal species)
- large enterprises of industrial-scale livestock production (e.g. poultry fattening, laying batteries, pig fattening, feedlots for cattle)

Fisheries and aquaculture are covered in a separate environmental brief.

Livestock farming is **possible wherever** arable farming is practised. It is also the only form of agriculture in semi-arid and arid regions as well as in high mountain regions in the zone beyond the arable farming limit up to the vegetation limit.

2.   Environmental Impacts and Protective Measures

2.1   Types of husbandry

2.1.1 Pasture use in general

The most noticeable consequence of **grazing** is the **defoliation of plants** by the animals, which influences the structure of the pasture vegetation and the variety of species which it contains. The precise nature of this influence depends on the type of animal concerned, the stocking rate (grazing pressure) and possibly also the time of year. Cattle and sheep tend to eat grass, whereas camels and goats prefer leaves.

An **ideal sheep or cattle pasture** will thus consist primarily of **grass and herbaceous plants**, while an **ideal camel or goat pasture** will contain more **trees and bushes**.

Grazing can **stimulate plant growth** and encourage the growth of creeping ecotypes of a particular plant species rather than those which grow upright. In grass/legume pastures, grazing often favours the legume component, because animals generally prefer grass during the early part of the vegetation period; with competition reduced in this way, **legume growth is promoted**. However, the young stages of some legumes are also popular with animals. While light grazing and browsing on bushes and trees can stimulate growth, removal of vegetation by livestock on a larger scale can **reduce growth** or even cause plants to die off and may **hinder regeneration** of fodder bushes from seeds and suckers.

The **effects resulting from trampling** of the vegetation by livestock depend primarily on the type of animal concerned, the stocking rate, the soil condition and the topography. **Damage caused by trampling** can **increase soil erosion;** however, the roughening-up of the ground can also create better conditions for germination and thereby **promote plant regeneration**. Where the soil in humid regions is heavily waterlogged, the vegetation cover can be destroyed as a result of trampling.

The **seeds of many pasture plants** are very small and can pass through the animals' digestive tract without any impairment of their germination capacity. Certain plants are thus dispersed with the animals' dung. Hard-shelled seeds are also scarified and the seeds are **redistributed** and **sown** by the animals.

Only a small proportion of the **nutrients and energy** intake by livestock actually finds its way into the animal products used by man. The remainder is excreted via dung, urine and, in the case of ruminants, methane (a gas which plays a part in the greenhouse effect). The breakdown of organic matter in the digestive tract of ruminants gives rise to energy and nutrient losses similar to those resulting from microbial breakdown in the soil; as the breakdown process in the stomach of ruminants is considerably faster, however, the grazing animals **accelerate the nutrient cycle**. If the animals are penned overnight, the excretion of dung in the pen means that the pasture is deprived of nutrients. Although the dung collected in pens can be used in **arable farming and horticulture** or for **production of biogas** and can thereby contribute to **improving soil fertility**, the loss of nutrients can accelerate **degradation of the pasture vegetation**.

In semi-arid and arid regions, the considerable fluctuations in annual rainfall mean that vegetation growth varies greatly not only according to the time of year but also from one year to another. The herbaceous vegetation layer in particular will thus not exhibit consistent growth. In drought years there may be so little vegetation growth that all the herbage is eaten by the animals. **If shrubs and trees are not to suffer permanent damage, the amount of their vegetation consumed as fodder must not exceed a specific proportion of the annual growth**, otherwise their capacity for survival and regeneration will be jeopardised.

Permanent damage is generally considered to have occurred when the **vegetation's capacity for regeneration** has been impaired and the surface of the ground has suffered erosion by wind or water. In view of the differences between plant communities and the differing regeneration capacities of the various species, it is not possible to lay down any **universally valid standards** specifying how much land can be used without impairing the productivity of the vegetation and what **stocking densities** are possible. American estimates work on the basis that 50% of the vegetation can be used, while studies from West Africa take figures of 30-50% (le Houerou 1980). Others graduate permissible vegetation use according to rainfall and take different levels of permissible utilisation for the bush/tree layer (25-50%) and the grass/herbaceous layer (30-50%) (Schwartz 1989). **Factors** which can assist in assessing degradation include the age structure and species composition of the tree and shrub community, seed reserves in the soil for the herbaceous plants and possibly also soil cover as well as depth and condition of the A-horizon.

The distribution of animals in an arid pasture area is determined primarily by the **availability of water. Deep wells containing plenty of water** supply a large number of animals and may thus give rise to **serious overgrazing** in their immediate vicinity. The size of the area around a well that can be used by animals for grazing depends among other things on the dry-matter content of the fodder, the type of animal and the animals' physiological status. Inadequately protected wells and watering places can easily be contaminated by dung and may also constitute a **health risk** for the local population if **drinking water becomes contaminated**. The concentration of animals around wells can promote the **spread of epizootic diseases**. Around every watering place there is a certain area which, although it contains an accumulation of nutrients by virtue of the dung produced by numerous animals, is almost totally devoid of vegetation as a result of trampling. The size of this area depends on the design of the watering place (e.g. troughs on hard ground) and the way in which access to it is controlled (e.g. fencing-in of watering places). Use of **fertiliser** in arable farming and horticulture in the vicinity of the watering place will not give rise to any problems.

Pastureland comprises natural pastures, fallow land and harvested fields. **Forested areas**, which in some cases are under the control of forest administrations, can also **be used as pastures**. In many cases, for instance in North Africa, the major proportion of the forest yield is derived from livestock farming. Fodder production is an integral part of agroforestry. It must be pointed out, however, that **forest pastures** are often **over-used**. If this is to be prevented, **a wide variety of measures are necessary**: reduction of tensions between forest administration and local farmers; employment of an adequate number of appropriately motivated personnel in order to enforce the regulations limiting pasture use; provision of alternative fodder resources for local livestock owners; steps to prevent use of pastures by non-local livestock owners not engaged in agriculture; reasonable charges (where payment is made for use of forest pastures) by comparison with the price of other fodder resources; and involvement of the local population in pasture-use planning. Both the dry and humid tropics offer examples of **balanced pasture management** which takes forest growth dynamics into account.

2.1.2   Pasture use with supplementary feeding

The environmental impacts of **supplementary feeding** depend on the context and the type of feed. Where fodder is of poor quality but available in large quantities, supplementary feeding of minerals can improve utilisation of the "standing hay". Provision of supplementary feed in the form of feed concentrate or high-quality roughage soon leads to a **reduction in the amount of fodder consumed** per animal **during grazing**, which benefits the pastureland. If, however, the **number of animals is increased** on account of the improved fodder supply and the natural pasture continues to be used, there is a greater risk of **degradation**. In some cases (e.g. in North Africa) livestock are given so much supplementary feed that this feed covers not only their performance requirement but also part of their maintenance requirement. Another reason for overgrazing is the desire to **improve the quality of the animals' meat**, since this will be reflected in higher meat prices. Meat quality is influenced in particular by the fact that the animals move around more and by the improved basic fodder supply.

2.1.3   Fodder production

Erosion-control strips can be used for fodder production. Appropriate planting of permanent fodder crops (such as sulla in North Africa) can serve as a form of "soft" erosion control. Fodder growing within a crop rotation system can have positive effects on soil structure and soil fertility (see Plant Production). The possibility that fodder crops may **compete** for land with crops that can be used as food for human beings must be borne in mind.

In the case of certain fodder crops, a large quantity of **nutrients** is **taken** from the soil together with the green matter. If these nutrients are not replaced, or if the dung is not returned to the field, there is a danger that **the nutrient balance may be disturbed**. If mineral fertilisers and herbicides are used in fodder production, there is a risk that **surface water and groundwater may become contaminated** and that the **diversity of species may be reduced** at the same time.

2.1.4 Confinement

While pasture use primarily involves ruminants, chickens, pigs and small livestock such as rabbits and guinea pigs are generally kept in confinement.

The **environmental impacts** of keeping livestock in confinement depend on the number of animals, the type of animal, the nature and origin of the feed and whether the livestock housing is open or closed. The **environment prevailing in the animal-sheds** (temperature, humidity, light, presence of noxious gases, dust and germs) has an effect on the animals, while livestock housing itself has an effect on its immediate environment through **odours, liquid manure and noise**. Where ruminants are kept, **methane** (a gas which plays a part in the greenhouse effect) is also released.

If livestock are kept in confinement, the **vegetation suffers far less damage** than if the animals are allowed to graze. However, use of cut fodder means that the soil is **deprived of nutrients** on a considerable scale; if these nutrients are not replaced, there is a danger that **soil fertility may be reduced**.

The enormous quantities of **liquid manure** produced where a large number of animals are kept can **impair drinking-water quality** and **contaminate both surface water and groundwater**. Large-scale chicken farms located near cities give rise to particularly **adverse environmental impacts** on account of their need to dispose of dead birds and droppings. Liquid and solid manure represent a major **potential source of infection** - especially for children - in many developing countries, particularly if no measures are taken to prevent contact with them. When used as **fertilisers**, liquid and solid manure can have a **beneficial** effect on **soil fertility** and **soil structure**, provided that they are not applied in excess.

2.2 Farming systems

2.2.1 Ranches

Ranches permit uniform management of comparatively large areas. Large-scale farms of this type nevertheless **do not guarantee conservationist use** of pasture resources (Harrington et al. 1984). In dry years a ranch too requires alternative fodder resources or the number of animals must be reduced in good time, otherwise heavy losses are likely. Supplementary feeding can lead to **over-use of pastureland** and thus increases the **risk of erosion**. When a large farm with "rational" stocking rates or a pasture reserve with a controlled stocking rate is established in an area where traditional livestock husbandry predominates, it should be borne in mind that although the reduced stocking rate on the land concerned may be more appropriate to site conditions than the original rate, the exclusion of animals from this land will increase the **grazing pressure** in the surrounding area.

Particularly in humid regions, large-scale **land clearance** to create pastures for ranches substantially **reduces the diversity of species found in the vegetation.** Apart from the resulting **erosion problems**, there may also be a risk of **climatic changes over a wide area.** The fact that ranches generally keep only cattle gives rise to **one-sided utilisation of resources**, which either permits only very low stocking rates or calls for sizeable inputs to preserve the pastureland. There is also a danger that the **pastureland may be acidified** as a result of waterlogging. **Damage caused by trampling** can have an adverse effect on the soil structure, leading to **increased surface-water run-off** and **a greater risk of erosion.**

Although ranches can improve the urban population's food supply, their carrying capacity per unit of area is smaller than that of traditional farming systems (e.g. Cruz de Cavalho 1974, de Ridder & Wagenaar 1986).

Environmental protection measures are difficult to realise **where ranching** is concerned. Attempts to standardise the **carrying capacity** of pastures are the subject of considerable **dispute** on account of the complex interrelationships and numerous variables involved, particularly in assessment of vegetation (e.g. Sandford 1983).

Some systems, such as those found in Australia, are based on detailed long-term studies and official determination of the permissible maximum stocking rate. As the land in Australia is generally not in private ownership, but is instead leased out by the government on a long-term basis, specific conditions can be imposed and the lease revoked if need be. In many countries the necessary **data** are not available and monitoring **institutions** are either non-existent or not equipped to perform the essential tasks. Rules aimed at preventing erosion should be worked out together with the ranch managers concerned.

2.2.2 Pastoral systems

In such systems, animal husbandry is the sole or principal economic occupation. Herding and a high degree of mobility make it possible to utilise resources in a manner complementing arable farming or to utilise areas that can be used for grazing only at certain times of year.

Pastoralists often keep mixed species herds, which permits intensive use of a wide variety of fodder resources. The products derived from the herds include milk, meat, traction power and manure.

- **Integration of grazing and arable farming**

Where pasture resources are used in combination with arable farming, the amount of land available for grazing varies greatly in the course of the year. During the growing season only natural pasture and fallow land can be used for grazing, while during the dry season harvested fields are also available for this purpose. Grazing has a variety of effects on fallow land and natural pasture. The **species composition of the vegetation** may change in such a way that a larger proportion of the vegetation can be used as fodder or for other purposes; at the same time, however, intensive grazing can also lead to **degradation**. If herded animals are penned at night, nutrients accumulate in the night paddock as a result of the droppings and urine produced by the livestock. These nutrients can be used to preserve soil fertility on arable land (dung), but are thereby removed from the nutrient cycle on the land used for grazing. Leaching from the night paddock can lead to **contamination of surface water and groundwater**. Use of crop residues as fodder may **accelerate the nutrient cycle** and result in **redistribution of nutrients** in a particular field or among fields. If crop residues are used on a large scale the **soil cover may be reduced** and this can lead to **erosion**. Rights to use fodder resources must be established through agreements between pastoralists and arable farmers.

- **Mobility**

A high degree of flexibility and mobility is required on the part of the pastoralists in order to permit ecologically appropriate and economically sound **use of arid regions**. Mobility in turn calls for large herds. In the course of their **migration** the pastoralists and their families must for the most part live on the products which they can derive from their herds. Reduced mobility generally leads to **overgrazing**, accompanied by increased **soil erosion**, in the area around the newly created settlements and to **under-use** of other areas. Under-use can also give rise to changes in the balance of species and reduce the vegetation's productivity.

As herds and people become increasingly sedentary and concentrated in specific areas, use of green twigs and branches to construct livestock paddocks and as domestic fuel leads to **destruction of the woody vegetation**.

- **Grazing rights**

Land and pasture use rights may comprise seasonal rights of use in specific areas and grazing rights in areas located a long way from one another. Apart from creating an opportunity **to use land resources for grazing as well as for arable farming**, this also helps to balance risks, as rainfall in arid regions is often highly localised. Communal grazing rights predominate in such areas. Communal pastures are traditionally used by clearly definable groups of livestock owners. Depending on the group's structure and effectiveness, this makes it possible to stipulate **stocking rates and times at which the pasture is not to be used**. In regions such as East Africa, controlling access to water is an important means of controlling stocking rates. Open access pastures - frequently equated with common pasture - offer virtually no opportunity for such a step. In such a context, **creation of watering places** outside the traditional structures can encourage **opportunistic** use and thus contribute to **overgrazing**. The **secondary consequences** of such a development will be degradation of the vegetation, reduction of the soil's rainwater infiltration rate and increased soil erosion.

- **Changes in ownership**

Changes in the herd ownership structure can likewise adversely affect the pastoralists' resource management. When herdsmen look after cattle owned by other people, for example, they are often allowed only to use the milk. **In order to have a secure livelihood** they require **large herds of their own** if they are not to become impoverished. Moreover, the owners' desire to keep a check on their property may cause them to restrict the herdsmen's mobility and thus also their flexibility where pasture management is concerned. This too can result in **over-use of the vegetation** (disturbance of the balance of species within the flora, disturbance of the water balance, soil erosion).

- **Division of labour**

In pastoral systems, the men are generally responsible for management and marketing of the largestock, while the women frequently tend the small ruminants and have responsibility for milk processing and marketing. The women's role is often underestimated, as it is the men who represent the family vis-à-vis other people. The decentralised processing and marketing of milk ensures a relatively reliable milk supply in rural areas, even though a woman may process and market only a few kilograms of milk a day. When **milk is processed** at household level, consideration must be given to possible **hygiene risks** (e.g. danger of infection).

- **External influences**

Pastoral land use frequently necessitates agreements between various population groups. External influences - and that includes government programmes - may disturb the often **fragile equilibrium**. If, for example, arable farming is expanded onto land used by pastoralists for dry-season grazing or as reserve pasture, the loss of this pastureland can **increase the pressure** on other areas and lead to overgrazing. Should the arable farmers start to keep livestock on a larger scale, the pastoralists may find themselves **driven out into marginal areas**. This not only has consequences in terms of grazing management and livestock productivity but can also affect the welfare of the population groups concerned.

If their mobility is restricted, pastoralists may be forced to make intensive long-term use of marginal areas on a scale which exceeds the natural carrying capacity.

The resultant **degradation process** intensifies competition for the decreasing fodder resources. By promoting over-use of the available land it also reduces the number of species found locally and marginalises large sections of the pastoral population.

2.2.3 Smallholder livestock husbandry

The number of livestock owned by a smallholder can range from a few small animals (e.g. chickens) to large herds, e.g. twenty goats or ten head of cattle. **Livestock management** normally **takes second place to the interests of arable farming**. Many smallholders keep more than one type of animal.

Smallholders generally use **pastures with supplementary feeding** (at least on a seasonal basis) or keep their livestock **in confinement**. Large herds - such as village herds - may be mobile (cattle placed in the charge of a herdsman by their owners).

The animals may be allowed to graze freely, or may be herded, tethered or kept in fenced pastures. The practice of **fencing off pastures with wooden posts** - which may have to be replaced at frequent intervals on account of termite damage - can have **adverse effects** on the **species composition** and **density of the tree stand**. By contrast, use of "living fences" or hedges to subdivide pastureland has **positive effects** on the tree stand but requires a considerable amount of labour.

Clearing land to create improved pastures can increase the **erosion risk** and thus have an adverse influence on soil fertility. Creation of improved pastures, particularly with legumes, can be integrated into ley farming (seeded pasture rotation) and will improve soil structure and fertility. **Competition** for use of fodder resources may arise between livestock owners, above all between pastoralists and smallholders as well as between the smallholders themselves, and can thus **impose increased pressure** on the available land.

As in pastoral systems, **management** of largestock is frequently the men's responsibility, while the women are in charge of the smallstock. As women in many rural societies have no land ownership rights, livestock husbandry plays an extremely important part in enabling them to accumulate capital. The income earned from animal husbandry can be used to finance necessary expenditure for arable farming (fertiliser, seed, hired labour, creation of erosion-control strips), while the animals' dung can be used to **preserve soil fertility**. Livestock perform a particularly important function as a form of "risk reduction" in regions where arable yields tend to be unreliable. If the harvest is insufficient to meet the family's subsistence requirements, animals can be sold to permit the purchase of staple foods. Without this means of offsetting risks it would be necessary to extend the area under cultivation, which would have negative effects in terms of soil erosion, soil structure, nutrient balance and diversity of species.

A **changeover from pasture use to keeping livestock in confinement** can have beneficial effects on the diversity of plant species and assist in preventing erosion. The **increased concentration of liquid manure and dung** may lead to greater pollution of surface water and groundwater. Keeping livestock in confinement requires more labour than pasture use and it is generally women who are called upon to perform the extra work.

High-performance animals have more demanding requirements in terms of fodder supplies and veterinary care. If chemoprophylaxis is necessary, **pathogen strains resistant** to the chemotherapeutic agents used can develop (see environmental brief Veterinary Services). Introduction of high-performance animals frequently does not lead to a reduced number of livestock; it does not lessen the burden on the available fodder resources either.

The actual and potential advantages of indigenous breeds and species are often underestimated. With a one-sided promotion of the use and importation of high performance animals, there is a danger of losing genetic resources adapted to the natural environmental conditions.

Urban livestock husbandry can be regarded as a special category of smallholder livestock husbandry. As urban livestock owners purchase far more fodder than those in rural areas, their existence can encourage **fodder growing** in the vicinity of towns. This can have positive effects on soil structure and fertility, besides boosting the fodder growers' income. Dairy cattle are kept in urban areas to supply the urban population with fresh milk. While other animals are kept primarily to meet their owners' food requirements, they can also serve as a form of "savings bank" and as a means to accumulate capital. The **dung** produced by the animals can help to improve the soil structure and nutrient balance, but may well give rise to direct and indirect health risks if it is used or disposed of incorrectly. As in rural areas, women make an important contribution to urban livestock husbandry, although it can be assumed that the division of labour between the sexes is less strict than in rural society.

Smallholder animal husbandry also includes beekeeping. Apart from producing honey, **bees** can substantially increase fruit yields by pollinating the blossoms and help to **preserve the diversity of species within the flora**. Modern intensive beekeeping involves chemical control of pests (mites etc.); such measures can create health risks for humans if the chemical agents are incorrectly used and if **residues find their way into the honey**. Importing of higher-performance strains of bee can eradicate indigenous species. Production of honey and beeswax, which is predominantly a male domain, can be a highly profitable source of income in rural areas.

Environmental protection measures in the field of pastoral and smallholder livestock husbandry may involve steps to change framework conditions or direct intervention. Examples of measures aimed at **changing framework conditions** include discontinuation of subsidies for feed grains - in North Africa such subsidies contributed to widespread overgrazing - and changes in land law (land reform). Where **direct intervention** in pastoral and smallholder production systems is envisaged, it is essential that the groups affected be involved in the measures right from the planning stage. The **measures** planned may relate to a wide variety of different areas, e.g. water resources management, erosion control, fodder growing or - where smallholders are involved - promotion of confinement. Simply demanding a reduction in the number of animals - as was frequently done in the past - reflects an inadequate understanding of the way in which pastoral and smallholder production systems function.

2.2.4 Large enterprises with intensive animal production (commercial farming)

Large-unit animal production generally does not depend on the availability of land to provide forage, as fodder is imported from other parts of the country or from abroad. For the purpose of supplying the urban population, large-scale livestock production focuses on pigs and poultry.

Large farms **consume** far more **fossil energy** per product unit than traditional farms. If growth stimulants such as **antibiotics or hormones** are added to the fodder, there is a danger that **residues may be found in foods** of animal origin and that **resistant pathogens may develop**.

The **high water consumption** of large farms is also likely to lead to excessive utilisation of scarce water resources.

In confinement housing, the **prevailing in-house-micro-climate** (temperature, humidity, noxious gases such as ammonia, hydrogen sulphide and methane, dust and germ content of the air) can have adverse effects both on the animals and on the farm workers (health risks). The mere size of the farms means that the danger of **surface water and groundwater being contaminated** by liquid manure and farm effluent is far greater than in the case of smallholdings. The problems attached to the **disposal** of dung and animal carcasses are also likely to be greater, as is the related hygiene risk. Use of disinfectants can endanger water, soil and possibly also health.

Where cattle are involved, sizeable quantities of **methane** - a gas which plays a part in the greenhouse effect - are produced in the animals' stomachs and released.

If large farm enterprises are in **competition** with smallholdings, they can have an adverse effect on the smallholders' income. This may force smallholders operating under marginal conditions to engage in arable farming instead of livestock husbandry. Apart from giving rise to undesirable consequences with regard to the balance of species and soil fertility, such a step also increases the **danger of erosion** in the region concerned. Some large farms, such as commercial cattle-feed lots or large dairy farms (agricultural combinates), may also compete directly with smallholdings for agricultural land (e.g. in irrigated areas) and thereby force the smallholders into marginal areas. However, this risk is greater in the case of plantations and other crop-growing farms than in the case of livestock farms.

Environmental protection measures on large farms with intensive livestock management focus primarily on technical aspects: housing design, whole farm layout, ventilation, distance from settlements, precautions during storage and disposal of liquid manure and dung, hygiene measures such as disinfection, ban on use of growth stimulants, fencing-in of livestock housing etc. Technical standards in Central Europe are well documented (e.g. the German DIN standard 18910 on controlled environment in livestock housing, specifications laid down by the Association of German Engineers [VDI], maximum workplace concentration values (MAK), construction specifications issued by the German committee on technology and construction in agriculture [KTBL]).

3.   Notes on the Analysis and Evaluation of Environmental Impacts

There are no generally applicable guidelines for analysing the environmental impacts of livestock farming. Useful background information concerning the **impacts of large farm enterprises** on water and the environment prevailing in the livestock housing can be obtained from German guidelines (e.g. DIN standard 18910, VDI specifications, planning documentation for livestock housing). Australian studies (e.g. Harrington et al. 1984, Squires 1981) can yield valuable pointers concerning **ranches and pasture use** in general. Collection of data to ascertain the impacts of livestock farming must be conducted on a long-term basis and can involve a variety of methods such as soil and plant monitoring, investigation of herd composition and livestock productivity, interpretation of aerial photographs (series) and possibly also interpretation of satellite images. An ecosystem analysis provides a sound basis for determining the carrying capacity of the ecosystem in question.

The ecological, economic and technical rationale of **pastoral and smallholder livestock farming** has sometimes been the subject of heated debate in recent years (Sandford 1983, Galaty et al. 1979; see also articles published within networks such as Nomadic Peoples and ODI Pastoral Development Network or by CRSP). The current state of knowledge does not permit any form of definitive assessment; the information sources cited above should instead be seen as offering pointers.

4. Interaction with Other Sectors

Livestock farming is interlinked above all with **plant production** and **forestry** and constitutes an element of **general resource management**. One link with plant production lies in the "transformation" of feedstuffs such as green forage, crop residues and cereals. Production and spreading of dung has beneficial effects on plant production, while the role played by livestock farming as a form of "savings bank" and as a means of accumulating capital can also permit investments in crop growing. The land requirements of pasture use are most likely to conflict with those of crop growing where the latter involves cash crops such as cotton and other crops cultivated in large-scale monoculture systems. Livestock farming also has a certain bearing on **rural water supplies.**

As natural pasture is often the major source of fodder for ruminants, the interests of livestock farming and pasture use must be taken into account in **regional planning**. Failure to understand livestock husbandry systems and the way in which they function can give rise to serious conflicts.

Food production and the related hygiene risks have an influence on the population's nutrition and health. Direct competition regarding product use can arise if cereals and other products that could be competition by humans without further processing are fed to livestock. Indirect competition occurs wherever feedstuffs (e.g. soya beans) are grown on a large scale for export, since this is to the disadvantage of smallholders engaged in livestock husbandry.

Livestock farming supplies raw materials for further processing by dairies, slaughterhouses, tanneries and spinning mills and is thus a source of raw materials for **agro-industry**.

Where draught animals are kept, livestock farming supplies "products" required in agricultural engineering; large farms are customers in this sector by virtue of their need to purchase items such as equipment for installation in livestock housing. **Veterinary medicine** essentially performs a service function for livestock farming. **Fishery** yields fish meal and thus also supplies feedstuffs for other forms of intensive livestock farming, while **aquaculture** can utilise wastes and by-products from livestock farming.

In the **processing sector**, environmental impacts depend on the nature and size of the enterprises concerned. With regard to slaughterhouses, see the environmental briefs Veterinary Services, Slaughterhouses and Agro-industry.

5.   Summary Assessment of Environmental Relevance

The environmental impacts of livestock farming are determined by the **intensity** of the production operations.

The following **critical influencing factors** are to be found in all farming systems and forms of animal husbandry:

-    land clearance for the purpose of pasture improvement or to permit forage growing

-    stocking rate, which is influenced by the number of animals, the herd composition in terms of species and classes of animals, and the availability of fodder

-    availability of water as a function of the number of watering places per unit of area, the distribution of watering places in the region and the design of the watering places.

However, the **extent of the environmental hazards** created by these critical influencing factors depends on the **farming system** in question. Stocking rate, for example, becomes less important in intensive livestock farming systems; at the same time, an increasingly significant role is played by critical factors in fodder growing such as type of fodder, form of use and fertiliser application, as well as by dung removal and possibly also residues in feedstuffs and animal products (which may also be the result of veterinary measures).

The greatest environmental hazards are caused by **industrial-scale animal production**. Apart from the considerable risk of water and air pollution through noxious gases and disposal of dung and liquid manure, its energy and water requirements can also be seen as having adverse impacts on the environment.

6. References

Cruz de Cavalho E., 1974: "Traditional " and "modern" patterns of cattle raising in southwestern Angola: a critical evaluation of change from pastoralism to ranching. Journal of Developing Areas 8, pp. 199-226.

DIN 18910, 1974: Klima in geschlossenen Ställen. Berlin: Beuth-Vertrieb.

Galaty J.G., Aronson D., Salzman P.C., Chouinard A. (Eds.), 1980: The future of pastoral peoples. Ottawa: International Development Research Centre (IDRC).

Harrington G. N., Wilson A. D., Young M. D. (Eds.), 1984: Management of Australia's rangelands. Melbourne: Commonwealth Scientific and Industrial Research Organisation.

Jahnke H. E., 1982: Livestock production systems and livestock development in tropical Africa. Kiel: Kieler Wissenschafts-Verlag Vauk.

King J. M., 1983: Livestock water needs in pastoral Africa in relation to climate and forage. ILCA Research Report No 7. Addis Ababa: International Livestock Centre for Africa.

Kotschi J., Adelhelm R., Bayer W., von Bünau G., Haas J., Waters-Bayer A., 1986: Towards control of desertification in African drylands: problems, experiences, guidelines. GTZ special publication No. 168. Eschborn: GTZ.

Niamir M., 1990: Herders' decision-making in natural resources management in arid and semi-arid Africa. Community Forestry Note 4. Rome: FAO.

Pastoral Development Network Discussion Papers. London Overseas Development Institute.

de Ridder N., Wagenaar K. T., 1986: Energy and protein balances in traditional livestock systems and ranching in eastern Botswana. Agricultural Systems 20, pp. 1-16.

Sandford S., 1983: Management of pastoral development in the Third World. Chichester: John Wiley & Sons.

Squires V., 1981: Livestock Management in the arid zone. Melbourne: Inkata Press.

VDI-Richtlinien 3471, 1986: Emissionsminderung Tierhaltung Schweine. Berlin: Beuth-Verlag.

VDI-Richtlinien 3472, 1989: Emissionsminderung Tierhaltung Hühner. Berlin: Beuth-Verlag.

Agriculture

Veterinary Services

31. Veterinary Services

Contents

1. Scope

Veterinary services are of even more immediate relevance to the environment than is the case for sectors such as plant or animal production. Their principal purpose is to **preserve or restore animal health** and their environmental impacts are thus essentially positive. However, the possibility of negative impacts - generally of an indirect nature - cannot be precluded. The veterinary sector primarily performs a **service function** for livestock farming and fisheries, as well as playing an important role in **food inspection**.

Activities in the veterinary sector cover the following areas:

- diagnosis and control of diseases, involving treatment, prophylaxis, vector control and epizootic-disease control
- artificial insemination and embryo transfer
- laboratory activities, comprising laboratory diagnostics, vaccine production and residue analysis
- food inspection, above all meat inspection in slaughterhouses and food hygiene.

In the fields of disease diagnosis, treatment and vector control, a distinction can be made between **"modern" measures** carried out by formally trained veterinary surgeons and **traditional practices** employed by the animal owners themselves or by healers.

In the **agro-industry** sector (meat and milk processing, fodder hygiene), veterinary services perform a **monitoring function**. Veterinary medicine is also closely linked with the pharmaceutical industry by virtue of its need for drugs and vaccines.

2. Environmental Impacts and Protective Measures

Veterinary services perform a vital function through their key tasks of combating animal diseases and ensuring that foods of animal origin comply with the necessary hygiene regulations. Measures to protect health and the natural environment are necessary above all wherever veterinary drugs and pesticides are liable to have side-effects, leave residues or be used incorrectly or negligently, as well as in laboratory work and vaccine production. Disposal of wastes and of possibly infected carcasses (or parts thereof) unfit for human consumption is discussed in the environmental brief Slaughterhouses and Meat Processing.

- For drugs, the following principles should be applied: **strict controls** on sale and use; monitoring of production if necessary; livestock owners to be **advised** on potential side-effects; greater **emphasis on use of traditional** remedies. Although traditional remedies derived from plants may not be totally free of environmental hazards, they are in general likely to have less environmental impact than "modern" pharmaceuticals. Changes in livestock husbandry systems can also help to reduce the need for drugs.

-        In prophylaxis and vector control, the following measures are essential: **refrainment from use of products** which are **broken down in the environment** only very slowly or not at all (e.g. DDT); greater emphasis on epidemiological aspects and **promotion of forms of livestock husbandry** likely to reduce parasite infestation.

Veterinary measures may **interfere with established social structures**, with adverse effects on the producers' rights and income. Women are particularly liable to be affected, since in many societies they play an important role as traditional healers, as livestock owners and in the processing and marketing of animal products.

2.1   Disease control

2.1.1 Diagnosis and treatment

Clinical diagnosis and treatment are carried out on the one hand by the animal owners themselves or by traditional healers, and on the other hand by formally trained veterinary surgeons. Clinical diagnosis has little direct impact on the environment (see environmental brief Analysis, Diagnosis, Testing).

Traditional methods of treatment often involve user-prepared plant extracts, although modern drugs are also used on a growing scale. Use of plant extracts (generally in aqueous form) can have undesirable effects on the **diversity of species** within the flora if medicinal plants are gathered in such large quantities that their existence as a whole is jeopardised. It can be assumed that "natural" remedies leave few residues.

Improper storage of **modern drugs** (chemotherapy) can have harmful effects on the environment. Certain drugs such as potent antibiotics are used too frequently or in incorrect doses; this can cause pathogens to become **resistant** to the antibiotic used and necessitate administration of a number of different antibiotics in rapid succession.

There is also a danger that drugs or drug residues may **accumulate** in products destined for human consumption - thereby giving rise to health risks - if the prescribed waiting periods are not observed before animals are slaughtered or used for other purposes (e.g. to obtain milk).

Practices such as using waste oil to treat dermatophiliasis can bring animals short-term relief but may cause contamination of water and soil.

Getting rid of disposable cannulas and containers made of **plastic and other synthetic materials** can give rise to problems. Incineration pollutes the air (e.g. with dioxins), while incineration residues may contaminate water and soil.

Successful treatment of sick livestock can lead to an increase in the number of animals; this in turn may result in **over-use of fodder resources**, giving rise to a greater risk of erosion and general degradation of fodder bushes, fodder trees and pastureland.

Where malnutrition is a contributory factor in disease, control measures should be combined with **improved feeding**.

2.1.2 Prophylaxis

- **Immunoprophylaxis**

Isolated immunoprophylaxis for infectious diseases (vaccination) can result in an increase in the number of livestock, leading to **overgrazing**. A shortage of fodder can in turn weaken the animals and eventually lead to their death.

The **disposable equipment** used (syringes, cannulas, vaccine containers) has direct impacts on the environment. Improper disposal creates a risk of injury for human beings and animals (cannulas), while disposal on landfill sites can contaminate water and soil. Waste incineration causes air pollution and incineration residues may accumulate in soil and water.

- **Chemoprophylaxis**

Chemoprophylaxis involves **preventive treatment** such as daily subtherapeutic doses of a vermifuge or prophylactic administration of trypanocides. Such treatment can also help animals **adapt** to new surroundings, for example **new pastures**, by enabling them to develop premunity. Chemoprophylaxis enables particular species or breeds to use pastureland on which they could not be kept previously, e.g. by making it possible for zebu to graze in areas infested with the tsetse fly.

However, chemoprophylaxis can cause pathogens to become **resistant** to the drugs used. It may also adversely influence the development of immunity or premunition, with the result that **mortality will rise** after chemoprophylaxis is discontinued until the animals have developed immunity of their own.

To prevent tensions from developing between population groups, veterinary measures must **pay equal attention to the needs and interests of all groups concerned**.

- **Preventive management measures**

Preventive livestock management measures that can reduce the animals' risk of infection include the following:

- **appropriate herd distribution**: depending on the varying spread of diseases specific to particular types of animal, certain areas are used only by cattle and small ruminants, or only by camels.

- **avoidance of specific pastures** (at particular times of the day or year, or throughout the year): if pastureland is not used in the early morning when the grass is wet, invasion by infectious larvae of gastro-intestinal parasites will be reduced. Areas with a large population of mosquitos and biting flies during the rainy season are only used for grazing during the dry season or not at all. Areas infested with worm eggs and larvae or ticks in various stages of development (e.g. abandoned paddocks) are avoided for a number of months.

- **keeping livestock away from moist pastureland**: this prevents worm infestation (liver fluke) and reduces the risk of such parasites being transmitted to man.

- during **migratory** herding, areas infested with parasites (worm larvae, tsetse flies, ticks) are avoided at the times of year when the parasite population is at its largest (Sutherst 1987, Sykes 1987).

These preventive practices have long been employed by ethnic groups engaged in traditional animal husbandry. They have extensive beneficial effects on biodiversity and pasture resources, as they ensure that pastureland is not over-used.

Drainage of land for the purpose of creating specific forms of landscape and vegetation can lead to the **loss of wet biotopes**. Both biodiversity and the landscape will benefit if wet areas are fenced off and not used for grazing.

A changeover from pasture farming to confined livestock raising in the interests of animal health (see environmental brief Livestock Farming) will increase the livestock owners' workload. At the same time, however, the fact that grazing is replaced by growing and cutting of roughage may help to **reduce the risk of erosion**.

The resistance of productive livestock can be enhanced by **improved feeding**, in particular by giving the animals high-energy and protein-rich feedstuffs along with minerals. The environmental impacts of pasture farming with supplementary feeding are treated in the environmental brief Livestock Farming.

2.1.3 Vector control

Vector control involves attempting to change the balance of species so as to hinder the transmission of diseases by intermediate hosts and vectors or interrupt the cycle of transmission to man and livestock.

Chemical control of vectors includes measures such as use of insecticides in dips and the like to combat ticks, large-scale or targeted spraying of insecticides to control flies and mosquitoes, and application of molluscicides to kill snails. Long-term use of such methods can cause **resistant strains of parasite** to multiply, with the result that in tick control, for example, the agents used (acaricides) must be changed at frequent intervals. There is also a danger that other arthropod species may be affected as well. Pesticides may **contaminate soil and water** and, if the specified **waiting periods** are not observed, leave **residues** in milk and meat. The acute and chronic toxicity of the insecticides used thus creates direct risks for both man and animal. Large-scale vector control measures, such as aerial spraying of insecticide to combat the tsetse fly, involve an additional problem, namely **disposal** of the insecticide containers. Such containers must be treated as hazardous waste and must not be used for storing or processing food.

A further disadvantage of chemical vector control is that indigenous animal populations may **lose their natural resistance** or premunity with respect to numerous diseases. If the continuity of chemical vector control is not guaranteed, man and animal frequently have an increased risk of contracting the disease transmitted by the vector.

Unlike large-scale chemical control measures, **use of attractants and insecticide-impregnated traps** - for example in tsetse control - admittedly does not lead to eradication of the vector but at the same time ensures an almost total **absence of insecticide residues**. There is also virtually no danger that livestock will lose their premunity. Biological control methods such as use of sterilised flies to combat the tsetse fly and screw-worm fly generally do not entail any risks apart from those attaching to the necessary radiation treatment in the laboratory.

Targeted attempts to **eradicate wild animals serving as a "reservoir" for pathogens causing particular epizootic diseases** destroy the diversity and balance of species within the wild fauna. By reducing opportunities for hunting, they can moreover jeopardise the income and food supply of specific population groups.

Land clearance has far more complex impacts. By destroying the habitat of tsetse flies and other insect pests it reduces the infection risk for both man and livestock. The balance of species will change, with grasses and herbaceous plants becoming dominant; at the same time there is a **risk of increased soil erosion** and a **reduction in the soil's water retention capacity**. Local clearance techniques which - like those used in West Africa - leave 30 to 50 trees standing per hectare and allow the topsoil to remain largely intact have considerably less environmental impact than technically sophisticated methods. The pastureland created through clearance is **highly susceptible to erosion if overgrazed**. However, land clearance can also help to alleviate the pressure on overgrazed areas, thereby reducing their erosion risk and enabling the vegetation to recover.

Bush fires are seldom started with the aim of improving animal health. A reduction in the presence of vectors such as ticks (West 1965) is merely a **side-effect** of such measures, which have complex impacts on flora and fauna. Fire can also help to keep a savannah open and thus ensure that the insect-pest population remains low. As a result of **interference with the species composition**, however, insect pests may penetrate into hitherto unaffected areas and subsequently multiply here.

Theoretically speaking, **breeding of livestock with particularly high resistance** to a disease or vector (e.g. ticks) makes it possible to introduce a particular species into areas where it could not be kept in the past (Sutherst 1987). Indigenous livestock, however, already possess a high degree of resistance. For example, West African zebu can acquire a certain "trypanotolerance" if they have lived in tsetse areas for a number of generations and are regularly exposed to the pathogen.

2.1.4 Epizootic-disease control

The purpose of epizootic-disease control measures is to **prevent diseases from spreading**. Such measures are necessary in connection with the export and import of animals and animal products. They comprise general control measures (e.g. export or import bans), compulsory vaccination, ring vaccination in the event of acute outbreaks of disease, quarantine measures, compulsory slaughtering of sick animals and directives governing disposal of the carcasses of animals that have died or been compulsorily slaughtered.

Compulsory vaccination is a means of keeping certain diseases effectively under control for a lengthy period of time.

Ring vaccination is often accompanied by quarantine measures. The resultant restriction of herd mobility may lead to overgrazing in some places, creating tensions between sedentary and nomadic livestock owners. To ensure their acceptance, government quarantine measures should also take account of traditional practices that help to curb the spread of epizootic diseases.

Compulsory slaughtering is the most radical control measure, but is seldom used. It gives rise to severe financial losses for the farms affected and may oblige them to change their livestock management practices. For example, pastoralists may be forced to become less mobile if their herds fall below the critical size necessary for migration and this can create an increased risk of local overgrazing.

Disposing of dead animals by **burning the carcasses** creates unpleasant odours and pollutes the air. If wood is used, it also increases fuelwood requirements and thus women's workload wherever women are responsible for procuring wood (see also environmental brief Meat Processing).

Compulsory slaughtering is an emergency measure which prevents the spread of epizootic diseases and has beneficial effects on the health of both man and animal.

2.1.5 Zoonosis control

Through treatment of sick animals, prophylaxis, vector control and epizootic-disease control, veterinary services help to reduce the incidence of zoonoses and thus **improve human health**. Epizootic-disease control measures such as banning the keeping of dogs to curb the spread of echinococcosis and reduce the risk of rabies can restrict herding or make it difficult for nomads to guard their camps and thus have far-reaching socio-cultural implications. They may necessitate changes in livestock management practices and, by **reducing** mobility, can lead to **overgrazing** in certain areas.

2.2   Laboratory activities

2.2.1   Laboratory diagnostics

Preparation, transportation and handling of infected specimens in connection with laboratory work can give rise to environmental hazards. Improper handling and disposal of infectious specimens can endanger **human health** and contribute to the spread of disease.

In addition to the problems involved in **disposing** of non-reusable materials, there is also a risk that air, water and soil may be contaminated during transportation, storage and disposal of chemicals and reagents. Incineration of specimens no longer needed likewise causes air pollution.

To protect the environment, it is essential that **safety regulations be strictly observed** and that glass and plastic containers, reagents, chemicals and the specimens examined be collected, recycled where appropriate and properly disposed of (see OECD 1983). Use of toxic chemicals can sometimes be reduced by selecting appropriate **analytical methods**.

2.2.2 Vaccine production

Apart from the usual environmental risks attaching to laboratory work, **vaccine production** also involves all the hazards that can arise when live pathogens are being handled.

The most essential environmental protection measures are strict **compliance with safety regulations**, improvement of safety facilities where necessary and **appropriate disposal precautions**.

2.2.3 Residue analysis

By bringing to light undesirable environmental impacts, residue analysis also helps to safeguard human health and can thus be seen as a form of environmental protection. Detailed residue analyses can often be conducted only in **specially equipped laboratories** (see also environmental brief Analysis, Diagnosis, Testing).

2.3   Artificial insemination and embryo transfer

Artificial insemination (AI) and embryo transfer (ET) are modern techniques for **importing high-performance breeds** (primarily cattle) into tropical and subtropical countries. Animals produced in this way and born in the importing country are **better adapted** to the environmental conditions there than those imported live. Artificial insemination is also a means of **controlling the spread of venereal diseases**.

AI and ET do not have any direct environmental impacts. By curbing the spread of venereal diseases they may contribute indirectly to improving livestock fertility and may thus lead to higher productivity and an increase in livestock numbers. The resultant effects on the environment depend on the prevailing husbandry system.

Importing of high-performance livestock calls for strict control of vectors and ectoparasites; it may also be necessary to step up chemoprophylactic measures (see Section 2.1 above). There is a danger that the contribution of AI and ET to raising animal production may be overestimated and existing production systems consequently neglected.

2.4   Food inspection

Veterinary control and inspection of foods of animal origin is intended to **prevent human health being endangered** by tainted or infected foods.

2.4.1 Meat inspection

Meat inspection has hitherto often been confined to large modern slaughterhouses. It is a prerequisite for the export of animal carcasses and thus contributes to improving the income of livestock dealers and producers.

Attempts to **introduce** and apply meat inspection regulations from other countries without creating the necessary infrastructure (monitoring services, analysis facilities) can lead to a loss of income. The activities of small village slaughterhouses may be considerably **restricted** if they too are required to comply with such regulations and this can have an adverse effect on the rural population's meat supply. As women in some countries play an important role in slaughtering and meat marketing (particularly where small livestock are involved), such a development would have particularly serious consequences for the women's income and economic status.

However, meat inspection and proper disposal of products seized by the authorities prevent the **spread of epizootic diseases and zoonoses**. Hides contaminated with anthrax pathogens, for example, can be a highly dangerous source of infection for tanners.

2.4.2 Food hygiene

Milk hygiene plays a particularly important role in this field. Bacteriological monitoring is intended to prevent the spread of diseases such as tuberculosis and brucellosis, while analysis of milk composition helps to ensure high product quality. Milk testing and **sales bans** imposed as a result can have **far-reaching social consequences** if they extend to smallholdings which generally process only a few litres of milk a day and on which there is little danger that large quantities of milk may be contaminated. **Direct marketing** of milk and other dairy produce is often a major **source of income for women**. Sour-milk products have the advantage that the souring process kills pathogenic germs. **Boiling the milk** to kill germs increases energy requirements.

Legislation on milk hygiene could conceivably be misused to force small-scale processing and marketing operations out of the milk production sector.

The possibility of health risks can be counteracted by **advising and informing** women about hygiene precautions to be taken during processing of dairy produce.

3.   Notes on the Analysis and Evaluation of Environmental Impacts

The environmental impacts of **traditional veterinary medicine** have not yet been the subject of any summarising assessment. Use of traditional practices is generally confined to specific groups. Relevant references are contained in a number of annotated bibliographies (e.g. Mathias-Mundy and McCorkle 1989).

The OECD guidelines on sound laboratory practice provide pointers concerning the **environmental impacts of laboratory testing and analysis**. Information on this subject will also be found in the environmental brief Analysis, Diagnosis, Testing.

The environmental impacts of **residue analysis** are discussed in the relevant literature (e.g. Barke et al. 1983, DSA 1984, Rico 1986, Großklaus 1989).

4.   Interaction with Other Sectors

Through treatment of disease, epizootic-disease control and vector control, the environmental impacts of veterinary measures are linked to those of **animal production** and fisheries. By **monitoring hygiene** in food production and processing, veterinary services contribute to environmental protection in other sectors (e.g. agro-industry, slaughterhouses and meat processing). Veterinary medicine is dependent on the pharmaceutical industry for its supply of modern drugs. **Wastewater and solid waste disposal** is of relevance to laboratory activities, which also have links with the **chemical industry** by virtue of the need for reagents and chemicals.

5.   Summary Assessment of Environmental Relevance

The principal tasks in the veterinary sector are disease control and food inspection. However, disease control measures and laboratory work may have adverse effects on health and the natural environment, either directly or indirectly. Regulations governing epizootic-disease control and food hygiene can interfere with the social structures forming the basis of the livestock owners' existence.

The traditional remedies used by livestock owners are often based on **plant extracts**, which play an particularly important role in the treatment of small animals.

Therapeutic and prophylactic measures may cause pathogens to **develop resistance** and can give rise to **residues** in food.

While traditional forms of treatment have few negative impacts on the environment, this is not true of **modern drugs**, particularly if they are used improperly.

The activities of veterinary laboratories may give rise to water and air pollution; disposal of laboratory waste can cause contamination of air, water and soil.

Improved animal health is reflected in lower mortality and higher productivity, providing the producers with a more secure livelihood. However, consequent expansion of livestock husbandry can increase the danger of overgrazing if veterinary measures are not accompanied by improvements in the fodder supply and appropriate livestock management measures.

Epizootic-disease control, measures to curb the spread of zoonoses and monitoring of compliance with food hygiene regulations all have essentially positive effects on human health and the livestock owners' income. In some cases, however, incomes can be adversely affected. Imposition of excessively stringent standards for food hygiene can lead to the disappearance of small-scale slaughtering and milk-processing operations. Such a development could adversely affect the supply situation in rural areas and the income of those engaged in such activities, particularly women.

31

6. References

Barke, E. et al. 1983: Rückstände in Lebensmitteln tierischer Herkunft. Situation und Beurteilung. Verlag Chemie.

DSA 1984: Safety and quality in food. Proceedings of a DSA symposium "Wholesome food for all". Views of the animal health industries. Brussels 29/30.03.1984. Amsterdam: Elsevier.

Großklaus, G. 1989: Rückstände in von Tieren stammenden Lebensmitteln. Berlin: Parey Verlag.

Mathias-Mundy, E. & McCorkle, C. M. 1989: Ethnoveterinary medicine: an annotated bibliography. Bibliographies in Technology and Social Change No. 6. Ames: Technology and Social Change Program, Iowa State University.

OECD-Grundsätze zur guten Laborpraxis: Bekanntmachung im Bundesanzeiger (Federal Gazette) Nr. 42 dated 2. März 1983, pp. 1814 ff.

Putt, S.N.H., Shaw, A.P.M., Matthewman, R.W., Bourn, D.M., Underwood, M., James, A.D., Hallam, M.J. & Ellis, P.R. 1980: The social and economic implications of trypanosomiasis control: A study of its impact on livestock production and rural development in Northern Nigeria. Reading: Veterinary Epidemiology and Economics Research Unit, Study No. 25.

Rico, A.G. 1986: Drug residues in animals. London: Academic Press.

Sutherst, R. W. 1987: Ectoparasites and herbivore nutrition. In: Hacker, J.B. & Ternouth, J.H. (Eds.) The nutrition of herbivores. Sydney: Academic Press, pp. 191-209.

Sykes, A.R. 1987: Endoparasites and herbivore nutrition. In: Hacker, J.B. & Ternouth, J.H. (Eds.) The nutrition of herbivores. Sydney: Academic Press, pp. 211-232.

West, O. 1965: Fire in vegetation and its use in pasture management with special reference to tropical and subtropical Africa. Hurley: Commonwealth Agricultural Bureaux (CAB), mimeographed publications 1/1965.

Agriculture

Fisheries and Aquaculture

32.   Fisheries and Aquaculture

Contents

1. Scope

Activities for the purpose of obtaining food and other products from water bodies involve **catching and gathering** as well as **farming and raising aquatic organisms** (above all fish, crustaceans, molluscs and algae). Annual worldwide production in the fishery and aquaculture sector amounts to around 95 million tonnes.

The **principal forms of activity** are:

- capture fisheries
- aquaculture
- stocking and ranching

All three types of activity can be carried out in **seawater, brackish water and fresh water** and in both **coastal and inland waters. Deep-sea operations** primarily involve **capture fishery**, with **aquaculture** playing only a **very small role.** Stocking and ranching may include use of deep-sea areas in that fish released near the coast (e.g. salmon) may spend their growth phase in the open sea.

While **inland and inshore fisheries and aquaculture** are predominantly artisanal, deep-sea operations are primarily on an industrial scale where capture fisheries are concerned and exclusively so in the case of aquaculture.

Capture fisheries utilise **natural stocks of aquatic organisms.** Such activities influence the stocks not only by **catching** them but also by means of **conservation measures** (closed seasons, protected areas, catch quotas, use of selective gear). In **aquaculture** measures are taken to **directly influence** at least the **growth stage** and if possible also the **reproductive stage,** above all by **controlling water quality** (through the conditions under which the organisms are kept), **nutrition** (through feeding and pond fertilising) and **health** (by means of prophylactic and therapeutic measures). The **reproductive stage can be controlled by influencing maturation, egg and sperm production, hatching and larva raising.** The **characteristics of the organisms bred** can be **genetically influenced** (e.g. by means of selection, crossing or genetic engineering).

Stocking and ranching combine aquaculture with **fishery** (culture-based capture fisheries). Natural or artificial bodies of water are **stocked** with young organisms which were **hatched under supervision** and spent the particularly critical early stages of their life cycle under **controlled conditions.** When the **stocks** created or augmented in this way reach **the end of their growth stage,** they are fished using normal **capture-fishery** techniques.

Between the "production" process - carried out under natural conditions (fisheries) or controlled conditions (aquaculture) - and consumption of the products there are a number of other stages which may likewise have environmental impacts: **keeping fresh, processing, packing, transporting** and **marketing**.

Fisheries and aquaculture can be divided into **five main areas**:

- artisanal small-scale fisheries
- small-scale aquaculture
- fisheries and aquaculture in artificial lakes
- fishery in the 200-mile exclusive economic zone
- fisheries and aquaculture in mangrove swamps

In the first two areas, emphasis must be on supporting **low-income groups of the population** and ensuring that **appropriate technologies** are applied. These two aspects likewise form the focus of attention in the use of artificial lakes for fisheries and aquaculture. By contrast, activities involving **fishery in the 200-mile exclusive economic zone** - predominantly at **industrial scale** - centre on preservation of resources and on managing and monitoring their use. **Particular importance** must be attached to **environmental protection and resource conservation** when the intention is to **utilise mangrove swamps for fisheries and aquaculture,** as measures involving the use of this **fragile ecosystem** should aim from the very outset to ensure that **adverse environmental impacts** are **avoided** altogether or kept to an absolute minimum.

2.   Environmental Impacts and Protective Measures

2.1   Artisanal small-scale fisheries

The actual **fishing activities** have the **greatest bearing on the environment,** as the long-term availability of the resources depends on the extent to which these activities are geared to the resource situation and to the conditions prevailing in the ecosystem fished. Through centuries of experience, traditional artisanal **small-scale fisheries** based in a specific location have made sure that they **do not over-fish the available resources.** Any attempt to **increase production** can **jeopardise** this well-established **equilibrium**.

It may nevertheless **be possible to increase production** without endangering the resources. Such an opportunity exists in cases where the stocks fished are **utilised at a level below** that guaranteeing optimum yield and sustainability. The same applies if **fishing activities** are **extended** to those components of the biocoenoses within the ecosystem that were previously utilised very little or not at all.

However, utilisation of additional species may be **limited** by the **food relation-ships** between various components of a biocoenosis. If the prey of a predatory fish starts to be utilised in addition to the fish itself, the potential yield that can be de-rived from the predatory fish is automatically reduced, as the food supply has been curtailed. Since **many such relationships** exist, it is essential that they should be carefully reflected in the **management models** if it is intended to simultaneously utilise a variety of different organisms within a single ecosystem.

In **management of fishery resources, a key role** is played by the nature of the **gear** used as well as by **when and where it is used. Modern fishing gear** can be highly **efficient** (i.e. may jeopardise the existence of stocks if no restrictions are imposed on its use) and highly **selective**. Fishing gear is considered selective if it catches only particular species or size categories of organisms. Its selectivity can be determined by **net mesh size, hook size,** or the **depth of water or depth zone** in which it is used. The **most important** fishery management **measures** include **closed seasons, protected areas, minimum mesh and hook sizes, limits on the number of sets of gear, boats or ships and on the times when they may be used, and stipulation of catch quotas and size categories for the organisms to be caught.**

Stock management calls for a high level of **training in fishery biology** and adequate knowledge of **fishery economics. Stock regulating measures** should be discussed, agreed upon and implemented by the **local fishermen** acting on a **collec-tive** basis.

Apart from the need to conserve the resources themselves, it is also essential to protect their **living environment** against influences that could raise problems in the short or long term; to this end, the **physical, chemical and biological condition of fishing areas** must be **monitored. Product quality** depends on the chemical and biological **conditions of the water** and on the **sanitary conditions prevailing ashore** (village hygiene). The destructive effects of **using wood resources for smoking fish** can be curbed in two ways: by employing energy-saving kilns which permit **more rational use of wood** and by ensuring appropriate **management** of the forest resources concerned. The **amount of wood required for boat-building** can be reduced by replacing dugout canoes with **boats made of planks** and by **using alternative construction materials.**

32

Where it is likely that **infrastructure** for landing places used in artisanal fishery can be modified or removed only with difficulty, such facilities should not be constructed unless their necessity and expediency have been **thoroughly reviewed**. Concrete structures can also mar the aesthetic value of their surroundings (tourism).

2.2   Small-scale aquaculture

Aquaculture offers considerably **greater options** than capture fishery as regards both the type of organisms to be produced and the production sites. The **natural stocks** of organisms suitable for aquaculture can be **most effectively conserved** if aquaculture **controls the entire life cycle**, beginning and ending with the reproductive stage, and does so not just for one or two generations but on a long-term basis. As yet, however, this is possible only in the case of a few aquatic organisms. The only way of overcoming this problem is to **promote applied basic research** in the fields of reproductive physiology and reproductive ecology.

The **production site should be chosen** with the aim of **conserving natural ecosystems and scarce water resources**. The **choice of the type of organism to be produced** can contribute to conserving heavily used food resources if preference is given to species whose **food requirements** can be met by **waste products or by-products** from other sectors. Such products can either be **fed directly** to the fish or be used to **fertilise the water** and thereby promote the multiplication of food organisms (algae, microfauna). This could, for example, reduce demand for fish meal as a constituent of fish food. However, **producers have a tendency** to concentrate on **expensive organisms** (e.g. certain species of predatory fish) which generally require food of extremely high quality.

Water quality within and downstream of an aquaculture facility is determined by **management practices**. Efforts must be made to ensure that **as little leftover food as possible** remains in the water and that **the quantities of nutrients and pollutants washed out** of the installation are kept to a minimum. The amount of **leftover food** can be **minimised** by gearing the quantities of food given and the frequency of feeding to the absorption capacity and appetite of the fish. If sizeable quantities of waste are nevertheless discharged (e.g. from intensively operated through-flow ponds), they can be caught in **settling ponds** and thus largely prevented from entering rivers and lakes.

Drugs for preventing and treating disease and for combating parasites should **not be used in running water** (through-flow ponds) for reasons of effectiveness and economy and should **not be used at all in open systems** (cages, pens), even if the fish then have to be transferred to **special containers** for treatment and are thereby exposed to stress situations.

The main way of saving **energy** in aquaculture is to **obviate the necessity of pumping** for the purpose of water renewal. Introducing new water benefits the **oxygen supply** and helps to **wash out wastes**, besides **compensating for evaporation and seepage losses**. The **extent** of the necessary water replacement depends largely on the stocking density. Pumping energy can be saved wherever **natural gradients** can be used to create a water flow. **Artesian springs** are sometimes also available.

Considerable **ecological advantages** are offered by ponds in which **wastes can be utilised by plants and microfauna** which for their part are suitable as **food** for productive aquatic organisms. Such ponds can be fertilised by livestock (poultry, pigs) kept above or next to them. The profitability of this type of **integrated aquaculture** depends on the ecological appropriateness of the **aquatic organisms kept,** their **popularity** with consumers, the **production costs** and **market prices**. A role is also played by the way in which aquaculture is integrated into the overall production system, which usually involves other forms of production requiring labour. It is important, however, to know what constitutes the basis of the microfauna's food supply (there is a risk that pesticide residues could find their way into the food chain).

When **setting up ponds** in **tropical countries**, it is essential to bear in mind the risks originating from **diseases** whose pathogens spend at least one stage of their life cycle in water or in aquatic organisms (malaria, schistosomiasis etc.).

Cage farming not only involves high **feeding costs**, but also gives rise to problems in **procuring the necessary materials** for making the cages, as nets, support rods and floats are **expensive**. Only in forested regions is the use of wood unlikely to present any problems.

Elimination of potential health risks attaching to consumption of aquaculture products must be given particular attention wherever human excrement and domestic wastewater are used for fertilising ponds. In **wastewater aquaculture systems,** the critical factors in this respect are the number of pond stages, the degree of dilution and the period for which the water is retained before it enters the fish ponds. **Accurate management**, along with regular checks on **sanitary conditions and water quality,** are essential in such cases.

2.3   Use of artificial lakes in fisheries and aquaculture

As use of artificial lakes involves a **combination of fish farming and fishing** (and can thus be placed in the category of "culture-based capture fisheries"), the environmental protection measures described in both 2.1 and 2.2 above are of relevance in this connection. However, the fact that **an artificial lake is a man-made entity** creates a substantially **different** situation, both in limnological and ecological terms as well as from the sociological and economic viewpoints. Man-made lakes differ from natural ones by virtue of their artificial nature, the fact that they are subject to continuous **management** to enable them to fulfil their primary purposes (drinking-water supplies, energy generation, irrigation), their **initial biological "void"** which - depending on the actual and to some extent random sequence of colonisation by flora and fauna - can offer scope for **a variety of biological development possibilities,** and last but not least the new options which this may offer in terms of fisheries and aquaculture. While an **artificial lake** thus allows man a considerable degree of **freedom in shaping ecological conditions**, it nevertheless confronts him with **far-reaching social and economic problems** when it comes to developing and establishing the ways in which it is to be used.

Two important **principles** should be observed when determining **how a new artificial lake is to be used**:

-        **Organisms that are foreign to the ecosystem and region concerned** should be introduced only with strict observance of internationally recognised precautionary measures or not at all.

-        No attempt should be made to **regulate fishery activities** until local traditions have been studied in detail; regulation measures should be realised in consultation with existing local fishermen and those willing to settle in the area.

When a **new dam is being planned,** consideration should be given to the various **options for fisheries and aquaculture** which the newly created lake will offer. Where appropriate, such aspects should be taken into account when deciding on dam design.

2.4   Fishery in the 200-mile exclusive economic zone

Optimum fishing of the **200-mile exclusive economic zone** (EEZ) calls for **use of advanced technology**. This will inevitably lead to **conflicts** in the **transition area** between industrial deep-sea fishing and artisanal inshore fisheries unless depth conditions and coastline configuration create a natural division between the two. Such conflicts are often to the detriment of the available **resources**, causing them to be **over-exploited** or even **destroyed**. They may also adversely affect the economic and social position of the **artisanal inshore fishermen**, who usually **come off worst** in such conflicts if their interests are not effectively safeguarded through government intervention.

While **old-established, traditional fishing communities** have developed fishing practices **designed to ensure that resources are preserved in the long term**, the technical potential of **modern deep-sea fishing** - which can totally exhaust resources within a short time - means that use of resources must be **strictly limited and monitored**. **Minimum mesh and hook sizes** must be laid down to make sure that the gear does not catch young organisms which are not yet mature enough to reproduce and thereby preserve the existence of the fish stocks. Such regulations can also **reduce the pointless destruction of small food organisms** caught in the nets together with the fish.

The only way to prevent **trawls** with "ploughing" structures from causing serious damage to entire communities of sea-bed organisms is to **ban** the use of such gear. Depending on local conditions (sea-bed conditions, reproductive cycle and migration of fish or other organisms), use of such nets must be banned either **completely**, or in specific areas or at certain times of year.

Complete bans must be imposed on catching certain types of organism while they are still going through their development phase in the "nursery areas". As such bans are often impossible to enforce, efforts are being made in many places to create **artificial refuges** - in the form of submerged concrete blocks, for example - to which fish and other aquatic organisms can retreat and from which they can repopulate areas which have been subject to adverse influences or whose stocks have been exhausted. However, the **effectiveness** and cost-benefit ratio of these "artificial reefs" are still the subject of considerable **debate**.

The **death of numerous fish and large marine fauna** (dolphins, turtles, birds etc.) in lost **drift nets** made of plastic that does not decompose in water can be prevented by using **degradable thread** to attach the net sections to the floats. The net sections would then collapse after a time and sink to the bottom. However, this method appears to be **too complicated** for general use and it is not known what **damage** the nets might cause on the **sea bed**.

Considerable problems are still posed by the question of what to do with the **"by-catch"** (of non-target species), in other words the organisms with little financial value that are caught together with the highly lucrative species (e.g. prawns or shrimps) constituting the intended catch. These organisms are large or bulky enough to be retained by the net together with the main catch even if the minimum mesh sizes are adhered to. However, their **market value** is so **low** by comparison with that of the main catch that it is not worthwhile landing them, despite the fact that a considerable proportion of this "by-catch" would often be **suitable for human consumption**. If a worldwide solution to this problem could be found, for example by having the by-catch **continuously collected** by special boats at sea or by means of other methods, **several million additional tonnes of fish** would become available as **food each year**.

As is generally the case with motorised seagoing shipping, **deep-sea fishing vessels'** high consumption of **fossil fuel** necessitates special measures to **dispose of residues on land**. Environmental problems on land as a result of fishing stem primarily from industrial processing of the catch. Mandatory standards regarding **disposal of solid wastes and wastewater** must be observed; in some places such standards have still to be introduced. Some of the **solid wastes** can be made into **fish meal**, while valuable constituents of liquid waste can be recovered in the form of extracts and used as **feed additives** (cf. environmental briefs Inland Ports, Shipping on Inland Waterways, Wastewater Disposal and Solid Waste Disposal).

2.5   Use of mangrove swamps in fisheries and aquaculture

The **traditional ways of using** the flora and fauna in mangrove swamps can be viewed in the same light as artisanal small-scale fisheries in other areas: they take into account the regeneration capacity of the resources and are thus **ecologically sound**. However, this is not true of **modern aquaculture** on large fish farms whose construction necessitates **complete clearance of the mangrove vegetation**. One example of this type of operation is the large-scale **raising of brackish-water prawns**. As **production** of these much sought-after crustaceans can yield **high profits**, the potential suitability of **mangrove swamps** as sites for brackish-water ponds has given rise to **dangerous pressure** on these areas. Since mangrove areas are subject to the daily ebb and flow of the tide, the water has the **necessary salt content** and **water replacement** can be achieved at relatively **little expense** because the **tidal cycle can be used** to **minimise the amount of energy** required for pumping.

Efforts should be made to counter the pressure on the **mangrove swamps** in as realistic and flexible a manner as possible. The paramount principle should be that **no form of use is to be permitted without thorough advance planning**. The principal purpose of such planning is to completely **rule out non-traditional use** of areas which are irreplaceable as **nature reserves, genetic resources, nursery areas** for important aquatic organisms or **protective belts guarding against coastal erosion**. Clearance of mangrove swamps for the purpose of aquaculture can also be prevented by making **areas immediately upstream of the mangrove belt** available for the creation of ponds. Provided that the installations are well-managed, the necessary pumping costs could be offset by earnings.

Where **use of mangrove swamps** appears **unavoidable** for economic reasons, activities should be concentrated in areas with **clayey soils**. In such areas the mangrove vegetation can easily re-establish itself if the ponds (or swamp-rice fields) are abandoned at a later date, whereas areas with sandy and peat soils will be nothing more than wasteland for a long time afterwards. Continuing efforts should also be made to find ways of **utilising the natural productivity** of suitable mangrove areas for semi-intensive small-scale aquaculture, without clearing them of all vegetation and without major additional expenditure on feed or fertiliser. The success of such experiments will depend on whether or not it proves possible to keep costs down to a level ensuring that even low yields per unit of area offer attractive economic prospects.

3. Notes on the Analysis and Evaluation of Environmental Impacts

The **environmental aspects** of fisheries and aquaculture fall into **five categories**:

- impacts on the natural environment which have adverse effects on aquatic organisms but which do not stem from either fisheries or aquaculture (pollution of water through disposal of wastes from industry, agriculture and households or caused by nutrients, pesticides and residues being washed out of soil on land; water-resources management measures); such impacts may affect both fisheries and aquaculture.

- influences on the existence and renewal of fish resources resulting from their use (such influences relate only to natural stocks and not to those maintained and controlled by man, i.e. aquaculture is affected only where it is dependent on young organisms from natural stocks).

- environmental impacts caused by fisheries and aquaculture (disturbance of ecological equilibrium, impairment of water quality etc.).

- influences on use of resources (and thus on the resources themselves) caused by changes in the social and socio-economic situation of producers and consumers (e.g. as a result of population growth).

- effects of fishery and aquaculture activities on the social and socio-economic situation of producers and consumers (e.g. in the event of local overproduction without sufficient access to more distant markets).

Computer-aided simulations of both the ecological and economic situation, using a standard model, can help to ensure that **natural fish resources are optimally utilised** in a manner which **preserves their capacity for renewal.** Such **models** are **essential** for developing a **reliable long-term utilisation strategy** which takes into account the **economic interests of both the fishermen and the country concerned without jeopardising** in the long run the **natural resources on which fishing depends.**

There as yet exist **no summarising overviews or evaluations of the serious impacts which various modern techniques** can have on resources (use of explosives and pesticides, bottom trawling, use of drift nets etc.).

Considerable efforts are currently being devoted to **studying and evaluating the environmental impacts of aquaculture activities**. In September 1990 the International Centre for Living Aquatic Resources Management (ICLARM) held a **symposium** on **environment and aquaculture** (results to be published in 1992).

4.   Interaction with Other Sectors

Activities in the fisheries and aquaculture sector can be combined with agricultural production and with water resources development. The following are examples of the ways in which **fisheries and aquaculture can be integrated with agricultural production**:

-       combining fish farming (or artisanal fisheries) with plant production and animal husbandry in an agricultural production system without physical integration of the individual components
-       combining fish farming in ponds with keeping of poultry, pigs or other livestock above the ponds
-       fish farming in swamp-rice fields

The following are examples of the ways in which **fisheries and aquaculture can be combined with water resources development**:

-       fishing in artificial lakes of all kinds (including those designed to provide drinking-water supplies)
-       fish farming in small, shallow irrigation reservoirs
-       fish fattening in large irrigation canals
-       cage fish farming in adequately large and deep artificial lakes not used to provide drinking-water supplies

(cf. environmental briefs Large-scale Hydraulic Engineering, Irrigation and Rural Hydraulic Engineering).

Fisheries and aquaculture also have extensive **links with agriculture** through the **use of waste products, by-products** and (in exceptional cases) **main products** of agriculture as **food or fertiliser** in aquaculture and through the use of fish meal in the **production of livestock fodder** (cf. environmental brief Livestock Farming).

Links with the forestry sector exist by virtue of the **wood required** for making boats and fishing gear, for preserving and processing fish by means of **smoking** and for **making cages**. The **close ecological links between forests and waters** are of particularly far-reaching significance and must be taken into account in both forestry and fishery activities.

Fisheries and aquaculture also have links with the **energy sector** through the **operation of boats, ships** and sophisticated **fishing gear, fishing ports, refrigera-**

tion plants and **industrial processing facilities, technically complex aquaculture installations** and **vehicles** for transporting people, equipment, supplies and products.

Attention has already been drawn in the text to links with other sectors.

5. Summary Assessment of Environmental Relevance

32

Fisheries and aquaculture are **dependent** on the existence of an **environment** which is **intact or at least not permanently damaged**, but can themselves **have negative effects** on the **environment and on resources.** As **fisheries rely** on continuous **natural renewal of fish resources**, activities in this field must **treat these resources and their habitats with care.**

Wherever **resources** have already been overfished and their **habitats** adversely affected by environmental changes, they must be rehabilitated if possible. Once fishing reaches a certain level of intensity, **improved utilisation of catches** is the only way of **raising production**. To this end, particular efforts should be made in the future to promote **consumption** of **types of fish** which are currently still **unattractive** from the commercial viewpoint and are simply used for making fish meal, and to ensure that fewer fish are lost as a result of spoilage.

Aquaculture is for the most part still a relatively new field of activity in the fishery sector. To promote its future development it requires **tailor-made strategies** which take particular account of the fact that most of the **natural resources employed** in aquaculture (water, land, feedstuffs; spawn in the case of most cultivated species) are already **used for other purposes** and can thus become **sources of conflict.** One of the most important strategic principles must therefore be to **avoid such conflicts** or **resolve them with a minimum of adverse consequences in the ecological, economic and social spheres**. This means, for example, that

- the **impacts of aquaculture** must initially be **compared with those of other ways of utilising resources** (e.g. mangrove forests as a means of preventing coastal erosion, tourism), with aquaculture activities then being designed as far as possible such that they complement use of water resources for other purposes;

- **by-products or waste products** that cannot be put to beneficial use elsewhere should be used as far as possible for **feeding aquaculture organisms** and fertilising the water. It is vital, however, that such products should be free of contamination (e.g. by pesticides).

Observance of the basic principles can be **encouraged** if the long-term advantages are demonstrated by **effective examples** and appropriate **political and ecological conditions** create a balanced combination of incentives and restrictions.

When involved in the development strategy and given appropriate training, women can play a key role in helping to **prevent, reduce and eliminate environmental and health risks**. Particular importance must be attached to awareness-raising measures that take religious considerations and cultural aspects into account.

6. References

Alabaster J.S. & Loyd R., 1980: Water quality criteria for freshwater fish. FAO/Butterworths.

Beveridge M.C.M., 1984: Cage and pen fish farming: carrying capacity models and environmental impact. FAO Fishery Technical Paper 255.

DANIDA, 1989: Environmental issues in fisheries development. Copenhagen.

Deutsche Forschungsgemeinschaft, 1980: Forschungsbericht: Methoden der Toxizitätsprüfung an Fischen; Situation und Beurteilung. H. Boldt Verlag.

Dwippongo A., 1987: Impacts of trawl ban on fisheries and demersal resources in the Java Sea. Ph.D. Thesis, Nihon University, Tokyo.

FAO, 1975-1983: Manual of methods in aquatic environment research. Part 1 - 9.

FAO, 1981: Conservation of the genetic resources of fish; problems and recommendations. FAO Fish. Tech. Paper 217, 43 pp.

FAO, 1981: The prevention of losses in cured fish. FIIU/T219.

FAO, 1987: The economic and social effects of the fishing industry - a comparative study. FIP/C314/Rev. 1.

ICLARM, 1982: Mismanagement of inland fisheries and some corrective measures. Contribution 110.

Johannes R. E., 1981: Working with fishermen to improve coastal tropical fisheries and resource management. Bulletin of Marine Science 31.

Lasserre & Ruddle, 1983: Traditional knowledge and management of marine coastal systems. Biology International, Special Issue 4.

Metzner G., 1983: Fischtests im Rahmen nationaler und internationaler Regelungen. In: Untersuchungsmethoden in der Wasserchemie und -biologie unter besonderer Berücksichtigung des wasserrechtlichen Vollzugs. Münchner Beiträge zur Abwasser-, Fischerei- und Flußbiologie 27, 47-69.

Mienno J. L. & Polovinci J. J., 1984: Artificial Reef Project - Thailand. ADB.

32

Nauen C. E., 1983: Compilation of legal limits for hazardous substances in fish and fishery products. FAO Fishery Circular 764.

Pauly D. & Tsukayama I., 1987: The Peruvian anchoveta and its upwelling ecosystem: three decades of change. ICLARM Studies and Reviews 15. (IMARPE/GTZ/ICLARM).

Pauly D., Muck P., Mende J. & Tsukayama I., 1989: The Peruvian upwelling ecosystem: dynamics and interactions. ICLARM Conference Proceedings 18. (IMARPE/GTZ/ICLARM).

Rosenthal H., Weston D., Gowen R. & Black E. (Ed. committee), 1988: Report of the ad hoc Study Group on "Environmental impact of mariculture". ICES Cooperative Research Report 154, Copenhagen.

Ruddle K. & Johannes R. E., 1985: The traditional knowledge and management of coastal systems in Asia and the Pacific. UNESCO/ROSTSEA Regional Seminar. UNESCO Regional Office, Jakarta, Indonesia.

UNDP/FAO, 1982: Fish quarantine and fish diseases in South-East Asia. UNDP/FAO South China Sea Fisheries Development and Coordination Programme and IDRC.

UNESCO, 1984: Coastal zone resource development and conservation in S. E. Asia with special reference to Indonesia.

UNIDO, 1987: Environmental assessment and management of the fish processing industry. Sectoral Studies Series 28.

World Bank, 1981: Socio-cultural aspects of developing small-scale fisheries: delivering services to the poor. World Bank Staff Working Paper 490.

World Bank, 1984: Harvesting the waters - a review of Bank experience with fisheries development. Report 4984.

World Bank, 1985: Integrated resource recovery, aquaculture: a component of low coast sanitation technology. Technical Paper 36.

Agriculture

Agricultural Engineering

33. Agricultural Engineering

1. Scope

The fundamental components of agriculture are **plant production** and - based upon this - **animal production**. **Agricultural machinery and implements** are **used by man for the purpose of influencing** the natural process of plant and animal growth. Such **mechanical aids** can be divided into three categories on the basis of their **energy source**:

- hand-held implements
- animal-drawn implements
- motorised implements (with internal combustion engine or - less commonly - electric motor)

Agricultural engineering covers all aspects of **using and manufacturing technical aids for agricultural production**, the **upstream and downstream sectors**, and **decentralised generation and use of energy** in rural areas.

It is in **plant production** that **agricultural engineering** plays by far its **most important role**, although it is also becoming increasingly significant in **livestock farming** (intensive livestock husbandry). Mechanical aids are **most commonly used** in **tillage** and **transportation**, as well as in **threshing** and - where appropriate - for **supplying water**. As an area of project activity, **agricultural engineering** can thus be viewed in particular as an **extension of the plant production sector**; **links** frequently also exist with **animal production, irrigation and agro-industry**. The comments made in the relevant environmental briefs regarding objectives, impacts and protective measures apply by analogy.

2. Environmental Impacts and Protective Measures

2.1 Man, ecosystem and agricultural engineering

2.1.1 Man and agricultural engineering

Agricultural operations are generally **mechanised** for reasons of **labour efficiency**, namely

- to raise per-capita productivity (worker performance) and
- to reduce the burden imposed by physical labour.

The **changeover to a different source of power** - i.e. from manual labour to animal traction to motorisation - brings major new **technical and economic factors** into play. This means that operation, maintenance and management must all meet correspondingly **higher standards**.

While the burden of heavy physical labour is **reduced**, work may subsequently become **one-sided or monotonous**. Animals or machines determine the pace of the work. **Noise** prevents communication and can adversely affect health, as can engine **exhaust emissions**.

Operators and other people can be **endangered** if **machines get out of control**. **Moving parts** (shafts, belts, rods) **increase the accident risk**.

Operation of machinery generally enjoys **higher status** than manual labour or handling of animals. Mechanisation can lead to **changes in the division of labour and distribution of income**, with "women's work" becoming "men's work" (seldom the reverse).

The **way in which technical aids are used** is generally the crucial factor determining whether they have **positive or negative impacts**. As **motorised techniques** tend to **magnify errors**, however, such methods can have **considerably more serious negative effects** than may result, for example, when hand-held implements are used.

It is **particularly important** that the **right equipment be selected** for a particular operation and that **machinery and implements be used properly and at the right time**. This can be achieved above all by **training and advising** the operators and by imposing **legislative requirements** (accident prevention, technical inspection etc.).

2.1.2   Ecosystem and agricultural engineering

As the **degree of mechanisation increases, cropping areas** and **roads and tracks are geared to the demands of the machines and implements**. Use of **tractors** and **self-propelled machines** such as combine harvesters - or indeed even the use of animal traction - calls for **large cropping areas**, which should be as **free as possible of obstacles** such as stones, trees and tree stumps.

Intercropping - i.e. simultaneous cultivation of several different crops in a single field - offers **few opportunities for mechanisation** and **single-cropping systems** therefore **predominate**. Following tillage, the **surface of the ground remains unprotected for several weeks** and may be exposed to the risk of **erosion by wind and water**. **Broadcast sowing is replaced by row seeding; rows which follow the slope of the terrain can increase the danger of erosion by water**. Roads and bridges, as well as irrigation and drainage channels, are often designed to meet the requirements of mechanisation. **Ecologically valuable areas** such as forests, hedges and fallow land are increasingly **lost**.

The spectrum of **flora and fauna** in a region may be **diminished or altered**; ecological diversity is reduced. An **absence of windbreak vegetation** in arable-farming areas increases the risk of **erosion by wind**.

Top priority must be given to promoting mechanisable **land use systems** which take both economic (including labour efficiency) and **ecological aspects** into account. Such production and farming systems have already been designed for certain regions (particularly in temperate climates) and their use is to be encouraged. Applied research and development work is still needed in other regions. **It is not enough, however, to provide purely technical training and advice** on proper use of machinery and implements. A **new awareness** is required on the part of all concerned (from agricultural workers to decision-makers) if **the inherent potential of mechanisation is to be utilised** and **risks are to be recognised and reduced**.

It is important to **preserve refuges**, forests, hedges, wetlands and other niches for flora and fauna. Such areas do not hinder large-scale mechanised farming, as there are few labour-related advantages in having cropping areas larger than 20 hectares. **Row seeding** is **essential** in order to permit **mechanical weed control techniques**, for example, to be used instead of chemical methods.

2.2 Agricultural engineering in general

2.2.1 Energy sources, drive systems, fuels and lubricants

The **principal sources of power** are **manual labour, animal traction and engines or motors. Wind and water power are used** all over the world to drive stationary machines (mills and pumps).

In many countries, **biomass** (particularly wood, but also straw and dung) is the major **source of energy** for cooking in rural areas (see also environmental brief Renewable Sources of Energy).

The demands of **agriculture** may sometimes **compete** with those of rural **households**. If draught animals are kept, land must be used for cultivating forage crops and is thus not available for growing food; use of dung as fuel deprives the cropping area of nutrients; both stationary and mobile engines (e.g. those of tractors) are still fuelled for the most part by **non-renewable forms of energy**, particularly petroleum products.

As such engines are not required to power vehicles travelling long distances, they emit only limited quantities of **nitrogen oxides and carbon monoxide. Use of clean fuel** and **correct engine tuning** can nevertheless help to **minimise harmful emissions**.

Where **internal combustion engines are used** (e.g. in tractors and water pumps), **surface water** may be contaminated by **fuels and lubricants**. The risk is particularly high in **parking areas and workshop yards** where **fuel tanks are filled and oil changes** performed.

33

The **technical facilities** used for transporting and storing fuels and lubricants are often **in need of improvement**. Tanks must be checked for leaks and contamination. Oil collectors for use during oil changes are to be provided and a **used-oil processing system is to be set up**. Negligence in handling fuels and lubricants (which can create fire hazards and lead to contamination of soil and water) can be reduced only by means of **long-term training measures** and **appropriate technical facilities**. Efficient governmental or private **monitoring institutions** (e.g. like Germany's water authorities and technical inspectorates (TÜV)) must be established.

Use of biodegradable oils is to be **promoted**. Where power saws (whose chains require a great deal of oil) are used in water conservation areas, for example, only vegetable-based lubricating oils (rape-seed oil) should be employed as is the case in Germany. This regulation is to be extended to cover **hydraulic oils** for vehicles operated in water conservation areas. It is recommended that use of such oils in agriculture also be encouraged.

2.2.2 Production of technical aids

Hand-held implements and simple animal-drawn implements are often made by the family wishing to use them or by **local craftsmen**. Such production activities have **few** if any **environmental impacts**. The comments made in the environmental brief Mechanical Engineering apply in the case of industrially manufactured agricultural machinery and implements.

2.3   Specific aspects of plant production

2.3.1 Tillage

Loosening the soil with the aim of **improving conditions for the crops** plays a crucial part in arable farming. One purpose of **turning the soil** is to **eliminate plants competing with the crop**.

The extent to which the **soil structure** is **changed** depends on the **form of tillage**, which may involve

- loosening the soil with hooks or tines,
- turning the soil with the plough, or
- crumbling and breaking up the soil with a powered rotary cultivator or harrow.

If **greater drive power becomes available**, this may lead to **selection of implements** (e.g. rotary cultivator rather than plough) which **modify the soil structure to a greater extent**. In addition, less suitable **(marginal) sites** may be "put to the plough". Both of these steps increase the **risk of soil degradation**, which may involve reduction of pore volume, water absorption capacity and water storage capacity, the danger of puddling and crusting and a loss of organic matter. Crusting impedes water penetration and plant growth.

The **optimum degree of soil moisture** for tillage lies within **narrow limits**. If the soil is tilled when it is **too wet** it will be **compacted**, while tilling a soil which is **too dry** will result - depending on its clay content - in formation of **clods** or **pulverisation**. Compaction may assume sizeable dimensions if **heavy tractors and implements are used**.

Among other things, **compaction of the soil** has effects on **plant growth, soil organisms** and the **availability and breakdown of plant nutrients and pesticides**. On **slopes**, soil layers above the compacted horizon may **slip**.

Loosening of the soil and **introduction of organic matter** have **positive** impacts on the **soil fauna**. By contrast, compaction and puddling, frequent disturbance of the soil by tillage and application of pesticides and fertilisers all adversely affect the development of soil organisms.

Measures for **preventing soil erosion and compaction** include the following in particular:

- The soil should if possible have permanent vegetation cover in the form of living crop plants (permanent cropping, intercropping, alley cropping) or dead plants (mulch).
- Crops should be sown directly into the remnants of the preceding crop (without the soil being turned).
- Crop residues should be left on the surface and should not be ploughed under.
- Coarse soil structures are to be created or preserved through appropriate crop rotation measures and use of suitable implements.
- Contour ridges or terraces are to be created on slopes. This may be a somewhat complicated operation; there may be no vehicle access to the land in question.
- Windbreaks are to be planted at right angles to the direction of the prevailing wind.
- Organic matter is to be preserved and increased if possible.
- The pressure to which the soil is subjected when vehicles are driven over it is to be reduced as far as possible by using small/light-weight tractors and machines or larger tyres.
- Where possible, the soil should be driven over and tilled only when the moisture content is optimum.
- Shallow and deep tillage implements are to be used alternately over the course of time.

Emphasis must be placed on promoting a **willingness** to abandon excessive exploitation of the soil in favour of **sustainable, site-appropriate forms of cultivation**.

2.3.2 Sowing/planting, crop tending and fertilising

Sowing and planting, which are performed after the soil has been tilled, are intended to create **optimum conditions** for the growth of the seed or young plant.

After tillage, the soil may be totally or partly without cover until the crops have fully developed. It is thus exposed to the risks of **erosion and to puddling** as a result of heavy rainfall or severe **evaporation**.

With large cropping areas, mechanical aids are virtually essential for distributing **chemical pesticides**. Such equipment calls for **highly skilled** operators. Unsuitable, defective or incorrectly operated equipment can result in **overdosage** of fertilisers and pesticides, which will have negative effects on soil, plants and water as well as on the equipment users.

Application of highly concentrated liquid pesticides using the ULV (ultralow-volume) technique may lead to **severe air pollution**, with drift **causing contamination over wide areas**.

Pesticide users can be exposed to **serious health hazards** by touching or inhaling the chemical substances. It is often difficult to empty the containers completely and water used to rinse them out can contaminate surface water and drinking water. Pesticides and the equipment used to apply them are often **stored** improperly and are frequently kept in the same place as food for want of any other lockable rooms (see also the detailed remarks in the environmental brief Plant Protection).

Under unfavourable conditions, **mechanical weed control measures** (hoeing) can destroy the soil structure and encourage erosion; despite this fact, they should still be **preferred to chemical methods**.

Agricultural engineering can make a major contribution to ensuring that **pesticides and fertilisers** are used and applied **correctly**. Apart from **selection of the right equipment**, which is determined in part by the **formulation of the agents used** (e.g. powder or liquid), **correct operation** is equally important. If the **time of application is appropriately chosen** and **economic-threshold concepts are used**, quantities can be cut, drifting reduced and **risks thereby minimised**. Protective clothing, including face masks, is to be provided.

Under the climatic conditions prevailing in the tropics and subtropics, however, wearing of such clothing imposes considerable **physical stress**.

The **stresses and risks** for the equipment operators can be **reduced** to a large extent if the **work is appropriately organised** (e.g. operators proceed in the direction in which the wind is blowing).

2.3.3   Harvesting, threshing, processing, preservation, storage

The **technical aids** used in harvesting and threshing are intended to **enable the work to be performed more easily and rapidly**, besides **minimising losses and risks**. Harvesting and processing of "dry goods" (e.g. grain, burned-off sugar cane) can give rise to **dust emissions** which affect only a small area but are highly intensive. Such emissions affect people working and living in the vicinity, as well as animals. They can be **reduced** by taking **technical measures at their source** or their effects mitigated through the wearing of **face masks and protective clothing**.

Threshing and processing may yield **by-products** (glumes, hulls etc.). However, the quantities occurring at farm level **do not constitute a serious environmental hazard**, since the farm can generally **utilise such products itself**.

When **crops are harvested**, various substances **are removed from the natural cycling systems on the site concerned**. Efforts should be made to ensure that at least the **by-products** are **returned to the soil** either directly, after use for other purposes (e.g. as livestock fodder) or after composting.

Technical measures for preserving and storing produce at farm level **seldom have any environmental impacts, unless chemicals are used**. Crop drying calls for an **energy supply**, however, and in certain cases this may lead to over-use of local forest resources. Efforts to alleviate this problem should focus on **ways of reducing energy consumption** (heat sources).

2.3.4   Supplying and distributing water

To supplement the environmental brief Irrigation, attention must be drawn here to a number of important **interrelationships** and **areas of overlap** between these two sectors.

The **water application and distribution** system (gravity-flow method with open channels, pressure method with pipes or hoses) has a considerable **influence on mechanisation**:

-        Channels determine the field size and bridges are needed to cross them.
-        Small embankments and ditches in the field are damaged when driven over.
-        Pipes have to be removed before a field can be tilled or urgent plant protection measures carried out.

2.4   Aspects of animal production

Pasture farming has always been the **traditional form** of livestock husbandry. The use of **technical aids** is confined to **protective and security measures** (pens etc.); the same applies to the keeping of small animals (e.g. poultry, rabbits, bees). It is not until **livestock farming is intensified**, with livestock kept in confinement, that technical aids become **increasingly important.** In industrialised countries where intensive animal husbandry is practised (particularly in Europe), technical equipment has come to play just as vital a role in livestock farming as it does in crop growing.

In poorly ventilated livestock housing, **heat, dust and gases** - particularly ammonia - can create stresses for both man and animal.

Sizeable quantities of **ammonia** can escape into the air when **animal excrement is stored and applied to fields as manure.** In industrialised countries, ammonia is one of the principal factors causing the gradual death of the forests in regions where intensive livestock husbandry is practised. One effective **countermeasure** is to make sure that solid or liquid manure is **immediately worked into the soil.**

Improper storage and spreading of animal manure can lead to **over-fertilising** (eutrophication) of both surface water and groundwater.

Protective measures must often focus first of all on bringing about **changes in awareness,** for only when this has been achieved can technical measures contribute to reducing negative environmental impacts. **Animal excrement** must be regarded and treated as a **valuable fertiliser** and not as waste. It is also important that this fertiliser be **spread** over the cropping area **precisely** in line with the plants' **nutrient requirements.** Only if such measures can be realised will it be possible to develop intensive livestock farming methods that are environmentally sound in the long term.

3.   Notes on the Analysis and Evaluation of Environmental Impacts

Among the **negative consequences of cultivation** - magnified by use of mechanical aids - it is **erosion** that has by far the greatest significance worldwide. While there are numerous **ways of reducing erosion** (e.g. crop-growing measures such as mulching, technical measures such as terracing and planting of windbreaks), **standards** for evaluating the effect of erosion are largely confined to criteria for **recording and assessing the removal of soil.** Specific **cultivation bans or requirements** are occasionally imposed in the catchment areas of reservoirs particularly at risk from sediment.

Manufacturers of tractors and agricultural machinery in industrialised countries are called upon to fulfil widely varying national **environmental requirements**. These include

- standards and guidelines on design and durability;
- provision of safety devices, protective circuits etc., particularly for motorised vehicles and machines;
- provision of special equipment if the vehicles use public roads (danger to other road users);
- emission standards (exhaust emissions, noise).

Institutions performing official functions, such as agricultural-machinery testing stations, carry out **type approval testing** whose results are binding on manufacturers. Compliance with requirements is relatively easy to monitor.

It is far more difficult to ensure that **regulations are observed by users**. Safety devices may be removed, protective clothing and masks not worn or emission standards, speed limits and the like disregarded.

4. Interaction with Other Sectors

Agricultural engineering is closely linked to the following sectors:

- Plant production: agricultural engineering constitutes an extension of this sector in virtually every respect
- Plant protection: mechanical methods, application techniques
- Livestock farming: use of animal traction, intensive animal husbandry (still on only a small scale in developing countries, but extremely widespread in industrialised nations), livestock farming practices
- Irrigation: particularly supply, application and distribution of water, gravity-flow and pressure irrigation methods (sprinkling, drop irrigation)
- Agro-industry: primary products ("large-area products" such as grain and sugar), use of wastes
- Rural hydraulic engineering: correlation with plot size
- Renewable sources of energy (biomass)
- Mechanical engineering
- Mills handling cereal crops

5.   Summary Assessment of Environmental Relevance

The **fundamental elements** of agriculture are **plant production** and - based on this - **livestock farming**. Man uses **technical aids** in order to **influence the production systems** and **enhance their productivity**. **Agricultural engineering** is an **integral component** of these systems; its environmental impacts cannot be considered without reference to those of plant and animal production. Mechanical aids are most commonly used in **tillage and transportation**, with effects above all on soil, plants and man.

Among the **negative consequences of cultivation**, it is erosion that has the greatest significance worldwide. **All other effects** resulting from the use of technical aids in agriculture remain **limited to the locality or at most the region concerned**.

Improper storage and application of pesticides, mineral fertilisers and animal excrement can lead to **contamination and/or over-fertilising** (eutrophication) of both **surface water and groundwater**.

Most agricultural operations are **mechanised** for reasons of **labour efficiency**. Machines and implements call for a high degree of expertise in operation, maintenance and management if their use is not to have negative impacts. In many countries, responsibility for certain types of work passes from women to men.

Intensification of agriculture with the aid of agricultural engineering can lead to a **change or reduction in the spectrum of flora and fauna** found in a region.

The most important **protective measures** comprise

-       provision of training and advice, and
-       development and introduction of mechanisable land-use systems which take both economic (including labour efficiency) and ecological aspects into account.

6. References

Krause, R., F. Lorenz and W. B. Hoogmoed: Soil tillage in the tropics and subtropics. Schriftenreihe der GTZ No. 150, Eschborn 1984, p. 320.

Derpsch, R., C. H. Roth, N. Sidiras and U. Köpke: Erosionsbekämpfung in Parana, Brasilien: Mulchsysteme, Direktsaat und konservierende Bodenbearbeitung. Schriftenreihe der GTZ No. 205, Eschborn 1988, p. 270.

Zweier, K.: Energetische Beurteilung von Verfahren und Systemen in der Landwirtschaft der Tropen und Subtropen - Grundlagen und Anwendungsbeispiele. Forschungsbericht Agrartechnik des Arbeitskreises Forschung und Lehre der Max-Eyth-Gesellschaft (MEG), Band 115, 1985, p. 341.

GTZ: Sustainable agriculture in German and Swiss technical cooperation. Register Nr. 15 der "Working paper for rural development", GTZ, Division 4210, Farming Systems, Eschborn, Feb. 1989, p. 148.

UNEP: Agricultural mechanisation. No., UNEP environmental management guidelines, United Nations Environment Programme, Nairobi, 1986, p. 17.

FAO: Agricultural mechanisation in development - guidelines for strategy formulation. Agricultural Services Bulletin 45, Rome 1984, p. 77.

World Bank: Agricultural Mechanisation - Issues and Options. A World Bank policy study. Washington D. C., June 1987, p. 85.

33

Examples of German regulations and standards:

BBA: Regulations laid down by the Biologische Bundesanstalt (BBA - German Federal Biological Research Centre for Agriculture and Forestry) concerning use of pesticides.

Berufsgenossenschaften: Accident prevention regulations (Unfallverhütungsvorschriften) laid down by agricultural and industrial employers' liability insurance associations ("Berufsgenossenschaften").

DIN: German Standards and regulations on construction and design.

STVZO: Provisions of the German road transport licensing regulations.

TA Lärm: Technical Instructions on Noise Abatement, 1968/1974.

TA Luft: Technical Instructions on Air Quality Control. First general administrative regulation accompanying the Federal Immission Control Act, 27 February 1986.

Agriculture

Irrigation

34. Irrigation

Contents

1. Scope

It is not just in arid climatic zones that irrigation and drainage are today increasingly coming to play an essential role in **agriculture**. Sprinkler systems or other types of irrigation schemes are also used in **rain-fed farming** to raise production and/or provide a safeguard against unfavourable weather conditions. Irrigation is the only way of permitting arable farming in some places at all; it has made it possible, for example, to reclaim what was once desert or steppe in countries such as Egypt, Israel, India and Mexico.

Apart from the demands of the market and the progressively more monetary nature of rural trade, it is above all rapid population growth that is making it necessary to introduce or improve (artificial) irrigation as one means of **raising production** on land which is in some cases in increasingly short supply. High growth rates are thus likely in this sector, which means that the **importance of providing a water supply** and the **quantity of water required** will both **increase dramatically**.

While in many places totally **unutilised water resources** are still to be found or existing resources are used on only a moderate scale, the **provision of a water supply** has already led elsewhere to immense and generally **irreversible ecological damage**.

Just as **wastewater disposal** plays a significant role in drinking-water or process-water supply systems (cf. environmental briefs Wastewater Disposal and Urban Water Supply), irrigation must always be accompanied by **drainage** measures. Although efficient drainage is often guaranteed simply by the natural structure of the terrain, planning of water conveyance systems frequently also has to address the question of drainage.

Failure to implement drainage programmes immediately after the introduction of all-year irrigation can lead to **irreversible damage** - primarily as a result of soil salinisation - and to a rise in the groundwater. Even small-scale irrigation projects have given rise in many countries to **salinisation problems** (= adverse influence on the soil's nutrient balance) in cases where no drainage system exists. Depending on soil type, between 10% and 20% of the irrigation water should be drained off in order to prevent long-term damage from salinisation.

In the light of the increasing demand for irrigation water and the related **water supply and conveyance costs**, there is a risk that drainage measures may be spread over a lengthy period of time or realised on as small a scale as possible. There is also a tendency for water-saving but more expensive conveyance systems to be over-hastily rejected on the grounds of cost in favour of open, unlined or over-strained low-tech systems. **Insufficient use has been made to date of "appropriate" solutions** which are not only inexpensive but also effective and thus help to conserve resources.

34

Irrigation covers the following **areas**:

-       provision of a water supply
        through storage in small reservoirs, use of river water and tapping of
        groundwater

-       conveyance and distribution
        of irrigation water by open channels and pipelines

-       application
        of irrigation water by means of flooding, basins, border strips, rills, sprin-
        kling, drop irrigation and subsurface systems

-       drainage
        by means of open and concealed systems

This environmental brief is concerned only with small and medium-sized irrigation
projects. It deliberately excludes large-scale dam projects and irrigation schemes for
entire regions with measures involving entire river systems.

2.   Environmental Impacts and Protective Measures

Given that water resources are limited, water consumption is rising and irrigation
and drainage systems are often inappropriate to their context, priority should be
given to

-       considering the question of water supply, as projects entailing large-scale
        utilisation of natural resources generally involve major environmental risks;
-       making sure that irrigation and drainage measures are well matched;
-       establishing whether the technology of the measures implemented is geared
        to the financial capacity of the country concerned and to other specific
        conditions (e.g. available technical know-how) and thus ensures that poten-
        tial environmental hazards can be reduced or ruled out.

2.1 Impacts on components of the natural environment

2.1.1 Supply, conveyance and distribution of water

Depending on the activity involved, every aspect of the environment (soil, water, air/climate, species, biotopes/landscape) may be affected. Impacts on the **soil** vary in nature. The embankments of small reservoirs and open channels for conveying water can create **erosion risks**. All construction measures change (destroy) the soil structure, while irrigation itself alters the soil dynamics. The risk of erosion can be counteracted by stabilising embankments, for example with ground-covering plants having a dense root system.

A wide variety of impacts on **water** can be observed. Although small reservoirs improve the **availability of surface water**, they may also - depending on the subsoil - cause groundwater resources to become contaminated. In addition, small reservoirs too are liable to exhibit impairment of **surface-water quality** and the **nutrient balance** (particularly as a result of warming and eutrophication). It should be borne in mind that impounding measures in a particular area can reduce the available water supply in the lower reaches of the watercourse concerned. If rainfall is highly seasonal, however, the opposite effect is likely. If **river water** is used for irrigation the amount of available surface water will be reduced, while if the groundwater is tapped groundwater resources will be depleted. In the case of groundwater the quantity withdrawn depends not least on the **tapping method**. The easier (or in economic terms, the less expensive) it is to raise the water, the more wasteful the use of water resources may be.

The effect which **tapping of groundwater** has on the **dimensions of water resources** is of particular significance. This may apply even to small-scale schemes or micro-projects (e.g. where cropping areas are situated primarily on geological basement formations, often with few water reservoirs, or in wadi systems on the fringes of the Sahara). Tapping of **fossil groundwater** with no natural inflow will by definition exceed the available quantity. It thus constitutes **destructive exploitation** of a vital resource and should be permitted only in exceptional justified cases.

There is a **danger** that the groundwater may become **contaminated** if the **sites where water is raised are left unprotected** and/or if substances such as faecal matter or oil are discharged into the water.

In addition to having effects on the **microclimate, small reservoirs** also have an influence on the range of **species** found in the area. However, the precise nature of their impacts in the latter sphere is **not clear**. Certain species of flora and fauna may be destroyed or displaced, while the water and its surroundings may favour other species or indeed attract them. A (negligible) **reduction in dry biotopes** must be set against **the creation of new aquatic biotopes**. Wetlands may increase (above all around the edges of the reservoir) or decrease (as a result of reduced flow in the lower reaches of the watercourse). Increases and decreases in the presence of particular species may have **both positive and negative consequences** for man and nature. Particular attention must be paid to the effects of **fluctuations in the reservoir's water level**. It can be assumed that small reservoirs make for a more varied landscape.

Open water conveyance and distribution systems lead to **water losses** on account of **evaporation** and have a (slight) influence on the **microclimate**. Water conveyance systems in the form of earth cross-sections may have effects on flora and fauna; as is the case with small reservoirs, however, the precise nature of these effects is not clear. Depending on context, open water conveyance and distribution systems may enhance or mar the varied nature of the landscape.

Unless installed above ground, **enclosed systems** generally have only **minor impacts** on the natural environment.

2.1.2 Water application and drainage

Depending on the method used, water application - in other words the actual process of **irrigation** - can affect the **soil** to varying degrees. It is also likely to have impacts on **water, species** and the **microclimate**. The main problem encountered with many irrigation methods is that of **soil salinisation**, particularly if the system is poorly managed and there is no drainage. In simplified terms, salinisation can be defined as an **extreme nutrient imbalance** (excess of salts) and **damage to the soil structure** (puddling, crusting, compaction).

Traditional irrigation methods often involve **water dosage problems** (e.g. flood, basin, border-strip and furrow irrigation). The possibility of **erosion** cannot be ruled out where such techniques are used. Sprinkling and in particular drop irrigation may also lead to **salinisation** if not carried out properly.

Particular attention should be paid to methods in which **modern components** have been inappropriately **added** to **traditional techniques**. Water conveyance systems or application methods that gave rise to no problems in the past can cause **erosion or scouring** if the **introduction of power pumps** changes the way in which the water is supplied. It may be necessary for the entire system to be modified at considerable expense.

All irrigation methods can have adverse effects on the soil microflora and microfauna. When **geared to local conditions and properly managed**, however, irrigation can also **contribute to the nutrient balance** and **benefit microflora and microfauna**.

Drainage can do much to counteract the problem of salinisation. It thus **contributes to the nutrient balance** and to **stabilising the soil structure**. Water application methods can be used to achieve at least partial desalinisation.

Drainage ditches in the form of earth cross-sections create a **risk of erosion**. Impacts affecting water are likely to take two forms. Traditional irrigation methods, sprinkling and open drainage systems cause **surface water to be lost** through **evaporation**. However, traditional methods and drainage ditches in the form of earth cross-sections can also induce **recharging of the groundwater**. Where over-irrigation recharges the groundwater, the crops may be adversely affected because the groundwater level is too high.

In arid regions, **seepage** represents a **waste of water** and can lead to over-exploitation of resources. Priority should therefore be given to **lining the water conveyance systems**. Evaporation losses in conveyance systems tend to be negligible (e.g. 1 - 2% in desert regions compared to seepage losses of up to 85% from unlined water conveyance systems in sandy terrain). Traditional irrigation methods, sprinkling and open drainage systems can all have an influence on the **microclimate**. Depending on local conditions, their effects may be **beneficial** (e.g. as regards oasis ecology) or **detrimental**.

All water application methods are likely to have an influence on **flora**. The natural balance of species will generally be disturbed, while the **number of species** may either **increase** or **decrease**.

As only relatively **small irrigated areas** are involved, there are still enough **refuges** available to the local **fauna** to prevent permanent changes in the balance and number of species. The fauna are more likely to be affected by the enlargement and use of the cropping area per se and by the type of crop growing practised (cf. environmental brief Plant Production).

Open drainage ditches in the form of earth cross-sections can have **influences on flora and fauna**. As is the case for water conveyance systems and small reservoirs, however, the **precise** nature of these impacts cannot be defined. The same applies to the potential influence of such drainage systems on the diversity of the landscape.

34

2.2 Impacts on the socio-economic environment resulting from water supply, conveyance, distribution and application as well as from drainage

2.2.1 Factor requirements, labour, income and distribution

General assertions regarding impacts on the socio-economic environment are bound to be fairly vague, if indeed they are possible at all. In order to reach any conclusions, it is essential to **analyse the circumstances of the particular case in question**.

Technically sophisticated systems generally not only call for a sizeable input of capital but may also require a great deal of **energy**. Attention must be drawn to the possibility of using small reservoirs and water conveyance systems in generating energy and of meeting energy requirements by using renewable energy sources. One way of reducing the amount of external energy required is to make use of the available **water power** in cases where irrigation water is obtained from rivers (water wheels with a lift ranging from 0.5 m to over 20 m).

The major problem encountered in **operating irrigation schemes** involving new technologies is generally that of meeting the considerable **training and management needs**. Introduction of irrigation systems is usually also accompanied by a move in the direction of technically more sophisticated and more intensive forms of agriculture, which are not automatically accepted everywhere. A great deal of **advice and encouragement is required** if this difficulty is to be overcome.

Women are often **excluded** from discussion, extension services and training measures, even though they may be responsible for certain areas of farm work or may be farmers in their own right. This factor is of particular significance when traditional **technologies** are to be **replaced** by new ones.

Construction and operation of irrigation systems necessitate a considerable amount of **extra work**, particularly when labour-intensive techniques are used, and in many societies it is primarily women who bear this additional workload. Income levels are satisfactory, however, especially in the case of capital-intensive methods. **Social disparities** may be increased.

The introduction of irrigation frequently brings **financial disadvantages for women**. It is often only the **men** who are registered as the **owners** of the land covered by irrigation schemes; in other cases, men may simply appropriate the irrigated land, which is considerably more valuable than that used for rain-fed farming.

Farmers may run into **serious economic problems** on account of the fact that **operating, maintenance and monitoring costs** and expenditure on **renewal of irrigation systems** are often inadequately calculated at the planning stage, or as a result of sudden changes in government support policy (cuts in extension services, equipment subsidies and even water subsidies). It should be established whether the technical design and dimensioning of irrigation systems allow the systems to be **used profitably** by the farmers even under **changed conditions**.

It can generally be assumed that irrigation makes for **more reliable yields and incomes.** This is not the case, however, where workers are paid only for work performed over a **limited period,** for example during system construction or for **seasonal work,** the volume of which varies considerably. If women participate in this seasonal work their **workload** may be **increased** at the expense of other activities (feeding the family etc.).

Irrigation is likely to influence the **distribution of income** (and not just the relative incomes of men and women). Capital-intensive methods can place less prosperous farmers at a disadvantage and cause income distribution to become **more unbalanced.** Women are often excluded where conversion of land to irrigation is carried out on the basis of a loan scheme. **Social distinctions** generally increase in proportion to the technical complexity and cost of an irrigation system. **Titles to land** should therefore be **distributed** as widely as possible or **upper limits** set for ownership of land within specified areas covered by irrigation schemes.

It is important to make sure that **women's traditional land-use rights** are taken into consideration, for example by making certain that women too are entered in the cadastral register as **land owners.**

2.2.2 Health

Irrigation schemes are likely to create a variety of **health risks.** The main problems are caused by **waterborne diseases,** particularly schistosomiasis and onchocercosis, whose foci may be located at different points within the irrigation system (stagnant/flowing water). By virtue of the way in which it is transmitted (via human excretion), schistosomiasis in particular may well occur in areas being irrigated for the first time. Irrigated farming can also **promote** the spread of hookworms (Ankylostoma duodenale) and eelworms (Ascaris lumbricoides).

Malaria, which often spreads in areas where large irrigation schemes are being realised, can also constitute a problem in small-scale projects using open reservoirs and water conveyance systems. The possibility of rheumatic ailments and accident risks must likewise be taken into account. Health risks are liable to arise in cases where irrigation systems are also used to provide a **drinking-water supply** (see environmental brief Rural Water Supply). It is particularly important to raise women's awareness of these risks by means of targeted **information and education measures,** as it is usually women who are responsible for providing the family's drinking-water supply. **Vector control** measures (using chemicals) in turn create environmental hazards.

2.2.3   Subsistence, housing and leisure

Unless the land is used exclusively for **growing non-food crops**, irrigation schemes generally **contribute to subsistence** in that land owners grow food **for their own consumption** or workers are **paid in kind**. Particular efforts must be made during crop planning to **ensure** that food crops are grown (cf. environmental brief Plant Production). Irrigation in arid regions generally increases the range of food crops that can be grown.

Irrigation can cause damage to the **fabric** of houses where construction materials such as lumps of clay, tamped earth, air-dried clay bricks or materials of plant origin have been used. Houses on irrigated land can be protected against rising damp by being built with stone **foundations**.

Irrigation projects may have effects on leisure if they considerably **increase** the workload of the land owners and their families. This applies in particular in areas where only rain-fed farming was practised previously. It is often the **women and children** who are called upon to perform the extra work. In extreme cases this may prevent the children attending school or force the women to abandon other important activities.

Irrigation systems should not unnecessarily ruin the **natural landscape** or disrupt **communications**. The population should not be obliged to make long detours on account of changes in the landscape (e.g. pipelines installed above ground on supports or in/on embankments, or wide open channels). Adequate **crossing facilities**, including routes for driving livestock, should be provided (e.g. routes passing underneath system components, bridges).

2.2.4   Training and social relationships

Many irrigation methods or activities lend themselves to on-the-job **training**, although they often call for a high prior level of skill and know-how.

If activities can be organised and carried out on a **communal** basis, they can **encourage participation and social interaction**. Although irrigation can as a whole be seen as a communal task, it does not necessarily always help to consolidate social relationships. In many regions, irrigation establishes **unrestricted private ownership** of land **for the first time**, with the result that neighbourly cooperation is increasingly replaced by a system of hired labour.

It is **above all women who are affected** by the decline in communal activities (e.g. fetching water or washing clothes together, possibly also communal field work). In Islamic countries, for example, such activities give women an important opportunity for communication not afforded in any other way because of the restrictions imposed by social norms.

3. Notes on the Analysis and Evaluation of Environmental Impacts

General guidelines on **quantitative water management** exist in the Federal Republic of Germany. With the exception of technical guidelines for hydraulic engineering measures, however, there are **no standards** governing activities in connection with irrigation schemes. Standards could nevertheless be laid down to cover aspects such as

- permissible changes in the groundwater level resulting from tapping (lowering), seepage (rise) and drainage (lowering);

- **reduction of flow** where river water is used for irrigation purposes;

- **limits on use** of surface water, in order to prevent adverse effects on and/or destruction of aquatic organisms (defining minimum water quantity and depth etc.);

- the **quality of the irrigation water**, e.g. in order to prevent soil salinisation;

- the **degree of salinity** of flowing water where it receives discharges from drainage systems, etc.

The following could also serve as the **starting points** for standards governing measures affecting the water balance:

- The **quantity of groundwater used** must not exceed the medium-term recharge rate (often difficult to ascertain).

- **Fossil groundwater** may be **tapped** only in cases of extreme need.

- The **low-water flow** represents the **critical factor** for surface water quality when water is drawn off.

4.   Interaction with Other Sectors

The environmental brief **Plant Production** should be additionally consulted in order
to assess the impacts originating from crops grown on irrigated land.
The **individual areas involved in irrigation** also **interact** with other agricultural
subsectors, including the following:

-       **Plant protection**, in respect of the need to ensure that irrigation and drainage
        water is free of pollutants, for drainage water particularly in cases where it is
        discharged into surface water or groundwater

-       **Livestock farming**

-       **Fisheries and aquaculture**

-       **Agricultural engineering**, e.g. in connection with use of organic manures
        and mineral fertilisers and their possible polluting effects

Use of water resources for irrigation purposes may **conflict** with other interests,
above all in the light of the general demand for **conservation of natural resources**.
Utilisation of artesian and/or fossil groundwater represents just one example of such
a conflict. Other conflicts may arise with respect to the **wastewater and rainwater**
subsector, leading to impacts on **health** in particular.

In certain cases there may also be links with

-       **large-scale hydraulic engineering**, in connection with dams and weirs;

-       **rural hydraulic engineering**, above all in connection with weirs (use of
        water for irrigation), contour canals and small earth embankments forming
        part of water storage facilities;

-       **wastewater disposal**, in connection with disposal of wastewater by means of
        discharge onto agricultural land or into receiving waters (surface waters).

5. Summary Assessment of Environmental Relevance

Irrigation systems of **virtually every degree of technical complexity** can be planned and constructed with a minimum of environmental (and social) impacts provided that the scheme incorporates **measures appropriate** from the ecological, technological, economic and social viewpoints. Caution must be exercised during assessment, as financial constraints and other criteria often restrict essential measures to a minimum. The **technical practicality** of an irrigation system must be established, since it represents an important prerequisite for success. Although raising technical standards may have impacts on the natural environment, it is above all within the context of the socio-economic environment that problems are most likely to arise.

The **small-scale irrigation projects** discussed here are bound to have **fewer impacts** than measures which involve large-scale hydraulic engineering schemes or raising large quantities of groundwater. The potential **technological solutions** are often **interchangeable**; in other words, a number of different options may produce the same result, making it possible to choose the soundest alternative from the environmental viewpoint. It should be remembered that **traditional irrigation technologies** may well be geared to the natural environment, but can cause **environmental problems** if used **in combination** with "modern" technologies. Where appropriate combinations of old and new technologies are used, however, they can help to prevent negative impacts on both the natural and social environment.

34

6. References

Basic literature

Achtnich, W. (1980): Bewässerungslandbau. Stuttgart.

American Society of Agr. Engineers (1981): Irrigation Challenges of the 80's. The Proceedings of the Second National Irrigation Symposium. Lincoln/Nebraska.

Baumann, W. et al. (1984): Ökologische Auswirkungen von Staudammvorhaben. Forschungsberichte des BMZ Nr. 60, Cologne.

Biswas, A. K. (1981): Role of Agriculture and Irrigation in Employment Generation. ICID-Bulletin 30, 46-51.

Böttcher, J.-U. (1983): Umweltverträglichkeitsprüfung und planerisches Abwägungsgebot in der wasserrechtlichen Fachplanung. Bonn (Jur. Diss.).

Deutsche Stiftung für Internationale Entwicklung DSE [German Foundation for International Development]/UNEP (1984): Environmental Impact Assessment (EIA) for Development. Feldafing.

Feachem, R. et al. (1977): Water, Wastes and Health in Hot Climates. New York.

Framji, K. K. (1977): Assessment of the World Water Situation. Irrigation Systems in Total Water Management. ICID-Paper, UN Water Conference, Argentina.

Framji, K. K. and Mahajan, I. K.: Irrigation and Drainage in the World. A Global Review (ICID). New Delhi.

Fukuda, H. (1976): Irrigation in the World. Comparative Developments. Univ. of Tokyo Press. Tokyo.

Hübener, R. (1988): Entwicklungstendenzen der Beregnungstechnik im internationalen Vergleich, in: Zeitschrift für Bewässerungswirtschaft 23, 111-143.

Hübener, R. (1988): Verbesserte Methoden der Wasserverteilung im Bewässerungslandbau, in : Der Tropenlandwirt, 89 Jg., 143-163.

Hübener, R. and Wolff, P. (1990): Fortschritte in der Technik der Oberflächenbewässerung, in: Z. für Kulturtechnik und Landentwicklung 31, 34 - 43.

Huppert, W. (1984): Landwirtschaftliche Bewässerung. Ein konzeptioneller Rahmen für problembezogene Projektansätze, Vorentwurf. Band 2, Anlagen. Anhang 2 - Umweltverträglichkeit von Bewässerungsmaßnahmen, GTZ Division 15. Eschborn.

Jenkins, S. H. (1979): Engineering, Science and Medicine in the Prevention of Tropical Water-Related Diseases. Progress in Water Technology 11.

Jensen, M. E. (1981): Design and Operation of Farm Irrigation Systems. ASAE Monograph. St. Joseph.

Larson, D. L. and Fangmeier, D. D. (1987): Energy in Irrigated Crop Production. Trans. ASAE 21, 1075-1080.

McJunkin, F. E. (1975): Water, Engineers, Development and Disease in the Tropics. AID, Dep. of State, Washington D.C.

McJunkin, F. E. et al. (1982): Water and Human Health. AID, Dep. of State, Washington D.C.

Mann, G. (1982): Leitfaden zur Vorbereitung von Bewässerungsprojekten. Forschungsberichte des BMZ 26, Cologne.

Rudolph, K. U. (1988): Die Umweltverträglichkeitsprüfung bei der Planung und Projektbewertung wasserbaulicher Maßnahmen, in: Wasser, Abwasser 129 Heft 9, pp. 571-579.

Tillmann, G. (1981): Environmentally Sound Small-Scale Water Projects: Guidelines for Planning, Cooel/Vita.

Tillmann, R. (1981a): Environmental Guidelines for Irrigation, prepared for USAID and US Man and the Biosphere Program, New York.

U.S. Environmental Protection Agency (1978): Irrigated Agriculture and Water Quality Management. Washington D.C.

White, G. (ed.) (1978): L'irrigation des terres arides dans les pays en développement et ses conséquences sur l'environnement. UNESCO, Notes Techniques du MAB 8.

34

Wolff, P. (1978): Bewässerungstechnik in der Evolution, in: Z. für Bewässerungs-
    wirtschaft 13, 3-20.

Wolff, P. (1985): Zum Einsatz von neuen Wasserverteilungssystemen - eine Be-
    trachtung aus bodenkundlich/kulturtechnischer Sicht, in: Z. für Bewässe-
    rungswirtschaft 20, 3-14.

Zeitschrift für Bewässerungswirtschaft (cf. various recent volumes). DLG-Verlag,
    Frankfurt.

Zonn, I.S. (1979): Ecological Aspects of Irrigated Agriculture, ICID Bulletin 28
    No. 2.

Institutional guidelines

Asian Development Bank (AsDB): Environmental Guidelines for Selected Agricul-
    tural and Natural Resources Development Projects, 1987.

BMZ: Compendium of Environmental Standards, 1987.

Commission of the European Communities (CEC): The Environmental Dimension
    of the Community's Development Policy (84) 605 Final, 1984.

Federal Republic of Germany (BMZ): Environmental Guidelines for Agriculture,
    1987.

Food and Agriculture Organisation (FAO): The Environmental Impacts of Irrigation
    in Arid and Semi-Arid Regions: Guidelines, 1979.

FAO: Man's Influence on the Hydrological Cycle. Irrigation and Drainage Paper,
    Special Issue 17, 1973.

FAO: Irrigation and Drainage Paper 31: Groundwater Pollution, 1979.

FAO: Irrigation and Drainage Paper 41: Environmental Management for Vector
    Control in Rice Fields, 1984.

FAO: Preliminary Operational Guidelines for Environmental Impact Studies for
    Watershed Management and Development in Mountain Areas, 1979.

FAO: Environment Papers No. 1: Natural Resources and the Human Environment for Food and Agriculture, 1980.

FAO: Soils Bulletin No. 44: Watershed Development - with Special Reference to Soil and Water Conservation, 1985.

FAO: Soils Bulletin No. 52: Guidelines: Land Evaluation for Rainfed Agriculture, 1983.

FAO: Soils Bulletin No. 54: Tillage Systems for Soil and Water Conservation, 1984.

FAO: Soils Bulletin No. 55: Guidelines: Land Evaluation for Irrigated Agriculture, 1985.

FAO: Conservation Guides No. 1: Guidelines for Watershed Management, 1977.

FAO: Conservation Guides No. 2: Hydrological Techniques for Upstream Conservation, 1976.

FAO: Conservation Guides No. 8: Management of Upland Watersheds: Participation of the Mountain Communities, 1983.

FAO/UNESCO: Irrigation, Drainage and Salinity. London, 1973.

International Union for the Conservation of Nature and Natural Resources (IUCN): Ecological Guidelines for the Use of Natural Resources in the Middle East and South West Asia, 1976.

Organisation of American States (OAS): Environmental Quality and River Basin Development: a Model for Integrated Analysis and Planning, 1978.

United Nations Educational, Scientific and Cultural Organisation (UNESCO): MAB Technical Notes Series No. 8: Environmental Effects of Arid Land Irrigation in Developing Countries, 1978.

UNESCO: MAB; Expert Panel on Project 4: Impact of Human Activities on the Dynamics of Arid and Semi-Arid Zone Ecosystems with Particular Attention on the Effects of Irrigation, 1975.

United Nations Development Programme (UNDP): Environmental Guidelines for Use in UNDP Project Cycles, 1987.

34

World Health Organisation (WHO): Establishing and Equipping Water Laboratories in Developing Countries, 1986.

WHO: Environmental Health Impact Assessment of Irrigated Agricultural Development Projects, 1983.

World Bank: Environmental Policies and Procedures of the World Bank (Operational Manual Statement OMS 236), 1984.

Mining and Energy

Reconnaissance, Prospection and Exploration of Geological Resources

35

35. Reconnaissance, Prospection and Exploration of Geological Resources

<u>Contents</u>

1.   Scope

This brief describes the environmental consequences and potential means of pollution control in connection with the reconnaissance, prospection and exploration of geological resources.

Geological resources in the present context comprise mainly **raw minerals** and **groundwater,** with attention to soil being restricted to the reconnaissance aspect. Reconnaissance, prospection and exploration are the terms used **for steps taken in preparation for the commercial extraction, i.e., utilization, of geological resources**. The environmental consequences of extracting, dressing, refining and distributing such resources are not dealt with in this brief. The entire petroleum/natural-gas exploration complex has also been excluded. Those areas and related sectors are investigated in separate briefs.

The purpose of **reconnaissance**, including stock taking and mapping of resources, is to obtain regional overviews and to identify and demarcate mineral prospects and/or pedological location factors.

Prospection aims to locate prospects and exploitation areas by way of geological, geophysical and geochemical methods of field investigation.

Exploration - the detailed study of prospective areas - uses the same methods as those employed for prospection, but also involves direct disturbance of the environment.

While there are various basic types of reconnaissance, prospection and exploration projects, their respective environmental impacts depend primarily on the individual activities involved.

Both **direct and indirect geoscientific methods** are employed in the reconnaissance, prospection and exploration of geological resources. As a rule, the **indirect** methods yield less accurate results but offer the capacity for covering large areas at low specific expense. More precise, and substantially more expensive, **direct** methods applied preferentially to prospective zones and already identified anomalies or deposits enable the refinement of data bases. The **raw minerals sector**, for example, employs the **following methods of investigation** (listed in the order of increasing exactitude):

–   interpretation of satellite photographs
–   interpretation of aerial photographs
–   interpretation of thematic geoscientific maps
–   interpretation of geophysical test data

35

- interpretation of borings with the help of geochemistry and well logging; analysis of core samples
- investigation of explored deposits via shafts and tunnels
- interpretation of dressing tests

Groundwater prospection investigates the demand for water, its quantitative management, quality and protection, and the ecological consequences of its extraction (for details, cf. section 4). The protection-worthiness and sensitivity of the existing ecosystems, the volumetric and pollution-load capacities of the receiving waters, the effects of relevant road-building measures, and the social, sociocultural and ecological impacts of the anticipated settlement effects must be duly considered and assessed (for details, cf. section 4).

The **"soils"** subsector involves the evaluation and assessment of soils on the basis of soil surveys and the appraisal of soil utilization potentials. Moreover, measures designed to protect the soil from erosion, salinization and the effects of fertilizers and plant phytopharmaceutical products demand appropriate illumination.

2.   Environmental Impacts and Protective Measures

The environmental consequences of the subject sectoral measures are extensively limited, and the relevant protective measures are for the most part uncomplicated and inexpensive. Unavoidable damage of tolerable extent demands settlement by material compensation.

Reconnaissance, prospection and exploration activities can impose various hazards on the environment. The environmental consequences tend to increase as the activities progress from reconnaissance to prospection to exploration. In the first two cases, the impacts are usually modest and temporary. Exploratory measures are more elaborate and expensive, and the cost factor therefore helps retard their excessive implementation.

The main purpose of protective measures is to minimize the environmental consequences and to prevent environmental damage with respect to both time and space. The **avoidance of permanent damage** is especially important. Since geological resources are immobile, field investigations are usually limited to a **particular site**. Proper consideration of seasonal weather conditions can help avoid damage to the environment, e.g., by performing such work outside of the growing/breeding season.

POTENTIAL FORMS OF ENVIRONMENTAL DAMAGE RESULTING FROM GEOLOGICAL RECONNAISSANCE, PROSPECTION AND EXPLORATION

X = high risk to the environment
(X) = conditional risk to the environment

Activities		Landscape (scenery)	Flora	Fauna/humans	Surface water	Groundwater	Soil	Air	Climate	Material assets	Cultural heritage	Land consumption	Infrastructural development	Noise	Gaseous waste	Liquid waste	Dust/particulates	Sedimentation	Radiation	Setting
		With effects on ... possible detriment to ...											With effects by ... possible detriment due to ...							
Work-area access	Acess roads, clearings, lanes	X	X	X	(X)	(X)	(X)	(X)		(X)	(X)	X	X	(X)	(X)	(X)	(X)			
Topographical and geological mapping	Field surveys, interpretation of aerial photographs, remote sensing by satellite	(X)	(X)	X										X	X	X			(X)	
Camps and support facilities	Lodging, roads, workshops, field laboratories, storeyards, infrastructure	X	X	X	X	X	(X)	(X)				(X)	X	X	X	X	X			
Geophysics	- airborne techniques			X										X	(X)					
	- prospection seismics (land and water)	(X)	(X)		(X)	(X)				(X)	(X)			X						
	- nonseismic techniques[1]			X	X	(X)	(X)							X	X					
	- Well-shooting[2]						X									X			(X)	
Hydrogeological investigations	Pumping tests, injection tests, water sampling tracing				(X)	X	(X)									(X)			(X)	(X)
Exploration	Trial pits, shafts, tunnels, dumps, boreholes	X	X	X	X	X	(X)	X					X	X	X		X	X	X	
Sampling	Surface samples, exposure samples, marine sampling	(X)	(X)		X		(X)													
Laboratory investigations	Diverse techniques[3]				X			X							X	X	X	X	X	(X)

1) e. g., gravimetry, magnetometry, geoelectricity, radiometry

2) e. g., X-ray fluorescence logging, neutron activiation, measurement of density, radioactivity, self-potential, resistance, magnetic susceptibility and magnetic field, well-shooting, mechanical and thermal measurements

3) e. g., microscope, wet-chemical, X-ray diffraction and sedimentary-petrographical investigations, isotope analysis and dressing tests

35

Damage to the environment can be **extensively avoided**, or at least limited, by:

—   careful execution of exploratory work - e.g., by avoiding the use of heavy (and accordingly expensive) equipment - inclusive of soil- and water-protection measures, stabilization measures, recultivation, etc.,

—   choosing environmentally benign (micro-)sites for (prospecting) lanes in order to minimize the environmental burden, e.g., through dissection; the same applies to the choice of locations for camps and support facilities,

—   taking measures to prevent environmental mishaps, e.g., by installing traps for oil and chemicals.

Environmental consequences can also be limited **by recovering and recycling materials and substances**. Recycling is preferable to (controlled) disposal. The ultimate goal is to restore the site as closely as possible to the state it was in prior to commencement of the work, or at least to preclude lasting detriment to the environment.

2.1   Access to work area

2.1.1   Access roads

It is frequently necessary to fell trees and move earth to make way for access roads. The damage resulting from such activities can by far exceed that caused by reconnaissance, prospection and exploration. Moreover, establishing **access to a previously inaccessible area** can lead to such social consequences as public unrest and land speculation. Controlling access to such roads can help prevent the subsequent uncontrolled generation of settlements.

2.1.2   Lanes

Geophysical investigations may require the cutting of narrow lanes as footpaths. This can cause temporary damage to the **vegetation** and expose the soil and subsoil to **erosion.**
In the tropics and subtropics, much more so than in semi-arid regions, the vegetation is normally able to close off such lanes within a year or two, so that **no permanent damage** remains. Protective measures are rarely necessary. The inadvertent provision of general access to the area in question must be avoided.

In areas characterized by a very fragile balance of nature (marginal locations, slopes) it may be necessary **to impose certain restrictions and to carefully accommodate the local situation**, e.g., by disturbing as small an area as possible and reducing the felling of trees to a minimum. If farmland is involved, the competent authorities and those affected must be consulted with regard to compensation.

2.2   Topographical and geological mapping

Unless **mapping activities** are **intensive** and require extensive **field checking**, little **impairment of flora and fauna** need be anticipated.

2.3   Camps and support facilities

In many cases, **permanent camps** comprising lodgings, workshops, field laboratories, storeyards, etc. can be required. The attendant land use, sealing of soil and general detriment to and disturbance of local flora and fauna are disadvantageous. Controlled disposal of liquid and solid wastes must be ensured.

2.4   Geophysics

2.4.1   Airborne techniques

The noise caused by **flyovers**, most notably in connection with helicopter-assisted methods of surveying, is **disturbing** to local animal populations.

2.4.2   Prospection seismics

The environmental consequences of prospection-seismic activities (blasting) on lanes can be extensively minimized by carefully **plugging** the blasting charges in the boreholes. Such activities cause no permanent damage to the environment.

2.4.3   Nonseismic geophysical investigations

All nonseismic geophysical investigative methods involve the use of **portable measuring instruments** at or slightly above ground level (≤1.5 m). The hauling and handling of the requisite equipment and the movements of personnel within the areas of interest can be expected to cause modest impairment of the local environment.

The **electricity** required for a camp and, possibly, for one or the other electric-powered piece of equipment may necessitate the use of a diesel- or gasoline-fueled generator. Environmental damage can result from improper or careless **handling and storage of fuels and lubricants**.

2.4.4 Well-shooting

Well-shooting is the term applied to measurements conducted in an **existing borehole** according to radiometric, electric, magnetic, acoustic, mechanical and thermal techniques to gain information on the immediate surroundings of the hole itself. Consequently, any **effects** on the environment are limited to the **immediate vicinity of the measuring point** - one exception to the rule being **radiometric measurements** performed with the aid of active radiation sources that require certain precautionary measures in connection with calibration and introduction of the probe - the loss of which must be avoided - into the borehole. Radioactive cores must be duly marked, and appropriate **protective measures** up to and including the services of a radiological safety officer must be taken in case of high-level radiation.

2.5   Hydrogeological investigations

2.5.1 Long-time pumping tests

The **sustainable yield** and/or groundwater permeability of wells and boreholes is determined by long-time pumping tests. **Lowering of the groundwater** level in the vicinity of the tested well can cause temporary detriment to other **wells situated nearby**.

2.5.2 Injection tests

Long-time injection tests serve in determining the sustainable injectivity of **drainage wells**. Such tests can **temporarily alter the groundwater regimen**. Care must be taken to ensure that the injected water is environmentally compatible.

2.5.3 Tracer tests

In karst areas, tracer tests are conducted to locate and determine **watercourses and groundwater retention times**. The methods employed rely on fluorescent dyes (1), radioactive substances (2), salts (3) and pollen (4). Tracers (1) and (4) have no environmental consequences, although fluorescent dyes could be perceived as a visual infringement. The initial activity and concentration levels of tracers (2) and (3) must be kept low enough to avoid detrimental effects on the environment.

2.6 Exploratory work

Exploratory work serves to enable sampling activities. Depending on the depth of the planned sampling point and on the geological situation, different opening operations are appropriate:

2.6.1 Trial pits

The main environmental consequences of establishing a project result from **removal of the local vegetation** and soil. It is sometimes necessary to penetrate more deeply into the exposed rock, although such measures normally involve depths of a few meters at most. Cutting into a steep slope causes **erosion**. Upon completion of the exploratory work, the prospect must be refilled with the excavated material and separately stored topsoil to prevent aggravated erosion and accidents. Additional case-specific measures may be necessary to preclude erosion.

2.6.2 Shafts/tunnels

If boreholes and trial pits are insufficient for the envisioned scope of exploratory work, horizontal or slightly inclined tunnels and/or vertical shafts can be dug to enable **underground reconnaissance**, including sampling. Due consideration must be given to the fact that tunnels require appropriate entrances and that they tend to **collect groundwater**, possibly resulting in the **dewatering/drainage of overlying rock**. Special protective measures may be required in connection with the location and exploration of **uranium deposits**. In the absence of appropriate national directives, the radiation protection ordinance of the Federal Republic of Germany should be applied accordingly.

Major exploratory operations quickly equate to regular mining operations, the environmental consequences and relevant protective measures of which are dealt with in detail in the respective sections of this handbook.

Tunnel faces and shaft mouths must be **closed off** for safety reasons whenever the work is interrupted and following its completion.

Any shaft or tunnel that interrupts the flow of **groundwater** can simultaneously jeopardize its quality. Consequently, when the work is finished, all such **holes** should be completely **refilled**. As long as the work is ongoing, shafts must be secured to prevent unauthorized access and accidents. If regular measures are not possible, a sturdy cover must be installed.

Dug wells providing potable water in rural areas of dry-climate regions are especially important. If the exposed groundwater is not effectively protected against pollution, such wells can have negative qualitative effects on the environment. The same applies in essence to groundwater trial pits, while groundwater stemming from an adit is, as a rule, hygienically unobjectionable.

2.6.3 Drilling

Drilling serves as a means of **subterranean geological exploration**. It allows geological surveys, geophysical measurements and sampling. Pumping tests are conducted for hydrogeological purposes (cf. 2.5.1). Drilling can cause a substantial **noise nuisance**, with attendant disturbance of the local populace and animal life. Thus, all requisite active and passive means of noise control must be adopted, and the applicable work safety directives in particular must be followed. Depending on the climatic zone, some extent of **land may have to be cleared** around the drilling site.

Wells and boreholes are **potential hazards for groundwater**. In the absence of protective measures, detrimental effects can result from cutting through confined groundwater (artesian, for example), from **interconnecting** different groundwater stories (possibly of divergent quality), and/or from **piercing** the bases of multiaquifer formations.

Bleeding artesian wells are a waste of groundwater reserves and can do damage to the borehole environs, e.g., by causing the soil to salt up. The hydraulic interconnection of different groundwater stories can detract from both the quantity and the quality of the entire resource. Intermediate groundwater stories can drain out to such an extent that **wells run dry** and the work of **fetching potable water** - mainly by women - for household purposes increases in relation to the distance to the next intact well.

Appropriate technical measures, however, can be taken for **drilling operations** (pressure regulating valves, special flushings, packers, clay seals) to prevent such damage. In areas with a relevant hazard potential, the geological and technical aspects of drilling must be carefully planned in all detail, with all the appropriate equipment provided and properly serviced. (For details, please refer to the environmental brief Petroleum and Natural Gas.)

In semi-arid regions, drill bits often encounter aquifers filled with fossil, nonrenewable groundwater. Thus, in any such case the **demand forecasts** and **proven reserves** must be carefully balanced in order to avoid both an unprofitable investment and consequential damage to the ecology.

Drilling operations can also have negative environmental consequences as a result of **drill cuttings, chemicals, process water and improper fuel storage procedures**. The incidental drill cuttings and flushings must be collected and properly clarified at the end of the drilling operations, so that only cleansed wastewater is returned to the environment. The drilling site must be cleaned up and restored as closely as possible to its original condition.

2.6.4 Solid waste/dumps

Solid waste can derive from laboratory work as well as from exploration and production operations. All scraps, e.g., worn drill rods, must be collected and either **properly disposed of or recycled**. The same applies to sludgy residues from flushing operations.

Trial pits, tunnels and shafts yield **excavated material** that requires temporary or permanent storage. The size of the requisite storage area depends both on how much material is excavated and on the local topography. Wind, precipitation and percolation can erode, leach, scour and elutriate excavation dumps and cause **water pollution** in the process. In particularly severe cases, the dump may **slump or slide**. The storage of any material with a high **hazard potential**, e.g., due to radioactivity, requires appropriate measures for:

- preventing washout and dust deflation,
- collecting wastewater/effluent (packing and possible percolation drainage) with subsequent clarification, and
- monitoring its discharge.

Environmental detriment attributable to erosion can be extensively avoided - and the dump's stability enhanced - by **turfing, greenbelting or otherwise covering it**.

2.7 Sampling

2.7.1 Surface sampling

Sampling for analytical purposes often requires either the **removal** of near-surface strata or the extraction of material from special-purpose exposures. In some **rare cases**, sampling can impose a burden on the environment in the form of noise given off by jackhammers. As a rule, though, such problems are short-lived and not particularly serious. By comparison, the work involved in establishing exposures is more likely to have negative environmental impacts, as described in section 2.6.

2.7.2 Marine sampling

Marine sampling operations can have environmental consequences for ecosystems in **shelf waters** as well as in the deep sea: alteration of the seafloor morphology, disruption of pediment, destruction of marine life, turbidity.

The following techniques and technology must therefore be employed for minimizing such effects:

—   exploration of the seafloor via TV probes in order to confine and delimit the sampling area,

—   selective sampling with TV-guided grabs,

—   no large-scale clearance or scavenging of the seafloor,

—   separating the sludge and fine slush from the liquid phase (undissociated suspensions can jeopardize marine fauna, especially if they get into the photic zone.),

—   avoiding the local release (into the seawater) of acidic processing residues.

In some cases, in-situ analytical instruments use radioisotopes as a **source of excitation**, with a possible attendant (normally harmless) increase in radioactivity.

2.8   Laboratory testing

2.8.1   Laboratory analysis

Activities in connection with chemical and physical laboratory testing and analysis can yield substantial amounts of **solid, liquid and gaseous wastes**, some of which may contain **toxic reagents**. Exhaust air and exhaust gases may require filtration or scrubbing, while **liquid wastes and effluents** can be neutralized, precipitated, clarified, separated, etc. Organic solvents must be collected and the escape of noxious fumes and vapors to the atmosphere prevented. Additionally, appropriate measures must be adopted to ensure either the orderly disposal (incineration, dumping, ultimate storage) or recycling of liquid and solid waste products. The environmental brief Analysis, Diagnosis and Testing contains pertinent information in detail.

2.8.2 Dressing tests

Deposit exploration projects sometimes necessarily include dressing tests. The incidental **wastewater** must be collected in settling tanks and appropriately treated to the extent that it contains substances capable of polluting the recipient body or groundwater; cf. environmental brief Minerals Handling and Processing.

3.   Notes on the Analysis and Evaluation of Environmental Impacts

Distinction must be drawn between the environmental consequences dealt with in section 2 and those which may result from **follow-up measures**. Any study, expert opinion or commentary prepared in connection with such projects should include references to the potential **environmental impacts** of subsequent project **implementation**. Even at the initial reconnaissance and exploration stage, such consequential effects should be appraised. If necessary, pertinent studies must be conducted in parallel with the prospecting activities. Such preliminary studies should focus on the **data requirements** of the subsequent environmental impact assessment. Section 4 and other environmental briefs offer further-reaching information on the scope, evaluation and possible countermeasures.

4.   Interaction with Other Sectors

The following other environmental briefs are also of relevance:

- Spatial and Regional Planning
- Water Framework Planning
- Urban Water Supply
- Rural Water Supply
- Road Building and Maintenance, Building of Rural Roads
- Rural Hydraulic Engineering
- Large-scale Hydraulic Engineering
- Surface Mining
- Underground Mining

- Petroleum and Natural Gas - Exploration, Production, Handling, Storage
- Minerals - Handling and Processing
- Cement and Lime, Gypsum
- Glass.

The **groundwater domain** is a focal point of interest in that connection. Regional planning, in particular for rural development, is heavily dependent on access to properly protected groundwater, and timely evaluation of the potential environmental consequences of project measures is therefore of major significance. Diverse cross-links also exist between the groundwater domain and the mineral resources and mining sector, since potential environmental impacts often become apparent at the feasibility-study stage.

5.   Summary Assessment of Environmental Relevance

The **project goals** encompass preparations for the environmentally appropriate satisfaction of basic needs (e.g., access to potable water), the protective use of resources (such as water), appropriate use of soils, self-sufficiency in the environmentally sound exploitation of mineral and fuel resources and, as a result, improved employment perspectives in conjunction with the extraction and exportation of resources. Two of the most important project objectives are the transfer of know-how and the enhancement of environmental awareness.

As long as the project is carefully planned and executed with due regard for the described consequences and protective measures, activities in connection with the reconnaissance, prospection and exploration of geological resources can be expected to have only **limited impacts on the environment**.
Appropriate and available **means of controlling or remedying environmental damage** can be implemented **with relatively modest inputs**.
The subject studies and investigations also serve to supply the environmentally relevant **data and information** needed to achieve sustainable utilization of soil and groundwater and the protective extraction of nonrenewable mineral resources.
In connection with the implementation of protective measures, the interested and concerned parties must be made aware of environmental concerns. The analysis and evaluation of potential environmental impacts must be considered an **integral part** of the project appraisal phase.

Pertinent directives serve to ensure that:

- intervention in the environment is limited to the smallest possible, essential scope;

- unavoidable encroachments are accommodated to the natural situation;

- resultant damage is remedied or, if that is not possible, at least controlled;

- permanent damage is avoided to the greatest possible extent.

The necessary measures and corresponding responsibilities must be defined and established during the project planning phase.

Controls aimed at ensuring the success of the protective measures should be **conducted during and at the end of the project**.

Attention must be drawn to potential environmental consequences resulting from the continuation or expansion of a project.

35

6.   References

Bender, F. (Ed.): Geologie der Kohlenwasserstoffe, Hydrogeologie, Ingenieurgeologie, Angewandte Geowissenschaften in Raumplanung und Umweltschutz. In: Angewandte Geowissenschaften III: Stuttgart (Enke) 1984.

Der Bundesminister für Wirtschaftliche Zusammenarbeit [BMZ - German Federal Minister for Economic Cooperation and Development] (Ed.): Sektorkonzept Mineralische Rohstoffe. Bonn 1985.

Der Bundesminister für Wirtschaftliche Zusammenarbeit [BMZ - German Federal Minister for Economic Cooperation and Development] (Ed.): Umweltwirkungen von Entwicklungsprojekten, Hinweise zur Umweltverträglichkeitsprüfung (UVP), Bonn 1987a.

Deutsche Gesellschaft für Technische Zusammenarbeit (GTZ) GmbH (Ed.): Consultant-Tag 1985, "Umweltwirkungen von Infrastrukturprojekten in Entwicklungsländern". Sonderpublikation 1981, Eschborn 1986.

Doornkamp, J.C.: The Earth Sciences and Planning in the Third World. - Liverpool Planning Manual, 2. Liverpool (University Press & Fairstead Press) 1985.

Ellis, D.V.: A Decade of Environmental Impact Assessment Marine and Coastal Mines. - Marine Mining, 6, 4, New York, Philadelphia, London 1987.

FINNIDA: Guidelines for Environmental Impact Assessment in Development Assistance, Draft 15, July 1989.

Gladwell, J.S.: International Cooperation in Water Resources Management - Helping Nations to Help Themselves. - Hydrological Science Journal, 31, 4, Oxford 1986.

Loucks, D.P. & Somlyody, L.: Multiobjective Assessment of Multipurpose Water Resources Projects for Developing Countries. - Natural Resources Forum, 10, 1, New York, 1986.

McPherson, R.B., et al: Estimated Environmental Effects of Geologic and Geophysical Exploratory Activities, Office of Nuclear Waste Isolation (ONWI), Technical Report, December 1980.

Meyer, H.J.: Bergrecht und Geoforschung in Entwicklungsländern. - Studien z. int. Rohstoffrecht, 10, Frankfurt am Main (Metzner), 1986.

Overseas Development Administration (ODA): Manual of Environmental Appraisal, without address, without year of publication.

Schipulle, H.P.: Umweltschutz im Rahmen der entwicklungspolitischen Zusammenarbeit der Bundesrepublik Deutschland - In: Tagungsbericht "Eine Umwelt für drei Welten", 23.02.1988, Dortmund (Inst. f. Umweltschutz) 1988.

Urban, K.: Bewässerung in Sahel - Eine kommentierte Literaturübersicht. - GTZ-Sonderpublikation, 217, Eschborn, 1988.

Zimmermann, G.: Strahlenschutz. 2. Aufl., Stuttgart (Kohlhammer) 1987.

Mining and Energy

Surface Mining

36. Surface Mining

Contents

1. Scope

Surface mining is the term used to describe diverse forms of **raw-material extraction from near-surface deposits**. It involves the complete removal of non-bearing surface strata (overburden) in order to gain access to the resource. Depending on the physical characteristics of the raw material and on the site-specific situation, **various surface-mining techniques** are applied:

Dry extraction of loose or solid raw materials: In hardrock mining, the product must first be "worked" (loosened). Then, it can be loaded, hauled and processed by mechanical means similar to those employed in loose-rock mining. Accordingly, dry surface mines require appropriate dewatering.

In **wet-extraction**, or dredging, operations, loose raw materials are mechanically or hydraulically extracted and transferred to a processing facility. The entire extraction equipment is normally located on/in the water, often floating on a river or artificial lake.

Offshore, or shelf, mining is the term used to describe the extraction of loose material from nearshore deposits (marine beach placers). Like in wet extraction, the material is excavated and conveyed by mechanical or hydraulic means.

Deep sea mining is a - future - form of mining in which raw materials are extracted from ocean beds; not to be dealt with in the present context.

The various surface mining techniques are applied to different types of raw material reservoirs.

Table 1
Forms of surface mining and major raw-material products

Hardrock mining		Loose-rock mining			
dry extraction		dry extraction		wet extraction	
				terrestrial	offshore
building stones diamonds gems feldspar gypsum limestone/ raw materials for cement	metalliferous ores (copper, iron, silver, tin) oil shale hard coal uranium ore	brown coal diamonds gold kaolin/ china clay phosphates sand, gravel	heavy minerals (ilmenite, rutile, RE-minerals[1], zircon) clay tin ore	diamonds gold heavy minerals tin ore sand, gravel	diamonds heavy minerals (ilmenite, rutile, zircon, monazite) tin ore

[1] RE-minerals = rare-earth minerals

Surface mines vary in size according to the **nature of the deposit and the employed techniques of extraction**. Among terrestrial workings, one encounters mines ranging in size from small **one-man operations** to huge **strip mines** measuring several kilometers in diameter. Due to the elaborate, expensive technology required, marine workings always strive toward minimum dimensions.

Since mining amounts to a **site-bound activity**, new and expanding operations often have to compete with other **potential users** of the premises in question, and the infrastructure required for surface mining operations may still have to be established. As regards the demarcation of surface mining activities, it is inherently difficult to separate them from the required mineral dressing facilities, because such processing normally takes place directly at the place of extraction.

2.   Environmental Impacts and Protective Measures

The environmental consequences of surface mining operations are strongly dependent on the project type. Consequently, this section distinguishes between impacts and control measures.

2.1   Potential environmental consequences of surface mining

Common to all surface mining activities is that their environmental impacts are both **size-dependent and location-dependent,** particularly with regard to climatic, regional and infrastructural contexts. For the sake of simplicity, the potential environmental impacts of surface mining operations are categorized in the following sections according to the employed type of raw-material extraction.

Table 2
Forms of surface mining and their main environmental impacts

	DRY EXTRACTION	WET EXTRACTION	NEARSHORE EX-TRACTION	DEEP-SEA MINING
earth's surface	areal devastation; altered morphology: danger of falling rocks at the faces; destruction of cultural assets	areal devastation; altered morphology and river course; formation of large dumps	altered ocean-floor morphology; coastal erosion	
air	noise; percussions from blasting; dust formation due to traffic, blasting, wind; smoke and fumes from self-ignited dumps; blast damp, noxious gases; vibrations	noise due to power generation, extraction, processing and conveying; exhaust gases	noise, exhaust gases	noise; exhaust gases
surface water	altered nutrient levels (potential eutrophication); pollution by contaminated wastewater; pollution by aggravated erosion	denitrification; burdening of recipient with large quantities of muddy wastewater; pollution by contaminated waste-water	turbidity; oxygen consumption; wastewater pollution	turbidity; oxygen consumption wastewater pollution
ground water	recession of groundwater; deterioration of groundwater quality	altered groundwater level; altered groundwater quality		
soil	denudation in the extraction area; loss of (agric.) yield, dryout, ground sag, danger of swamping due to local groundwater recovery, soil erosion	denudation in the worked area	altered seafloor; deterioration of seafloor nutrient content	deterioration of seafloor nutrient content
flora	destruction in worked area; partial destruction/alteration in surrounding area due to altered groundwater level	destruction in the worked area		
fauna	expulsion of fauna	expulsion of fauna	destruction of stationary marine life (corals)	destruction of stationary marine life (corals)
humans	land-use conflicts; induced settlement, destruction of recreation areas	land-use conflicts; social conflicts in boom times; induced settlement	impaired fishing (destruction of spawning grounds)	impaired fishing (destruction of spawning grounds)
structures	water damage due to groundwater recovery			
miscellaneous	potential modification of microclimate	modification of microclimate; growth of pathogens in still-water areas		

36

2.1.1 Dry extraction

Differentiation is made between loose-rock and hardrock mines. Wherever necessary, the following sections include reference to specific influences. The environmental consequences are broken down according to physical, biological and social
effects.

- **Physical environmental impacts of dry surface mining**

In essence, the foremost environmental impact of surface mining is the **extraction
of nonrenewable resources**. The processes and activities involved in the extraction
of a raw mineral can involve mining losses, free-standing ore pillars, presently
uneconomical sections of deposit, overcutting, etc., **with resultant destruction of
sections** to the extent of their becoming inaccessible for future extraction. The strip
mining of carbonizable or combustible raw materials such as coal or peat can lead
to the destruction of resources by **fire (seam fires)**.

The **space requirements** of surface mining operations can be quite substantial,
comprising the **quarry** itself and **dumps for overburden**, which can be very siz-
able for deep hardrock mines (e.g., open-cast ore mines), **tailings heaps**, which also
can become very large for low-grade ore, and **room for infrastructural facilities**
(miners' lodgings, power supply, transportation, workshops, administration building, processing equipment, etc.). Since surface mining operations are inherently
bed-bound, their size and location are determined by the given geological conditions of the bedding and associated strata. And since major **disruption of the
earth's surface** is unavoidable in connection with surface mining operations, the
question of tolerability under the prevailing conditions must be given due consideration prior to commencing with any extractive processes.

In and around the mine and its dumps, some of the soil has to be **removed**, and
some gets **covered over**. Nearly all **industrialized countries** have regulations
governing the treatment of cultivable soil (topsoil). As a rule, its **removal** and
temporary storage prior to the beginning of direct mining activities is mandatory.
In addition, subsequent replacement of the topsoil and recultivation of backfilled
ground may also be prescribed.

Surface mining operations also alter the **morphological makeup** of the mine site as
a (temporary) result of shaping the quarry and its dumps and heaps. Once an
abandoned mine has been recultivated, some such changes remain behind in the
form of **permanent, residual (submorphic) hollows**, the size of which depends on
how much material has been extracted from the mine. Morphological changes can
be particularly pronounced in **hardrock mines**, which tend to have very steep
slopes and for which little material is left for refilling (e.g., in stone quarries).

By comparison, the morphological changes occurring in **loose-rock** mines consist
primarily of the **overburden dumps** established at the time of opening the mine,
and **ground subsidence** caused by dewatering.

Surface mining activities also interfere with the **surface water regimen**. Relevant intentional intervention aims to keep surface water and groundwater out of the workings by **collecting and channelling the water** from around the perimeter as well as from the mine proper. Riverbeds are bypassed around the mine, and runoff water from precipitation and drained slopes is collected in ponds and discharged into the natural hydrographic network. Increased sedimentation and altered chemism resulting from such measures can cause qualitative **degradation of the recipient water body**.

Loose-rock surface mining can also interfere with the **groundwater regimen**, with resultant **loss of groundwater quality** due to the infiltration of contaminated wastewater and in washout and leaching of dumps, heaps and the mine itself. If the **groundwater level** is not lowered in time, groundwater will flow into the pit. Consequently, all around and within the mine, wells are sunk to below the lowest pit bottom in order to enable dry extraction while enhancing the stability of both the slopes and the floor by relieving the effective hydraulic pressure. The well water is generally unpolluted and can be fed directly into the natural river system. Lowering the groundwater level has major **consequences** for the surrounding area, e.g.:

— drying up of nearby wells,
— settlement/subsidence,
— disturbance of the vegetation due to altered groundwater supply.

When the mine is closed down, **hollows** resulting from extraction of the resource and removal of overburden during the opening phase remain behind. The hollows eventually form groundwater-fed ponds and lakes reflecting the return of the groundwater level, which may proceed very slowly, depending on the depth of the erstwhile mine and on the given hydrogeological situation. Indeed, it may take more than 50 years for a new state of equilibrium to be achieved. If the zone of contact between the water and the soil contains soluble substances, power-plant ash and/or industrial residues, the **water quality** may suffer. The most well-known problem in that connection is an excessively low **pH** in the lakewater. A lack of affluxes and effluxes aggravates the problem, promoting **eutrophication**, particularly if the surrounding areas are intensively farmed.

36

The extraction activities impose a **noise nuisance** on their surroundings, with major noise sources including the machines and devices required for getting, loading, hauling, reloading, etc. In hardrock mines, drilling and blasting constitute two additional sources of noise. In addition to the **sound of the explosion**, the attendant **vibrations and reverberations** amount to an additional dynamic burden on the environment that not only annoys the neighbors, but can also cause damage to structures.

Finally, dry surface mining activities also lead to **air pollution**, the causes and effects of which are multifarious:

—   Blasting in hardrock causes **dust pollution** in that rock dust becomes entrained in the blast damp. The wind can stir up any and all exposed materials, especially during loading, reloading and dumping operations, all of which adds to the dust nuisance;

—   Air pollution in the form of **gases** results from the exhaust of vehicles and engines, which tend to be diesel-driven, as well as from the escape of blast damp. Open-pit coal mines are susceptible to still other, deposit-specific hazards: the extraction of deep-lying coal can give rise to the escape of **methane**, and spontaneous combustion can release other noxious gases.
Hot, dry weather poses a considerable **fire hazard** - by spontaneous combustion - for exposed coal at the bottom of the pit and at the loading and unloading points.
Additionally, self-ignition can cause hard-to-extinguish **smoldering fires** in overburden dumps and feigh heaps containing small amounts of coal. Such fires can pollute the environment with odors and noxious gases for years or even decades.

—   **Radiation exposure** can occur in special cases, i.e., in connection with the mining of uranium ore or rare-earth pegmatites.

•   **Interference with the biological environment by dry surface mining**

The surface extraction of raw materials necessitates **areal exposure** of the deposit. Removal of the soil in and around the mine itself, the surrounding dumps and the requisite infrastructure **destroys the local flora**.

In turn, **fauna is driven out** of the area by the destruction of its natural habitat.
Aquatic ecosystems can be disrupted by **qualitative and quantitative changes in surface water conditions**, while **wetlands** can be emburdened by an **altered groundwater level**, e.g., its lowering or recovery with subsequent lake/swamp formation. Fragile ecosystems in extreme locations are particularly susceptible to permanent damage or destruction.
Terrestrial ecosystems are also affected by mining-induced situational changes (in connection with the groundwater level, for example). Even after the mine has been abandoned and recultivated, the residual changes in soil physics and chemistry, available water resources, etc. can lead to the appearance of different plant and animal associations constituting an **irreversible alteration** stemming from the original disruption.

- **Effects of dry surface mining on the social environment**

The areal nature and deposit dependence of surface mining activities engender the presumably most serious **effects on human living conditions**. Frequent consequences include:

— the necessity of **resettling** the inhabitants of the area to be mined. Surface mining operations demand the relocation of settlements as well as traffic routes and communication infrastructure. The consequences range from economic loss to sociological and cultural disruption. The latter will be all the more serious, where the local population feels strongly attached to a limited natural environment, cultural or religious localities, established tribal structures, territorial sovereignties, etc.;

— **land-use conflicts** when the area to be mined is being used for agricultural or forestry purposes or contains significant cultural monuments, recreation areas/facilities or the like that stand to be destroyed or negatively affected by the mining operations.

If, due either to the large area to be affected by a surface mining operation and/or to attendant damage to the local flora and fauna, farmland and, hence, **income potentials** are lost, or even the **relocation of entire settlements** necessitated, those responsible and those affected must investigate in advance which special **consequences and impacts** the project can be expected to have for existing groups - women in particular. Likewise, the extent to which **women** will be able to **partake of the economic** advantages the region stands to gain from the mining operation must be duly investigated.

Moreover, the environmental effects of mining operations can affect the health of both the miners themselves and the people living in the surrounding area.

Finally, the establishment of mining infrastructure can inadvertently induce the uncontrolled generation of **settlements** in areas which otherwise may have remained undisturbed.

2.1.2   Wet extraction

With regard to the environmental consequences of wet surface mining, the previous subdivision according to physical, biological and social impacts is maintained. In case of identical consequences, the reader is referred to section 2.1.1.

- **Physical environmental impacts of wet surface mining**

Since the wet extraction of raw materials is a function of site- and mineral-specific factors such as a low degree of consolidation, certain particle-size spectra, well-balanced, shallow topography and adequate quantities of water, the number of potential locations and, hence, the **scope of environmental consequences** are **more limited** than for dry extraction.

The differences begin with the **space requirement**. Wet extraction normally involves a very limited extraction area. Precious-metal and tin dredgers, for example, rarely require more than one hectare, unless overburden has to be removed in advance. On the other hand, the extraction area wanders more or less rapidly over the entire **explorated field**, which eventually becomes completely modified: when dry land is being worked, the soil is removed, but when a river is being worked, the entire riverbed is altered, and the entire course of the river is likewise affected. Cutting and winning leaves behind rubble containing large amounts of **classified material** that is extensively lacking in fine and superfine contents. Consequently, pedogenisis, or **soil formation**, as an essential prerequisite for recolonization by plant associations, **is seriously impeded**. Meanwhile, the fine and superfine fractions emburden the river with large quantities of muddy wastewater. Such wet-extraction **sludge plumes** sometimes develop into water pollution loads that remain clearly visible over hundreds of kilometers before the clay fraction finally settles out of suspension. The situation can be additionally aggravated by **contaminated wastewater**. The escape of mercury from gold-placer processing activities, for example, or the uncontrolled disposal of used oil, constitute serious pollution potentials.

With regard to resources, noise and air, the reader is referred to the hazards discussed in item 2.1.1.

- **Effects of wet surface mining on the biological environment**

Like dry extraction, wet extraction also **destroys flora and drives away fauna**. Also and in particular, however, wet extraction disturbs the **aquatic ecosystem**. The aforementioned mining-induced sludge contamination of affected rivers degrades the water quality, alters the river bed by depositing fine and superfine material, and disrupts the nutrient balance of the rivers, with consequential effects on river fauna and flora. Frequently, such pollution leads to **lower fish populations** due to dying and migration away from the affected sections of the river.

In tropical areas, wet extraction of mineral resources poses an additional serious environmental hazard in that resultant still waters can serve as **breeding places for pathogenic agents** such as malaria-carrying mosquitos. Indeed, it can happen that regionally eradicated tropical diseases flare up anew.

- **Effects of wet surface mining on the social environment**

In otherwise infertile areas, the loss of fertile flood plains or easily irrigated areas to wet surface mines can lead to bitter **land-use conflicts**. Even if the areas in question are recultivated afterwards, irreversible damage may remain behind on a location- and situation-specific basis. The **impairment of fish-farming activities** by the aforementioned sludge pollution of rivers counts more as **a temporary effect**. By contrast, **health impairment** resulting from the contamination of rivers with mercury, for example, counts as **irreversible, permanent damage**.

Social conflicts in connection with wet mining activities become particularly serious when boom times (a local gold rush, for example) draw large numbers of small miners (diggers, garimpieros, pirquineros) into a particular area. Many such newcomers lack legal mining titles and either breed or intensify diverse problems (crime, speculation, exploding prices, disease, social tension among the native population, etc.). As the originally rich deposits become harder to work and eventually depleted, such problems tend to intensify.

36

2.1.3   Nearshore marine mining

In dealing with the environmental consequences of marine mining, deep-sea mining is not gone into separately, because it does not yet actually contribute to the
production of raw materials. The environmental effects of deep-sea mining are
comparable to those of nearshore marine mining, with the latter limited by definition to the use of bucket chain (scoop) and suction dredgers in waters with a
maximum depth of about 50 meters.

*   **Physical environmental effects of marine mining**

The most serious effect of extracting minerals from the ocean is that such activities
alter the ocean floor. The ground is removed by mechanical or hydraulic means in
order to separate it from its ore in an on-board processing facility. Altering the
morphology and composition of the ocean bed amounts to its total restructuring,
since **natural classifying processes** take place when the oversize, tailings and
perhaps overburden is re-deposited - assuming, of course, that the raw material in
question contains low-grade ore (e.g., heavy mineral sand) and that processing
leaves behind large quantities of nonbearing materials. When a large percentage of
the material being extracted is commercially valuable (sand, gravel), its **removal in
large volumes** also modifies the seafloor morphology, possibly resulting in
intensified **coastal erosion** and **accumulation of sediments**, since the "new" ocean
floor is less compact and lacking in fine and superfine particles.
The fine and superfine fractions that are left over from ore processing and which
swirl up from the ocean floor remain in suspension for a long time, causing **turbidity** that can be carried off by ocean currents to pollute areas as far as 10 km away
from the source.
If the water flows slowly, the fine and superfine particles settle out, covering the
ocean floor with a layer of clay.

Moreover, by way of analogy to dry mineral extraction, the mining equipment,
machines and apparatus generate noise and pollute the air and water.

- **Effects of marine mining on the biological environment**

The altered seafloor **interferes with the natural ocean-bottom nutrient balance**, both within the mined area and in the emburdened vicinity. The **effects** are particularly **devastating for immobile marine organisms** such as corals, which can be partly or completely destroyed by the combination of high turbidity and fine-particle sedimentation.

The clouds of turbidity also impair marine life in the water itself, e.g., by reducing insolation, lowering the available oxygen level due to oxidation of stirred-up particles, obstructing the respiratory passages of marine organisms, and possibly even poisoning them with trace metals.

Mobile marine fauna can evade the polluted environment by moving off, but are nonetheless unable to prevent the destruction of their spawning grounds.

- **Effects of marine mining on the social environment**

Since marine mining has no direct impact on the human environment, its social effects are limited to **usufructuary conflicts**, most notably **with fish farmers**, whose livelihood can be prejudiced by such mining, **and with the operators of recreation facilities** that can be adversely affected by mining-induced pollution.

2.2 Measures for limiting the environmental consequences of surface mining activities

A selection of **technical options** for use in limiting the pertinent environmental impacts prior to, during and subsequent to surface mining activities are pointed out below. Naturally, the limitation of environmental consequences (= pollution control) entails a suitable **institutional basis** and the **existence, adherence to and monitoring of appropriate directives**.

2.2.1 Measures prior to commencement of mining activities

The most important precommencement measure is to ascertain the **momentary condition of the environment** as a basis for evaluating subsequent environmental impacts. The relevant studies should give due consideration to cultural and historical monuments, soil conditions, groundwater and surface water qualities and quantities, as well as flora, fauna, land use, etc.

In the case of marine placers, the marine flora and fauna, prevailing currents, seafloor gradients, etc. also should be determined in advance.

36

Careful planning of operational sequences enables significant limitation of environmental consequences even before mining activities begin. For example, a suitable time schedule with provisions for the archiving and conservation of archeological finds, the harvesting of standing timber in the area to be worked, and/or keeping the mine open only as long as necessary is extremely useful. Likewise, careful separation and separate storage of humus and the upper soil horizons of the overburden ensure that suitable material will be available for subsequent recultivation. Selective dewatering according to a time scale and the use of modern drainage techniques and/or sealing methods can help minimize the problems arising from groundwater recession.

With a view to precluding potential social tensions, all relevant planning must - in order to **protect their interests** - involve the **groups of persons who will be affected** either directly, e.g., by having to resettle, or indirectly, e.g., by impaired fishing conditions. It is particularly important that all parties concerned and affected, as well as the local authorities, be allowed to appropriately **participate** in the planning and execution of relocating measures, compensation and possible resettlement.
Finally, both the decision makers and the miners must be **instructed and sensitized** in and toward the environmental and health aspects of surface mining activities prior to their commencement.

2.2.2   Measures in the course of mining activities

In order to **avoid excessive land consumption,** inside dumps should be established, i.e., the overburden should be stored within the open spaces of the mine.
The **noise nuisance must be limited** by appropriate soundproofing of individual pieces of equipment. Whole units of equipment can be **encapsulated** or equipped with special exhaust systems (mufflers). Additionally, the miners must be required to wear **personal noise-protection gear**, e.g., ear protectors. Finally, **time limits can be imposed on noise emissions**, e.g., by limiting blasting operations to once a day. Moreover, the propagation of acoustic waves in the near vicinity of noise emitters can be reduced by such measures as noise-control embankments.
In hardrock mines, **optimized blasting methods** can substantially reduce noise and dust emissions. By optimally matching the explosive quantities to the drilling pattern and by stemming the holes, the overall quantity of explosives and, hence, the magnitude of the explosion (vibrations), the incidence of microfine dust, and the intensity of the blasting noise can be substantially reduced.

Dust control measures in surface workings can encompass such **individual measures** as sprinkling water on the roads and other conveying routes, washing down transport equipment (trucks, etc.), irrigating and turfing dumps and exposed areas, and applying dust bonding agents as necessary. Also, individual pieces of equipment such as crushers over belt feeders can be **encapsulated and surrounded with trees or hedges** that filter out dust and reduce the overall drift (deflation). Drills and boring tackle can be fitted with wet or dry dust precipitators.

Wastewater can be cleansed of suspended solids, neutralized and clarified in wastewater treatment facilities to meet **minimum quality standards** for release into a recipient body. For each and every solution or suspension, there are **appropriate liquid/liquid and solid/liquid separation processes** for use in purifying contaminated water. For metal-polluted acid mine drains (a.m.d), electrolysis is indicated, while an ion-exchange technique is more suitable for radioactive wastewater. In general, all means of countering the **causes of pollution** should be exploited. For example, the use of bypass microfilters in engine lubrication systems can reduce the incidence of used oil by up to 90 % by prolonging its useful life.

The **dredges** used for working nearshore marine placers should be equipped with **long rubbish chutes** for use in covering the tailings/trash and oversize with overlay shelf in order to restore a close-to-natural particle-size spectrum to the seafloor.

Wet extraction from an artificial lake is preferable to working directly in a river, because it involves much less of a sludge load for the latter. **Wells and other large boreholes** that are no longer needed but could disturb groundwater barriers (aquitards) **should be sealed**.

Particularly for fragile working faces, the **angle of slope** around the perimeter of the mine must be designed to preclude major flank movements (slides and falling rocks).

In dry coal mines, care must be taken at the planning stage to protect **coal-bearing dumps** from spontaneous ignition by appropriate **surface compaction and air-exclusion measures**. The same applies to coal pillars and abandoned working faces, which also require sealing to prevent smoldering fires.

Such **special measures** as the posting of trespassing notices, installation of fences and blocking of roads can help protect and preserve adjacent ecosystems.

36

Persons likely to be affected can and must be afforded appropriate protection through, say, the appointment of environmental affairs and/or safety officers and occupational physicians. Since damage to the environment cannot be limited exclusively to the mining area, the **right to medical services** should also be extended to persons living in the general vicinity.

Continuous monitoring of all important factors must accompany all surface mining activities and attendant pollution-control measures. Such factors include exhaust gases, noise levels, vibrations, water pollution, particulate emissions, slope movement/stability, ground subsidence and groundwater levels.

2.2.3 Measures following termination of mining activities

As soon as any section of a deposit has been fully exploited and refilled with waste from new operations, appropriate **rehabilitation measures** can and must be taken. Since surface mining operations tend to be quite expansive, ongoing mining operations in one area can be accompanied by rehabilitative measures in another. The same is true of wet mining operations outside of riverbeds. To rehabilitate means to immediately transform the areas concerned into **as natural a landscape as possible**. Following wet extraction, particularly in tropical locations, all worked-out areas must be drained and graded to eliminate all **open bodies of water** that could serve as breeding grounds for pathogenic vectors like malaria-transmitting mosquitos. On the other hand, **bodies of water created by surface mining activities** can, on a case-by-case basis, also be utilized as **dry-season water reservoirs or** for such commercial purposes as **fish-farming**.

Dumps, open-pit perimeters, outside dumps and erstwhile extraction areas require immediate greenbelting or planting with **indigenous vegetation** in order to limit or prevent erosion, especially in humid, tropical climates, and deflation in arid climates. Special **erosion control** methods such as drainage and consolidation must be employed in particularly vulnerable areas.

The ultimate aim must be to fully **recultivate** the worked out areas to enable appropriate and corresponding use, or to renature them for another purpose. To **reclaim the land**, it must be graded, compacted and covered with **soil and humus** to allow immediate oversowing and subsequent soil management. It should be borne in mind, however, that recultivation is not the only means of limiting environmental detriment. Recultivation is very **time-consuming**, and the ultimate **success is usually uncertain**. Especially recultivation in tropical areas has not yet been adequately researched and developed with regard to planting sequences, site-appropriate species, etc. Moreover, **successful recultivation** entails extensively natural soil physics (permeability, granularity/type of soil) and soil chemistry (pH, nutrients, absence of pollutants). Otherwise, the soil would not be able to fulfill its diverse functions as a water reservoir, a biotope for plants and animals, and a basis of agricultural production.

3.   Notes on the Analysis and Evaluation of Environmental Impacts

The principal **regulations** governing mining activities and pertinent environmental protection **in Germany** are the *Bundesberggesetz* BBergG [Federal mining law] dated August 13, 1980, and the *UVP-VBergbau* (ordinance on the environmental impact assessment of mining projects) dated July 13, 1990, the *TA-Luft* (Technical Instructions on Air Quality Control), the *TA-Lärm* (Technical Instructions on Noise Abatement), the *BImSchG* (Federal Immission Control Act) and its various implementing provisions, as well as the respective mining regulations of the various states and their laws governing landscape, preservation of nature and excavation. In addition, the *Verein Deutscher Ingenieure* (Association of German Engineers) has issued a number of guidelines dealing primarily with the relevant mechanical equipment.

Other industrialized countries like the USA, Canada and Great Britain have similar, in part more stringent, laws and regulations - including, for example, the U.S. "Clean Water Act" (1977) and the "Surface Mining Control and Reclamation Act" (public law 95/87, 1977), with supplementary provisions drawn up by the Office of Surface Mining Reclamation and Enforcement (OSM) and by the Environmental Protection Agency (EPA).

A precommencement **status quo study** with thorough investigation of all matters relevant to the physical, biological and social environment provides a **crucial basis** for evaluating the environmental consequences of surface mines and planning recultivation measures; cf. environmental brief Reconnaissance, Prospection and Exploration of Geological Resources.

Growing awareness of the environment and the will to protect it is emerging in many parts of the world. To some extent, however, that new awareness has not yet found its expression in appropriate national laws. But even where laws protecting the environment already are in place, their enforcement is frequently neglected for a **lack of control and monitoring options.** The absence of an appropriate legal basis and/or of its proper implementation has **serious** large-scale and small-scale **consequences for the environment,** whereas mining regulations could be adopted with which to hold mine operators responsible for the consequences of their own mining activities. For **small mines that are difficult to monitor,** a pertinent recommendation was proposed at the UN-sponsored International Round Table on Mining and the Environment Congress in Berlin: **recultivation guarantee funds** can be set up, e.g., included in the concession fee. If the mine operator fails to rectify substantial environmental damage when he leaves his concession, financial reserves will be available to pay for recultivation. Otherwise, the withheld funds could be returned to the mine operator following satisfactory inspection of the properly recultivated areas.

Illegal mining is the biggest problem with regard to environmental destruction and recultivation. When large numbers of gem seekers and gold diggers intrude into and begin working an area in a completely uncontrolled manner - especially in developing countries - their activities are bound to cause areal destruction, often accompanied by pollution of the soil and rivers (with mercury and cyanide in the case of gold diggers). **Legal measures have proven totally inadequate** as a means of control, because the form of mining involved requires very little equipment, thus promoting a high level of mobility and, hence, good chances of **evading control.** Moreover, **supervision becomes nearly impossible** when large numbers of such people converge on an area and are **willing to use force in defense of their interests.** Consequently, damage to the physical and biological environment is accompanied by pronounced social tensions between the various interest groups.

4.   Interaction with Other Sectors

In sparsely populated and undeveloped regions, mining tends to serve as a **pacemaker for infrastructural development.** Frequently, mining projects have to carry the major share of the relevant cost of building access roads and establishing rail links to deposits for hauling away the mineral products and of building homes for the miners and their families, including all the requisite supply and disposal facilities. The new infrastructure can act as a catalyst for extensively **uncontrolled settlement and economic development of the area in question.**

Ore mines in particular tend to include an **initial processing stage** for local first-step product enrichment. Frequently, the purchaser and the mine operator agree to share the storage and supply facilities. In many brown-coal and hard-coal strip mines, the raw (possibly upgraded) coal is used directly for fueling **thermal power plants.** Accordingly, power generating facilities and distribution systems tend to be installed near such mines. **Storage grounds for disposing of residues** can be established in worked-out parts of the mine for some future use. Fly ash from power plants, for example, is often used for consolidating mine roads.

Land-use conflicts can quickly arise due to the **space requirements** of surface mining. The various interests must be reconciled within the context of appropriate **regional planning**.

While **land-use problems** occur less frequently in sparsely populated countries, legal problems nonetheless may arise. **Property rights**, for example, may not have been duly registered, and the boundaries may have been inaccurately mapped. Such problems intensify when those concerned happen to be groups of people with no lobby and a life-style and social status that afford them few options for preserving their traditional habitat. The very existence of such groups can be threatened. What is needed is a form of **regional and development** planning that duly accounts for ecological and ethnic concerns alongside of economic interests.

The following sectors, the environmental consequences of which are described in other environmental briefs, can be affected by surface mining activities and therefore require consideration:

- Spatial and Regional Planning
- Planning of Locations for Trade and Industry
- Overall Energy Planning
- Water Framework Planning
- Wastewater Disposal
- Transport Planning and Traffic
- Reconnaissance, Prospection and Exploration of Geological Resources
- Underground Mining
- Minerals - Handling and Processing
- Thermal Power Stations.

5.   Summary Assessment of Environmental Relevance

Surface mining of mineral resources involves different methods: wet and dry, marine and terrestrial. Common to all, however, is that they have **serious environmental consequences**.

Although **most mining activities are temporary by nature** (approx. 20 - 50 years), they often cause **permanent damage to the environment through irreversible disruption**. The earth's surface and the groundwater and surface-water regimens tend to sustain the most serious direct damage. Mineral extraction by surface mining methods also causes air pollution, noise nuisance, alteration of the soil, flora and fauna, and social problems arising from land-use conflicts, resettlement, etc. Such impacts are invariably dependent on the area involved, the location and the climate. Additionally, **points of law and control options** play major roles in determining the extent of environmental damage caused by surface mining activities and/or its limitation by such means as recultivation or renaturation. **Recultivation**, however, always amounts to substituting a new ecosystem for the original one in the affected area. Moreover, the ultimate success of such measures can rarely be guaranteed, especially in locations for which no relevant empirical data is available.

The **extent of damage can be limited** through careful planning, preparation and implementation of the mining activities. A thorough **analysis of the actual situation** in the region is an indispensable prerequisite and basis for planning with due consideration of the mining activities' anticipated effects in the form of environmental impacts and structural modification of the subject region. This must include the regulation of compensation and the planning of resettlement measures as well as the elaboration of recultivation plans.

As a flanking measure, concerned organizations, institutions and individuals must be **sensitized and informed** to prepare the way for the ecologically oriented implementation of the project.

The need to minimize costs must not be allowed to induce the promotors and others responsible for the project into cutting back on expenditures for environmental protection. Consequently, project desk officers should see to it at the project appraisal and authorization stage that the project encompasses **adequate landscape and environmental protection measures**, including the optimal use of resources, and that a sustainable structure with the **appropriate control and regulatory** functions is in place.

6. References

Agbesinyale, P.: Small Scale Traditional Gold Mining and Environmental Degradation in the Upper Denkyira District of Ghana, Spring Phase 1, Universität Dortmund 1990.

Bender, F. (Ed.): Geologie der Kohlenwasserstoffe, Hydrogeologie, Ingenieurgeologie, Angewandte Geowissenschaften in Raumplanung und Umweltschutz. In: Angewandte Geowissenschaften III: 674 pages, Stuttgart (Enke) 1984.

Chironis, N.P. (Ed.): Coal Age Operating Handbook of Coal Surface Mining and Reclamation, McGraw Hill, New York 1978.

Crawford, J.T.; Hustrulid, W.A.: Open Pit Mine Planning and Design, SME/AIME, New York 1979.

Cummins, A.B.; Given I.A. (Ed.): SME Mining Engineering Handbook, Vol. 1 & 2, SME/AIME, New York 1973.

Down, C.G.; Stocks, J.: Environmental Impact of Mining. Applied Science Publishers Ltd. London 1977.

E.I. du Pont de Nemours & Co. (Inc.) (Ed.): Blasters' Handbook, 16th ed., Wilmington, Delaware, USA, 1977.

Günnewig, D.: Die Umweltverträglichkeitsprüfung beim Abbau von Steinen und Erden. Inaugural-Dissertation, Institut für Mikrobiologie und Landeskultur der Justus-Liebig-Universität Gießen, 1987.

Hermann, H.P.: Schwerpunkte der Verwaltungsvorschrift zur Änderung der TA-Luft, in Braunkohle 35, 1983, Heft 6, p. 190 - 194.

Hofmann, M.: Bundesforschungsanstalt für Naturschutz und Landschaftsökologie: Dokumentation für Umweltschutz und Landespflege. Bibliographie. Abgrabung (Bodenerosion, Tagebau, Gewinnung oberflächennaher mineralischer Rohstoffe und Landschaft). Deutscher Gemeindeverlag, Köln 1988.

Hößlin, W. v.: Technische und rechtliche Probleme bei der Schaffung von Tagebauseen der Bayerischen Braunkohlen-Industrie AG in Schwandorf, in Braunkohle, 1980, Heft 9, p. 273 - 277.

Jung, W. et al.: Überblick der aus der bergbaulichen Tätigkeit resultierenden Umweltauswirkungen in der ehemaligen DDR. In: Z. Erzmetall 43, 1990, Heft 11, p. 478 ff.

36

Karbe, L.: Maßnahmen zum Schutz der Umwelt bei der Förderung metallischer Rohstoffe aus dem Meer. Vortrag gehalten auf der Tagung Meerestechnik und Internationale Zusammenarbeit. Tagungsband erschienen im Verlag Kommunikation und Wirtschaft. Oldenburg 1987.

Knauf (Ed.): Praktizierter Naturschutz. Dokumentation über Rekultivierungsverfahren abgebauter oberflächennaher Lagerstätten, 1987.

Koperski, M.; Musgrove, C.: Reclamation Improves With Age, in Coal Age, 1980, no. 5, pp. 162 - 169.

Kries, O. v.: Braunkohle und Landesplanung, in Raumforschung und Raumordnung, 1965, Heft 3.

Kröger, K.: Theoretische Grundlagen von Lärmemissionen und -immissionen bei Fördersystemen des Braunkohlenbergbaus, in Braunkohle 30, 1978, Heft 9, p. 260 - 266.

Krug, M.: Angewandte Planungsmethoden beim Aufschluß des Tagebaues Hambach 1978/79, in Braunkohle, 1980, Heft 4, p. 71 - 81.

Pfleiderer, E.P.: Surface Mining, 1st ed., AIME, New York 1968.

Robinson, B.: Environmental Protection: A Cost-Benefit Analysis, in Mining Magazine 151, 1984, no. 2, pp. 118 - 121.

Salomons, W.; Förstner U. (Eds.): Environmental Management of Solid Waste. Dredged Material and Mine Tailings. Berlin, Heidelberg, New York, London, Paris, Tokyo (Springer-Verlag) 1988.

Schultze, H.J.: Braunkohlenbergbau und Umwelt im Rheinland, in Erzmetall, 1985, Heft 2, p. 65 - 72.

Seeliger, J.: Eine europäische Umweltverträglichkeitsprüfung, in Umwelt- und Planungsrecht, 1982, Heft 6, p. 177 - 185.

Seeliger, J.: Kohlennutzung und Umwelt, in Glückauf 121, 1985, Heft 14, p. 1103 - 1107.

Sengupta M.: Mine Environmental Engineering, Volumes I and II, Boca Raton, Florida, 1990.

Stein, V.: Anleitung zur Rekultivierung von Steinbrücken und Gruben der Steine- und-Erden-Industrie, Köln (Deutscher Instituts-Verlag) 1985.

Thiede, H.-J.: Immissionsschutz in den Braunkohletagebauen des rheinischen Reviers, Energiewirtschaftliche Tagesfragen, 1979, p. 535 - 540.

Welch, J.E.; Hambleton, W.W.: Environmental Effects of Coal Surface Mining and Reclamation on Land and Water in Southeastern Kansas, Kansas Geological Survey, Mineral Resources Series 7, 1982.

Yundt, S.E.; Booth, G.D.: Bibliography. Rehabilitation of Pits, Quarries, and other Surface-Mined Lands. Ontario Geological Surveys Miscellaneous Paper 76, Ministry of Natural Resources, 1978.

Zepter, K.-H.: Schutz der natürlichen Umwelt - Möglichkeiten und Grenzen, in. Erzmetall, 1979, Heft 9, p. 357 - 418.

36

No Single Author

Bundesberggesetz (BBergG) und Verordnung über die Umweltverträglichkeitsprüfung bergbaulicher Vorhaben (UVP-V Bergbau - federal mining law and
ordinance on the environmental impact assessment of mining projects),
Verlag Glückauf GmbH, Essen 1991.

Environmental Aspects of Selected Non-ferrous Metals (Cu, Ni, Pb, Zn, Au) Ore
Mining: A Technical Guide, Draft Report, unpublished UNEP/IEO.

Environmental Protection Agency (EPA), USA 1986.
Part 11 - Natural Resource Damage Assessment
Part 23 - Surface Exploration, Mining and Reclamation of Lands
Part 434, Subpart E - Post Mining Acres

Mining and Environment Guidelines, International Round Table on Mining and
the Environment, Berlin 1991. UNDTCD/DSE.

Reclamation, in Mining Magazine, 1982, no. 11, pp. 449 - 451.

Texas Water Commission, 1985. Instructions and Procedural Information for Filing
Applications for a Permit to Discharge, Deposit or Dispose of Waste.

Update on Reclamation Regulations, in Coal Age, 1981, no. 7, pp. 68 - 73.

World Bank Environmental Guidelines: Mining and Mineral Processing, Draft
Report.

Mining and Energy

Underground Mining

37. Underground Mining

Contents

1. Scope

Mining is defined as the **extraction of mineral resources from the earth**. **Underground mining** is the extraction of raw materials below the earth's surface (deep mining) and their conveyance to the surface. Access to the vein or lode is by **shafts and tunnels** with links to the surface. (The subsequent stages of **raw material processing** are dealt with in a **separate brief**: Minerals - Handling and Processing.) The present brief examines only the underground extraction of solid mineral resources.

There are some 70 individual types of **useful minerals** that occur in minable concentrations either alone or in combination with other minerals, frequently as **natural mixtures** (aggregates).

Underground mining includes all **work involved in the winning of raw materials** by people using technical contrivances. Apart from the actual extraction and conveyance processes, the term underground mining also covers **development of the deposit** and provision of the requisite **infrastructure** (transportation/handling, storage facilities, surface plant, e.g., administration building, workshops, etc.) and all measures devoted to ensuring the **safety** of the miners. This includes:

- working
- loading
- conveyance
- drainage
- ventilation
- support

Small-scale mining activities in many countries frequently include a **transitional form** of extraction referred to as **trench mining**, or burrowing.

In **special cases**, the **mineral** can be made transportable and hauled off from its natural surroundings with no need of **exploratory work** (brine mining, in-situ leaching and in-situ gasification of coal).

Deep mining creates underground spaces in which people work. Their **working conditions** with regard to air temperature and humidity, presence of harmful or explosive gases or radiation, as well as moisture, dust and noise, **can be specific to the mined mineral** and/or **the surrounding rock, the depth of the mine, and the type of machinery** in use.

The **locations** of deep mines are dictated by the presence of potentially profitable raw materials. Underground extraction is practiced in all climate zones, in remote areas as well as under large cities, on the ocean floor and in alpine regions. The **size**, or output, of such mines ranges from less than 1 to more than 15 000 tons a day, and the **depth** at which extraction takes place ranges from a few meters to more than 4 kilometers.

37

2.　Environmental Impacts and Protective Measures

Deep mining impacts the environment in **three different areas**: in the deposit it-self and the surrounding rock, in the underground spaces created by and for the mine, and aboveground. Optimal exploitation of the resource with attendant limi-tation of environmental effects is **dependent on detailed planning** of the sequence of operations and on the mining methods and technology to be employed.

2.1　Environmental impacts on the deposit and the surrounding rock

2.1.1　Exploitation of resources

The most important **environmental consequence** of underground mining is that it involves the exploitation of a **nonrenewable resource**. The process of extracting the raw material necessarily also involves mining losses and impairment of other parts of the deposit. The best way to counter the latter effects is to **carefully plan** the extraction operations, stowing measures, etc.
Some raw materials (coal and several sulfidic ores) can under certain circumstances ignite spontaneously and cause **mine fires**.

2.1.2　Disruption of rock structure

The opening up of underground workings creates **cavities** and leads to **stress and motion** in the surrounding rocks. The **effects** of mining on the rock structure can include:

—　　subsidence due to cave-ins in the cavities. The resultant settling can propa-gate to the surface, possibly causing damage to structures and facilities (subsidence damage; cf. section 2.3.3 for protective measures);

—　　destruction of hanging parts of the deposit (most likely as a result of inade-quate extraction planning).

2.1.3　Disruption of groundwater flow

The opening up of underground workings **modifies** the formerly **stable water bal-ance** of the rock structure by creating new water conduits. Water drainage, for ex-ample, can cause significant **recession of the groundwater level** with substantial attendant detriment to vegetation within the affected area (cf. section 2.3.2).

2.1.4 Alteration of groundwater quality

Mining activities can pollute groundwater in several ways: **mine waters** (cf. item 2.2.4), for example, can enter the groundwater system, and various **alkaline and other solutions** used in in-situ dressing processes, as well as **leakage losses of refrigerants** used in the sinking of shafts, all can contaminate the groundwater, just as the **leaching of dumps** produces percolating water that can alter the character of groundwater. Effective preventive measures include the **sealing off of soils**, shafts and worked-out parts of the deposit, **drainage** and/or **canalization**.

2.2 Underground environmental impacts

Man, machine, rock and climate all interact underground, whereas man is impacted most significantly. Matters concerning the health and **safety of miners** are therefore given **priority consideration**.

2.2.1 Air / climate

The underground climate is influenced by the elevated **temperature of deep rock** and by the **gases and liquids** it contains.

37

Table 1

Factors influencing the atmosphere in underground mines

Potential hazard / Reference values	caused by ...	danger of ...	Preventive measures
Oxygen deficiency (O_2) --------- 19 % min.	displacement by irrespirable (black) damps and fire-damps, respiration, open mining lamps, mine fires	fatigue, asphyxia	ventilation
Radiation	radioactive rock components, measuring probes	radiation affection	limited exposure time with dosimetric control
Radon	gas evolution from surrounding rock	radiation affection	ventilation, limited exposure time
Methane (CH_4) --------- 5 - 14 % = explosive	gas evolution from coal	explosion	gas extraction, ventilation, flameproof equipment
Coal dust	mining, handling of coal	explosion	dust precipitation, flame-proofing
Carbon monoxide (CO) --------- > 50 ppm	exhaust, gas evolution in abandoned hard-coal mines	poisoning	ventilation
Carbon dioxide (CO_2) --------- > 1 %	gas eruption in salt, exhaust, gas evolution from thermal waters	asphyxia	ventilation
Hydrogen sulfide (H_2S) --------- > 20 ppm	gas evolution from mine and thermal waters	poisoning	ventilation
Oxydes of nitrogen (NO_x) and blast damp	blasting	poisoning	ventilation, specification of blasting times
Exhaust gases	engine exhaust	poisoning	ventilation
Low-temperature carbonization gases, smoke	mine fires	poisoning	extinguishment, damming off, precautionary measures
Aerosols of oil	pneumatic equipment	poisoning	oil precipitation
Heat	elevated rock temperatures, off-heat from engines	fatigue	ventilation, air cooling

2.2.2 Noise

In underground workings, noise is generated by **drilling and blasting,** by **internal-combustion engines and pneumatic and hydraulic motors,** and by various **means of conveyance** (conveyor belts, trains, vehicles) and **fans.**
Machine-generated noise can be reduced by various **design measures,** and ear protectors are mandatory beginning at certain sound intensity levels.

2.2.3 Dust

Exposure to dust (stone dust in coal mines, for example) must be limited to minimize the **incidence of related diseases,** the most dangerous of which is silicosis resulting from the inhalation of silica particles. Dust forms when rock is destroyed by mechanical means (drilling, blasting, crushing, handling, etc.).
Dust consisting of the following mineral substances **poses a hazard to human health**: asbestos, beryllium, fluorspar, nickel ores, quartz, mercury, cinnabar, titanium dioxide, manganese oxide, uranium compounds and tin ores. Pulverized asbestos and respirable dust containing nickel ore and/or beryllium, as well as soot from diesel engines, are **carcinogenic.** Coal dust can cause **dust explosions.**
Countermeasures against dust pollution include its **consolidation** during drilling and conveying, either by spraying it with water or by saturating the face through appropriately arranged boreholes prior to extraction. **Gas masks** prevent the inhalation of dust, and **filters** on engines bond soot particles.

37

2.2.4 Mine waters

Mining activities **alter the characteristics** of mine waters.
Appropriate safety clothing protects miners against aggressive mine waters, and appropriately resistant materials prevent corrosion of material goods.

Table 2
Pollution of mine and surface waters

Type of pollution	Typical polluting substances	Preventive Measures
Altered pH		neutralization
Soluble inorganic substances	heavy metals, salts, sulfur	precipitation
Insoluble inorganic suspended solids	mud	agglomeration and settling
Organic substances	oil, grease, lubricants, emulsifying agents	precipitation in settling tanks
Heat		cooling, mixing

2.3   Aboveground environmental impacts

The aboveground environmental consequences derive from communication between the mine and the surface in the form of **ventilation, mine pumping and conveyance of the product**, in combination with establishment of the requisite aboveground mining **infrastructure**. **Vibrations caused by blasting and ground movement** are also perceptible aboveground.

2.3.1 Air / climate

The harmful effects of air pollution, particularly on nearby vegetation can be alleviated by **filtering the outgoing** air from the shafts and tunnel faces. Dumping and wind-induced erosion of dumps can cause substantial air pollution, most notably in the form of **dust**.
Dust evolution can be controlled by appropriate **sprinkling in connection with dumping** and by immediate **greenbelting**, oversowing and **protective dams**. In arid regions where land planting is hardly possible, preventive measures must be taken in the form of **restricted use in the prevailing wind direction**.
Coal mining releases large quantities of **methane** (CH_4), one of the most notorious **"greenhouse gases"**. The best way to control methane is to **"drill and extract"** (with subsequent utilization). Particulate solids in the vitiated air from underground mines can be extensively eliminated by filtration.

2.3.2 Water

The **pH** of mine waters, particularly in the presence of sulfidic ores, can range below 5.5 (**acidic**). Adherence to the **limits prescribed** for sulfates, chlorides and metals is essential.

If the groundwater is being used as **drinking water** and ore is being discharged into a body of surface water, the relevant values must be **monitored**. It is important to know which anions and cations can occur in mine water and which of them constitute **potential hazards** on the basis of their concentration or toxicity.

It is also important to mention that heaps of material extracted from an underground mine are liable to contain **high concentrations of chlorides and sulfates** and that, in a humid climate, such salts can be leached out by precipitation.

Whenever **minewater** is **discharged into a body of surface water**, care must be taken to avoid damaging any sensitive ecosystems and to ensure that no long-term accumulation of pollutants occurs in the sediment and that overall use of the water in question, e.g., for fishing purposes, is not impaired.

Marine pollution and alteration of the ocean floor or fishing/spawning grounds can result from the **conveyance of polluted water** through rivers leading to the coast.

Finally, underground mining **consumes water** for such activities as drilling, gobbing/stowing, hydro-mining, etc.

The **measures** described in section 2.2.4 (table 2) should be adopted to prevent pollution of surface and groundwater by mine waters.

2.3.3 Subsidence

For the day surface, the most frequent danger resulting from underground mining activities is **subsidence**, or settling. Subsidence-induced tilt, curvature, thrust, stretch and compression of the day surface can cause **damage to buildings and infrastructural facilities** as well as to the **natural environment**. Watercourses such as canals and rivers - and **rice paddies**, for example - react very sensitively to the slightest change in ground inclination.

Protective measures begin with **early regional planning** with due consideration of the potential mining-induced consequences of ground subsidence.

Settling can also be avoided or at least reduced **by** properly **lining the mine with support material** and backfilling the face workings with rejects and/or the use of certain suitable extraction techniques. Well-planned and controlled extraction allows slow areal settling that is unlikely to damage buildings or public utility lines and facilities.

2.3.4 Dumps, land consumption, landscape

Underground mining activities are usually accompanied by the appearance of large **rubbish heaps** within the immediate vicinity of the mine, where rejects and other useless material are dumped. The **residual metal contents** of such material should be ascertained, even though the metal burdens emanating from dressing heaps can be expected to be higher. Frequently, rubbish dumps are **difficult to recultivate**, and appropriate measures therefore should be included in the working plans.

Underground mines require a certain extent of surface area for the requisite **infrastructure** (hoists, buildings, workshops, storage areas, power generating equipment, access road, etc.). The aboveground facilities can impair the appearance of the **landscape**, and relevant architectural measures have limited effects. The establishment of any such industrial complex is bound to alter the landscape in the vicinity of the mining facilities. To the extent that **resettlement** is necessary, the affected parties must receive appropriate compensation.

Lowering the groundwater level can have detrimental effects on the local vegetation, including the **drying out of ponds, streams, etc.** Moreover, the local fauna and human population can be adversely affected by a **diminishing supply of drinking water** as a result of the altered water regimen.

Adequate protection of wetlands against such negative impacts may require the **artificial recharge of groundwater**, particularly since receding groundwater tends to cause settling, with damage to structures as one likely result.

Finally, **vibrations** caused by blasting and ground movement are also perceptible aboveground.

2.4   Other consequences of underground mining

Establishing mining operations in remote areas can have the inadvertent effect of **opening the area** up to uncontrolled settlement and land use. Appropriate **planning-stage backup measures** are therefore called for.

The intensive use of wood for timbering mines can trigger the **large-scale felling of trees** and, hence, **erosion** of the exposed soil. **Orderly silvicultural activities** in the area around the mine can help prevent such problems, especially if fast-growing species of trees are planted. Nonetheless, long-term effects on the ecosystem remain unavoidable. The use of **anchoring techniques and steel supports** in underground mines can extensively reduce wood consumption.

The world over, underground mining provides **employment almost exclusively for men**, because cultural and traditional conceptions forbid women to work underground. If at all, jobs for women are to be found in the areas of mineral processing, marketing and attendant services. Children should never be allowed to work in underground mines. Other **social problems** can arise in connection with mining if the housing for the miners and their families is either inadequate or not accompanied by the appropriate infrastructure (water, markets, schools, etc.) and if the miners are not covered by social insurance.

3. Notes on the Analysis and Evaluation of Environmental Impacts

3.1 Air / climate

The **gas contents** of air in underground mines is regulated in Germany by pertinent laws such as the **mining ordinances** *(Bergbauverordnung)* BVOSt and BVOE of the North Rhine-Westphalian mining inspectorate *(Landesoberbergamt LOBA)* and its pertinent and specific directives.

For **methane** (CH_4), the following limits apply to free airflow:

more than 0.3 %	:	tram shutdown
more than 0.5 %	:	recorded monitoring
more than 1.0 %	:	electrical equipment shutdown
more than 2.0 %	:	monitoring equipment shutdown

Gas extraction equipment is subject to measures in accordance with the relevant gas extraction directives.

Carbon monoxide (CO) in concentrations of 50 ppm and higher calls for special rescue, recovery and security measures according to a life-saving plan (Hauptstelle für das Grubenrettungswesen der Bergbau-Forschung GmbH, 1982).

Mines must be evacuated if the **carbon dioxide** (CO_2) level reaches 1.0 % or higher.

Nitrous gas levels of 300 ppm NO_x, including 30 ppm NO_2, allow a maximum exposure time of 5 minutes. A level of 100 ppm NO_x (including not more than 10 ppm NO_2) extends the maximum exposure time to 15 minutes per shift.

37

The **oxygen content** must amount to at least 19 %.

The **hydrogen sulfide** (H_2S) concentration must not exceed 20 ppm.

All **gas measurements** must be performed using **calibrated** commercial-type **instruments**.

The **airflow velocity** should amount to at least 0.1 m/s in large spaces and at least 1.0 m/s in fast-line sections. The air velocity in levels used for travel (tram levels) should not exceed 6.0 m/s.

Minimum **air volumes** amount to 6 m^3/min per person, plus 3 - 6 m^3/min per diesel horsepower for CO levels ranging from 0.06 % to 0.12 %.

Airflow velocities are **measured** with anemometers, and the airflow volumes are **calculated** by multiplying the velocity by the cross-sectional area.

The regulations governing gas contents, air volumes and airflow velocities differ from country to country (hard-coal mines in India, mines in Chile, the People's Republic of China, etc.).

3.2   Noise

Underground noise limits can be drawn up along the lines of rules issued by the **North Rhine-Westphalian Mines Inspectorate (LOBA)** in Dortmund.

The **sound intensity level** of noise generated by **drills** should not exceed 106 dB (A) at a distance of 1 m (LOBA *Rundverfügung*).

Transgression of a certain reference intensity calls for the use of **ear protectors**. The 1988 EC directive on noise in mining came into force in Germany in 1992. **Noise measuring specifications** have been developed by the Westphalian miners' union fund *Westfälische Berggewerkschaftskasse* in Bochum, and the appropriate measuring instruments are commercially available.

3.3   Dust

In the Federal Republic of Germany, the German Research Foundation *(DFG - Deutsche Forschungsgemeinschaft)* publishes yearly dust emission limits/standards in the form of **occupational exposure limits (*MAK-Werte*), technical exposure limits (TRK)** and **biological tolerance values for working materials (BAT)**. To the extent that the limit values in question are directly **relevant to human health**, the above or comparable guidelines, e.g., from the World Bank or other international organizations, should be adhered to.

The most important occupational exposure limit, or MAK-value, is that pertaining to **fine silica dust**, which amounts to 0.15 mg/m^3. The corresponding value for **siliceous fine dust** is 4 mg/m^3. In hard-coal mining, the limits for fine silica and siliceous dust presently (as of this writing) amount to 0.60 mg/m^3 and 12 mg/m^3, respectively, and were scheduled for reduction in 1992. Fine dust is referred to as siliceous if it contains more than 1 % quartz.

The maximum personal **dust exposure**, measured in mg/m^3 x number of shifts worked in five years, shall not exceed 2500. All **underground work** is classified according to different **dust-exposure categories**.

Workers suffering from incipient pneumonoconiosis (or anthracosis) may not be exposed to more than 1500 (mg/m^3 x number of shifts worked) in the span of five years. In North Rhine-Westphalia, the German *land* with the largest number of mines, the mining ordinance for hard-coal mines *Bergbauverordnung für Steinkohlebergwerke*, section 44 - 48, version dating from February 19, 1979) governs the measurements and interpretation.

Table 3
Miscellaneous dust limits (MAK-values) with mining relevance

	Fibers/m³	mg/m³
Asbestos, crocidolite	0.5 x 10^6*	0.025*
All other types of asbes-	1 x 10^6*	0.05*
tos-laden fine dust	--	2.0*
Beryllium	carcinogenic	
Iron-oxide powder	--	6
Fluorspar	--	2.5
Nickel-ore dust (sulfid.)	carcinogenic	
Mercury		0.1
Cinnabar		0.01
Titanium dioxide		6
Manganese oxide		1
Uranium compounds		0.25

Determined by means of atomic absorption analysis and X-ray fluorescence analysis. Application to projects in developing countries in accommodation of local measuring techniques and analytical methods (cf. references) is recommended.

* technical exposure limit (TRK)

3.4   Water

The discharge of industrial process water and mining effluent is strictly regulated in Europe. The EC Council Directive 80/778 relating to the quality of water intended for human consumption, dated July 16, 1975, supplemented July 15, 1980, lists **three water categories** requiring less extensive (category A1) or more extensive (categories A2 and A3) treatment. The **guideline values** (G) and **imperative values** (I) for the third category are listed in the following table along with the threshold values (TV) and limit values (LV) stipulated by the North-Rhine Westphalian State Agency for Water and Waste (*Landesamt für Wasser und Abfall Nordrhein-Westfalen*) in the draft ordinance on potable water *Trinkwasserverordnung* (TVO) dated July 26, 1994, selected on the basis of relevance to deep-mine waters.

Table 4
Potable water obtainment guidelines

Element	EC-values		NRW (North-Rhine/ Westphalia)		Ele- ment	EC-values		NRW	
[g/l]	G	I	TV	LV	[mg/l]	G	I	TV	LV
Fe	-	0.2	-	0.2	Cr	-	0.05	0.03	0.05
Mn	-	0.1	-	0.1	Pb	-	0.05	0.01	0.04
Cu	1	-	0.03	-	Se	-	0.01	-	-
Zn	1	-	0.1	2.0	Hg	0.0005	0.001	-	-
B	1	-	-	-	Ba	-	1	-	-
Mg	-	-	25	50	NO_3	25	50	5	11
Na	-	-	50	150	SO_4	150	250	120	240
K	-	-	5	12	Cl	200	-	25	-
Ni	-	0.05	0.03	0.05	F	0.7/1.7	1	-	-
As	-	0.1	0.006	0.04					
Cd	-	5	2	5	pH	5.5-9		6.5-8	

3.5   Soil

Oversown dumps are rarely used for agricultural purposes. In the event that such a use is envisioned, the applicable heavy-metal tolerance values for soils are to be found in the guidelines and directives issued by the Darmstadt-based *Verband Deutscher Landwirtschaftlicher Untersuchungs- und Forschungsanstalten* (German association of agricultural research and analysis stations) and by the Biologische *Bundesanstalt für Land- und Forstwirtschaft* (Federal Biological Research Centre for Agriculture and Forestry) in Berlin. It is generally necessary to determine the

constituents of the dump and any **leaching behavior** that could impose limits on the available soil utilization options.

4. Interaction with Other Sectors

With regard to environmental consequences, underground mining is closely linked to a number of other sectors, including in particular:

- **prospection and exploration** of deposits in preparation for the actual underground extraction activities;

- **processing** of the raw materials to obtain marketable products, with such processing normally taking place in centralized plants situated directly at or near the mine;

- **conversion into electricity** in thermal power stations, many of which are located in the near vicinity of brown-coal mining operations;

- building construction and civil engineering as sectors pertinent to establishment of the requisite mining **infrastructure** and means of transportation to the market. (Mines tend to be found in isolated locations, accordingly intensive construction activities are required.);

- **waste disposal**, e.g., for thickener sludge, hydraulic oil, spent oil and the like, and problems concerning ultimate disposal;

- **water management**, since natural water is quantitatively and qualitatively altered by the discharge of mine water into surface waters or groundwater as well as by the extraction of water for use as process water;

- **forestry** as a bulk provider of timbering wood;

- and, finally, **regional development**, which consistently derives strong impetus from mining activities.

37

5.   Summary Assessment of Environmental Relevance

In sum, underground mining can be referred to as an activity with **substantial impact on the environment**. The consequences can be very detrimental to the environment, especially through the extraction of resources, alteration of the rock structure and groundwater regimen, pollution of the air, the effects of noise and dust, pollution of surface water and alteration and disruption of the landscape. **Compared to surface mining, underground mining has modest surface area requirements**, both for the winning of raw materials and for other industries. With the exception of leftover rubbish dumps, the area in question is **only needed for as long as the deep mine remains in operation**.

Among the most significant environmental effects of underground mining is its **impact on the miners themselves**, whose **health and safety** are quickly and seriously jeopardized, if the protective rules, regulations and measures are not systematically adhered to.

Finally, underground mining has **social consequences**, especially in connection with speculative forms of mining, e.g., for precious metals or gems.

Many environmental consequences can be moderated but not prevented. Extensive data is needed as a basis for assessing the environmental impacts and designing protective measures; the uncertainty levels are accordingly high. Even the preparatory activities (reconnaissance, prospection and exploration) necessitate good coordination between the relevant environmental impact assessments and their data requirements.

The stipulation, enforcement, monitoring and control of **limit values** and underground mining operations has, to a certain extent, evolved to exemplary levels. **Direct application** of limit-value enforcement and monitoring to other countries is only conditionally possible, since the **basic prerequisites** usually differ. Nevertheless, every attempt should be made to apply and meet standards designed to preclude detrimental effects on man and the environment. Probably the **biggest problem** from an environmental standpoint are the uncounted "informal" **small-scale mining** activities employing uncontrolled, inadequate, unsafe methods that also tend to be hazardous to the environment.

Proper and orderly mining operations require **stringent supervision** (routine measurements, data collection and monitored adherence to essential limit values). That, in turn, calls for **competent executing agencies**.

6. References

General Literature

Arndt, P., Luttig, G.W.: Mineral resources, extraction, environmental protection and land-use planning in the industrial and developing countries. Stuttgart 1987.

Bender, F. (Ed.), 1984: Geologie der Kohlenwasserstoffe, Hydrogeologie, Ingenieurgeologie, Angewandte Geowissenschaften in Raumplanung und Umweltschutz. - In: Angewandte Geowissenschaften III: 674 pages; Stuttgart (Enke).

Bundesberggesetz (B Berg G) - 2. Auflage, Glückauf-Verlag, Essen 1989.

Deutsche Forschungsgemeinschaft: Maximale Arbeitsplatz-Konzentrationen und Biologische Arbeitsstofftoleranzwerte, Weinheim 1990.

Down, C. G.; Stocks, J.: Environmental Impact of Mining. Applied Science Publishers Ltd., London 1977.

EEC 85/337: Council Directive of 27 June 1985 on the assessment of the effects of certain public and private projects on the environment - Off. J. no. L175, 05/07/85, p. 0040.

Environmental impact of iron ore mining and control. Jain N.C.J. Mines Metals Fuels, vol. 29, no. 7/8, July/Aug. 1981.

Environmental monitoring and control. Wld. Min. Equip., vol. 10, no. 5, May 1986.

Franke, H., Guntermann, J. und Paersch, M.: Kohle und Umwelt, Glückauf-Verlag, Essen, 1989.

Inter-American Development Bank - Environmental Checklist für Mining Projects.

Johnson, M.S., Mortimer A.M., comps.: Environmental aspects of metalliferous mining. A select bibliography. Letchworth, Herts.: Technical Communications, 1987.

Jones, S.G.: Environmental aspects of mining developments in Papua New Guinea. Prepr. Soc. Min. Engrs. AIME, no. 88 - 155, 1988.

37

Kelly, M.; assisted by Allison, W.J., Garman, A.R., Symon, C.J.: Mining and the freshwater environment. (Elsevier Applied Science)

Klima-Bergverordnung (Klima Berg V), Glückauf-Verlag, Essen 1983.

Lambert, C.M., comp.: Environmental impact assessment, a select list of references based on the DOE/DTp. London, Department of the Environment and Department of Transport Library, 1981.

Rawert, H.: Die Erschließung neuer Abbrauräume als landes- und regionalplanerisches Problem - das Beispiel Haard. In: Markscheidewesen 86 (1979), Nr. 2, p. 31 - 41.

Schmidt, G.: Umweltverträglichkeitsprüfung bei Projekten des Bergbaus. Glückauf 125 (1989) Nr. 5/6.

Sengupta, M.: Mine Environmental Engineering, Volume I and II, CRC Press, Inc., Boca Raton, Florida, 1990.

Servicio Nacional de Geologia y mineria - Chile: Reglamento de Seguridad Minera. Decreto Supremo No. 72 of October 21, 1985, Ministerio de Mineria, 1988.

Solving environmental problems. World Min. Equip., vol. 9, no. 6, June 1985.

Stein, V.: Bergbau und Umwelt, Erzmetall 37, 1984 Nr. 1, p. 9 - 14.

United Nations Department of Technical Cooperation for Development (UNDTCD): Proceedings International Round Table for Mining and Environment, DSE Berlin, 1991.

World Health Organisation: Environmental pollution control in relation to development, report of a WHO Expert Committee. (World Health Organisation technical report series, no. 178). Geneva 1985.

<u>Specialized Literature</u>

<u>Methane</u>

Landesoberbergamt Dortmund: Rundverfügung 33-111.15/7455/64-17.2.65; Bergbauverordnung Steinkohle (BVOSt) § 158, § 150; Bergbauverordnung Erz-
   bergwerke (BVOE), § 97; Sonderbewetterungsrichtlinien; Gebirgs-
   schlagrichtlinien; Gasausbruchrichtlinien; Gasabsaugrichtlinien

<u>Carbon Monoxide</u>

Landesoberbergamt Dortmund: BVOSt § 150.

Plan für Grubenrettungswesen, Hauptstelle für das Grubenrettungswesen der
   Bergbau-Forschung GmbH, Essen, 1982.

<u>Carbon Dioxide</u>

Landesoberbergamt Dortmund: BVOSt, § 150.

<u>Hydrogen Sulfide</u>

Landesoberbergamt Dortmund: BVOSt, § 150.

<u>Oxides of Nitrogen</u>

Landesoberbergamt Dortmund: Sprengschadenrichtlinie.

<u>Air Velocity</u>

Landesoberbergamt Dortmund: BVOE, § 19; BVOSt, § 151; Sonderbewetterungsrichtlinien.

<u>Airflow</u>

Landesoberbergamt Dortmund: BVOSt, § 150.

<u>Temperatures</u>

Landesoberbergamt Dortmund: Klima-Bergverordnung, § 3.

37

Noise

Landesoberbergamt Dortmund: Maßnahmen für den Lärmschutz Kleinkaliber-
    Bohrgeräte (Bohrhammer, Drehbohrmaschinen), Rundverfügung 12.21.11-
    4-7 (SB1.A 2.4).
Westfälische Berggewerkschaftskassen, Bochum: Geräuschmeßvorschriften DIN
    45, 365; 52 Gruben-Diesellokomotiven; 53 Dieselkatzen; 54 Gruben-Gleis-
    los-Fahrzeuge; 55 Rangierkatzen.

Dust

Landesoberbergamt Dortmund: BVOSt, § 44 bis 48, mit Plan für die Staubmessun-
    gen an ortsfesten Meßstellen zur Feststellung und zur gravimetrischen Beur-
    teilung der Feinstaubbelastung, MAK und BAT.

Water

Landesamt für Wasser und Abfall Nordrhein-Westfalen: Grundwasserbericht
    84/85, Düsseldorf 10/85.
EEC 75/448: Council Directive of 16 June 1975 concerning the quality required of
    surface water intended for the abstraction of drinking water in the Member
    States - Off J. no. L194, 25/07/75, p. 0026
EEC 80/778: Council Directive of 15 July 1980 relating to the quality of water in-
    tended for human consumption - Off. J. no. 2229, 30/08/80, p. 0011.

Dumps, Soil

Kloke, A.: Orientierungsdaten für tolerierbare Gesamtgehalte einiger Elemente in
    Kulturböden, Mitteilungen des Verbandes deutscher landwirtschaftlicher
    Untersuchungs- und Forschungsanstalten, Heft 1 - 3. Januar, Juni 1980.
Kloke, A.: Die Bedeutung von Richt- und Grenzwerten für Schwermetalle in Böden
    und Pflanzen, Mitteilungen der Biologischen Bundesanstalt für Land- und
    Forstwirtschaft, Berlin-Dahlem, Heft 223, Oktober 1984.
Stein, V.: Anleitung zur Rekultivierung von Steinbrüchen und Gruben der Stein
    und Erden Industrie, Köln, Deutscher Institutsverlag, 1985.
Über die Schwermetallbelastung von Böden, Pflanzenschutzamt Berlin, 1985.

Clarifying Ponds

Davis, R.D.; Hucker, G.; L'Hermite, P.: Environmental Effects of Organic and
    Inorganic Contaminants in Sewage Sludge, Commission of the European
    Communities, 25./26.05.1982, Reidel D. Publishing Company Dordrecht,
    Boston, London.

Mining and Energy

Minerals - Handling and Processing

38. Minerals-Handling and Processing

Contents

1.   Scope

Processing constitutes the technological **link** between the **extraction**, or mining, **of raw minerals and their conversion** into industrially useful working materials. The techniques applied are designed to separate the valuable from the barren material while upgrading, or concentrating, the former. The **large variety of raw materials** and the many different types of deposits in which they are found naturally necessitate an accordingly **broad array of processing routes**, from the simple classification and washing of sand and gravel to the more elaborate methods of processing hard coal, and on to the material beneficiation of disseminated metal ores. Ores processing (dressing) does not, however, include the various stages of metallurgical processing described in the brief dealing with the production of nonferrous metals.

In many cases, the environmental relevance of a given stage of processing increases in relation to its scope and/or degree of difficulty. The present brief therefore focuses on the **environmental aspects** of ore processing facilities as the source of most damage potentials.

It must be noted in that connection that **no account is made of special cases** such as uranium ore processing, which is already subject to special statutory regulations around the world. Likewise, no processes are dealt with that serve in the **reclamation or reprocessing** of spent merchandise such as worn-out batteries, scrap glass, etc.

2.   Environmental Impacts and Protective Measures

2.1   Handling

The **loading and unloading** of trucks and railroad cars **can generate large amounts of dust**. During transportation, fine dust is lost to relative (head) wind, while trucks emit pollutant-laden exhaust gases, and both trucks and trains are noisy. Transportation by truck or rail entails the **consumption of land area** for roads and railways. The construction and use of traffic routes can have **detrimental effects** on nature and residential quality; cf. briefs dealing with transport and traffic planning, provision and rehabilitation of housing, and road traffic.

In the interest of environmental protection, the **mineral processing plant** should be located either **directly on or in the immediate vicinity of the mine premises**. That way, the ore can be moved from the mine to the processing facility by conveyor belt instead of by truck or rail. If transportation by truck is unavoidable, the **haul roads** should be provided with a course of bituminous road-building material or concrete and **kept clean** at all times. A wheel-washing stand and/or routine washing of the vehicles helps **reduce dust emissions**. Low-emission, noise-abated trucks are designed to help reduce overall emissions of carbon monoxide, hydrocarbons, oxides of nitrogen, soot and noise. **Other in-transit protective measures** include moistening the load with water, tarping it over, or using closed containers. **Dust extraction and control devices are required for loading and unloading operations,** i.e., on loading equipment such as downcomers, and on unloading equipment such as dumping chutes. When filling closed containers with dust-generating products, the displaced air must be dedusted. The **required degree of dust extraction depends on the hazardousness** of the dust in question. Cyclone separators and fabric filters are inherently suitable.

Conveyor belts should be encapsulated as a pollution-prevention measure (not for maintenance purposes), i.e., as a means of restricting dust and noise emissions. The conveyor drives at the corners (diversion points) emit sound intensity levels reaching as high as 120 dB (A). Any **sound insulation** employed should be harmonized with that used for other noise sources within the processing plant. The use of noise locks on bunkers also helps reduce noise emissions, since the size of the opening is decisive for the amount of sound radiated during unloading.

2.2   Crushing, screening, milling, classifying

The rock material is preferably **rough crushed** in jaw crushers and subsequently screened, with the oversize being returned for recrushing. The normal fractions are collected in a surge bin. A conveyor transfers the material from there to the fine crusher. **Classification** to standard sizes involves continuous feedback of the oversize and interim storage of the standard-size fractions. Additional classification and particle-size reduction can be effected in rod or ball mills, with separation of the desired size fractions and raw materials.

All of the above processing steps involve **dust and noise emissions that can emburden both the workplace and the environment.**

There are **no generally applicable values** for the dust quantities encountered, because they depend on the crystalline structure of the minerals and of their geological association, requisite extent of crushing and various engineering factors. However, in view of now-common ore throughputs of up to 50 000 t/d, even minimal proportional dust emissions can put pressure on the soil and vegetation around ore processing facilities. In particular, the attendant **deposition of heavy metals can jeopardize human health** by way of the food chain, and the presence of fibrogenic dust at the workplace can cause silicosis or asbestosis.

In order to minimize dust pollution, the machinery should be **encapsulated**. Wherever that would be unfeasible for technical reasons, the dust-laden exhaust air should be collected and put through a **dust precipitator**. The type of filter to be used depends on the composition and particle-size distribution of the dust. Generally, cyclone filters are used for coarse filtering, while fabric filters serve to remove **fine dust particles**. Such equipment can achieve residual dust contents (clean gas dust loads) of less than 10 mg/m^3. Equipment operators at dusty workstations must be required to wear **dust masks** (particle respirators). Masks designed for use in very warm climates should have appropriately large filtering surface areas.

In the interest of noise control, such facilities must have enclosures with a minimal number of openings. Since processing plants operate around the clock, suitable noise control measures in the form of **safety distances**, embankments, shielding walls and the like must be planned in at an early stage to preclude excessive prejudice to adjacent residential areas.

The only real options for **limiting the workplace noise nuisance** is to automate and install **control centers**. The operators of noisy equipment generating high acoustic intensities must be provided with **ear protectors** and made aware of their importance for preventing noise-induced deafness.

2.3   Separation, flotation

Ore processing facilities **use water for separating** buoyant and nonbuoyant, i.e. floating and nonfloating, materials: in cyclones and screen **classifiers** for grading by gravimetric separation or for **pulp preparation**, where water serves as a working medium for separating the useless material by gravimetric means and for eliminating suspended solids from the concentrate. The overall **water requirement varies widely,** depending on the type of raw material, the nature of the deposit, and the processes employed.

Dense-medium techniques are used exclusively for the **coarse-size range**, with medium solids consisting of magnetite, lead glance (galena), ferrosilicon and, occasionally, heavy spar (barium sulfate). Between 0.3 and 1 g of sodium hexametaphosphate can be added per liter of pulp to reduce its consistency. The water used in heavy media separation processes should be **recirculated**. Accordingly, the entrained solids have to be separated out in settling tanks, irrigated electrostatic precipitators or hydrocyclones. Even if the water from pulp regeneration is recirculated, the **fresh water requirement** can still amount to 0.5 - 1.5 m^3/ton of crudes.

38

Concentration by flotation is achieved with the aid of flotation agents. **Special chemicals** induce physicochemical surface reactions that are useful for **separating and separately concentrating** mixed and disseminated ores that have been sufficiently comminuted to eliminate most intergrowth between the constituents of interest. Consequently, the solid contents of **flotation slimes** in part occupy the microfine to colloidal size range. Since such slimes **sediment out very slowly**, part of the process water can be recovered more quickly by dewatering the flotation products in thickeners. The still-wet mining wastes (tailings) are then pumped into settling tanks and given ample time - perhaps a week - for extensive sedimentation of solids. The liquid phase can be recaptured as **gravitation water**.

Among the various **flotation agents**, distinction is made between collectors, frothers and modifiers. **Collectors**, or collecting agents, are **surface-active substances** that make the surface of the ore water-repellent. Organic compounds serving as collectors are selectively employed according to the type of ore. In the flotation of sulfide ore, for example, between 10 and 500 g of xanthate is needed per ton of ore, while anywhere from 100 to 1000 g of sulfonates or unsaturated fatty acids are consumed per ton of nonsulfide ores.

Frothers, or frothing agents, which **influence** the size of air bubbles and help **stabilize** the froth in the flotation apparatus, include terpenes, cresols, methyl isobutyl carbinol, and monomethyl esters of various propylene glycols. Consumption levels run between 5 and 50 g/t for flotating crude sulfide ores.

The **modifiers**, or modifying agents, include chemicals for **regulating the pH**: lime, soda and caustic soda for adjusting the alkalinity, and predominantly sulfuric acid for acidification. Passifiers and actifiers, which are used to **intensify the differences between the water-repelling properties** of the ores to be separated, include copper sulfate and zinc sulfate. Alkali cyanides serve in the **selective flotation of sulfide ores**. Cyanides can only be added to an alkaline pulp; otherwise, hydrogen cyanide could evolve and be released to the atmosphere. The amounts required range from 1 to 10 g/t ore. Sodium sulfide, dichromate, water glass and complexing agents also belong to the group of selective flotation agents.

Many flotation agents and other chemical additives constitute **a hazard to water**. Consequently, carefully monitored **dosing apparatus** is required to preclude overdosing, and **special safety requirements** must be met by plant and equipment used for storing, decanting, handling and using such hazardous-to-water flotation agents. The facilities must be designed to safely **preclude contamination of surface water and groundwater** to an extent reflecting both the pollutive potential of the substances in question and the protection requirements of the relevant locations, e.g., potable water protection areas. Impervious, chemical-resistant, drainless collection and **holding vessels** must be provided to the extent necessary for intercepting in a controlled manner any media that may escape as a result of leakage, overfilling or accidents. The retention volume must suffice to hold back the escaped substances until such time as appropriate countermeasures can be brought to bear. Additional safety precautions include **double-walled storage tanks, overflow prevention devices and leakage sensors**.

All requisite measures and precautions for avoiding hazards due to potential water pollutants in the form of flotation agents should be stipulated and communicated via appropriate **handbooks**. Plans pertaining to monitoring, repair and alarm response to malfunctions should also be compiled in handbook form. In addition, **occupational safety measures** must be instituted and monitored in connection with the handling of potentially dangerous flotation agents.

Sensitization and training measures are of essential importance, because the inexpert handling, storage and transportation of working agents are frequent sources of environmental pollution.

Along with the depleted material, small amounts of **flotation agents, leaching chemicals and/or heavy medium** can get into the tailings ponds. The **gravitation water** collecting in the drains should be tested for the presence of flotation agents and chemicals prior to its return to the process water circuit. Most of the agents and chemicals remain in the floated concentrate. When the concentrate is dewatered, the **agents and chemicals are washed out and re-injected** into the fine-grinding cycle. Once the concentrate has been thickened, filtered and dewatered, its **residual moisture content** will amount to roughly 8 %. Thus, the **freshwater requirement** for such processing facilities can amount to about one third of the overall process water consumption rate of about 5 m^3/ton of ore. The **water consumption** of a given concentration plant must be **carefully attuned** to the existing original water budget, i.e., to the available volumes of groundwater and surface waters, in order to avoid both detrimental effects on the environment and problems with the supply of drinking water.

The **process water** should be appropriately treated and **recirculated**. Processes in which the water is discharged into a recipient body on a once-through basis can cause silting and contamination of the receiving water due to high sediment contents and residual chemical additives.

The **disposal** of barren rock and tailings is also problematic in that it **consumes land area**. As the percentage of valuable material diminishes, the throughput quantities increase, and the long-term areal requirement rises proportionately. An ore processing facility with a throughput of approximately 45000 tons/d, for example, requires a **settling basin** measuring some 400 to 500 hectares in area and 300 to 350 million m^3 in volume for 20 years of operation. In some cases, the tailing ponds can be kept somewhat smaller by extracting dried material for use in **refilling** underground mines. Due to the altered material properties, however, this option is only **conditionally appropriate** and would never be able to fully replace tailings ponds and rubbish dumps.

Large settling basins should never be constructed prior to painstaking pertinent investigation including precise specification of the physical and chemical com-

positions of the tailings as well as of the geological and, above all else, the hydrological set-up. The permeability of soil strata, for example, and natural drainage
systems are very important with regard to **groundwater protection**. Since many
tailings ponds stay in service for decades on end, building up all the while, the
relevant accident analysis must consider a possible **dam failure** due to excessive
surface runoff.

Rubbish dumps must be established with due attention to the fact that precipitation
can induce **leaching processes** with attendant **pollution of surface and gravitation
water**. Any mining waste containing large amounts of water-soluble substances or
heavy metals can jeopardize the groundwater, unless the soil under the dump is
sufficiently impermeable. Thus, the essential **protective measures** include an
adequately dense subgrade, minimal sprinkling and the collection of runoff water.
Before the first load of material is dumped, **observation wells** should be sunk for
monitoring the groundwater.

It would be impossible to **preclude all dust generation** in connection with dump
operations, but it can be minimized by keeping the **discharge heights** of dry tailings
as low as possible and by **encapsulating the transfer points. Wind erosion can be
limited** by compacting the surface, sprinkling the pile, applying suitable,
environmentally benign binders to the surface, or planting the windward side of the
heap. The equipment required for dump operation (pumps, dump trucks, conveyor
belts, bulldozers, ...) can be quite **noisy**. Noise control measures in the form of quiet
tools and vehicles, acoustical barriers, etc. are called for whenever sensitive
legitimate residential areas are located nearby.

The **surface and gravitation water** (percolation) from rubbish dumps should be
collected by way of an impermeable peripheral trench and tested before being
released to a recipient body. Moreover, before the water is discharged, its settleable
solids content must have been ascertained as appropriate to the outlet channel's own
sensitivity and intended use. Depending on the material composition of the tailings
in the pond and/or of the rubbish in the dumps, additional testing for the presence of
environmentally relevant pollutants such as heavy metals and processing
chemicals may be necessary. The treatment required for the impounded water may
consist merely of settling in an appropriate basin or, depending on the entrained
substances, of physicochemical processes (precipitation, flocculation, chemical
oxidation, evaporation, ...).

Long-term, if not permanent, **monitoring** of the surface runoff and gravitation water
is called for, because the nature and extent of discharge can change over time due to
weathering (surface disintegration).

In addition to flotation, **leaching and amalgamation** also serve as separation processes. In gold mining, for example, the gold is extracted from the gravity-separated concentrate by making it react with metallic **mercury** to form amalgam. The concentrated residue is then leached with a cyanide solution. Both processes have negative environmental impacts that are very difficult to control. The **mercury** content of the effluent **is particularly problematic, if the wastewater is discharged to the outlet channel without having been treated**. It is still an open question as to whether or not the new ion-exchanger resins will, in the long run, be able to bind enough mercury to meet the residual concentration requirements. **Leaching** involves the use of numerous different chemicals. In gold processing, for example, these include cyanide, lime, lead nitrate, sulfuric acid and zinc sulfate. The processes themselves also **jeopardize the air, water and soil**. All measures and precautions that would apply to the concerns of environmental protection and occupational safety in connection with an **industrial-scale inorganic chemical process must be allowed for** at the planning stage. This would include, for example, capturing the exhaust vapors from the reaction tanks and vessels and installing vapor scrubbing equipment (vapor stacks) to prevent harmful emissions. The aqueous solutions emerging from filter presses should be recirculated, and the waste sludge from suction filters must be tested for disposability and treated as necessary. The **wastewater** from amalgamation and leaching processes requires periodical monitoring.

2.4   Roasting

38

The processing of sulfide ores includes **roasting**. The **roasting gases** contain large amounts of **sulfur dioxide** and therefore require gravitational separation (inertial impaction) and electrostatic precipitation. Further processing of the incidental sulfur dioxide should be obligatory, because release of the unprocessed roasting gases would unavoidably **destroy most of the vegetation** around the roasting plant. It is particularly important that the feed and discharge devices on the roasting furnace be airtight. Fabric filters mounted on the roasted-ore silo can extensively **preclude dust emissions**. To the extent that the blowers give off too much **noise**, their **encapsulation** is recommended. A chlorinating roasting process may involve the formation of polychlorinated **dibenzodioxins** and furans in the exhaust gas, the roasting residue and/or the slag, depending on the operating conditions and on the nature and extent of organic substances. Whenever the formation of any such harmful substance is detected in connection with a chlorinating roasting process, the **operating conditions** must be altered such as to minimize the level of emissions.

2.5   Storage and handling of concentrate; recultivation

If concentrates are stored outdoors and unprotected, wind- and precipitation-induced erosion can pollute the air, the soil and the waters.
The **ground in the storage area should be sealed** to prevent contamination of the
topsoil. Continuous maintenance of adequate surface moisture and/or covering the
ground with mats does not always suffice to prevent all wind erosion. Consequently, the concentrate **storage area should be roofed over and enclosed, and
appropriate measures**, e.g., low dumping heights, **should be taken to minimize
dust generation** during loading and unloading.
The measures to be taken in connection with hauling correspond to those described
in section 2.1.
The extent to which planned heaps and sedimentation facilities would occupy the
former **life space**, i.e., the **habitats**, of local flora and fauna must be ascertained on
a case-by-case basis. The possibility of promptly recultivating slopes should also be
examined as a means of preventing wind- and water-induced erosion while
achieving a certain degree of ecological compensation. The nature and extent of
early **recultivation** must be discussed and coordinated with those responsible for
regional/landscape planning and defined in **a catalogue of measures**. If the area in
question is to be used for agricultural or horticultural purposes, the anthropogenic
pollutive burdens in the stored material and their mobility (pollutant transfer factors) must be accounted for by appropriate measures such as sealing or compacting
of the subsoil to interrupt the paths of emission. Even at the planning stage, information should be gathered on the **availability of cultivable materials** fit for
land restoration.

3.   Notes on the Analysis and Evaluation of Environmental Impacts

The processing, handling and transportation of raw minerals can cause **substantial
environmental pollution by dust evolution**. The most effective available means of
dust collection and precipitation must be applied to dust containing cadmium,
mercury, thallium, arsenic, cobalt, nickel, selenium, tellurium or lead. Quartzose
dust (silica dust) can cause **silicosis** and therefore must be allowed for as an occupational safety consideration. Depending on the mass flow, the material must be
analyzed for the presence of the aforementioned heavy metals, and clean-gas limits
need to be defined, whereas those for cadmium, mercury and thallium should be
lower than those pertaining to the other heavy metals. The workplace dust
concentrations must be monitored as a basis for controlling the silicosis hazard.
Industrial medical care must be provided for the workers.

The **local vegetation is liable to be destroyed by the caustic effects** of mineral constituents dissolved by rain. Also, a thick layer of dust can so strongly impede the plants' natural assimilation process that they die off. The **soil** around processing facilities for ores containing heavy metals can eventually become **contaminated**. The geogenic contents of the soil should be determined prior to erection of any such facility.

Well-proven dust collecting and precipitating devices are available **for** use in **controlling dust emissions**. Their adequate separation efficiency in continuous operation must be monitored. The nature and extent of **inspections, preventive maintenance and repair** of precipitators should be specified in a service manual.

Under certain unfavorable conditions, an accumulation of heat, an overheated bearing or a spark can trigger the **ignition or fulmination of fine dust**. **Good ventilation**, possibly in combination with **inertization, pressure-surge-proof encapsulation** and/or the use of **pneumatic drives,** can substantially reduce the hazard.

Substances constituting a hazard to water in connection with ore dressing processes can lead to **soil and water pollution** due to leakage, carelessness, accidents, etc. Consequently, all facilities required for storing, decanting, handling and using potentially water-polluting substances must be designed and operated such as to avoid contamination of the soil and water. Appropriate precautionary measures also must be taken for the transportation and disposal of the chemicals, and pertinent **occupational safety measures** must be specified for handling them. The potential environmental hazards emanating from the chemicals (cyanides, mercury, etc.) and from the acidic roasting gases involved in **separation and concentration processes** based on leaching, amalgamation and roasting can be particularly severe. Thus, appropriate measures must be taken to hold back the mercury, cleanse the roasting gases, control the leaching process, and otherwise contribute toward the minimization of emissions.

Tailings ponds, settling basins and rubbish dumps for the residues of dressing processes all have substantial **space requirements**. Knowledge of the subsoil structure is important for properly assessing the effects of harmful emissions. With a view to ensuring the long-term protection of groundwater and surface waters, relevant **special studies and analyses** must be conducted at the planning stage. There are as yet no official limit values for acceptable levels of ground contamination by mill slurries from the processing of raw minerals. Consequently, planners of new facilities have to rely on **experiential values** gleaned from settling basins for similar dressing plants. In the case of coal mud heaps, good compaction is required to prevent spontaneous combustion.

To the extent that **farmland** and, hence, income potential must be sacrificed for processing activities, the **consequences** for the affected subpopulation, women in particular, must be **investigated** and suitable alternatives developed as necessary. Early **involvement of the local populace** in the dissemination of information and decision-making processes is an effective means of **avoiding or alleviating conflicts** in advance.

38

The **effluent** from mineral processing activities and the **gravitation water** emerging from tailings ponds and rubbish dumps may contain heavy metals or potentially water-polluting chemicals that pose a hazard to surface water, groundwater and the soil. Special attention must be given to the possible jeopardization of **potable water supplies**. In case of excessive sediment contents, the **river bed** is liable to silt up and **accumulate harmful substances**. The wastewater from ore processing plants therefore has to be **continuously monitored**. Depending on the nature and extent of settleable solids, heavy metals or chemicals posing a hazard to water, the **effluent** will require appropriate **treatment**.

Properly sized equipment **enclosures** with adequate **acoustic insulation** properties are very important for reducing the amount of **noise** emitted by processing plants. Appropriate **safety clearances** should also be planned in for between the plant and neighboring residential areas. Suitable noise control measures also should be applied to the operation of tailings ponds and rubbish dumps located in the near vicinity of residential areas.

Permissible **noise emission levels** are specified in Germany's *TA-Lärm* (Technical Instructions on Noise Abatement). The site surroundings - e.g., an industrial zone, commercial area or residential district - are decisive for the maximum allowable noise intensity level.

As in Germany, ore processing plants should have **immission-control, water-pollution-control and waste-management officers**, whose positions should be independent of the production division. A **safety officer** and an **occupational physician** should be available for matters concerning occupational safety.

4. Interaction with Other Sectors

As a rule, mineral processing plants are attached directly to the relevant mining operations. The environmental **briefs pertaining to mining** therefore apply.

The large area required for a processing plant necessitates its coordination with present and planned regional land use. As such, the environmental briefs Spatial and Regional Planning, Planning of Locations for Trade and Industry should also be consulted.

If the processing plant cannot be installed directly at the mine, appropriate road-building measures become necessary, in which case important details can be found in the briefs Road Building and Maintenance, Building of Rural Roads.

In arid regions, the water neeeded for operating the processing equipment is a very important resource, and its judicious use must be incorporated into Water Framework Planning.

5. Summary Assessment of Environmental Relevance

If the **planned site** of a processing plant is located in a thinly populated area, it must be brought **in line with the goals of regional development planning**. In selecting the site, importance should be attached to choosing a location with a relatively **low level of ecological sensitivity** and which is not crucial to the vitality of the regional natural household.

Most processing plants emit **large amounts of material- and process-generated dust and noise**. Within the **plant premises**, such nuisances/pollution can be reduced to tolerable levels by the use of suitable enclosures and dust retention devices. Dust emissions from **dry heaps** and dumps, however, is more difficult to control, particularly when finely comminuted material is exposed to wind and weather. Such material must be kept moist and/or covered, and the surface should be consolidated or sown over.

Large volumes of low-grade material accumulate at processing plants and have to be pumped into **tailings ponds** for sedimenting. Before any such settling basin is established, its long-term environmental impacts must be **carefully analyzed**, because it could well remain in operation for several decades, becoming larger and larger all the while. The analysis must cover important aspects of protection for the soil and groundwater, stability (e.g., in case of flooding) and subsequent reculti-vation, including definition of appropriate measures.

Prior to establishing and operating **rubbish dumps**, the site-specific hazard potentials for the subsoil, groundwater and surface waters must be carefully investigated. The subgrade must be sealed and a means of collecting all surface runoff and gravitational water provided.

Old tailings ponds and dumps should be given **close-to-natural shapes** prior to their **recultivation** in order to **fit them into the landscape** in a manner appropriate to their planned future use.

Process effluent and gravitational water from processing plants, tailings ponds and rubbish dumps must be put through **wastewater treatment facilities**, the nature and extent of which depend on the sensitivity and manner of utilization of the recipient body. Silting should be avoided, of course, and pollution by mercury and other heavy metals should be minimized. Observation wells should be sunk to allow **monitoring of the groundwater**.

Bulk transportation by road or rail can have negative impacts on the environment: through construction of the required hauling routes (and attendant erosion potential) and in the form of airborne dust and noise. Dust emissions can be avoided by hauling the material in **closed containers**. Quiet, low-emission trucks should be given preference. **Fine-grained material should not be stored in the open air** for any length of time. Otherwise, wind- and precipitation-induced erosion could cause pollution of the soil and water. Frequently, the **cost** of relevant environmental protection measures **is more than offset** by resultant reductions in the loss of resources.

6.   References

<u>Rules and Regulations</u>

Anforderungskatalog für HBV-Anlagen: Anforderungen an Anlagen zum Herstellen, Behandeln und Verwenden wassergefährdender Stoffe (HBV-Anlagen) Ministerialblatt für NRW, Nr. 12, 1991, p. 231 - 234.

Deutsche Forschungsgemeinschaft: Liste maximaler Arbeitsplatzkonzentrationen (MAK-Wert-Liste), 1990, Mitteilung XXVI, Bundesarbeitsblatt 12, 1990, p. 35.

EC Directives: Protection of workers from the risks related to exposure to noise at work, May 12, 1986 - 86/188/EEC, and of June 14, 1989 - 89/392/EEC - on the approximation of the laws of the Member States relating to machinery.

Environmental Protection Agency (EPA): Standard of Performance for Nonmetallic Mineral Processing Plants, EPA 40, Part 425 - 699 (7-1-86 Edition) 60, Subpart 000, Preparation Plants and Coal Preparation Plants, EPA 40, Part 425 - 699 (7-1-86 Edition) 434.23.

Katalog wassergefährdender Stoffe: Lagerung und Transport wassergefährdender Stoffe. LTWS Reihe No. 12, 1991, Umweltbundesamt [German Federal Environmental Agency] Berlin.

Technische Anleitung zum Schutz gegen Lärm: *TA-Lärm*, (Technical Instructions on Noise Abatement) vom 16.07.1968, Beilage BAnz. (supplement to the Federal Gazette) Nr. 137.

Unfallverhütungsvorschriften: Hauptverband der gewerblichen Berufsgenossenschaften, Bonn - u.a. UVV-Lärm, VBG 121 v. 01.01.1990.

VDI Guideline 2560: Personal Noise Protection, December 1983.

VDI Guideline 2058, sheet 1: Assessment of Working Noise in the Vicinity, September 1985.

VDI Guideline 2263, sheets 1 to 3: Dust Fires and Dust Explosions; Hazards, Assessment, Protective Measures, November 1986, November 1989, May 1990.

Erste Allgemeine Verwaltungsvorschrift zum Bundes-Immissionsgesetz, vom 27.02.1986 (Technische Anleitung zur Reinhaltung der Luft - TA-Luft), GMBI. (joint ministerial circular) 1986, Ausgabe A, p. 95.

38

16. Allgemeine Verwaltungsvorschrift: Mindestanforderungen an das Einleiten von Abwasser in Gewässer. Steinkohleaufbereitung und Steinkohlebrikettfabrikation. GMBI. (joint ministerial circular) No. 6, 1982.

27. Allgemeine Verwaltungsvorschrift: Mindestanforderungen an das Einleiten von Abwasser in Gewässer, Erzaufbereitung. GMBI. (joint ministerial circular) No. 8, 1983, p. 145.

Wiedernutzbarmachung von Bergehalden des Steinkohlebergbaus Bellmann Verlag, Dortmund, Verlags-No. 614, 1985.

Zulassung von Bergehalden: Richtlinien für die Zulassung von Bergehalden im Bereich der Bergaufsicht, MBI. NW. (North-Rhine/Westphalia ministerial circular), p. 931, dated July 13, 1984.

Technoscientific Contributions

Alizadeh, A.: Untersuchungen zur Aufbereitung von Golderzen. Aufbereitungstechnik 5, 1987, p. 255 - 265.

Alizadeh, A.: Grundlagenuntersuchung zur mathematischen Beschreibung der Flotation von oxidischen Eisenerzen. Aufbereitungstechnik 2, 1989, p. 82 - 90.

Atmaca, T.; Simonis, W.: Freistrahlflotation von oxidischen und sulfidischen Erzen im Feinstpartikelbereich. Aufbereitungstechnik 2, 1988, p. 88 - 94.

Diesel, A.; Lühr, H.P.: Lagerung und Transport wassergefährdender Stoffe, Erich Schmidt Verlag, 1990.

Kirshenbaum, N.W.; Argall, G.O.: Minerals Transportation, Proceedings of First International Symposium on Transport and Handling of Minerals, Vancouver, 1971.

Sciulli, A.G. et al: Environmental approach to coal refuse disposal, Mining Engineering, 1986.

Ullmanns Enzyklopädie der technischen Chemie, 4. Auflage: Band 2 Verfahrenstechnik I, 1972, Band 6, Umweltschutz und Arbeitssicherheit, 1981, Verlag Chemie, Weinheim.

Williams, R.W.: Waste Production and Disposal in Mining, Milling and Metallurgical Industries, Miller Freeman Publ., San Francisco, 1975.

Mining and Energy

Petroleum and Natural Gas
- Exploration, Production,
Handling, Storage

39

39. Petroleum and Natural Gas - Exploration, Production, Handling, Storage

Contents

1. Scope

In the year 2000, **petroleum and natural gas** together will cover **between 50 % and 70 % of the global energy requirement**, and the **energy-coverage ratio** between the two will be **about 2 : 1 to 1.5 in favor of petroleum**. Obviously, in view of the corresponding scale of petroleum and natural-gas production, **countries with major resources and corresponding development projects in the mining sector** will have continued exposure to **consequential environmental impacts**. Due to the **immobility of deposits** and to the **technical processes** required to **obtain the crude product, petroleum and natural-gas mining activities** have their own specific environmental consequences. According to prevailing international definitions, **a typical petroleum/natural-gas development project comprises the following three phases**:

–   **exploration**, on- and offshore, mainly by **geophysical methods** and **exploration drilling**, including a **test phase following any discovery**;

–   **production**, beginning with **field development drilling** as a precondition for actual production, in the course of which certain phases are run through, up to and including **basic conditioning of the raw material**. The production of crude oil and natural gas requires a certain infrastructure;

–   **handling and storage** directly following the production stage, i.e., prior to the further processing of crude oil and natural gas into energy-market products. These activities utilize part of the overall infrastructure.

39

2. Environmental Impacts and Protective Measures

2.1 Exploration

Exploration is the term used to define the **scientific prospecting and reconnaissance of raw material deposits** by means of

–   mapping/charting
–   geophysics
–   exploratory drilling.

Exploration for petroleum and natural-gas **is based on a large-scale, onshore-specific aerial mosaic** (photomap). In many regions of the world, **superficial analysis** of such maps can suffice to detect **promising areas**. Exploration then continues according to **geophysical and geochemical methods of prospection**. Finally, the **superficial geological, geophysical and geochemical reconnaissance** of promising structures requires **confirmation** by way of **exploratory drilling**, incl. **well shooting**, and the **interpretation of drill cuttings and cores**.

The **environmental consequences of exploration are relatively minor** on the whole, though the attendant **drilling** has a substantially **higher disruptive hazard potential**; **cf. environmental brief Reconnaissance, Prospection and Exploration of Geological Resources**.

2.1.1 Nature and ecology

Modern **airborne mapping techniques** employed at the beginning of the exploration phase pose **no direct threat to the environment**.

Depending on the applied techniques, the **environmental consequences of geophysical prospection can extend over a period of months or years. Distinction** must be drawn **between gravimetric techniques** and predominantly airborne **magnetic measuring methods** on the one hand, and **seismic measuring methods** on the other. The latter put **geophysicists** in a position to **locate geological bedding boundaries** at depths of several thousand meters by registering **reflected compression waves**. Indeed, the **seismic reflection method** is the **most important prospecting tool**, but it is not without **consequences for the environment**.

Even assuming a relatively brief disturbance, **negative environmental impacts must be limited**. Geophysical surveying teams, for example, live more or less self-sufficiently in **remote areas** for various lengths of time. Their **access and transportation routes** should preferably be **by air or water**, depending on the circumstances. **Overland routes** must include any deviations/detours necessary to **avoid ecological disruption**. With regard to **blasting, the magnitude of the explosions** used to generate pressure surges must reflect the **state of the art**. In some cases, **vibroseismics** may constitute a **less disruptive alternative**. Advanced receivers and amplifiers yield extensive information at lower pulse levels. **Offshore blasting has destructive effects on marine life, particularly in the littoral zone**. Alternative use of the **air-pulse technique effectively protects marine flora and fauna**.

On a regional basis, the **most pronounced environmental consequences** for nature and the ecology derive from **deep drilling**. If, however, **state-of-the-art drilling equipment** is (properly) used, the **environmental impact** frequently will remain **far below what laypeople would expect**. The **main objective for drilling operations** is to carefully plan, equip and conduct **terminal exploration projects** such as to either **avoid negative environmental impacts** altogether or at least **reduce them** to a tolerable level.

In connection with the **preparation of drilling sites and the construction of access roads**, due consideration must be given to subsequent **renaturation**, and **disruption of the surface** must be **limited** to the necessary minimum. The **topsoil** must be **protected** as well as possible (in covered heaps, etc.).

Drilling must be conducted such as to **preserve the intactness of the rock strata and water-bearing horizons** in their original, virgin separations through appropriate **casing and cementation programs**.

The **media required for drilling**, in particular the **drilling fluid**, should be chosen with attention to **low environmental impact** and subsequently **recycled** to the greatest possible extent.

Borehole safety, which is understood mainly as uninterrupted **control over the dynamic pressure situation and borehole stability**, must be ensured by **adequately sized casings** and **cementation** in combination with a **blow-out preventor** serving as a drilling-phase closure system (state of the art). **Preventive measures** in the form of **technical equipment and disaster plans** must be taken to **limit the consequences of blow-outs. Such precautionary measures can prevent major environmental damage**, which, while seldom irreversible, can be very **expensive to repair.**

Unavoidable, unrecyclable **mining refuse** like borehole cuttings and spent drilling fluid must be **properly disposed of**. With due regard for the environmental circumstances, preference must be given to **dilution, thermally optimized incineration and/or encapsulation.**

The **slim-hole drilling technique** must be considered as an **alternative to conventional deep-well drilling**. The technique is characterized by a **much-reduced diameter, minimal use of operating media, less technical inputs overall**, and such substantial **time savings** that the **cost of drilling can be cut in half**. Slim hole drilling does, however, **presuppose certain geological conditions** and is inherently unfeasible for deep wells.

The **exploration** drilling phase is not **complete** until all appropriate protective measures have been taken to **counter the adverse environmental consequences** of activities pursued in connection with successful exploration, i.e., the discovery of an exploitable field, which may last several years.

39

Any well that remains **dry** must subsequently be properly **plugged**, and the attendant **aboveground facilities**, including **access routes**, must be either **recultivated or surrendered to some other controlled form of utilization**.

2.1.2 Sociology

Exploration projects can **seriously alter the social fabric** of a country or region. Practically overnight, they **expose native social systems to** the activities and influence of multinational companies employing **modern technical know-how**. **Conflicts of interest** resulting from the **immobility of prospective petroleum and natural gas deposits** must be dealt with appropriately. The **project** must be **integrated into the prevailing social structure** as quickly as possible. That, in turn, requires the **involvement of all social groups**.

2.1.3 Human health and occupational safety

In general, urgent priority must be attached to **occupational safety** and to **preservation of the health of petroleum and natural-gas exploration workers**. The **effects** of such projects on **parties not directly involved** in the exploration work are **insignificant**.

The most obvious **problems** are those deriving from the **difficult, privative work of geophysical surveying crews**, especially in **remote regions**, and which continue through the end of exploratory drilling.

Since **local personnel** can be hired and trained relatively quickly **for some of the work**, appropriate individual **support** must be planned in. **Medical care, hygiene and occupational safety** must be guaranteed, and **acceptance of worker protection measures**, which demand a certain amount of training, must be ensured.

2.2 Production

Successful exploration is followed by the petroleum and/or natural-gas **production phase**, which includes:

— field development drilling, incl. complete preparations for production,
— aboveground installations and processing facilities,
— infrastructural measures.

A **substantial share** of the **petroleum and natural gas resources** that took millions of years to develop has been used up within a relatively **short** span of **time**. In the interest of **long-term utilization**, the human race must exercise a **sense of responsibility** in dealing with those natural resources - which are **only renewable in terms of geological time spans**. In reality, however, traditional oil-producing countries tend to adopt **volume-oriented production strategies**, accepting **substantial environmental consequences** as attendant phenomena. Production strategies in general are heavily influenced by the **demand situation** and as yet **inadequate alternatives**.

The **time between exploration and production** should be used to carefully **analyze the project's anticipated environmental impacts** for the duration of an average production field life (15 to 25 years for an oilfield and 50 to 100 years for gas) - and beyond. The analysis must be based on timely local and individual registration of the **sociological, cultural economic, climatic and ecological situation**, which, of course, differs widely around the world. The **results of analysis** must then be **incorporated into** each and every relevant **resource production project**.

The **beginning of the field development** drilling work should coincide with the establishment of requisite **infrastructure**, e.g., **access routes, incoming and local service lines**, and even the **aboveground pumping facilities, processing plant**, etc. With regard to the attendant environmental consequences, the reader is referred to the corresponding brief Road Building.

2.2.1   Nature and ecology

The long-term **production phase** of a typical petroleum/natural-gas project begins with the **first regular output. Field development drilling prepares a deposit for production** on the basis of the **production geological and field engineering targets** defined to reflect the underground conditions prevailing in the reservoir. The environmental consequences of exploration, as described in section 2.1.1 apply in full.

Especially in **sensitive areas with valuable biotopes**, the equipment used must be chosen with a view to **minimizing space requirements**. Thanks to **advanced drilling technology (directional drilling** with deflecting tools), a single onshore or offshore location now often suffices for tapping several square kilometers of a reservoir. And **horizontal drilling** can help to drastically reduce the overall number of boreholes.

39

Large-scale destruction or alteration of an area's flora and fauna (e.g., in a rain forest, tundra or coral reef) **need not occur** in connection with petroleum/natural-gas projects, which have **relatively modest aboveground space requirements** for technical equipment and infrastructure.

Through state-of-the-art **plant dimensions** in combination with necessarily redundant **automatic monitoring equipment, emissions** occurring under normal and disturbed operating conditions can be held at **low levels.**

Damage to the environment as a result of accidents, oil spills in particular, must be limited by **safety-relevant controls**, e.g., valving. **Oil-contaminated water and soil** must be rehabilitated by **chemicobacterial means of artificially accelerating the biodegradability of hydrocarbons.** A **properly managed oil** well causes no problems with regard to **groundwater protection.**

The **economically efficient exploitation** of natural energy vehicles must attach priority to **controlling the environmental impacts while conserving the resource itself.** For petroleum and natural gas, the conservation of resources covers both the **effective utilization** of their entire energy potential (by **avoiding such activities as the pollutive flaring off of surplus production** that cannot be directly utilized) and the alternative employment of **high-tech production techniques.**

2.2.2   Sociology and economics

The **productive phase of an oil or gas field lasts** on average about **as long as a normal person's working life** - if not even longer, as is frequently the case in gas production. That fact alone imposes a **major social responsibility** on the project. In continuation of the initial exploratory-phase measures, the **living conditions, nutrition, education, health and cultural environment**, including religion, of the personnel must be treated as **importantly** as the purely **technical production facilities. Ghettoization must be countered**, and the **growth of social fabric must be promoted. Industrialization must be conducted cautiously** and in a manner to allow **incorporation of the cultural heritage** of the aboriginal society.

2.2.3   Human health and occupational safety

One of the project executing organization's most important tasks is to **promote health care**, not only among the workers themselves, but also **throughout the entire project region.**

The same applies to **occupational safety**, which can and should be implemented **in imitation of measures applied in industrialized countries. The assignment of well-trained and qualified personnel** to such tasks is a **crucial prerequisite.**

2.3   Handling and storage

Handling and storage is understood here as the last step **following exploration and production. The rough-processed crude products are transported by pipeline, railroad car or road tanker, and by inland waterways and oceangoing vessels,** all of which requires special infrastructure. The products are **stored in aboveground tank farms and underground stores, cavities and pore spaces.**
The transportation/shipping, distribution mechanisms and finished-product storage scopes are not covered by this brief; please refer to the briefs relevant to adjacent sectors, e.g., shipping, ports and harbours, inland ports.

2.3.1   Nature

The requirements stated in section 2.3.1 apply analogously to transportation.
The large-scale **storage** of crude oil and/or natural gas requires **special environmental safety measures,** particularly with regard to the **prevention of fires and explosions. Special importance is attached to leakage detection, alarm sounding and catchment techniques. Underground tanks are preferable** to aboveground tanks, though they do **call for more sophisticated safety engineering.**
As an alternative to tank farms, **underground storage** in depleted mines, rock caverns, salt caverns and pore spaces has the **least extensive environmental consequences. Pore spaces** are only suitable for storing gas, and **salt caverns** demand appropriate utilization or disposal options **(proximity to the ocean)** for the brine. Both alternatives - pore spaces and salt caverns - presuppose the appropriate geological formations.

2.3.2   Human health and occupational safety

The **large-scale handling and storage** of oil and gas poses **hazards** such as the **escape of hydrocarbons** and the possibility of **accidental explosions. Technical transportation monitoring measures and redundant storage safety engineering** substantially **reduce** the **risks** involved. **Pipeline safety** can be ensured via **monitoring stations, self-acting pressure control devices and aerial line inspections. Storage tanks and piping** must be **protected against corrosion.**

39

3.   Notes on the Analysis and Evaluation of Environmental Impacts

The **environmental impacts** must be **evaluated** with due consideration of the individual, **project-specific situation**. Project planning should be conducted with **emphasis** on the potential sociological consequences and the **earliest possible involvement of local nationals.** Environmentally relevant **experience** drawn from comparable projects **must be duly considered.**
The **training of local manpower** for all levels constitutes an **important step** in the direction of **responsible management with a capacity for controlling environmental consequences.** The project must be **implemented in line with the pertinent laws, standards, codes, limit values and technical know-how of industrialized countries.**

4.   Interaction with Other Sectors

As **the profitability of exporting natural gas is limited by the long transportation distances** involved, many countries neglect its production. In that connection, the **technical utilization of liquid natural gas** (LNG) would appear **worthy of promotion**, since the transport problems would be relativized by the use of accordingly large tankers. By reason of its **high efficiency, natural gas** makes a good, nonpolluting **substitute for other primary energy carriers**.

The **petroleum/natural-gas sector has numerous points of contact** with other sectors, among the more important of which are:

-     regional planning
-     overall energy planning
-     water supply
-     planning of locations for trade and industry
-     mechanical engineering, workshops, shipyards
-     oil and fats.

References to adjacent sectors have been included in the appropriate passages of the above text.

5.   Summary Assessment of Environmental Relevance

Global experience shows that the **petroleum and natural gas industry** can maintain an **ecological orientation** with the aid of **modern science and technology.** **Environmental awareness** must be **promoted** by applying the **standards of the most advanced industrialized countries.**

Risks and undesirable environmental consequences must be minimized by responsibly implementing each project in accordance with its own **ecological and sociological significance. Interdisciplinary management** with the **direct involvement of all sections of the population is appropriate and advisable.**

Ecologically oriented operations presume the existence and adequacy of the requisite **control organs.** In that connection, an **environmental protection officer** carrying the responsibility for **training the workers and instilling them with environmental awareness** can be a major asset.

39

6.   References

ASUE: Erdgas als Beitrag zur Milderung des Treibhauseffektes, AG Sparsamer Umweltfreundlicher Energieverbrauch, Frankfurt/Main, 1989.

ASUE: Die Richtung stimmt - Erdgas als Brücke zur idealen Energie, AG Sparsamer Umweltfreundlicher Energieverbrauch, Frankfurt/Main, 1990.

BMFT (German Federal Ministry for Research and Technology): Schriftenreihe Risiko- und Sicherheitsforschung, S. Lange, Ermittlung und Bewertung industrieller Risiken, Berlin, 1984.

BMI (German Federal Ministry of the Interior): Beirat LTwS Lagerung und Transport wassergefährdender Stoffe, diverse publications.

CONCAWE: Methodologies for hazard analysis and risk assessment in the petroleum refining and storage industry, Den Haag, 1982.

CONCAWE: 1989 Annual Report, Brussels, 1990.

Deutsche BP: Das Buch vom Erdöl, Kleins Druck- und Verlagsanstalt, Langerich, 1989.

DGMK: Forschungsbericht zum Umweltschutz, Hamburg, 1974 - 1986.

Deutsche Shell: Neue Aspekte der Öl- und Gasförderung, Deutsche Shell AG, Hamburg, 1989.

Enquete-Kommission, Bundestag: Schutz der Tropenwälder, Economica Verlag, Bonn, 1990.

Enquete-Kommission, Bundestag: Schutz der Erde, Teilband II, Economica Verlag, Bonn, 1991.

Friedensburg/Dorstewitz: Die Bergwirtschaft der Erde, Ferdinand Enke Verlag, Stuttgart, 1976.

Hoffmann, Jürgen P.: Öl - vom ersten bis zum letzten Tropfen, Westermann Verlag, Braunschweig, 1983.

IMO: Inter-Governmental Maritime Organization, Results of International Conference on Tanker Safety and Pollution Prevention; with Regulations and Amendments, London, 1981.

Konzelmann, Gerhard: Öl, Schicksal der Menschheit? Sigloch Service Edition, Künzelsau, 1976.

Mayer, Ferdinand: Weltatlas Erdöl und Erdgas, Westermann Verlag, Braunschweig, 1976.

Müller, Karlhans: Jagd nach Energie, Sonderausgabe, Regel und Meßtechnik GmbH, Kassel, 1981.

OECD: Emission standards for major air pollutants from energy facilities in OECD member countries, Paris 1984.

OTA: Office of Technology Assessment of the Congress of the United States, Technologies and Management Strategies for Hazardous Waste Control, Washington, 1983.

UBA Materialien: Symposium Lagerung und Transport wassergefährdender Stoffe, 2/83.

UBS Texte 32/83: Vorhersagen von Schadstoffausbreitungen auf See - insbesondere nach Ölunfällen.

Ward, Edward: Öl in aller Welt, Orell Füssli Verlag, Zurich, 1960.

World Bank: Environmental guidelines, Washington, 1983.

World Bank: Environmental requirements, Washington, 1984.

Mining and Energy

Coking Plants, Coal-to-gas Plants, Gas Production and Distribution

40. Coking Plants, Coal-to-Gas Plants, Gas Production and Distribution

Contents

1.   Scope

This environmental brief covers various coal upgrading technologies, incl. coking and low-temperature carbonization as processes yielding the target products coke and gas plus tar products and diverse raw chemicals.

The relevant **facilities** can be installed and operated either **separately** or in **conjunction with neighbouring industrial technologies**; cf. environmental briefs Planning of Locations for Trade and Industry, Spatial and Regional Planning. **Interconnected operations** can be characterized by **proximity to either collieries or iron works**.

While in the former case the **working product coal** can be **conveyed directly to the upgrading facility** located only a short distance away, **close proximity to a metallurgical plant avoids** the **necessity of transporting** the **product coke** over long distances and enables direct **supply of the gas to the consumer** by way of a low-pressure network; **blast furnace gas** from the metallurgical plant can serve as **low-sulfur fuel gas** for, say, a coking plant.

If the coal upgrading facility is instead **located separately** and independently, **substantial infrastructural measures** will be required for conveying, loading, unloading and storing the working materials, process materials and products; cf. environmental briefs Transport and Traffic Planning, Railways and Railway Operation, Inland Ports.

In addition, the **generated gas** has to be compressed and **purified to piped-gas quality** before it can be supplied to the consumers.

The **low-temperature-carbonization and coking** processes, as applied to coal in the sense of this brief, are based on **heating** in exclusion of air in appropriate reactors.

Depending on the **temperature** at which the process takes place, distinction is made between:

- low-temperature carbonization (450 - 700°C)
- medium-temperature coking (700 - 900°C)
- high-temperature coking (>900°C).

While the above **processes do not differ at all in principle, the different temperatures** yield **different products** and **process conditions** and call for the use of different reactor systems.

a)   Low-temperature carbonization processes

Low-temperature carbonization processes, as applied primarily to **lignite (brown coal)**, take place in fixed-bed reactors, fluidized-bed reactors or entrained-bed reactors.

The **heat** supply derives from:

—   the use of hot coke as a heat transfer medium or
—   the direct supply of heat to the working material via hot cycle gas (circulation gas).

The **gas** produced by low-temperature carbonization is cooled (condensed), **detarred, compressed** and **purified** prior to delivery. **Residual coke** is **cooled** by means of either wet quenching or cold gas before being **supplied to the users.**
Low-temperature carbonization processes are used primarily for **obtaining tar products, diverse raw chemicals and low-temperature-carbonization gas**. The **incidental coke** is of **no particularly high quality and therefore not used for metallurgical purposes. Consequently, other uses must be found for such coke, preferably uses that do not require high crush resistance.**

b)   Coking processes

Hard coal is coked in batteries of regenerative **horizontal chamber ovens**. Distinction is made between **top charging and stamping operation**, depending on the extent to which the coal is likely to produce high-quality, adequately stable blast-furnace coke.
Coke ovens are heated indirectly **by fuel gas**, the heat of which is transferred to the charge (feed coal) via heating walls. The fuel gas can consist **of partially purified coke oven gas, blast furnace gas** or a **mixture of combustible gases.** Even if the entire operation is heated exclusively with coke oven gas, the plant will still yield **surplus quantities of gas** with calorific values ranging from roughly 16 000 to 20 000 kJ/m³. Following appropriate **purification**, that gas can be **supplied to diverse consumers.**
Oven service equipment is needed **to fill coal into the ovens** (charging cars), **transfer the coke to quenching stations** (quenching cars) and **convey the hot coke** to the wet quenching or dry cooling plant.
Coke oven gas is produced above the charge at **coking temperatures of 750 - 900°C**. Through **riser pipes**, the gas passes into a so-called **collecting main**, where it is **sprayed** with **recirculating water** to **induce cooling and partial condensation**, thus **precipitating** most of its **crude tar content**.

The next stage of treatment consists of additional **cooling** to approximately 25°C, followed by **final tar removal in electrostatic precipitators,** and the primarily absorptive **extraction** of such **constituents** as H_2S, NH_3, HCN, CO_2, benzene and naphthalene.

Those ingredients are **further processed** according to **various techniques** to obtain:

—   ammonium sulfate (after transformation of H_2S into sulfuric acid)
—   Claus sulfur (with simultaneous cracking of ammonia)
—   crude benzene and
—   crude tar.

If **the surplus coke oven gas cannot be injected into an LP (low-pressure) network,** it is put through a **gas compression stage** that **includes additional purification to remove H_2S and raw benzene/naphthalene and to lower its dew point.**
Wastewater deriving from condensation of the gas and from the H_2S/NH_3 scrubbing stage is treated in **multiple stages** including **distillation** in so-called strippers and **dephenolating processes** (extraction, biological elimination).
Modern coking plants can handle between 6 000 and 10 000 tons of feed coal a day for a **coke output** of 4 500 - 7 500 t/d.
The attendant **gas production** amounts to between 80 000 and 150 000 m³/h, and **process wastewater accumulates** at the rate of 80 - 150 m³/h.

c)   Classification of process options

From the standpoint of **environmental protection,** carbonization and coking are roughly equivalent.
With regard to **production capacities** and **technological applications,** however, coking takes priority over carbonization, as evidenced by the fact that **most laws, regulations and guidelines** pertaining to **the control of emissions** refer mainly to coking processes. Nonetheless, carbonizing facilities should be dealt with under the **same premises**.

40

2.   Environmental Impacts and Protective Measures

2.1   Environmental impacts

The **erection and operation of coking and/or coal carbonizing plants** at locations previously not used for industrial purposes alters the landscape and **consumes land** to an extent that depends on the size of plant.

In addition to ascertaining the potential effects of **emissive pollution**, it must also be determined to which extent the requisite **extraction of water** from given resources would **interfere with existing ecosystems**, since **make-up water** is required at various points of both operations. The required **volume of water** can range from 200 to 500 m³/h; cf. environmental briefs Water Framework Planning, Water Supply.

Particularly in connection with **coke oven batteries, emissions** must be anticipated both from **certain operating points** (e.g., exhaust stacks) as well as from **diffuse sources** such as leaky shutoff valves and cracks in the masonry of coke ovens.

The following emissions are deemed particularly relevant:

a)   Air pollutants

Including:

-    suspended solids (coal and coke dust)
-    gaseous and vaporous emissions, e.g.:
     - sulfur dioxide (SO_2)
     - hydrogen sulfide (H_2S)
     - oxides of nitrogen (NO_x)
     - carbon monoxide (CO)
     - benzene, toluene, xylene (BTX)
     - polycyclic aromatic hydrocarbons (PAH)
     - benzo(a)pyrene (BaP)

b)   Wastewater pollutants

Including:

- various nitrogen compounds
- phosphorus
- chemical and biological oxygen demand
- phenols
- polycyclic aromatic hydrocarbons
- cyanides
- sulfides
- BTX
- sum of all toxic substances classified as such on the basis of, say, toxicity toward fish (fish test).

c)   Noise emissions

Coking plants have numerous **noise emitters** at all points of the operation. Each and every **drive unit**, for example, constitutes a source of noise.
The **equipment used for mixing, crushing and screening coal and coke and for compressing gas** is particularly **noisy** and therefore requires comprehensive **noise-control** measures. Otherwise, some emitters are liable to develop **noise intensity levels** significantly above 85 dB(A).
In order to **preclude noise-induced detriment to human health**, both the sound sources themselves and the general vicinity of such equipment are subject to certain **emission and immission limits.**

d)   Soil and groundwater

The **handling and storage of coking products, crude tar, crude benzene, sulfuric acid and various purchased chemical additives pose a hazard potential for soil and groundwater.**
Environmental consequences result from emissions, the effects of which in the vicinity of the installations can be **damaging both to human health** and to nature and which are monitored by way of ground-level pollutant concentrations. Also, **near-source pollution** occurring directly at and around the **workplace** demands careful attention. In the interest of **occupational safety**, so-called maximum working-site concentrations (*MAK*-values), so-called occupational exposure limits, and technical concentration guidelines (*TRK*-values) have been specified.
Improper handling of substances constituting a hazard to water can cause con-tamination of the **soil and groundwater; contaminated effluent** can be **toxic** (toxicity as a common parameter), cause bad taste (phenols) and/or excessive fertilization and, hence, consumption of oxygen (nitrogen, phosphorus).

40

Consideration must also be given to the fact that the **construction and operation of technical facilities affects the living conditions of sundry groups**. The relevant **socioeconomic** and **sociocultural aspects** therefore have to be analyzed.

2.2   Protective measures

Environmental protection and **occupational safety** at **coking plants** are governed by **statutory provisions**; in **Germany**, these include the *TA-Luft* (Technical Instructions on Air Quality Control), the *Gefahrstoffverordnung* (**Hazardous Substances Ordinance**), and the Wasserhaushaltsgesetz (**Federal Water Act**).

In accommodation of amended laws with in part substantially more stringent provisions, **technical advances** have emerged which now enable **comprehensive protection of the environment**.

One such advance is the **development of large coke ovens**, so that the operation of new coking plants of equal capacity now requires less frequent **opening** (around 80%) and has a far smaller sealing area to be cleaned (approximately 65% reduction). At least the **following basic emission control measures** are now being implemented on **new facilities**:

a)   **Coal handling, including unloading, storage, conditioning (mixing, grinding) and hauling**

- erection of **stationary sprinkling systems** over coal storage areas to keep the coal moist, with provision for climatic boundary conditions;

- **minimized dumping heights** for mobile discharge and transfer stations;

- use of **enclosed conveyors**;

- installation of **dedusting facilities** on grinding and mixing equipment, plus dust silos.

b)   **Coke oven batteries**

- **collection of charging gas** and transfer by two separate routes to the crude gas, e.g., via so-called mini risers leading to the adjacent oven and then through the "real" riser into the collecting main;

- **charging gas aspiration** by means of stationary or mobile systems equipped for aftercombustion and **dedusting**;

- **mechanical cleaning of charging hole lids and frames**, plus sealing after each charge;

- **mechanical cleaning** (aspiration) **of the oven roof**;

- **mechanical cleaning of the risers** (closures equipped with water seals);

- installation of **mechanical cleaning devices for the doors and chamber frames** of the coke oven machines;

- **collection and treatment of emissions** emerging upon **removal of the doors**;

- use of special-purpose **door maintenance cars**;

- installation of **tight door sealing systems** with gas relief channels to avoid excessive gas pressure in the vicinity of the sealing elements;

- installation of **aspiration hoods** for the door and frame cleaning devices;

- **leakage gas aspiration** for emissions resulting from leakages around oven doors, with injection of exhausted air into the batteries' combustion air supply;

- use of **combustion gases with sulfur contents safely below 0.8 g/m³** to limit SO_2 emissions;

- gradual air feed and internal/external flue gas recirculation to **reduce NO_x emissions** in connection with heating of the ovens;

- use of highly heat-conductive **stone linings** for the heating walls;

- **collection and purification of emissions** from coke pushing.

c)   **Coke cooling**

- use of dry coke cooling technology, comprising:
    - moistening of the dry cooled coke to suppress dust evolution at the transfer points
    - dedusting of the delivered coke
    - dedusting of surplus gas by means of bag filter
    - intergas generation to replace cooling cycle gas based on the use of low-sulfur gas;

- emission control measures for wet quenching, e.g., provision of baffle plates for the wet quenching towers.

d)   Coke treatment

- installation of **enclosed coke conveying equipment**;

- **encapsulation** of **coke screening plant**;

- **collection and removal of particulate emissions**, e.g., at the feed bunkers, sieving lines, crushers, belt feeders, etc.;

- installation of **remoisteners** for dry cooled coke to limit dust generation at the coke transfer points.

e)   Gas treatment and coal-constituent recovery systems

- use of **effective sealing systems/elements** for pumps, valves and flanges;

- **forced ventilation** of tanks, water lutes, etc. and injection of the ventilation gases into the crude gas suction line;

- use of Claus systems with **injection of tail** gas into the crude gas (**tail gas recirculation**);

- provision of waste **gas filtration** and additional catalyst installation for the H_2SO_4 systems to extensively preclude SO_2/SO_3 emissions.

f)   Wastewater treatment; cf. environmental briefs Wastewater Disposal and Mechanical Engineering, Workshops, Shipyards

- use of **upstream strippers** employing alkaline additives (e.g., caustic soda solution) to reduce the so-called fixed ammonia compound burden of the coking plant's process water;

- installation of **multiple-stage biological wastewater treatment facilities**, including a nitrification/denitrification stage to enable elimination of nitrogen compounds in the coking plant's process effluent.

g)   Conservation of soil and water

- **separate drainage systems** for surface runoff and process wastewater (from gas treatment and coal-constituent recovery systems);

- placement of all tanks and apparatus used in the handling or treatment of substances constituting a hazard to water in **collecting tanks**; installation of **intercepting sewers for wastewater**, e.g., by way of biological water treatment;

- installation of **monitorable tank bottoms** (on strip footing); use of **overfill protection devices**;

- use of **suitable materials** and external **anti-corrosion measures** to substantially improve the availability of plant components.

h)   Noise control

- **noise control at the source**, e.g., encapsulation of machines, pumps, etc.;

- **noise control for structures**, e.g., solid construction, sandwich construction, use of vibration dampers, partitions;

- erection of **acoustical barriers**;

- **individual examination** of noise sources with a view to satisfying equipment-noise and neighbor's-rights requirements.

40

The **measures** listed under a) through h) are **technically tried and proven** and routinely implemented for **new facilities**.
Stated **in proportion to the total investment** for a new coking plant, the **cost of environmental protection** measures accounts for 30 - 40 %.
The **operational reliability and availability of environmental protection provisions** - like that of the entire coking plant - is highly dependent on the **qualifications of the operating personnel**. Consequently, appropriate **training** is required to put the personnel in a position to **operate and use the equipment in an expert, competent manner**.

3.   Notes on the Analysis and Evaluation of Environmental Impacts

3.1   General

Any evaluation of the detrimental effects of **coking plant emissions** must allow for **numerous factors**, some of which are **difficult to quantify**, e.g.:

-        baseline pollution by other emitters,
-        climatic influences, particularly of wind, on propagation behavior,
-        accumulative capacity of surrounding ecosystems.

It has been qualitatively determined that some such emissions, **BTX** and **benzo(a)pyrene** in particular, have **carcinogenic effects on humans and animals** and that **particulate emissions** and some gases can cause **diseases of the respiratory tract**.

The fact that coking plants, most notably in the near vicinity of the coke oven batteries, have both **definite and diffuse sources of emissions complicates the stipulation of tolerable emission levels**, e.g., in the form of emission factors. This problem is evidenced in **pertinent German directives and regulations** such as *TA-Luft*, which states **allowable concentrations** in the exhaust gases/air from **definite points**, but also applies **qualitative technical measures** to the erection of **numerous different systems and types of plant**.

Hence, only a **narrow selection of emission limits/factors** is available for reference.

That information, however, is supplemented by *MAK/TRK*-values which define **limits for airborne workplace pollution** and allow for the **registration and monitoring of emissions from diffuse sources**.

3.2   Summary of limit values and standards

Proceeding on the basis of Germany's Federal Immission Control Act *(Bundesimmissionschutzgesetz)*, the following **directives and regulations** apply in essence to the design and planning of coal processing in the Federal Republic of Germany:

-        Technical Instructions on Air Quality Control (*TA-Luft*) dated February 27, 1986,
-        Limit values for pollutants in coking plant wastewater according to the Federal Water Act, section 7a,
-        Limit values according to the Technical Instructions on Noise Abatement (*TA-Lärm*) dating from July 1984 (5th update),
-        Hazardous Substances Ordinance.

Additional directives and regulations to be heeded for planning purposes are listed in section 6 (References).

The following tables (1.1, 1.2, 2, 3 and 4) survey the presently valid **limit values** for emissions, pollutant concentrations and MAK/TRK-values **according to German standards**.

It must, of course, be kept in mind that **more stringent requirements may apply, depending on the baseline pollution** level (initial load) at the location in question. The tables compare the emission limits imposed by various **industrialized countries of Europe** with those reflected in German standards. **With few exceptions,** as quickly becomes apparent, **the German standards impose the most stringent environmental protection requirements**.

Table 1.1

Emission limits according to *TA-Luft* (general exhaust emissions)

Component	Definition of mass-flow range	German emission limits	European comparative values[1]	Re-marks
Dust	> 0.5 kg/h < 0.5 kg/h	50 mg/m³ 150 mg/m³	50 - 115 mg/m³, Ø 94 mg/m³ 150 mg/m³	
NO_x (as NO_2)	> 5 kg/h	0.5 g/m³	0.35 - 0.8 g/m³, Ø 0.55 g/m³	0.35: Belgian outlier
SO_2	> 5 kg/h	0.5 g/m³	0.5 - 0.8 g/m³, Ø 0.6 g/m³ alternative: limitation of annual load to 10 000 - 12 000 t	
H_2S	> 50 g/h	5 mg/m³	5 mg/m³	
HCN	> 50 g/h	5 mg/m³	5 mg/m³	
C_6H_6	> 25 g/h	5 mg/m³	5 mg/m³	
Benzo(a)py-rene	> 0.5 g/h	0.1 mg/m³	no emission limit defined	

40

---

[1]   The comparative values derive from the Netherlands, England, Belgium, France, Spain, Austria, Sweden and Finland; the Ø-values represent the arithmetic mean of the respective limit values.

Table 1.2

Emission limits for coking-plant waste gas/purified exhaust air

Component	Definition of mass-flow range	German emission limits	European comparative values[2]	Remarks
Coal plant	Dust	50 mg/m³	100 mg/m³	
Coal drying and preheating	Dust	100 mg/m³	115 mg/m³	
Coke screening	Dust	50 mg/m³		
Coal charging (filling process)	Dust PAH for mass flows > 0.5 g/h	25 mg/m³ 0.1 mg/m³	15 - 230 mg/m³, Ø 92 mg/m³ 0.1 mg/m³, alternative: limitation of daily load to 2 kg	15: Dutch outlier
Coke pushing (operation)	Dust	5 g/t coke	5 - 115 mg/m³, Ø 46 mg/m³ or 5 g/t coke	
Dry coke cooling	Dust	20 mg/m³	20 mg/m³	
Wet coke quenching	Dust	50 g/t coke	50 - 800 mg/m³, Ø 330 mg/m³ or 80 g/t coke	
Exhaust stack, coke oven batteries (new facilities)	NO_x as NO_2 SO_2 CO Dust	0.5 g/m³ 0.8 g sulfur in UF gas 0.2 g/m³ 10 mg/m³	0.2 - 0.8 g/m³, Ø 0.53 mg/m³ 0.5 - 1.7 g sulfur in UF gas Ø 9 g sulfur in UF gas or 0.5 g H_2S in UF gas 100 - 200 mg/m³, Ø 130 mg/m³	0.2: Dutch outlier 0.5: Spanish outlier
Exhaust stack of by-product plants (new facilities)	Dust CO NO_x SO_3 (H_2SO_4 systems) SO_2 (H_2SO_4 systems) H_2S (Claus systems) Sulfur (tolerable emission level) Production capacity: < 20 t S/d Production capacity: 20 - 50 t S/d Production capacity: > 50 t S/d	 60 mg/m³ SO_2-to-SO_2 conversion rate: > 97.5 % or 2500 mg/m³ 10 mg/m³ 3 % 2 % 0.5 %	50 mg/m³ 200 mg/m³ 0.1-0.35 g/m³, Ø 0.225 g/m³ 60 - 10 mg/m³, Ø 70 mg/m³ 500 - 3000 mg/m³, Ø 1750 mg/m³ 10 mg/m³ 3 % 2 % 0.5 %	 500: Austrian outlier

---

[2]   cf. table 1.1 footnote

Table 2

**General administrative framework regulation (Rahmen-Abwasser VwV) on
minimum requirements
for the discharge of wastewater from coking plants, Appendix 46
(draft dated August 1990), valid for direct discharge**

Parameter

Totals:		
NH_4-N NO_2 NO_3-N	40	mg/l
Phosphorus	2	mg/l
BOD_5	30	mg/l
Filterable substances	50	mg/l
COD	200	mg/l
Phenol index	0.5	mg/l
PAH	0.1	mg/l
BTX	0.1	mg/l
CN (volatile)	0.1	mg/l
Sulfide	0.1	mg/l
Fish toxicity	4	(dilution factor)

Note: The above emission limits apply to undiluted coking plant wastewater occurring at the rate of 0.3 m³/t of coal.
Coking plants equipped for HP gas treatment and with installations for
collecting and recycling contaminated rainwater, or with appropriate
supplementary process stages, can increase the specific wastewater
discharge to as high as 0.42 m³/t of coal.

40

Table 3

Standard immission values for noise emissions (July 1984)

The noise immission levels are specified for:

a) areas containing only commercial or in-
    dustrial facilities and living quarters for
    supervisory and standby personnel, to:

                                                            70 dB(A)

b) areas containing primarily non-residential   (daytime)   65 dB(A)
    buildings, to: (nighttime)   50 dB(A)

c) areas containing nonresidential and resi-
    dential buildings accommodating neither
    mostly nonresidential occupants nor pri-
    marily residential occupants, to: (daytime)   60 dB(A)
                                                 (nighttime)   45 dB(A)

d) areas containing primarily residential   (daytime)   50 dB(A)
    buildings, to: (nighttime)   40 dB(A)

e) areas containing exclusively residential (daytime)   50 dB(A)
    buildings, to: (nighttime)   35 dB(A)

f) areas containing health resorts, hospitals, (daytime)   45 dB(A)
    nursing homes, to: (nighttime)   35 dB(A)

g) residential buildings attached to indus-   (daytime)   40 dB(A)
    trial/business premises, to: (nighttime)   30 dB(A)

- **Comments**

- The acoustic engineering of industrial facilities and the requisite prognostical calculations rely heavily on the algorithms elucidated in VDI (Association of German Engineers) guideline 2714 (E) "Outdoor Sound Propagation" and VDI guideline 2571 "Sound Radiation from Industrial Buildings".

- If the above guideline values are exceeded and/or adulterated by superimposed extraneous noise to such an extent that accurate measuring becomes impossible, appropriate correction factors must be allowed for, as generally laid down in *TA-Lärm* (technical instructions on noise abatement). If measurements cannot be conducted, sound propagation calculations must be performed.

- "Nighttime" is understood as the eight hours between 10:00 p.m. and 6:00 a.m.

- In supplementation of the workplace ordinance *Verordnung über Arbeitsstätten* (section 15: protection against noise), the following provisions shall apply to workplace noise nuisance:

    (1)   The workplace noise level must be kept as low as possible for the type of operation in question. The workplace reference intensity, inclusive or exclusive of extraneous noise, shall not exceed:
      - 55 dB(A) for primarily intellectual work
      - 70 dB(A) for simple or mostly mechanized office work and comparable activities
      - 85 dB(A) for all other activities; to the extent that the prescribed sound intensity level cannot be adhered to by available, reasonable means, it may be exceeded by 5 dB(A).

    (2)   The sound intensity level prevailing in breakrooms, duty rooms, rest rooms and first-aid rooms shall not exceed 55 dB(A). The reference sound intensity level shall be set by taking into account only the noise generated by installations inside the rooms plus the extraneous sounds entering the rooms in question.

40

Table 4

MAK-values (occupational exposure limits) and TRK-values
(technical concentration guideline values)

Component	Limit values in Germany	Comparative values in Europe[3*]	Remarks
Dust	6 mg/m³	10 - 15 mg/m³, ∅ 11 mg/m³	
NO_x (as NO_2)	9 mg/m³	4 - 6 mg/m³, ∅ 5.3 mg/m³ 30 mg/m³ NO	
SO_2	5 mg/m³	<u>1.5</u> - 5 mg/m³ ∅ 4.7 mg/m³	<u>1.5</u>: Spanish outlier
CO	33 mg/m³	<u>29</u> - 57 mg/m³ ∅ 45 mg/m³	<u>29</u>: Dutch outlier
Benzene	16[4)] mg/m³	<u>3</u> - 32 mg/m³	<u>3</u>: Swedish outlier
Toluene	375 mg/m³	375 mg/m³	
Xylene	440 mg/m³	425 - 435 mg/m³, ∅ 430 mg/m³	
Benzo(a)pyrene	2-5[4] mg/m³[5]	2 - 5 µg/m³	
H_2S	15 mg/m³	14 - 15 mg/m³	
HCN	11 mg/m³	10 - 11 mg/m³	
NH_3	35 mg/m³	17 - 18 mg/m³	
Phenol	19 mg/m³	19 mg/m³	
Mercaptans	1 mg/m³	1 mg/m³	
Biphenyl	1 mg/m³	1 - 1.5 mg/m³	
Carbon disulfide	30 mg/m³	30 mg/m³	
Naphthalene	50 mg/m³	50 mg/m³	

[3*]   cf. table-1.1 footnote

[4]   TRK-values

[5]   In the direct vicinity of the batteries, higher levels (measured in µg/m³) can be tolerated but then call for supplementary organizational and/or hygienic measures in addition to personal protective equipment, e.g., breathing masks.

Adherence to the permissible emission levels and the **efficiency of the airborne emission control equipment** must be **monitored** by measurements.

Subsequent to commissioning of facilities, it shall be **determined by measurements** whether or not the **numerical data** assumed at the planning stage **correspond to the actual operating conditions. The measurements** shall be conducted by **neutral institutions,** authorities or the like **with due consideration of pertinent guidelines.**

The *TA-Luft* and relevant **VDI guidelines** describe in detail the **conduct of emission and ground-level pollution measurements.**

3.3   Evaluation of environmental impacts

The aforementioned **emission limits** can be complied with by **implementing the protective measures** described in section 2.2.

Compared to existing facilities, the following **reductions in emission levels** (referred to a complete coking plant, including so-called diffuse sources) are foreseeable:

SO$_2$	by	20 - 40 %
NO$_x$	by	20 - 40 %
CO	by	30 - 35 %
BTX	by	> 95 %
Dust	by	approx. 50 %
PAH	by	approx. 90 %
BaP	by	approx. 90 %

The following **emissions** can be **completely avoided by replacing wet quenching facilities with dry coke cooling systems:**

—   H$_2$S up to 80 g/t of coke, amounting to 160 t/a of H$_2$S for an annual coke output of 2 million tons;

—   NH$_3$ up to 15 g/t of coke, amounting to 30 t of NH$_3$ per year.

4.   Interaction with Other Sectors

Coking plants are closely allied with the **iron-producing industry** (cf. environmental brief Iron and Steel). However, some coking plants are located near mines (see briefs on mining), with a coal processing facility (coal washing plant) operating at the mine.
New developments in the steel-making industry such as those enabling the use of both oil and coal in the blast furnace process can help **reduce the specific coal requirement** of blast furnace operation. At present, however, there is **no sign of coke becoming dispensable** as a fuel and mainstay medium (reducing agent) for blast furnaces.
References to other **adjacent sectors** are to be found at the **appropriate text passages**.

5.   Summary Assessment of Environmental Relevance

Without the safe and sure operation of comprehensive antipollution devices, **coking plants can cause substantial pollution of the air, soil and water**.
In addition to **emission reduction measures** to avoid pollution from **definite sources**, the **avoidance of emissions from diffuse sources** is also important. Moreover, the **maximum allowable workplace concentrations** (occupational exposure limits) must be observed.
Systematic implementation of well-tested modern **environmental protection measures**, in combination with the **observance of pertinent rules and regulations**, can ensure that **coking plants, like other coal processing facilities** such as carbonization systems, **need not be classified as environmentally hazardous**.
With due regard to local circumstances, official **ordinances must ensure that** environmental protection **measures are properly executed**.
To that end, it is advisable to designate **environmental protection and environmental safety officers**, who, with **proper training and technical support**, are in a position to **assume supervisory functions** and attend to the **interests of environmental protection and occupational safety** in connection with all relevant industrial activities.
Early involvement of affected groups (women in particular) **in the planning and decision-making processes enables consideration of their interests and helps alleviate environmental problems**, e.g., in connection with the contamination of foodstuffs and/or health impairment in the vicinity of such undertakings.

6.   References

<u>Laws, Regulations, Directives</u>

AD-Merkblätter

Bergverordnung zum gesundheitlichen Schutz der Beschäftigten (Gesundheitsschutz - Bergverordnung - GesBergV) dated 31.07.1991.

Druckbehälterverordnung (pressure vessel code), Bundesarbeitsblatt Nr. 3, Teil Arbeitsschutz, March 1990.

Hinweise für das Ableiten von Abwasser in öffentliche Kläranlagen, ATV Arbeitsblatt 115.

MAK-Liste, Liste maximaler Arbeitsplatzkonzentrationen, 1990, Mitteilung XXVI, Bundesarbeitsblatt 12/1990.

Richtlinie für Rohrleitungsanlagen zum Befördern wassergefährdender Stoffe, Gemeinsames Ministerialblatt GMBL (joint ministerial circular) Nr. 8, 02.04.1987.

Technische Regeln zum Umgang mit brennbaren Flüssigkeiten, TRbF, BGBL III (Federal Law Gazette III).

Unfallverhütungsvorschriften UVV, Hauptverband der gewerbliche Berufsgenossenschaften, Bonn.

VDI-Richtlinie 2058, Blatt 1, Beurteilung von Arbeitslärm in der Nachbarschaft, September 1985.

VDI-Richtlinie 2058, Blatt 3, Beurteilung von Lärm am Arbeitsplatz unter Berücksichtigung unterschiedlicher Tätigkeiten, April 1981.

VDI-Richtlinie 2560, Persönlicher Schallschutz, December 1987.

VDI-/VDE-Vorschriften.

Verordnung über Arbeitsstätten (Arbeitsstättenverordnung), Ausgabe 1983, Bundesminister für Arbeit und Sozialordnung (German Federal Minister of Labour and Social Affairs).

40

Verordnung über gefährliche Stoffe (Gefahrstoffverordnung), BGBL (Federal Law Gazette), Aug. 26, 1986.

Wasserhaushaltsgesetz, insbesondere mit dem § 7a, Mindestanforderungen an Kokereiabwässer, BGBL I (Federal Law Gazette I) (1986)

Other applicable provisions cited in diverse rules and regulations.

Miscellaneous

Mitteilungen des Europäischen Kokereiausschusses zu Emissionsgrenzwerten und MAK-/TRK-Werten (unveröffentlicht).
Bernd Schärer, Article "US Clean Air Act" in Staub - Reinhaltung der Luft 52 (1992) 1 - 2, Springer Verlag.

Mining and Energy

Thermal Power Stations

41. Thermal Power Stations

Contents

1. Scope

Thermal power stations are facilities in which the **energy content** of an energy carrier, i.e., a fuel, can be **converted into either electricity or electricity and heat**. The **type of power plant** employed depends on the **source of energy** and the **type of energy** being produced.

Possible **energy sources** include:

— fossil fuels such as coal, petroleum products and natural gas
— residual and waste materials such as domestic and industrial refuse and fuel made from recovered oil
— fissionable material

Thermal power plants can be designed **for different fuel sectors** in the interest of greater **fueling flexibility** and/or **higher efficiency** - one example being a combination power plant with a gas turbine running on natural gas and an oil- or coal-fired steam generator feeding a steam turbine.
Renewable sources of energy such as wood and other forms of biomass are not dealt with here, as they are the subject of a **separate environmental brief. Nuclear thermal power plants** have also been **omitted** from this catalogue. The frame of reference concentrates extensively on **fossil-fueled power plants**, in particular types using coal and petroleum products, the present and near-future use of which is of eminent importance in most developing countries. With regard to **hydropower**, the reader is referred to the environmental brief Large-scale Hydraulic Engineering.

As far as the form of energy being generated is concerned, there are **three main types of thermal power stations**:

— condensing power plants used exclusively for generating electricity
— steam- or hot-water producing heating stations for domestic or industrial purposes
— district heating power stations, or cogenerating plants, for the simultaneous generation of electricity and available heat.

It is important to note that, **for economic reasons, process heat and heat for heating purposes should only be generated in close proximity to the users**. For thermal outputs ranging from 50 to 100 MW, the **distance** between the power plant and the user **should not exceed 2 to 5 km**. Conversely, electricity can be **economically transmitted** over very **substantial distances**; cf. environmental brief Power Transmission and Distribution.

41

The **unit power ratings of fossil-fueled thermal power plants** range from a **few hundred kW** (diesel stations) to **more than 1000 MW** (oil- and coal-fired stations). In **many countries,** preference is given to unit ratings of **200 - 300 MWel** with deference to **power system stability.** The better the boundary conditions, the higher the achievable capacities.

2.   Environmental Impacts and Protective Measures

The **environmental consequences** of any given plant are both **plant- and site-dependent.** Thermal power stations can impact the environment in different ways and at different locations. A typical thermal power plant is likely to comprise the following **principal components:**

- facilities for preparing and storing working materials
- facilities for burning fuel and generating steam
- facilities for generating electricity and available heat
- facilities for treating exhaust gases and solid and liquid residues
- cooling facilities

In **Appendix A-1,** a thermal power plant is reduced to a **block diagram** showing the most likely **material inputs, outputs and environmentally relevant flows of material.**

Table 1 surveys the **potential emissions** occurring at **different stages of the generating process:**

Table 1
Potential emissions from thermal power plants

Type of emission	Step of process					
	Fuel storage and processing	Combustion and steam generation	Flue-gas cleaning	Power generation	Cooling systems	Treatment of residue
Particulates	*	*			*	*
Noxious gases		*				*
Wastewater	*	*	*		*	*
Solid residues		*	*			*
Waste heat		*		*	*	
Noise	*	*	*	*	*	*
Groundwater contamination	*					

As the table indicates, thermal power stations can **affect** the media **air, water and soil**, as well as **human beings, plants, animals** and the **landscape**.

The **disposal of residues**, e.g., those associated with oil- and coal-fired facilities, is dealt with in section 2.3.

The **main environmentally relevant effects** of a thermal power plant derive from the **combustion process** and its **particulate and gaseous emissions**. As a rule of thumb, the **environmental impacts** of thermal power plants, i.e., pollution, spatial requirements and residues, tend **to increase** in severity for **gas, light fuel oil, heavy fuel oil** and **coal**, in that order.

Prior to examining the environmental consequences and **possible protective measures** for the various domains, let us begin with a few basic **introductory remarks**. The running text contains **information essential to the subject environmental consequences and protective measures**, while the **relevant technical measures are detailed in the appendix**.

In addressing the environmental consequences of a thermal power plant, **distinction** is drawn between **emissions**, i.e., the release of pollutants from various parts of the plant, the smokestack in particular, and **immissions**, i.e., the actual environmental effects of the pollutants, normally referred to ground level - as indicated by the terms **"ground level concentration/pollution"** and **"ambient air quality concentration/standards"**. Emissions and immissions are interlinked by a number of **factors**, e.g., **plant-specific technical parameters** (emission volumes, outlet velocity, temperature), **meteorological factors** (weather category, wind speed) and **range-specific data** (distance between the emitter and the ground-level pollution point). **Parameters** belonging to the first and last categories, e.g., height of stack and distance from residential areas, are more or less **freely selectable for new power plants**, while the **actuating variables** for **existing plants** all belong to the first category. According to the laws of physics (conservation of matter), practically all **noxious emissions** with the notable exception of CO_2, for example, eventually **fall to earth**. The **height of the smokestack, the outlet velocity of the exhaust**, and the **prevailing wind velocity** determine the size of area that will be affected. From a technical standpoint, it is relatively easy to **reduce ground-level concentrations** for a given area by **increasing** the **height of the smokestack**. Since, however, the specific **emission volume does** not change, but is simply distributed over a wider area, the extent to which such a measure would **aggravate** the **environmental impact outside of the subject area** would have to be clarified.

Measures aimed at **reducing the environmental consequences** of thermal power plants can be categorized as follows:

- alteration of boundary conditions
  - incentives for the efficient utilization and conservation of energy, e.g., cost-covering power rates and taxes
  - appropriate siting
- nontechnical protective measures
  - regulations dictating the mandatory use of piped energy (district heating) in congested urban areas
  - compensation models for the replacement of major emitters
- technical protective measures
  - reduction of ground-level concentration, e.g., by extending the height of the smokestack
  - emission-reducing measures
    * pollution-control measures to prevent or reduce pollution by combustion modification, e.g., choice of an appropriate low-impact fuel such as natural gas (in place of coal), homogenization of fuel to avoid peak emissions, efficiency-enhancing measures, and NO_x limitation by combustion engineering measures
    * post-combustion measures, i.e., flue gas clean-up.

The **order** in which the protective measures are taken is **subject to the principle of attaching priority to avoidance or reduction over subsequent rectification.** First of all, pollution-control measures must be taken to **preclude,** or at least **minimize,** the **occurrence of pollutants,** before any further-reaching post-combustion technical remedial processes are initiated.

One very **important** relevant **measure** is to achieve **higher efficiency,** e.g., by erecting **combination power plants** or opting for the **combined generation of heat and power** (cogeneration) in efficient heating power stations with an accordingly **low specific pollution level.** High efficiency is also the most effective way to **reduce CO_2 emissions** and, hence, their **greenhouse effect.** For additional means of reducing CO_2 emissions, e.g., through the use of **renewable sources of energy** for power generation, cf. the brief Renewable Sources of Energy.

With regard to **environmental consequences,** distinction is made between **direct consequences,** e.g., the emission of pollutants, and **indirect consequences** such as the transfer of pollutants from the atmosphere to water via scrubbing processes (assuming that the liquid effluent is not subsequently processed) or the environmental impact of limestone mining and attendant road traffic, e.g., the transfer of limestone by truck from the mine to the power plant. Moreover, **consequential problems** can arise, e.g., the need to properly dispose of the gypsum resulting from flue-gas desulfurization processes (FGD).

The environmental consequences and potential protective measures applicable to the aforementioned areas are explained below.

2.1 Air

The **particulate** and **noxious gas emissions** from thermal power plants primarily and directly pollute the air.

Eventually, the particulate emissions and, for the most part, the noxious gases and any atmospheric transformation products that may have formed (e.g., NO_2 and nitrate from NO) fall to earth either by way of **precipitation or dry deposition,** thereby imposing a **burden on the water and/or soil,** with resultant potential **damage to flora and fauna.**

Depending on the **fuel employed** (type, composition, calorific value) and the type of combustion (e.g., dry or slag-tap firing), given amounts of pollutants (particulates, heavy metals, SO_x, NO_x, CO, CO_2, HCl, HF, organic compounds) become entrained in the exhaust gases. **Table 2** shows the **potential concentration ranges** of different emissions for various fuels in facilities devoid of flue-gas emission control measures.

41

Table 2

Potential ranges of pollutant concentration levels in untreated gas

Type of fuel

Type of emission	Natural gas	Light fuel oil	Heavy fuel oil	Hard coal	Lignite (brown coal)
Sulfur oxides (SO_x) [mg/m^{3}STP]	20 - 50	300 - 2000	1000 - 10000	500 - 800	500 - 18000
Oxides of nitrogen (NO_x) [mg/m^{3}STP]	100 - 1000	200 - 1000	400 - 1200	600 - 2000	300 - 800
Particulates [mg/m^{3}STP]	0 - 30	30 - 100	50 - 1000	3000 - 40000	3000 - 50000

Table 2 lists the **noxious emissions** in mg/m^{3}STP, as prescribed by the applicable **German rules and regulations** [*TA-Luft (Technical Instructions on Air Quality Control),* and *Großfeuerungsanlagenverordnung (Ordinance on Large-scale Firing Installations)*]. SO_x and NO_x are postulated as SO_2 and NO_2. Some **emissions are** limited in terms of **mass flow**, e.g., in kg/h, or of **minimum separation efficiency** (cf. Appendix A-6). With a view to enabling conversion of the stated concentrations to other units such as ppm, g/GJ or lb. of pollutant per 10^6 BTU energy input, as commonly employed in the U.S.A., **Appendix A-6** includes an appropriate **conversion table**.

The **ranges quoted in table 2** for oxides of sulfur relate to differences in fuel-specific sulfur content, whereas many countries use large quantities of indigenous fuels like lignite with comparatively **low calorific values** and **high sulfur contents**. Such a combination naturally produces **relatively high SO_x concentrations** in the (untreated) flue gas.

The **lesser part of the NO_x concentrations** derives from the **nitrogen content of the fuel** (fuel NO_x). The **major share** results from the **oxidation of atmospheric nitrogen** at combustion temperatures exceeding 1200°C (thermal NO_x). Consequently, **high combustion temperatures** go hand in hand with **relatively high NO_x emission levels**. Appropriate **combustion engineering measures** that are relatively **inexpensive** for new plants can keep the **emissions** at the **lower end of the respective range**. However, care must be taken to ensure that a **high quality of combustion** is maintained. Otherwise, excessive combustion engineering measures aimed at **reducing NO_x emissions** could result in a **disproportionate increase in other emissions**, e.g., carbon monoxide and combustible (unburned) hydrocarbons.

In general, CO_2 **emissions** are mainly limited by **controlling the burnout process** such as to **minimize the discharge of CO** and the escape of **combustible hydrocarbons**. Unlike particulates, SO_2, NO_x and halogen compounds, CO and **combustible hydrocarbons** effectively defy retentive measures. **Combustible hydrocarbons** in particular include numerous **chemical substances** that can cause **toxicological problems**, e.g. benzpyrene.

Plants fueled with **coal or heavy fuel oil** also emit small amounts of **hydrogen chloride** and **hydrofluoric acid** (HCl and HF) ranging from 50 to 300 mg/m³$_{STP}$. As a rule, the concentrations stay well below the SO_2 levels and respond favorably to **desulfurization** processes, by which they are reduced even more than S_2.

There are many **combustion-stage and post-combustion alternatives** for use in **reducing air pollution** from **thermal power plants. Appendix A-2**, for example, sketches out an integral set of $deNO_x$, particulate-control and desulfurization measures for the **flue gas** of a **steam generating facility**. The various measures are individually described in the following subsections.

2.1.1 Dust control

Dust control for power plants can be based on **ordinary** and **multiple cyclone separators** and **electrostatic precipitators** or **fabric filters** - with the order of mention corresponding to their respective separation efficiencies: from 60 % - 70 % for cyclone separators to >99 % for electrostatic precipitators and fabric filters. To be sure, the **cost** of the various options **rises disproportionately for increasing separation efficiency**. The separation efficiency of electrostatic filters depends on the **number of consecutive** fields. Like fabric filters, they can achieve **extremely low residual emission levels**, i.e., about 50 and 30 mg/m³$_{STP}$, respectively. The **drawback of cyclone separators** is that they tend to **eliminate coarse particles much more efficiently than respirable** - and, hence, toxicologically critical - **microparticles. Fabric filters** are very **good** at separating out **fine dust** and its accumulated **heavy metals**. The **capital outlay** for **flue-gas dust control** depends on such **parameters** as the **type of fuel**, the **required separation efficiency** and the **technique employed**. As a rule, the **initial cost** ranges from 20 to 70 DM/kW$_{el}$, while the **operating expenses** amount to 0.1 - 0.6 DM/MWh. The **high-ash fuels** used in some countries makes flue-gas dust control a **difficult problem** - including the **proper disposal of the dust** yield, either by **recycling** it in, say, building materials, or by **depositing** it in a **landfill**. Certain characteristics of the **fly ash** may require the use of additives to obtain a solidified product that is less susceptible to leaching, this with a view to **preventing groundwater contamination**.

2.1.2 Desulfurization

SO$_x$ from combustion plants can be **reduced** either by **combustion modification measures** (use of low-sulfur fuel, direct desulfurization in the furnace, dry-additive method) or by **post-combustion clean-up measures** such as extracting the SO$_x$ from the flue gas.

The use of **low-sulfur fuels** is frequently **precluded by economic considerations. In each case, the lowest total-cost concept must be ascertained.** For example, while the **use of a low-sulfur fuel may increase the cost of operation**, it could **save the cost of installing and operating desulfurization equipment**, thus yielding **lower overall costs** for the power station. Of course, such considerations must also account for **other criteria**, e.g., using **indigenous fuels** in order to assure their **safe supply**.

Like solid fuels, **sulfurous petroleum products** are also amenable to **pre- and post-combustion measures. The pollution-control measure of choice** is to **hydrate the sulfur** by adding hydrogen in order to **extract the product from the oil**, e.g., as vacuum gas oil or a remanent of atmospheric or vacuum distillation. Such processes are only **cost-efficient for large capacities** and therefore only feasible for oil refinery applications. In a **thermal power station**, the appropriate measures for **reducing SO$_x$** emissions are **restricted** to the **use of a low-sulfur petroleum product, mixing different fuels and, primarily, flue-gas desulfurization** according to the same principle as that employed in solid-fueled facilities (described below and in Appendix A-3).

For **coal-fired power plants**, particularly in response to the pronounced compositional variance observed in the indigenous coals of many countries, appropriate **mixing and homogenization** can have the positive effect of **lowering the peak-value extremes** that have to be accounted for in the design of desulfurization systems. Consequently, major importance must be attached to a **conscientious analysis** of the calorific values and the water, ash and sulfur contents of fuel deriving from, say, different sections of a coal mine. It is also important to **ascertain the possible extent of spontaneous desulfurization** attributable to the presence of calcium compounds in the coal.

Coal can be desulfurized directly at the mine or pit as part of a process in which **sulfur and various inert constituents are extracted primarily by wet methods**. Depending on the type of coal and on the kind of sulfur linkage, the **sulfur content of the coal** (glance coal in particular) can be **reduced by 5 % to 80 %**. No such conditioning measures, however, can reduce the **organosulfur content. Sulfite** in the form of pyrite (FeS_2) can be **separated out** if it is freely present in the raw coal or is so coarse-grained in intergrowths that it becomes amenable to removal **following crushing**.

Direct in-boiler desulfurization is applied to **solid fuels in fluidized-bed combustion systems**. Separation efficiencies of 80 % to 90 % are achievable. **Dry additive techniques** remove between 60 % and 80 % of the sulfur from coal (cf. Appendix A-3 for details).

Flue-gas desulfurization techniques enable SO_2 **separation efficiency levels of 90 - 95 %**. Since **flue-gas desulfurization equipment is expensive to install and operate**, it is more judicious in some cases to install a **component-flow desulfurization system** in which only **part** of the flue gases are desulfurized, while the remaining, undesulfurized flue gases can be used for heating the treated gases.

Of all the described alternatives, **flue-gas desulfurization is the most expensive and elaborate**. In each case, particularly **for retrofitting projects, the spatial integration options must be carefully investigated** in advance.

A **comparison** of the aforementioned **pre- and post-combustion desulfurization measures** shows that the **former** offer the **lower separation efficiencies,** but are also **less expensive and more conducive to retrofitting**. Fluidized-bed combustion, however, is an exception to the rule, as it can only be implemented in new facilities (maximum capacity of commercial-scale systems to date: 150 MW_{el}).

All methods of desulfurization and dust control **involve the consequential problem of properly recycling or disposing of the residues and,** possibly, of **wastewater** resulting from operation of the equipment (cf. section 2.3).

41

Depending on the **size of the plant**, the **process employed, the separation efficiency** achieved, and other factors, the **investment cost** of a desulfurization system can amount to anywhere from roughly 30 to 550 DM/kW_{el}. Also the increase of auxiliary power consumption to run the system is unavoidable. **Dry-additive methods** are the least expensive, while **regenerative techniques** producing compounds of sulfur as their end product are the **most costly**.

The various desulfurization processes also and incidentally precipitate **halogen compounds** such as HCl and HF even more efficiently than sulfur.

2.1.3 DeNO$_X$

The available means of **nitrogen removal** also comprise **pre- and post-combustion alternatives**. With regard to sulfur content, a **careful choice of fuel** can do much to limit **NO$_X$ emissions**. On the other hand, the **NO$_X$ formation process** is **more complicated** than the conversion of fuel sulfur into SO_2, as described in section 2.1. The combustion modification measures aim to **reduce the rate of NO$_X$ formation during the combustion process**, essentially by **lowering the maximum temperature of combustion**. This can be achieved by **design measures**, e.g., the combustion-chamber geometry, burner design and configuration, staged air supply, reduced excess air, and such **operational measures** as reduced combustion-air-preheating temperature or the use of low-nitrogen fuel.

The **post-combustion deNO$_X$ measures** are concerned with **reducing the exhaust-side NO$_X$ emissions** by various means designed to remove the NO$_X$ either alone or together with SO$_X$.

The **only** process to have gained **commercial-scale acceptance** to date is the **selective catalytic reduction** of NO$_X$ (SCR method). In this process, ammonia (NH_3) reacts with NO$_X$ in a **catalytic converter** to form water and nitrogen. The process therefore produces **no residues** (like those from dust-control and desulfurization processes) that would require subsequent disposal. The **SCR process** takes place at temperatures of 300 - 400°C and **can be integrated** either on the raw-gas end, e.g., upstream of the air preheater (SCR → economizer) or on the clean-gas end behind a desulfurizing system (SCR → FGD).

SCR-base processes achieve NO$_X$ separation efficiencies of approximately 80 - 90%.

Another approach that is particularly well-suited for relatively low separation efficiencies of about 60 % or less is the **SNCR process** (selective non-catalytic reduction), in which NO$_X$ reduction is achieved by **spraying ammonia** into the boiler at a temperature of some 1000°C.

The **initial cost** of flue-gas deNO$_x$ equipment **depends on the size of the plant, the required separation efficiency, configuration**, etc. and ranges from roughly **120 to 250 DM/kW$_{el}$**.

2.1.4 Greenhouse effect

The **greenhouse effect**, i.e., the long-term warming of the earth's atmosphere due to the presence of **anthropogenic trace gases**, is chiefly attributable to the **accumulation** of gases such as **carbon dioxide** (CO_2), **methane** (CH_4), **chlorinated fluorocarbons** (CFCs), **tropospheric ozone** (O_3) and **nitrous oxide** (N_2O) - with the order of mention corresponding to the relevant significance of the gases. Their specific contributions to the greenhouse effect are widely variant. Methane, for example, has roughly 21 times the effect of CO_2, but occurs globally in much smaller mass volumes than does CO_2 as the end product of any combustion process involving carbonaceous (organic) fuel.

The **principal protective measure** to counter CO_2 emissions is to **ensure high combustion efficiency**, e.g., by way of a combination or cogeneration process.

Other measures like the use of **renewable sources of energy** - hydropower in particular - for generating electricity, in addition to measures aimed at steering the demand for electricity, are very important, but would never suffice to render superfluous the generation of electric power in fossil-fueled thermal power plants.

2.1.5 Diffuse emissions

In addition to the aforementioned types of emissions, most of which emanate from the smokestack, thermal power plants can also emit **pollutants from other areas** (cf. table 1). **Particulate emissions**, for example, can occur in connection with **fuel storage, handling and processing**. Such emissions can be extensively reduced by suitable measures such as **moistening with water or enclosing/encapsulating critical areas**. The same applies in effect to the **storage and handling of petroleum products**, i.e., via suitable contrivances on the tanks and pumping facilities, either to minimize **evaporation or to return the condensate to the system**. Such measures can be of major importance in **countries with a warmer climate** than that encountered in Central Europe.

41

2.2   Water

Most **water** in thermal power stations is used for **cooling**. After **absorbing enough heat to raise its temperature by 4 - 8°C**, the water normally is **returned to the extraction point**. Power plants designed for **non-circulating water cooling** require about 160 - 220 m³/h•MWel (with cooling water losses usually staying below 2 %).

In **pure power generation**, the cooling water absorbs approximately 60 % to 80 % of the fuel's energy content as **waste heat**. Less energy is wasted by plants with inherently higher efficiency, e.g., cogenerating facilities. Depending on local conditions, the **waste heat can impose a thermal burden on surface water**, e.g., cause an increase in the temperature of a river, with the **volumetric flow and/or water regimen** as an actuating variable. Particularly in **developing countries**, water bodies are subject to **pronounced seasonal variation. Oxygen depletion** therefore has two main causes: **accelerated consumption due to rapid metabolism**, and the **lower solubility of oxygen in warm water**. Oxygen deficiency can be seriously **detrimental to aquatic life**.

The **in/out temperature gradient of cooling water** can be limited by putting it through a **cooling tower** (once-through or circulation cooling) before it is returned to the river. Depending on the prevailing climatic conditions, however, such cooling systems involve **major evaporative water losses** and, hence, locally **elevated atmospheric dampness**. Such problems can be **avoided or minimized** by the use of **closed-loop cooling systems** in combination with **dry or hybrid cooling towers**. Natural-draft cooling towers are relatively expensive to build but comparatively inexpensive to operate, while induced-draft cooling towers have the disadvantage of operating on electricity, the generation of which increases the **overall ecological burden**.

Apart from their cooling-water consumption, **power plants have very modest water requirements** (0.1 - 0.3 m³/h•MWel) for topping up the steam cycle, cooling the ashes and operating certain types of flue-gas purification equipment (spray absorption, wet processes).

Water effluent from thermal power stations, particularly from coal-fired plants, can **pollute surface waters**.

The following **types of wastewater** can occur in power plants:

—   regenerate from the conditioning of makeup water and desalination of condensate
—   water used for washing condensate filters
—   effluent from coal handling and coal storage
—   sensitive wastewater, e.g., from pickling and conservation
—   ash-laden water (deslagging water) from liquid ash removal
—   water from the boilers, turbines and transformers
—   cooling-tower discharge and makeup-water conditioning
—   wastewater from flue-gas purification.

The quantities of such **wastewater** depend on the **type of fuel** and on **various plant-specific boundary conditions** and can be expected to range between 10 and 100 l/h for each MWel power output. Such effluent can be **polluted by entrained suspended solids, salts, heavy metals, acids, alkalies, ammonia and oil.**
Wastewater treatment can be based on **physical, chemical and thermal methods.** For some forms of wastewater, e.g., **filter backwashing water and coal-storage effluent,** physical treatment in the form of **filtration, sedimentation and/or ventilation** will usually suffice. Other forms of wastewater such as **regenerate from makeup-water and condensate polishing, flue-gas cleaning water or other wastewater streams** require **chemical treatment** - flocculation, precipitation, neutralization - or even **thermal treatment** - evaporation, drying - before they can be discharged; cf. environmental briefs Wastewater Disposal and Mechanical Engineering Workshops, Shipyards.
As mentioned in section 2, **wastewater** occurring as a **consequence of certain flue-gas desulfurization processes** can contain various **pollutants** deriving from the flue gas. The **composition** of such wastewater depends on a number of parameters, e.g., the type of fuel, the process water and the quality of the additives.
As a rule, **wastewater from flue-gas cleaning** requires **physicochemical conditioning** in the form of neutralization, flocculation, sedimentation and filtration to **remove heavy metals and suspended solids** (gypsum, etc.).
The amount of wastewater occurring in connection with wet desulfurization methods having **gypsum as a by-product** depends mainly on the **chloride content** of the coal and on the permissible concentration of **chloride in the washings.** In a typical hard-coal power plant, flue-gas desulfurization processes can yield wastewater in quantities between 20 and 50 l/h per MW of power output.
The high **water solubility of calcium chloride** ($CaCl_2$) entrained in the wastewater makes it an **unprecipitable saline emission.**

41

If **no salt** is allowed to be discharged into the receiving water, the **FGD wastewater** can be **evaporated** to yield dry, water-soluble salts requiring **controlled disposal**, e.g., in an **underground sensitive-waste depot**. Since the evaporation process requires high **energy inputs**, it should be ascertained for such cases whether or not an **effluentless method** (dry process, spray absorption) would be suitable.

Apart from the aforementioned **direct effects**, power plants can also have **indirect effects** on water. Consider, for example, the **"acid rain" phenomenon** involving the washout of airborne pollutants (SO_x, HCl, NO_x) from power plants in connection with natural precipitation.

2.3   Soil and groundwater

Thermal power plants can have multifarious impacts on soil and groundwater. The **soil quality**, for example, can be adversely affected by **dust sediment**, particularly in the near vicinity of the plant. The **seriousness of ground-level pollution** depends on the **heavy-metal content** of the dust. The **chemism of the soil** can be altered by acidic precipitation (acid rain) characterized mainly by the **acid formers SO_2 and NO_x**. Under unfavorable conditions, acidification can pass from the soil to both the groundwater and surface waters. The extent of **soil and groundwater pollution** does not depend on how much particulate matter and acid formers are contained in the exhaust, but rather on the **absolute quantities** emitted in the course of a year **(total annual emissions)** and on the **conditions of distribution**. Thus, it is important to limit such emissions by separation capacities commensurate with the size of the power plant.

The ground and, even more so, the groundwater in the immediate vicinity of the power plant are **threatened** by the escape of **water-polluting substances**, the main sources of which are various **weak points** in the collection and purification of wastewater, the leakage of oil and oil-containing liquids, and storage areas for oil, coal and residues.

The **deposition of residues** also has **consequences for the soil** and, even more so, for the **groundwater**. Power-generating residues consist primarily of **slag, fly ash, remanents from flue-gas desulfurization, and sludge from the treatment of raw water and effluent**. The **residual quantities** depend in part on the processes employed; in general, however, it may be said that, **the lower the quality of the coal, the higher the quantity of residues**.

Slag and fly ash can be put to various uses (for roadbuilding or as cement aggregate), depending on their composition. To the extent that they **cannot be recycled**, such substances must be **disposed of at suitable dumps** (e.g., above groundwater level). In Germany, these matters are regulated by *TA-Abfall* (Technical Instructions on Waste Management).

Part 1 in Appendix C of the catalogue of particularly sensitive wastes specifies **aboveground deposition** in the form of a **mono-type hazardous waste dump** for **solid reaction products** resulting from the **purification of combustion-plant exhaust gases, excluding gypsum**; cf. briefs Solid Waste Disposal, Disposal of Hazardous Wastes.

The nature of FGD residues depends on the method employed (cf. Appendix A-3) and may occur in **recyclable form, e.g., gypsum**. The quantities depend on the **sulfur content and the calorific value of the fuel, the degree of desulfurization, and the additives involved**. Prior to choosing a particular desulfurization process, it should be ascertained whether or not the respective **remanent substances** occurring as by-products of the different processes could be **marketed** in the respective country. This would require a **detailed local market analysis**, appropriately involving **local contractors/consultants**. Potential uses for the residues (as building materials) must be investigated; in their absence, it must be clarified whether or not and under which conditions the substances can be safely disposed of.

The following table compares the **quantified residues of flue-gas desulfurization in facilities** fired with heavy fuel oil and two different types of coal:

	Hard coal	Lignite	Heavy fuel oil
Calorific value [kJ/kg]	28 000	10 000	40 000
Sulfur content [weight %]	2.0	2.0	2.0
Degree of desulfurization [%]	85	85	85
SO_x in raw gas [kg/MW$_{el}$h]	14	38	9.5
[mg/m³STP]	4 000	8 600	2 850
SO_x in treated gas [kg/MW$_{el}$h]	2.1	5.7	1.4
[mg/m³STP]	600	1 300	427

Residual quantities [kg/MW$_{el}$] (process-dependent)	Hard coal	Lignite	Heavy fuel oil
Gypsum	32	87	22
Sulfite/sulfate	36	97	24
Sulfur	6	16	4
Sulfuric acid	18	50	12

41

When both **fly ash** and **desulfurization products** (gypsum or a sulfite/sulfate mixture) require **disposal**, it is advisable to **mix the products first**. A blend of fly ash and desulfurization products can be **hardened** to **stabilize the water-soluble constituents** (stabilizate) and **reduce their leachability**.

Desulfurization processes with useful end products require appropriate treatment of the **wastewater**. The resultant **sludge** contains **large amounts of heavy metals** and therefore should be treated as **sensitive waste**.

2.4   Human health

Adverse effects of thermal power plants on human health can derive from the **direct impact of noxious gases on the organism** and/or their **indirect impact via the food chain and changes in the environment**. Especially in connection with **high levels of fine particulates**, noxious gases like S_2 and NO_x can lead to **respiratory diseases**. SO_2 and NO_x can have health-impairing effects at concentrations below those cited in the German smog ordinance. The **duration of exposure is decisive**. **Injurious heavy metals** (e.g., lead, mercury and cadmium) can enter the **food chain** and, hence, the **human organism** by way of **drinking water and vegetable and animal products**. **Climatic changes** such as **warming and acidification of surface waters**, *Waldsterben* (forest death) caused by **acid rain** and/or the **greenhouse effect** of CO_2 and other trace gases can have **long-term detrimental effects** on human health. Similarly important are the effects of **climatic changes on agriculture and forestry** (and thus on people's standard of living), e.g., large-scale shifts of cultivation to other regions **and/or deterioration of crop yields**. Hence, the construction and operation of thermal power plants can have both **socioeconomic and sociocultural consequences**; appropriate **preparatory studies, gender-specific and otherwise**, are therefore required, and the state of **medical services** within the project area must be clarified in advance. **Early, comprehensive involvement of the concerned sections of the population in the planning and decision-making process** can help **reduce and avoid points of conflict**.

Noise, as an item of emission from thermal power plants, has direct effects on humans and animals. The main **sources of noise** in a power plant are: the mouth of the smokestack, belt conveyors, fans, motors/engines, transformers, flues, piping and turbines.

At least some of the **personnel** working in power plants are exposed to a more or less **substantial noise nuisance**.

Diverse noise-control measures can be introduced to **reduce immissions** to a tolerable level, whereas the primary goal must be to **protect the power plant staff**. To the extent possible, **power plants should be located an acceptable distance from residential areas**, and all appropriate **noise-control measures** must be applied to the respective sound sources at the **planning and construction stages**.

Two particularly effective measures are the use of **sound absorbers** to reduce flow noises and the **encapsulation of machines** and respective devices to reduce airborne and structure-borne sound levels. Appropriate enclosures constitute an additional means of simultaneously reducing both the emission and immission of noise. Incidentally, **enclosures** also provide weather protection and are therefore used widely in power plant engineering.

2.5   Landscape

Power plants have substantial **spatial requirements**. The extent of **land consumption** is generally **higher for coal-fired facilities than for gas- or oil-fueled plants**. With regard to siting, cf. environmental briefs on Spatial and Regional Planning, Planning of Locations for Trade and Industry.

The **landscape** is also affected by construction of the **roads** needed for delivering operating media and disposing of residues; cf. environmental briefs on railways, roads and waterways. The associated **mining activities** to obtain **coal** and, say, **limestone** (for desulfurization purposes) and for **disposing of residues not to be recycled** also tend to **alter the landscape**. In connection with the **disposal of residues**, priority should be given to **landfilling schemes** (e.g., in worked-out strip mines) or **land reclamation in coastal areas**. Both alternatives avoid the need for **separate dumping facilities** and put the residues to an **advantageous use**. The **residues**, of course, should be environmentally benign, either by nature or by reason of appropriate **treatment** to impart, for example, a **low level of leachability**. Additionally, it must be **ascertained** whether or not and which measures (**sealing, controlled drainage, conditioning of percolating water**) will be required to **keep soluble heavy metals and other substances** contained in the residues from passing into the **groundwater** or **coastal water** (cf. sections 2.1.1 and 2.3).

41

Moreover, attendant **pollution** can cause **damage to forests, lakes and rivers**, re-sulting in **serious permanent changes in the landscape**.

3.   Notes on the Analysis and Evaluation of Environmental Impacts

3.1   Immissions Limits for Air

As already explained in section 2, the **decisive atmosphere-specific environmental impact parameter is "ground-level pollution"**, i.e., the effects of air pollution on humans, animals, plants and inanimate objects. **In evaluating the environmental consequences of thermal power plants, air pollution is normally of central interest**. With the exception of CO_2, the **main pollutants** are increasingly regulated by the **particular immission limits** adopted by different countries. In actual practice, concrete projects must attach **primary importance** to **abiding** by the **applicable standards**. In some countries, those standards are even more stringent than those stipulated by Germany's *TA-Luft* (Technical Instructions on Air Quality Control). To the extent that the relevant standards have not yet been set or have been set too high, recourse should be taken to the **long-term standards prescribed by** *TA-Luft* with regard to impairment of human health and, in part, to the protection of vegetation, materials, water bodies, etc. (cf. **Appendix A-4**).

If in connection with a concrete project the **relevant standards** will obviously be **exceeded** by the **baseline pollution load** or foreseeable developments, then the **promotion of thermal power plants must be ruled out** from the beginning **on environmental grounds**. According to *TA-Luft*, exceptions can be made for new power plants if the **additional burden** attributable to the planned facility will **not exceed 1 % of the long-term immission limits** (irrelevance clause).

If an **existing power plant** contributes considerably toward **substantial transgres-sion of the relevant immission limits**, the first step to take is to investigate the pos-sibility of its - economically feasible - **relocation**. If the results of the study indicate retention of the existing site, the annual relevant pollutant concentrations attribut-able to the power plant must be significantly reduced in absolute terms by appropriate **rehabilitation** measures. If the contribution of the existing power plant toward the overall pollutive burden does not exceed 1 % of the standard values after rehabilitation, the irrelevance clause may be applied by way of analogy to the exemption provisions for new plants.

Whenever the **relevant standards are significantly exceeded**, care must be taken to prepare an appropriate **sanitation concept** for the affected sphere of influence. Such a concept must provide for the reduction of pollution from sources not standing in direct connection with the project of interest.

With regard to the immission limits listed in **Appendix A-4**, the reader's attention is called to the fact that the **particulate, sulfur dioxide and nitrogen oxide values** serve as vitally important indicators for the environmental consequences of **thermal power plants**. The **limit values for hydrogen chloride, cadmium and lead** gain significance, when those elements are **more abundantly present than normal in the fuel**. In such cases, **all considerations concerning the environmental relevance of the thermal power plant must be made subject to an analysis of the fuel** to be used.

As far as **German immission limits** are concerned, it should be noted that they only come to bear in **increasingly rare cases**, because **steady cuts in pollutant releases** have enabled **extensive compliance** in most areas in recent years. Any requirements exceeding the immediately prophylactic scope are substantiated on the basis of the pollution prevention principle. **Pollution limits are not schematically transferable to other situations and other countries**, because, for example, the **sensitivity of the local vegetation**, the **prevailing climatic and weather conditions**, and the **composition of the local soil(s)** can be **wholly different**, hence justifying either **more stringent or more lenient standards. Those specified in** *TA-Luft* give due account to the **protection of human health**. As such, they are **more stringent for clean-air areas than for regions in which high levels of baseline pollution already prevail**.

3.2 Emission limits for air

As explained in section 3.1, **the premier measure for limiting the environmental consequences of thermal power plants is adherence to the pertinent immission limits**. Nonetheless, power plant emissions also should be appropriately limited - **since an ounce of prevention is better than a pound of cure**. As mentioned in section 2, there are a number of **tried & tested commercial-scale pollution control technologies**, each with its own particular **benefits and drawbacks**. One frequent **drawback** is the relatively **high cost of efficient technology**. The extent to which a **less complex and therefore less expensive approach** could significantly reduce the adverse environmental impacts of a thermal power plant should be **ascertained in advance**.

41

For example, it certainly would make sense to eliminate particulate emissions with a relatively low-cost cyclone instead of a more efficient and accordingly more expensive electrostatic precipitator or fabric filter, particularly since the high cost of the latter could be regarded as prohibitive, with the result that, ultimately, no dust control effect whatsoever is achieved. According to that same line of reasoning, it would be better to install a single-field electrostatic precipitator than none at all on the grounds that a multiple-field unit would be too expensive. Moreover, the use of more **elementary processes** has the added advantage of simplifying the **operation, maintenance and repair** of the equipment while offering **a higher level of operational reliability.**

Appendix A-5 lists the **main laws, rules and regulations governing the release of power plant emissions** to the air, water and soil **in the Federal Republic of Germany.**

As a rule of thumb **for concrete projects, the emission limits adopted by the developing country** or countries in question **should be adhered to.** In some cases, of course, this could result **in transgression of the comparatively strict emission limits prevailing in the Federal Republic of Germany.** Depending on the general context, though, that still could be regarded as tolerable. Nevertheless, the **pollution prevention principle** dictates that every attempt be made to install **appropriate emission control technologies, even on a stage-by-stage basis** if necessary, e.g., by first installing a cyclone separator and leaving room for the eventual retrofitting of an electrostatic precipitator.

Appendix A-6 summarizes the **essential emission limits for airborne pollution from large-scale combustion plant in the Federal Republic of Germany.**

As the table shows, **the requirements differ according to type of fuel and size of installation** (the latter expressed in terms of thermal output), whereas the larger installations generally are expected to satisfy more stringent environmental protection standards.

Other European Countries go by emission limits similar to those applying in Germany, particularly by way of EC Directive 88/609, most notably for SO_2. The **Japanese and U.S. American** emission limits are also comparable, but how stringently they are enforced depends on local circumstances (competent authorities, baseline pollution levels, etc.). **Appendix A-6** also lists the **emission standards** for new, large-scale coal-fired power plants in selected countries, along with the corresponding EC standards, for the indicators SO_x, NO_x and particulate emissions. Also included is a **conversion chart for converting SO_2 and NO_x units from** $mg/m^3 STP$ to ppm or $lb/10^6$ BTU.

The **limit values** prescribed in Appendix A-6 can be achieved at **justifiable expense for favorable fuels,** i.e., for those with high calorific values and low sulfur contents. For **unfavorable fuels,** however the **stipulation of low emission limits** can be rather **problematic.** For example, according to table 2, it would take a separation efficiency of roughly 98 % to limit the SO_x emission level to 400 mg $SO_2/m^3 STP$ for a raw gas concentration of roughly 18 000 mg $SO_2/m^3 STP$. For such fuels, however, **stipulation** of an 85 - 95 % **degree of desulfurization** corresponding to the justifiable techno-economic expenditures would be more advantageous.

In **some** countries, the only available **fuels** are of such **inferior quality** that the **emission levels** listed in Appendix A-6 **cannot be adhered** to, and **higher levels are** therefore **permitted**.

It would be **inappropriate to simply transfer the emission limits** of, say, the Federal Republic of Germany to other countries, since **identical limitations** in combination with **inferior fuels** would call for **more sophisticated purification technology** than that required in Germany. To maintain a like level of expenditures, one must work from the given emission levels and automatically arrive at higher limit values. It should be noted in that connection, that some of the fuel used in the Federal Republic of Germany does not meet standard German specifications.

From the standpoint of environmental protection, **emission limits** serve merely as **expedients** denoting a certain state of technological development under a certain set of boundary conditions. The **primary purpose of environmental protection**, however, must be to **protect human health**, the vegetation, water bodies, etc. In other words, the **primary objective** of such provisions is to comply with the immission limits (cf. section 3.1). The factors governing ground-level pollution were discussed in section 2.

3.3   Monitoring of pollution levels

As a rule, it takes **very sensitive instruments** to accurately measure **pollutant concentrations**, since the levels in question can be situated several orders of magnitude below the emission concentrations. Still, certain **conclusions can be drawn concerning past pollution by studying the proposed site and its surroundings**. The **baseline pollution level** will be **all the higher**, of course, if **other power plants** and/or **emission-intensive industries** are located in the near vicinity or if the proposed site borders on a **major traffic artery. A conflict of purposes** could arise in that **cogeneration**, for example, as its high efficiency and accordingly low emission levels requires a **nearby consumer**, normally some form of industrial enterprise. If the consumer is characterized by relatively high emissions, the correspondingly high baseline pollution level could partially or even entirely counteract the environmental merits of cogeneration.

With regard to emission measurement, care should be taken to ensure that the scope of supply for the power plant includes instruments for measuring dust, SO_x and NO_x emissions. Such **pollutants** are relatively **easy to monitor** with the aid of **mobile local instruments** applied to **flues or breeching**. The requisite **gas analyzers** operate according to **different principles**. Differentiation is made between photometric and physicochemical measuring processes.

41

Photometric processes operate on a purely physical basis (nondispersive infrared process, nondispersive ultraviolet process), while the **physico-chemical processes** are based on a chemical reaction. Such instruments offer resolutions extending to 1 ppm.

Particulate concentration levels are monitored primarily by physical techniques, e.g., using graphimetric and radiometric instruments.

3.4   Emission limits for wastewater/effluent

In the **Federal Republic of Germany, effluent** from water treatment and cooling systems is subject to **discharge limitations** pursuant to section 7a *Wasserhaushaltsgesetz - WHG* (Federal Water Act) and Appendix 31 of the *Rahmen-Abwasser VWV* General Administrative Framework Regulation on Wastewater as listed in **table 3**.

Table 3

Discharge limitations for effluent from water treatment and cooling systems

Closed-loop systems of:

		Power plants	Industrial processes	Other steam-generating sources
		Random sample		
Settleable solids	mg/l	0.3	0.3	0.3
Available chlorine	mg/l	-	0.3	-
Hydrazine	mg/l	-	-	5.0
		2-hour composite sample		
Chemical oxygen demand (COD)	mg/l	30	40	-
Phosphorus (P_{tot})	mg/l	3	5	8
Vanadium	mg/l	-	-	3
Iron	mg/l	-	-	7

Source: *Rahmen-Abwasser VWV* (General Administrative Framework Regulation on Wastewater), Appendix 31 (Aug. 13, 1983)

To the extent that a **flue-gas desulfurizing system** produces wastewater, the **minimum discharge requirements** put forth in Appendix 47 of the General Administrative Framework Regulation on Wastewater as per section 7a Federal Water Act dating from Sept. 8, 1989, shall apply (cf. Appendix A-4).

The discharge of **effluents other** than those described in section 2.2 is governed by additional appendices to the General Administrative Framework Regulation as per section 7a of the Federal Water Act; its Appendix 49, for example, applies to **oily wastewater.**

The above requirements are in line with the **stringent provisions** of the **German Federal Water Act,** which stresses the importance of **prevention** and prescribes **limits** based on the **hazard levels of the respective substances.** Moreover, the *Abwasserabgabengesetz* (**Wastewater Charges Act**) rewards users who satisfy the requirements of section 7a, WHG (75 % lower wastewater charge) or who maintain existing facilities at least 20% below the prescribed limits (**setting off the cost of investment** against the past three years' wastewater charges.

For a concrete project, the type and nature of **tolerable water pollution** naturally depends on the **size, quality and manner of utilization of the receiving water. Weak, sensitive recipient bodies** must be **analyzed** in any case. Particularly in **tropical countries,** the **water flow rate can vary** widely on a seasonable basis - a fact that must be given due consideration. In that connection, consideration must be given to either **relocating** the plant or, as discussed in section 2.2, installing a **dry cooling tower.** Apart from the pollution load, the **tolerable thermal load** on the receiving body must be critically examined for each concrete project. According to the **recommendation of the German Länder working group on water *LAWA*,** the maximum temperature increase of a receiving body in a temperate climate zone should not exceed 3 K.

3.5   Noise

Depending on the **local situation,** the noise **immission requirements** for power plants can **differ widely.** According to the *TA-Lärm* (Technical Instructions on Noise Abatement) in the Federal Republic of Germany, the following noise **immission limits** (guide values) should be complied with:

41

	day dB (A)	night dB (A)
areas containing only nonresidential buildings	70	70
areas containing primarily nonresidential buildings	65	50
areas containing nonresidential and residential buildings	60	45
areas containing primarily residential buildings	55	40
areas containing exclusively residential buildings	50	35
areas containing health resorts, hospitals, nursing homes	45	35

The concrete-case values also depend on the baseline noise-immission levels.

As a rule, **power plants** should be located **as far as possible from residential areas**. According to the **North-Rhine/Westphalian spacing ordinance** *Abstandserlaß*, a distance of **800 m** or more means that the power plant can be expected to **cause no impairment**. In a number of German cities, power plants are situated much closer to residential areas, particularly in the case of cogenerating facilities, since the district heat produced by the power plant suffers substantial transmission losses with increasing distance to the consumer heat sinks.

The distance between a power plant and the nearest residential area depends primarily on the **noise immission levels** encountered at the points of interest, i.e., where the noise is measured. Noise immissions from the **boiler and turbine plant** can be substantially reduced by the application of **noise control** measures to the façade.

The **delivery of fuel and process materials** and the **hauling away of residues** (incl. the loading and unloading of trucks, railroad cars, barges, etc.) **contribute substantially to the overall noise pollution levels** from a power plant. For a coal-fired plant, the noise caused by the coaling system must also be allowed for. Consequently, **delivery and removal activities**, as well as **operation of the coaling system**, often have to be restricted to the **daytime** hours.

4.   Interaction with Other Sectors

Power plants release certain pollutants into the air, water and soil. If a **substantial number of small individual industrial furnaces with relatively poor pollution characteristics** can be **replaced** by a single **central thermal power plant**, or if such a plant is able to provide process heat as well as electricity to industrial enterprises, the resultant gain in efficiency and environment-friendly technology can yield a **relative improvement in the overall emission/immission situation**. Within that context, **cogeneration** appears as **a favorable** option, as long as the plant can be located in an industrial zone or integrated into an industrial complex with adequately large heat demand.

Power plants require diverse **operating media**. The relevant interaction with other industrial sectors is particularly pronounced in the case of coal-fired power plants. The sectors of essential relevance include **mining**, of course, as the coal source, and the **nonmetallic minerals industry** as a supplier of lime products for flue-gas desulfurization. If gas is used as fuel, the power plant will interact closely with the natural **gas industry**, and oil-fueled plants depend on **oil producers, refineries** and **petroleum-product storage** and **transport firms**. Reciprocity between a thermal power station and such other sectors involves the entire system catena, e.g., from the mining of the fuel to the disposal of residues (cf. section 5). Additionally, the power plant's water consumption must be viewed in context with the public **water supply system**, if both are competing for the same scarce water resources.

Relations with yet other industrial sectors can be entered into in connection with the disposal of residues. Fly ash and slag, for example, can serve as aggregates in the **cement industry**, and a number of byproducts from flue-gas desulfurization (gypsum, stabilizate and compounds of sulfur) can be useful in the **cement, plaster or chemical industry** (e.g., as **fertilizer**), depending on their properties and degree of purity. Such connections can help reduce the exploitation of natural resources like gypsum. Fly ash and desulfurization products (gypsum, sulfite, sulfate) can also be used in the **construction of roads and dams** or as **fillers** for purposes of **recultivation** (backfilling of mines).

5.   Summary Assessment of Environmental Relevance

As explained in sections 2 and 3, **thermal power plants have negative environmental impacts** in the form of **emissions** extending from **particulates, noxious gases** (SO_x, NO_x, CO, CO_2, HCl, HF, ...) **and waste heat to noise pollution**. Diverse measures such as **appropriate siting, the use of efficient, environment-friendly technologies** (cogeneration, i.e., the combined generation of heat and power) and the **avoidance or reduction of noxious emissions** can substantially alleviate such negative environmental consequences. Nonetheless, it is **not always possible to limit the environmental consequences to an acceptable scale**, particularly if **inferior fuel** is used, the **power plant is unusually large**, or the **surroundings** (human population, flora and fauna) are **particularly sensitive**.

For the purposes of an **environmental impact assessment, the entire system catena - from the production and transportation of fuels and chemicals to their in-plant combustion and on to the disposal of residues and the consumption of energy produced in other areas**, e.g., a user industry - must be given thorough consideration. Such a **holistic approach** helps **identify additional burdens** resulting from, say, transportation of the fuel or residues by truck, as well as **reductions**

41

ascribable to such aspects as **credits** granted for the replacement of older, less ecologically sophisticated combustion plant.

Since the **primary objective** in the erection of an environmentally compatible power plant must be to **reduce pollution of the environment**, the siting and baseline-pollution evaluation aspects are exceedingly important. However, a **conflict of goals** can arise by reason of the fact that the **positive effects of reduced emissions** - thanks to **cogeneration**, for example - can be partially or entirely **negated** by the necessity of **locating the plant in the near vicinity of an industrial complex** in which the **pollutant concentrations** already have contributed to **baseline pollution** in the area in question.

Regarding the **limitation of particulate, SO$_x$ and NO$_x$ emissions** by thermal power plants, various **well-proven commercial-scale techniques** are available. Since, for economic reasons, many countries prefer to fuel their power plants with **indigenous coal** characterized by **high ballast** and **sulfur contents**, special attention must be paid to reducing both of those pollutants. Depending on the local boundary conditions and in consideration of the overall situation, every attempt should be made to **reduce emissions to below 150 mg/m³$_{STP}$ particulates** and/or **2000 mg/m³$_{STP}$ SO$_2$**. Technically feasible measures for **low-NO$_x$ combustion** should be incorporated at the planning stage to ensure limitation of NO$_x$ emissions. Depending on the type of fuel in question, such pollution-control measures can **confine NO$_x$ emissions to the 200 - 600 mg/m³$_{STP}$ range** (excl. slag tap firing).

In general, **priority** should be **attached** to a **combination of avoidance and combustion-modification measures**, e.g., high efficiency, with favorable effects on CO$_2$ emissions. **Secondary measures** in the form of post-combustion flue gas clean-up, for example, should remain just what the name implies.

In assessing the environmental compatibility of a thermal power plant, proper **monitoring** is extremely important, since the best of all emission-control **measures** can only be as **efficient** as the attendant monitoring. One suitable approach would be to **appoint one or more in-house environmental protection officers**.

The following catalogue of criteria should be applied to the **planning and evaluation** of the **environmental relevance** of **thermal power plants**:

— **efficiency** in the **production** and **ultimate use** of electricity and/or heat (subsidized rates?);
— substantiable **necessity of the project** (size of plant, interaction with other sectors);
— description and analysis of the **project** and its impacts (technical concept, choice of fuel, emission sources, control systems, safety considerations);
— discussion of **siting alternatives** and determination of **baseline pollution levels** and the **prospective overall burden** at the selected location (ground-level pollution, ambient air pollution, effects on water, soil, flora, fauna, human health, physical and cultural assets);

— ascertainment of the **environmental relevance** of effects emanating from the anticipated overall burden, plus **measures aimed at reducing relevant environmental burdens** (siting, avoidance measures, pollution control by pre- and post-combustion measures).

41

6.   References

General

Asian Development Bank: Environmental Guidelines for Selected Industrial and Power Development, Projects, 1987.

Biswas, A.K.; Geping, Q.: Environmental Impact Assessment for Developing Countries, London: Tycooly Publ., Editor: United Nations Univ., Natural Resources and the Environment Series, vol. 19, 1987.

Deutsche Stiftung für Internationale Entwicklung (DSE - German Foundation for International Development): Environmental Impact Assessment (EIA) for Development; Proceedings of a joint DSE/UNEP International Seminar in Feldafing, Federal Republic of Germany, April 9 - 12, 1984.

Fleischhauer, M.; Friedrich, R.; Häring, S.; Haugg, A.; Müller, J.; Reuter, A.; Voß, A.; Wystrcil, H.-G.: Grundlagen zur Abschätzung und Bewertung der von Kohlekraftwerken ausgehenden Umweltbelastung in Entwicklungsländern, Institut für Energiewirtschaft und Rationelle Energieanwendung, Stuttgart, May 1990.

Storm, Bunge: Handbuch der Umweltverträglichkeitsprüfung, Berlin: E. Schmidt-Verlag, Umweltprogramm der Vereinten Nationen, Ziele und Grundsätze der Umweltverträglichkeitsprüfung, January 16, 1987.

World Energy Conference: Environmental Effects Arising from Electricity Supply and Utilisation and the Resulting Costs to the Utility, Report 1988, Oct. 1988.

Air Protection

Anton, P.; Elsässer, R. F.: Problemverschiebungen bei der Umweltpolitik zwischen Luft, Wasser und Boden, VGB-Kongreß "Kraftwerke 1985", p. 207 - 211.

Basu, P.; Greenblatt, J.; Wu, S.; Briggs, D.: Effects of Solid Recycle Rate, Bed Density and Sorbent Size on the Sulfur Capture in a Circulating Fluidized Bed Combustor, Proceedings from the 1989 International Conference on Fluidized Bed Combustion, San Francisco, Ca, pp. 701 - 707.

Baumüller, F.: Überblick über die Entschwefelungsverfahren, Sonderpublikation der BWK, Staub, Umwelt, p. 7 - 11, 1986.

Berman, I.M., Fluidized bed combustion systems: FBC presents a way to burn coal with minimal SO_2 and NO_x emissions. Development work is leading into demonstration units by a number of manufacturers, POWER ENGINEERING, November 1982.

Boardman, R.D.; Smoot, L.D.: Prediction of Fuel and Thermal NO in Advanced Combustion Systems, 1989; Joint Symposium on Stationary Combustion NO_x Control, March, San Francisco, Ca.

Davids, P.; Haug, N.; Lange, M.; Oels, H.-J. und Schmidt, B.: Luftreinhaltung bei Kraftwerks- und Industriefeuerung, BWK 39, Heft 4, p. 180 - 188, 1987.

EPRI Report, Inorganic and Organic Constituents in Fossil Fuel Combustion Residues, Volume 1: A Critical Review, EPRI EA-5176, Project Z4BS-8, Interim Report, August 1987.

Given, P.H.: An Essay on the Organic Chemistry of Coal, COAL SCIENCE, Volume 3, Edited by Gorbaty, M.L.; Larson, J.W. and Wender, I., pp. 63 - 252, 1984.

Graßl, H.: Anthropogene Beeinflussung des Klimas, VGB Kraftwerkstechnik 69, Heft 11, November 1983.

Haji-Javad, M.; Heinisch, M.; Hetschel, M.; Hutter, F.; Ludwig, H.: Konzeption eines Steinkohlekraftwerks aus umweltfreundlichen Komponenten, Forschungsbericht BMFT-FB-T 85 - 065.

Haßler, G.; Fuchs, P.: Verfahren und Anlagen zur kombinierten SO_2-/NO_x-Minderung, Sonderpublikation der BWK, Staub, Umwelt, p. 21 - 27, 1986.

Kalmbach, S.; Kropp, L.: Umweltrelevante Stoffe, Umweltmagazin, p. 53 - 55, May 1987.

Kanij, J.B.W.: The Emission of Polycyclic Aromatic Hydrocarbons by Coal-fired Power Stations in the Netherlands, Kema Scientific & Technical Reports 5, 1987.

Krolewski, H.: Maßnahmen zur Luftreinhaltung bei Kraftwerken und ihre Auswirkungen auf Wasser und Abfall, VGB Kraftwerkstechnik 65, Heft 9, pp. 801 - 806, 1985.

Leckner, B.; Amand, L.E.: Emissions from a Circulating and a Stationary Fluidized Bed Boiler: A Comparison, Proceedings from the 1987 International Conference on Fluidized Bed Combustion, Boston, Ma, Vol. 2, pp. 891 - 897.

41

Lee, Y.Y.; Hiltunen, M.: The Conversion of Fuel-Nitrogen to NO_x in Circulating Fluidized Bed Combustion, 1989 Joint Symposium on Stationary Combustion NO_x Control, March, San Francisco. Ca.

Leithner, R.: Einfluß unterschiedlicher WSF-Systeme auf Auslegung, Konstruktion und Betriebsweise der Dampferzeuger, VGB Kraftwerkstechnik 69, July 1989.

Natusch, D.F.S.: Final Report: Formation and Transformation of Particulate Polycyclic Organic Matter Emitted from Coal-fired Plants and Shale Oil Reporting, U.S. DOE Contract DOE-AC02-78EV04960, University of Colorado, April 1984.

Natusch, K.; Ratdjczak, W.: Meßtechnik zur Überwachung des Betriebsverhaltens von Rauchgasreinigungsanlagen. Sonderpublikation der BWK, Staub, Umwelt, p. 29 - 34, 1986.

Perhac, R.M.: Environmental Effects of Nitrogen Oxides, 1989 Joint Symposium on Stationary Combustion NO_x Control, March, San Francisco, Ca.

Smith, R.C.: The Trace Element Chemistry of Coal During Combustion and the Emission from Coal-fired Plants, Prog. Energy Combustion Science, 6 (1) pp. 53 - 119, 1980.

US EPA Report: Preliminary Environmental Assessment of Coal Fired Fluidized Bed Combustion Systems, EPA Report No. 600-7-77-05, May 1977.

US EPA Report: The Hydrogen Chloride and Hydrogen Fluoride Emission Factors of NAPAP (National Acid Precipitation Assessment Program) Emission Inventory, US EPA Report No. 600/7-85/041, October 1981.

US EPA Report: Locating and Estimating Air Emissions for Sources of Polycyclic Organic Matter, EPA 450/4-84-007P, September 1987.

Vernon, Jan L.; Soud, Hermine N.: FGD Installations on Coal-fired Plants, IEA Coal Research, EACR/22, London, April 1990.

Weber E.; Hüber, K.: Übersicht über rauchgasseitige Verfahren zur Stickoxidminderung, Sonderpublikation der BWK, Staub, Umwelt, p. 12 - 16, 1986.

Yeh, H.; Newton, G.J.; Henderson, T.R.; Hobbs, C.H.; Wachtner, J.K.: Physical and Chemical Characterization of the Process Stream for a Commercial Scale Fluidized Combustion Boiler, Environmental Science & Technology, Vol. 22, July 1988.

Residues

Hackl, A.: Vom Rohstoff bis zum Sonderabfall, Entsorgungspraxis 3, p. 81 - 83, 1987.

Pietrzeniuk, H.-J.: Rückstände bei der Verbrennung: Flugaschen, Filterstäube und REA-Gips, Umwelt Nr. 6, p. 455 - 458, 1986.

Verwertungskonzept für die Reststoffe aus Kohlekraftwerken, VGB Kraftwerkstechnik 66, Nr. 4, p. 377/385, 1986.

Wastewater / Effluent

Burfmann, F.: Betriebserfahrung mit der Abwasseraufbereitung hinter einer Rauchgasreinigungsanlage, VBG Kraftswerkstechnik 66, 1986 H. 9, p. 866 - 871.

Heitmann, H.G.: Chemische Behandlung von Abwässern aus Kraftwerken, BWK 38, Nr. 11, p. 499 - 509, 1986.

Ludwig, H.: Abwasserbehandlung, BWK Bd. 437, 1985, Nr. 9, p. 343 - 351.

Neumann, J.C. und Hofmann, G.: Behandlung und Aufarbeitung von Abwässern aus Rauchgaswäschen, BWK Bd. 437, 1985, Nr. 9, p. 352 - 355.

Sieth, I.: Abwasser aus Rauchgasreinigungsanlagen, Techn. Mitt. 78., Jahrg. 1985, H. 1/2, p. 71 - 73.

Laws, Directives

Wastewater Charges Act (*Abwasserabgabengesetz* dated Nov. 6, 1990; Federal Law Gazette, *BGBl.* I, p. 2432).

First General Administrative Provision Pertaining to the Federal Immission Control Law (*Erste Allgemeine Verwaltungsvorschrift zum Bundes-Immissionschutzgesetz*) Technical Instructions on Air Quality Control (*Technische Anleitung zur Reinhaltung der Luft - TA-Luft*) dated February 27, 1986, joint ministerial circular (*GMBl. Gemeinsames Ministerial-Blatt* p. 95, ber. p. 202).

Deutsches Umweltrecht, WLB, Verlag Technik GmbH, Berlin, 1991.

Act on the Prevention of Harmful Effects on the Environment Caused by Air Pollution, Noise, Vibration and Similar Phenomena (*Gesetz zum Schutz vor schädlichen Umwelteinwirkungen durch Luftverunreinigungen, Geräusche, Erschütterungen und ähnliche Vorgänge*) Federal Immission Control Act

41

(Bundes-Immissionsschutzgesetz - BlmSchG) as amended and promulgated on May 14, 1990 (Federal Law Gazette *BGBl.* I, p. 880).

Waste Avoidance and Waste Management Act *(Gesetz über die Vermeidung und Entsorgung von Abfällen (Abfallgesetz - AbfG)* dated August 27, 1986 (Federal Law Gazette *BGBl.* I, p. 1410, ber. p. 1501).

Act on the regulation of matters relating to water resources *(Gesetz zur Ordnung des Wasserhaushalts; Wasserhaushaltsgesetz - WHG)* dated September 23, 1986 (Federal Law Gazette *BGBl.* I. p. 1529).

Lärmbekämpfung 81, Entwicklung - Stand - Tendenzen, Umweltbundesamt (German Federal Environmental Agency), (Ed.), Berlin 1981.

General Administrative Framework Regulation on Minimum Requirements for the Discharge of Wastewater into Waters *(Rahmen-Abwasser-Verwaltungsvorschrift)* with Annexes 31 and 47 to section 7a WHG, dated November 25, 1992.

VDI guideline 2113 (12/76): Emission control, supplement units for solid fuels fired boilers.

Vernon, Jan L.: Emission Standards for Coal-fired Plants: Air Pollutant Control Policies, IEACR/11, IEA Coal Research, London, August 1988.

Verordnung zur Durchführung des Bundesimmissionsschutzgesetzes (Störfall-Verordnung, Ordinance for the Implementation of the Federal Immission Control Act = Hazardous Incident Ordinance), with
- Erster Allgemeiner Verwaltungsvorschrift zur Störfall-Verordnung (first general administrative provision on the hazardous incident ordinance) and
- Zweiter Allgemeiner Verwaltungsvorschrift zur Störfall-Verordnung (second general administrative provision on the hazardous incident ordinance).

Vierte Verordnung zur Durchführung des Bundes-Immissionsschutzgesetzes (fourth ordinance for the implementation of the Federal Immission Control Law) Verordnung über genehmigungsbedürftige Anlagen - 4. BImSchV (ordinance on installations subject to licensing) dated July 24, 1985 (Federal Law Gazette BGBl. I, p. 1586).

Zweites Gesetz zur Änderung des Bundes-Immissionsschutzgesetzes (second law amending the Federal Immission Control Act) dated October 4, 1985 (Federal Law Gazette BGBl. I. p. 1950).

Second General Administrative Provision on the Waste Avoidance and Management Act (Zweite allgemeine Verwaltungsvorschrift zum Abfallgesetz) Technical Instructions on Waste Management (TA-Abfall), Part 1: Technical Instructions on the storage, chemical, physical and biological treatment, incineration and storage of waste requiring particular supervision (Technische Anleitung zur Lagerung, chemisch/ physikalischen, biologischen Behandlung, Verbrennung und Ablagerung von besonders überwachungsbedürftigen Abfällen) dated March 12, 1991.

41

7.   Appendices

A-1   Flow diagram of energetically and environmentally relevant materials in a thermal power plant

A-2   Schematic diagram of a thermal power plant equipped with various flue-gas cleaning systems

A-3   Details of various desulfurization processes

A-4   Immission limits standards as per the German *TA-Luft*

A-5   German laws and regulations governing the limitation of emissions from thermal power plants

A-6   Emission limits for air pollutants from large firing installations (≥ 50 MW) in Germany; emission limits for new, large-scale, coal-fired power plants in various countries, plus pertinent EC and World Bank standards; conversion chart for SO_2 and NO_x emissions

A-7   Minimum requirements as per German Federal Water Act (WHG) section 7a Appendix 47: Scrubbing of flue gases from combustion plant, Sept. 8, 1989

41

Appendix A-1

Flow diagram of energetically and environmentally relevant materials in a thermal power plant

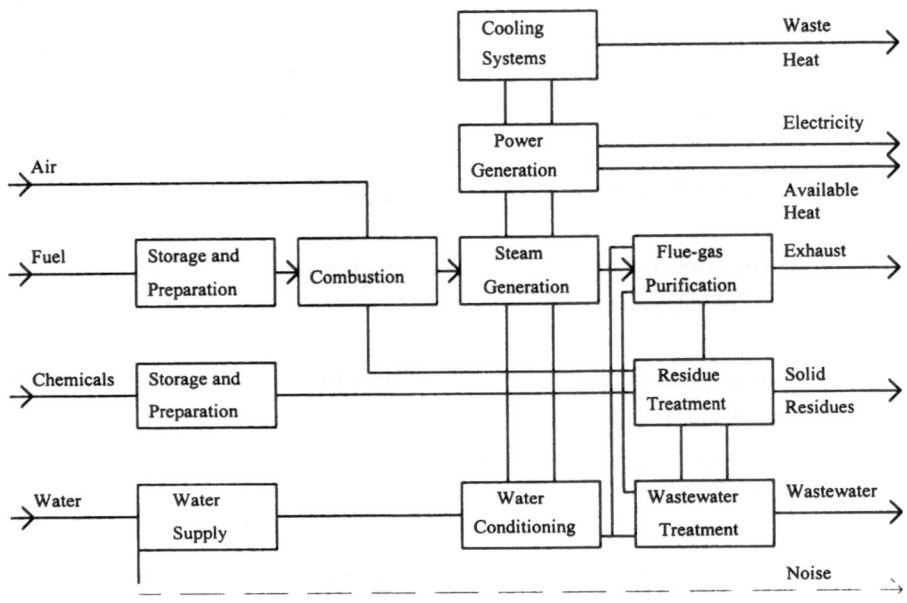

Appendix A-2

Schematic diagram of a thermal power plant equipped with various flue-gas cleaning systems

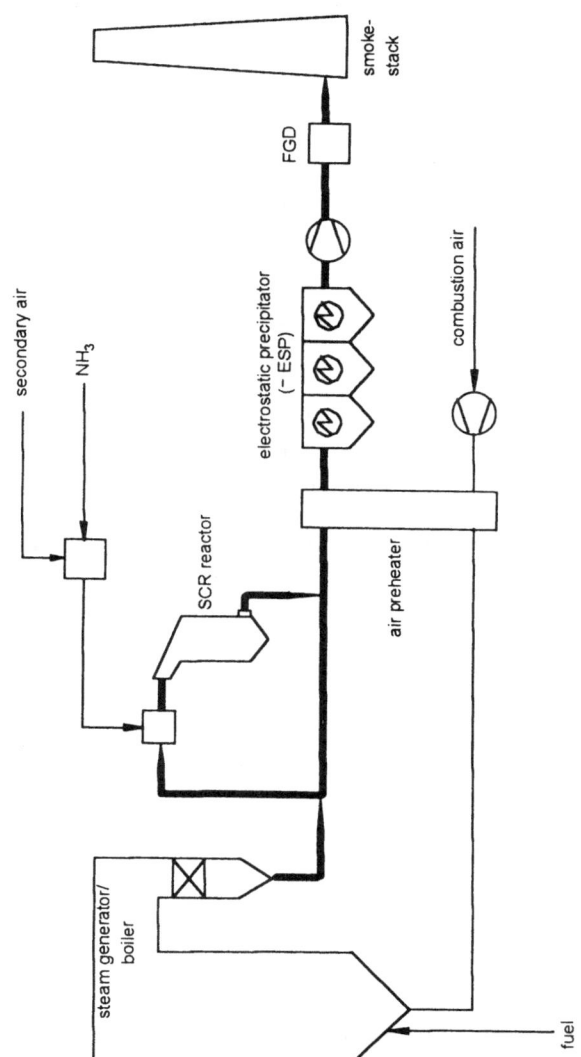

Appendix A-3

Details of various desulfurization processes

In-boiler desulfurization techniques are employed for **solid fuels**, e.g., in fluidized-bed combustion systems. The SO_2 forming in the flue gas combines with lime or limestone injected into the combustion system. Desulfurization therefore takes place simultaneously with fuel combustion at roughly 850°C. That relatively low combustion temperature helps limit NO_x emissions to 200 - 400 mg/m³STP. The **degree of desulfurization** ranges between 80 % and 90 %. Fluidized-bed combustion systems, which only can be used in new power plants, operate according to either the stationary or the circulating principle, with the latter achieving the lower emission levels under otherwise identical boundary conditions.

Dry additive processes can be applied to **coal-fueled**, grate- and dust-fired boilers. At a temperature below 1000°C, a pulverized lime product, e.g., slaked lime, is injected into the flue gas at a point above the combustion chamber, where it reacts with the SO_2 and precipitates. The requisite equipment is retrofittable and can **remove 60 - 80 % of the sulfur** from the flue gas.

The **residual product from fluidized-bed combustion and dry additive processes** - a mixture of coal ash, CaO or other unreacted additive, and various calcium salts ($CdSO_4$, $CaCl_2$, CaF_2) - is separated out in a downstream dust precipitator. In each case, it should be ascertained whether or not the residue can be put to some practical use, perhaps in the building materials industry (usually somewhat problematic due to the mixed salts), or will instead require safe disposal.

There are three basic types of **flue-gas desulfurization** processes:

- wet processes
- spray-dryer processes
- dry processes.

The **wet process** using limestone, lime or slaked lime as the additive and producing gypsum as a reaction product is the most widely employed **commercial-scale alternative**. It has yielded the largest worldwide empirical potential and is used in the majority of facilities. Appropriately processed via drying and pelletizing, for example, the gypsum can be used by the building materials industry, mixed with fly ash and landfilled, or used for land reclamation purposes in coastal areas (cf. section 2.5).

In the **spray-dryer process**, the sorbent (lime or slaked lime) is sprayed as an aqueous solution into an absorber at 60 - 70°C. As the water introduced with the suspension evaporates, the additive reacts with any SO_2 present to produce a fine-grained reaction mixture that precipitates out in downstream particulate removal equipment. Consisting of various calcium salts ($CaSO_4$, $CaSO_3$, $CaCl_2$, CaF), excess sorbent and residual fly ash, the reaction product can either be landfilled or used for land reclamation purposes. Potential preteatment requirements for the residue and additional measures for preventing groundwater or coastal-water contamination due to leaching are dealt with in section 2.5.

Other flue-gas desulfurization techniques, most notably **dry processes** using activated charcoal and **regenerative processes** involving sodium sulfite as the sorbent and sulfur dioxide as an intermediate product capable of further processing into sulfuric acid or sulfur, have become widely accepted in some areas and can be used for various other specific situations. As a rule, however, such processes are more elaborate and expensive than the limestone/gypsum techniques, and they impose particularly stringent standards on the quality of the end products, for which a corresponding market, e.g., the chemical industry, is needed.

Assuming otherwise identical constraints, the **quantities of residue** produced derive in descending order from the dry sorbent, spray drying, scrubbing with gypsum, and scrubbing with sulfuric acid or sulfur (cf. section 2.3).

41

Appendix A-4

Immission Limits as per the German *TA-Luft*

Pollutant		IW 1	IW 2
- Suspended particles	mg/m³	0.15	0.30
- Lead and inorganic lead compounds as components of the suspended particles indicated as Pb	µg/m³	2.0	-
- Cadmium and inorganic cadmium compounds and components of the suspended particles indicated as Cd	µg/m³	0.04	-
- Hydrochloric acid indicated as Cl	mg/m³	0.10	0.20
- Carbon monoxide	mg/m³	10.0	30.0
- Sulfur dioxide	mg/m³	0.14	0.40
- Nitrogen	mg/m³	0.08	0.20

The above table lists the immission limits for the prevention of health hazards as prescribed by the German TA-Luft, (Technical Instructions on Air Quality Control). The values **IW1 and IW2** are the **short-term** and **long-term limits**, respectively. In assessing the environmental compatibility of a thermal power plant, its IW2-value (continuous operation), for which *TA-Luft* specifies a monitoring period of one year, is significant.

As protection against substantial detriment and nuisance attributable to particulate precipitation, *TA-Luft* prescribes the following limits referred to as deposition values.

Pollutant		IW 1	IW 2
Particulate deposition	g/(m²d)	0.35	0.65
Lead	mg/(m²d)	0.25	-
Cadmium	µg/(m²d)	5.0	-
Thallium	µg/(m²d)	10.0	-
Fluorine	µg/(m²d)	1.0	3.0

The inorganic compounds in the above table are regarded as particulate constituents, with fluorine HF and the inorganic gaseous fluorine compounds counting as F.

Relatively **little information** is available on the **combined**, so-called synergistic, **effects** of pollution and on the interaction of atmospheric pollutants.
The **toxicological effects of the various pollutants** are listed in the **Compendium of Environmental Standards**.

The main **effects of the most important pollutants** are essentially as follows:

- Lead inhibits various globular metabolic enzymes in humans and other mammals, causing disruption of the oxygen balance and tidal volume. Sustained intake of less than 1 mg Pb/d has injurious effects. For plants, which take in lead primarily from the soil as opposed to the air, lead is only mildly toxic, the tendency being to lower the quality, but not the quantity, of the produce.

- Cadmium, a freely soluble metallic element, is resorbed into the digestive tract and stored in the liver and kidneys of humans and other mammals. Like its various compounds, cadmium has carcinogenic properties. In Asia, the so-called Itai-Itai and Aua-Aua syndromes were found to have been caused by Cd-polluted rice. Even low concentrations of cadmium in the soil can cause extensive damage to plants. Plants assimilate cadmium through their roots as well as their leaves and branches. Apart from reducing yields, cadmium contamination is hazardous to human health in that it enters the food chain as a cumulative toxin.

- Carbon monoxide, with its affinity for hemoglobin, the protein pigment responsible for the transport of oxygen, is toxic to humans and other vertebrates. Ingested exclusively by inhalation, carbon monoxide is odorless, colorless, tasteless and otherwise imperceptible to the senses. CO is nontoxic to plants, because it rapidly oxidizes to form CO_2, which plants need for photosynthesis.

- Sulfur dioxide causes corneal clouding, respiratory distress, inflammation of the respiratory tract, irritation of the eyes, disorientation, pulmonary edema, bronchitis and cardiovascular insufficiency in humans and other mammals.

Sulfur dioxide exposure causes both direct damage to the aboveground parts of plants and indirect damage primarily by way of soil acidification.

41

-     Nitrogen oxides resulting from combustion processes occur in the atmosphere mainly as nitrogen monoxide NO and nitrogen dioxide NO_2. The preferred generic term is nitrous gases (No_x)
Inhaled by a human or animal, an NO_x gas enters the lungs and irritates the
mucous membranes. While NO_2 leads to pulmonary edema, NO affects the
central nervous system.

In a photochemical smog situation, nitrogen oxides and hydrocarbons combine to form nitrate compounds that cause irritation of the eyes and mucous
membranes.

All nitrous gases are toxic to plants, as evidenced by brown to brownish-
black leaf margins and spots. The poisoning process culminates in dry with-
ering of the damaged cells.

Compared to nitrogen monoxide NO, nitrogen dioxide NO_2 is substantially
more toxic. For plants and animals, NO_2 is less hazardous than for human
beings. Atmospheric NO oxidizes to NO_2; consequently, NO is prevalent in
the near vicinity of combustion plants and is gradually supplanted by NO_2
with increasing distance from the source.

Appendix A-5

German laws and regulations governing the limitation
of emissions from thermal power plants

- Federal Immission Control Act
  (*Bundes-Immissionsschutzgesetz*)   (BImSchG)
  - Ordinance on Large Firing Installations
    (*Großfeuerungsanlagenverordnung*)   (GFAVO)
  - Technical Instructions on Air Quality Control
    (*Technische Anleitung zur Reinhaltung der Luft*)   (TA-Luft)
  - Technical Instructions on Noise Abatement
    (*Technische Anleitung zum Schutz gegen Lärm*)   (TA-Lärm)
  - Hazardous Incident Ordinance (*Störfallverordnung*)

- Resolution of the Conference of Ministers for the
  Environment concerning mandatory dynamization
  of the Ordinance on Large Firing Installations
  with regard to nitrogen oxide emissions
  (*Beschluß der Umweltministerkonferenz*)   (UMK)

- Federal Water Act (*Wasserhaushaltsgesetz*)   (WHG)
  - General Administrative Framework Regulations on ...
    Wastewater (*Rahmen-Abwasser-Verwaltungs-*
    vorschrift), Annexes 31, 47 to section 7a   (AbWVwV)
  - Provisions governing the handling of substances
    constituting a hazard to water (section 199)   (AbWVwV)

- Ordinance on the industrial sources of wastewater
  (*Abwasserherkunftsverordnung*)   (AbWHerkV)

- Waste Avoidance and Waste Management Act
  (*Abfallgesetz*)   (AbfG)
  - Technical Instructions on the storage, chemical,
    physical and biological treatment, incineration
    and storage of waste requiring particular
    supervision   (TA-Abfall)

41

Appendix A-6

**Emission limits for air pollutants from large firing
installations (≥50 MW) in Germany**

Type of fuel	MW*	Dust	NOx (as NO$_2$)	SO$_x$ (as SO$_2$)	CO	HCl	HF
Data stated in mg/m³STP, dry base							
Solid	≤300	50	400	2000 (400)[1]	250	200	30
	>300	50	200	400	250	100	15
Liquid	≤300	50	300	1700	175	30	5
	>300	50	150	400	175	30	5
Gas	≤300	5	200	35 (100)[2]	100		
	>300	5	100	35 (5)[3]	100		

* MW = megawatt thermal output

---

[1] for fluidized-bed combustion
[2] coke oven gas
[3] liquid petroleum gas

Appendix A-6

Emission limits for new, large-scale, coal-fired power plants in various countries, plus pertinent EC and World Bank Standards

Country	SO_2 emissions [mg/m³]	Size of plant	NO_x emissions [mg/m³]	Size of plant	CO emissions [mg/m³]	Size of plant	Dust emissions [mg/m³]	Size of plant
EC	400	> 500 MWt	650	> 50 MWt			50	> 500 MWt
World Bank	500 t/d or 50 µg/m³ additional immission over slight prior SO_2 burden (≤ 50 µg/m³) 100 t/d or 10 µg/m³ additional immission over high prior SO_2 burden (> 100 µg/m³)	> 200 MWt	858 (780 for lignite)				100 (150 in rural areas and when immission < 260 µg/m³ beyond power plant perimeter)	-
Australia	200		800	> 30 MWt	1000		80	
Austria	80 % (sep. efficiency)	> 200 MWt	800	> 50 MWt	250	> 2 MWt	50	> 50 MWt
Belgium	400	> 300 MWt	200	> 100 MWt			50	> 50 MWt
Canada	740		740				125	
Denmark	860	> 50 MWt	1150	> 50 MWt			57	> 5 MWt
Finland	140	> 150 MWt	200	> 300 MWt			57	> 50 MWt

41

Country	SO$_2$ emissions [mg/m³]	Size of plant	NO$_x$ emissions [mg/m³]	Size of plant	CO emissions [mg/m³]	Size of plant	Dust emissions [mg/m³[	Size of plant
France	1700 - 3400	(regional)					130	> 9.3 MWt
Germany	400	> 300 MWt	200	> 300 MWt	250	> 50 MWt	50	> 5 MWt
Great Britain	90 % (sep. efficiency)	> 700 MWt	760	> 700 MWt			97	> 700 MWt
India	height of stack > 500 MWt : 275 m > 200 < 500 MWt : 200 m < 200 MWt : (equation)		no limits				150 (350 for plants with < 200 MWt in unprotected areas)	
Italy	400	> 100 MWt	650	> 100 MWt			50	> 100 MWt
Japan	plant-specific		411	> 70000 m³/h			50	> 200000 m³/h
New Zealand							125 - 500	> 5 MWt
Netherlands	400	> 300 MWt	400	> 300 MWt			50	
Spain	2400						200	> 200 MWt
Sweden	290		430				35	
USA	740	> 29 MWt	740	> 29 MWt			37	> 73 MWt

/.../ The minimum size of plant to which the relevant limit applies is stated in MWt: the volumetric flue-gas flow is stated in m³STP/h

Appendix A-6

SO_2 and NO_x Emissions
Conversion Chart

To convert From ↓		To: (Multiply by) → mg/ m³	ppm NO_x	ppm SO_2	g/GJ Coal[A]	Oil[B]	Gas[C]	lb/10⁶ Btu Coal[A]	Oil[B]	Gas[C]
mg/m³		1	0.487	0.350	0.350	0.280	0.270	8.14×10^{-4}	6.51×10^{-4}	6.28×10^{-4}
ppm NO_x		2.05	1		0.718	0.575	0.554	1.67×10^{-3}	1.34×10^{-3}	1.29×10^{-3}
ppm SO_2		2.86		1	1.00	0.801	0.771	2.33×10^{-3}	1.86×10^{-3}	1.79×10^{-3}
g/GJ	Coal[A]	2.86	1.39	1.00	1			2.33×10^{-3}		
	Oil[B]	3.57	1.74	1.25		1			2.33×10^{-3}	
	Gas[C]	3.70	1.80	1.30			1			2.33×10^{-3}
lb/10⁶ Btu	Coal[A]	1230	598	430	430			1		
	Oil[B]	1540	748	538		430			1	
	Gas[C]	1590	775	557			430			1

A:- Coal:- Flue Gas dry 6 % excess O_2: Assumes 350 Nm³/GJ - ref IEA Paper 1986.

B:- Oil :- Flue Gas dry 3 % excess O_2: Assumes 280 Nm³/GJ - ref IEA Paper 1986.

C:- Gas :- Flue Gas dry 3 % excess O_2: Assumes 270 Nm³/GJ - ref IEA Paper 1986.

41

Appendix A-7

Minimum requirements as per German Federal Water Act (WHG*), section 7a

Appendix 47: Scrubbing of flue gases from combustion plant, Sept. 8, 1989

	COD	Filterable substances [4]	Fluoride	Sulfate	Sulfite	Lead	Cadmium	Chromium	Copper	Nickel	Mercury	Zinc	Sulfide
	Accepted engineering practice					State of the art							
General	80[5] 150[6] mg/l	30 mg/l	30 mg/l	20000 mg/l	20 mg/l	0.1 mg/l	0.05 mg/l	0.5 mg/l	0.5 mg/l	0.5 mg/l	0.05 mg/l	1 mg/l	0.2 mg/l
Hard-coal power plants (pollutant concentr., mg/kg)	see above					3.8 mg/kg	1.8 mg/kg	18 mg/kg	1 8 mg/kg	18 mg/kg	1.8 mg/kg	36 mg/kg	7.2 mg/kg
Lignite power plants with chloride contents up to 0.05 weight % (pollutant concentr., g/h) [7]	see above					0.2 g/h	0.1 g/h	1 g/h	1 g/h	1 g/h	0.1 g/h	2 g/h	0.4 g/h

* WHG = *Wasserhaushaltsgesetz*

4) via quicklime
5) via limestone
6) pollutant concentration in g/h per 300 MW installed electrical output
7) after subtraction of the prior COD pollutant concentration introduced with the service water

Mining and Energy

Power Transmission and Distribution

42

42. Power Transmission and Distribution

Contents

1. Scope

Adequate power supply systems constitute an essential part of any country's technical infrastructure. Such systems comprise facilities for the generation, transmission and distribution of electricity.

This brief deals with the planning, construction and operation of all technical facilities required for transmitting and distributing electric power.

Transmission is understood as the conveyance of electrical energy from its place of generation to its place of use. Power transmission is characterized by the conveyance of electric power over comparatively long distances with the aid of high-voltage and medium-voltage systems. Depending on the relative locations of the power generating facilities and the power consumers, many different forms of landscape and vegetation can be affected.

Distribution is understood as the delivery of electric power from the bulk power source to the consumer. As a rule, this involves relatively short distances within populated areas by way of medium-voltage and low-voltage systems.

The technical facilities required for transmitting and distributing electric power comprise mainly:

- overhead power lines
- cables
- transformer and switching stations.

2. Environmental Impacts and Protective Measures

Direct effects on the environment result from the erection and operation of such facilities, with the extent and intensity of the impacts depending substantially on the physical circumstances and the project planning.

This section describes and illuminates the **direct and indirect effects** of power transmitting equipment on the natural environment, i.e.:

- resources (water, soil, air) and
- ecological systems (flora and fauna, interlinked biotopes)

as well as on humans, i.e.:

- their health and safety, occupational and otherwise,
- their socioeconomic and sociocultural circumstances, as well as
- their visual perceptions.

42

2.1   Consequences for the natural environment

•   **Soil, water and air**

In wooded areas, the **erection and safe operation of overhead power lines** necessitate the **maintenance of unobstructed lanes,** the widths of which vary between 25 and 100 meters, depending on the size of the transmission line. Paved or unpaved access roads may be needed for installing and inspecting the lines and towers. That, in turn, entails the **permanent destruction of forest**; due to the loss of its original vegetation, the disturbed soil is at least temporarily unprotected and exposed to the climatic effects of heat, frost and rain, all of which promote erosion. Soil compaction resulting from project-site motor vehicle traffic intensifies the soil's **susceptibility to erosion.** Afterward, the affected area is only conditionally suitable for other forms of utilization. Any strip of land that cannot be used for forestry purposes due to its being located along a right of way (danger of grounding) should be greenbelted in order to combat erosion. The use of space-saving components helps reduce the space requirement quite substantially.
The erection of **towers and tower footings** on steep slopes demands detailed knowledge of the subsoil situation. Any mistake made in planning and executing the work can seriously impair the stability of the slope and lead to **slip erosion.**
The construction of **switching and transforming stations** permanently occupies certain areas and **jeopardizes the soil and groundwater through** the potential **leakage of coolants** and insulants (mineral oil or other liquids possibly containing toxic polychlorinated biphenyls - PCB) in large quantities from such components as transformers, capacitors, ground-fault neutralizers and underground cables.
Suitable collecting troughs/separators must be provided to prevent contamination of the groundwater and soil.

•   **Flora and fauna**

Due to the machines involved, **the erection phase** of power transmission lines and switch plant **imposes stress on - and causes potentially permanent damage to - the surrounding flora and fauna.**
The clearing of lanes in wooded areas **modifies the microclimate** by admitting more insolation and wind, thus altering the temperature distribution. Such changes can disrupt the local ecosystem.

Depending on the attitude of the transmission route, such lanes can seriously **enhance the windslash incidence** in the adjacent woods.

Frequently, **fire and herbicides** are used to create lanes and keep them clear of vegetation. Since such practices are very damaging to both flora and fauna, they should be dispensed with to the greatest possible extent.

Due consideration should also be given to **the danger of dissecting biotopes into small "islands"** that are not likely to survive beyond the medium term.

Overhead power lines are a fourfold hazard for **birds**:

— they debase breeding grounds;
— birds (particularly night-flying species) can fly into the wires;
— birds can be killed by simultaneously contacting two wires or a wire and the tower (medium-voltage lines);
— the "magnetic compass" effect can interfere with their navigating system.

In Germany, the populations of several species of large birds have been substantially decimated; some 70 % of all white storks are lost to electrocution.

In extremely rare cases, a **short circuit** or other defect in a transformer or switchgear has been known to cause a **fire** with resultant destruction of the surrounding flora and fauna.

The roads and lanes established to enable the erection and maintenance of overhead power lines can have the same environmental consequences as other traffic routes, in particular opening-up effects (cf. environmental briefs Transport and Traffic Planning, Road Building).

• **Minimization and avoidance measures**

The aforementioned **impacts can be minimized or avoided** by heeding the following points in connection with the planning and erection of overhead power lines:

— consider possible **alternatives** to new construction, e.g., conversion or more efficient/multiple utilization of existing lines;

— **adjoin** overhead lines to existing traffic routes and pipelines;

— **adapt** the right of way to existing landscape structures, i.e., avoid such exposed locations as hilltops and domes, ridges, razorbacks, etc.;

42

— significantly reduce the consumption of landscape and forest by installing high towers to permit **greater spans between towers**, thus traversing larger areas;

— **avoid** nature preserves and other protected areas, biologically and/or ecologically significant regions and recreation areas;

— use insulation sleeving, shrouding covers, perching and nesting platforms on towers for MV and LV lines as **protection for birds**;

— allow for the future installation of additional lines, i.e., for **multiple use of the transmission route** (multiple-circuit lines);

— significantly **reduce space requirements** by choosing suitable tower types (latticed steel, steel-pipe/concrete/wood) and configurations (size and arrangement of line-supporting cross-arms) and by using trefoil insulated LV and MV conductors;

— **reduce the ground-area-requirement** by using cables instead of overhead lines, even though the cable route also has to be kept free of tree growth (The use of cables can be problematic, however, for economic and technical reasons attributable to the high cost of investment and the need for highly qualified maintenance personnel.);

— **minimize the danger of soil and groundwater pollution** by performing routine safety inspections of pole impregnating facilities and by either replacing tar-base impregnants with more environmentally appropriate (salt-base) agents or opting for vacuum- or high-pressure impregnated wood;

— **prevent soil erosion** by mulching and greenbelting to cover all exposed ground. In climate zones with rainy seasons, this should be done at the beginning of the rainy season in order to prevent sheet erosion;

— **modify and strengthen existing lines** to save energy and additional cables;

— **restock** working areas under forest-traversing spans.

2.2   Human health, occupational safety and accident prevention

●   **Accidents**

The primary cause of accidents imperiling human life and limb (electrocution, serious burns) is **inadvertent contact with live components**, possibly **in inadequately protected facilities**, and the secondary cause is **fire resulting from short circuits**.

The danger of accidents is most acute when:

—   **technical specifications relevant to safety measures are disregarded** in the planning and erection of plant and equipment (use of low-quality components, inadequate sizing, negligent execution, nonobservance of safety clearances), so that the finished facilities are inherently unsafe;

—   **the operating personnel has not received sufficient training** in connection with safety measures and their observance;

—   **the local populace has not been properly educated with regard to electrical hazards**, which can lead to such misbehavior as climbing up on towers, trespassing on switching stations, lack of lightning conductors, illegal tapping of electricity, etc.

In the past, and to a certain extent even today, the use of **polychlorinated biphenyls** (PCB, askarel/chlophen) as flame-retarding dielectric liquids in transformers and capacitors has constituted a health hazard in its own right. PCBs are very toxic. They accumulate in the food chain, cause chronic disorders and are carcinogenic. Moreover, their incineration (due to, say, exposure to an accidental fire) produces highly toxic dioxins and furans.

With but few exceptions (e.g., for electric plants in underground mines), the use of PCB in electrical plants is now generally **prohibited** in many countries.

Tar-base impregnants for wooden poles constitute a health hazard in that they can cause skin ailments.

Such risks can be substantially **reduced or even completely avoided** by:

—   choosing plant components of the proper type and size,
—   precluding unauthorized access to electrical plant and installing anti-climbing guards on high-voltage towers,
—   reducing the danger of fire by using noncombustible dielectric liquids or dry transformers and refractory partitions,

42

— avoiding the use of dielectric liquids containing PCBs and of coolants in new installations; ensuring proper disposal and replacement of old transformers and the like,
— providing the operating personnel with appropriate safety clothing and suitable tools and test instruments,
— ensuring that the operating personnel receives the proper, duly qualified training,
— educating the local populace about the dangers of electrical installations.

• **Effects of electric and magnetic fields on human health**

According to information derived from prolonged observations and experiments in numerous countries, the electric and magnetic fields around power transmission and distribution facilities (exhibiting frequencies between 50 and 60 Hz) have **no harmful effects** on human health.

According to a WHO publication dealing with the effects of magnetic fields on human health, field strengths below 0.4 mT at 50 - 60 Hz induce **no detectable biological reaction**. The magnetic fields acting on the ground below overhead lines develop a maximum field strength of 0.055 mT for the above frequencies.

• **Noise nuisance**

Substation and distribution-system transformers generate a monotonous buzzing sound that can be annoying in residential areas. The **use of quiet-running transformers and/or appropriate structural measures** (incl. adequate distances) can avoid such problems.

2.3   Optical impairment of the landscape

Overhead power lines amount to an **optical disturbance**. The extent of disturbance depends on:

— the size, type and general configuration of the lines and towers,
— the concentration of overhead lines within a given area,
— the transmission route and/or visibility of the lines, i.e., how well the right of way has been accommodated to the landscape (color, "low profile"),
— the location (undeveloped/developed land, population density, industrial/residential areas, etc.).

The **recreational value** of landscapes and affected areas is diminished by the optical impairment.

The aforementioned preventive measures apply in equal measure to the avoidance of optical impairment.

2.4   Socioeconomic and sociocultural impacts

Any **direct consequences** the installation and operation of power transmission and distribution facilities may have on the socioeconomic and sociocultural environment are of minor significance. Radio and television reception, for example, can be substantially disturbed by corona discharges [luminous discharge along undersized and/or improperly arranged conductors (bundle-conductor lines)].

Indirect consequences derive from the purpose of such facilities, namely to improve living conditions by supplying electricity to a region or center. Access to electricity **increases comfort and convenience** in private life (e.g., time saved and work facilitated) and in the public domain. In combination with other technical infrastructural measures, it can initiate or stimulate economic activities aimed at creating new jobs (i.e., reducing unemployment) or rationalizing production processes[1].

On the other hand, past experience has shown that electrification and other forms of regional development can lead to a **loss of traditional ways of life, modes of behavior**, cultural peculiarities and sociocultural ties and structures. Moreover, it can have a whirlpool effect on neighboring regions, giving rise to emigration and new congested areas.

3.   Notes on the Analysis and Evaluation of Environmental Impacts and on Occupational Safety Standards

Numerous different authorities, associations, public and private bodies both corporate and individual must be **involved** in defining the rights of way and the locations of substations. The process must include appropriate consideration of environmental interests.

Suitable structural measures (e.g., to prevent erosion) and technical measures (e.g., to prevent the escape of transformer oil) must be taken to avoid **pollution of the soil and/or water**.

Optical impairment of the landscape is unavoidable but should be minimized. The extent of impairment depends both on how the land is used (work - recreation) and on its optical complexity. The right of way can be visually assessed with the aid of a computer.

---

[1]   The negative environmental effects of power generation can be aggravated by **excessive demand** resulting from artificially low (submarginal) supply tariffs.

Detriment to flora and fauna must be appraised with a view to the protection of endangered species and in consideration of local, national and international standards and regulations. Determination of the local and regional significance of biotopes must be based on a large-scale survey in which suitable measures for the protection of birds are included.

Internationally recognized and harmonized, detailed **standards on safety clearances, protective measures against contact with and entry to, in addition to working on, live systems** [e.g., the German Standards DIN 0800, DIN 0848, DIN 57106, Association of German Electrical Engineers' *VDE* guideline 0106, accident prevention provisions and implementing instructions for electrical equipment and operating equipment issued by the Verband gewerblicher Berufsgenossenschaften ("Elektrische Anlagen und Betriebsmittel" - VBG 4)] should be consulted in connection with the planning of power transmission and distribution facilities.

The use of **PCB** in closed systems (transformers, capacitors, etc.) has been prohibited in the EC since 1985, although the continued operation of existing PCB-filled equipment is permitted for the duration of its service life. In the interest of environmental protection, however, such equipment should be replaced and properly disposed of (sodium-base dechlorination of the oil). Its incineration would produce dioxins!

4.   Interaction with Other Sectors

The planning and installation of power transmission and distribution systems depend on decisions deriving from **higher-level** (national, regional) **planning** processes devoted to regional development, general energy development, town and country planning, general power supply measures, etc. (cf. corresponding environmental briefs).

There is a direct connection to the power generating sector (cf. environmental brief Thermal Power Stations). As soon as power transmission is correlated to a particular power source, the environmental impacts of the latter, i.e., of power generation, demand consideration; high transmission losses also have environmental consequences in that they necessitate the generation of additional power.

The rights of way for transmission lines are extensively determined by the **relative locations of the power plant and the power consumers**. Particularly valuable biotopes and landscapes must be protected by routing such rights of way around them.

Coordination with existing or still-to-be-installed technical infrastructure (roads, railways, waterways, other supply lines, etc.) is not only possible but even necessary for, say, crossing airports, waterways, roads, etc. and for the parallel routing of power transmission and telecommunication lines - all in order to ensure the safe, reliable operation of all facilities concerned.

With regard to the reprocessing and disposal of transformer oil (with or without PCB content), please refer to the environmental brief Disposal of Hazardous Waste.

5. Summary Assessment of Environmental Relevance

The aforementioned environmental impacts and their consequences are evaluated below, and potential means of minimization and avoidance are proposed.

Landscape consumption in the form of pressure on natural resources (soil, vegetation) and destruction of landscape is generally unavoidable, though adequate attention to environmental concerns at the planning stage can at least diminish its consequences.

Appropriate structural measures can be adopted to reduce, but not eliminate, the **hazard for birds** posed by overhead power lines.

The **danger of accidents for humans** emanating from transmission and distribution installations can be reduced by strict adherence to existing, recognized rules, regulations and standards. Relevant training and sensitization are crucial in this area.

The **emissions** (noise, corona conduction) of power transmission and distribution installations can be reduced to negligible levels by appropriate technical means. The **use of liquids containing PCBs in transformer substations** still constitutes a substantial hazard potential in that such liquids are liable to escape to the environment as a result of equipment malfunction or accident (leakage, fire). Consequently, the use of components and equipment containing PCBs should be globally prohibited, and existing equipment should be replaced.

Compared to other means of energy conveyance (road, rail, water, pipeline) the transmission of electricity involves a **modest, though by no means negligible, risk**. Whenever new facilities for transmitting and distributing electric power are deemed absolutely necessary (e.g., if there is no **possibility of opting for noncentralized power generation**), appropriate low-impact approaches should be sought out.

The easiest and most effective way to minimize or completely avoid harmful environmental impacts is to conscientiously **allow for environmental concerns from the planning stage on**.

42

6. References

Algermissen, W.; Lübben, W.; Kübler B.: SF_6-isolierte Lasttrennschalteranlagen, Eine Technik für die Netzstation von morgen, Elektrizitätswirtschaft, 1988, Heft 16/17.

Asian Development Bank: Environmental Guidelines for Selected Industrial and Power Development Projects; Manila 1988.

Biegelmeier, G.: Wirkungen des elektrischen Stromes auf den menschlichen Körper, etz., 1987, Heft 12.

Borris, D., v.: Umweltbelastungen durch Transport: Speicherung und Verteilung von Energie, Energie und Umwelt, Heft. 7/8, Bundesforschungsanstalt für Landeskunde und Raumplanung, Bonn, 1984.

Deutscher Bund für Vogelschutz, Landesverband Baden-Württemberg e.V. (Ed.): Verdrahtung der Landschaft: Auswirkungen auf die Vogelwelt, Ökologie der Vögel, Sonderheft 1980, Band 2, 1980.

Deutsches Institut für Normung e.V.: DIN 18005, Noise abatement in town planning, 1982.

Dreiser, Rolf: Ursachen und Folgen von Arbeitsunfällen in Elektrizitätsversorgungsunternehmen, Elektrizitätswirtschaft, 1983, Heft 13.

Fanger, U.; Weiland, H.: Entscheidungskriterien bei Projekten der ländlichen Elektrifizierung aus sozio-ökonomischer und entwicklungspolitischer Sicht. Kurzgutachten im Auftrag des BMZ, Arnold-Bergstraesser-Institut, Freiburg, 1984.

FINNIDA: Guidelines for Environmental Impact Assessment in Development Assistance; Draft 1989.

Groß, Markus: Graphische Datenverarbeitung in der Freileitungsplanung - Innovative Methoden mittels Sichtbarkeitsanalyse, Elektrizitätswirtschaft, 1990, Heft 6.

Haubrich, H.J.: Biologische Wirkung elektromagnetischer 50-Hz-Felder auf den Menschen, Elektrizitätswirtschaft, 1986, Heft 16/17.

Haubrich, H.-J.; Dickers, K.; Lange, G.: Influenzwirkung auf Personen und Fahrzeuge im elektrischen 50-Hz-Feld, Elektrizitätswirtschaft, 1990, Heft 6.

Jarass, L.: Hochspannungsleitung geplant - was is zu beachten?

Jarass, L.: Auswirkungen einer Dezentralisierung der Stromversorgung auf das Verbund- und Verteilungsnetz, in: Bodenbelastung durch Flächeninanspruchnahme von Infrastrukturmaßnahmen, Bundesforschungsanstalt für Landeskunde und Raumordnung (German Federal Research Institute for Regional Geography and Regional Planning (Ed.)), Bonn, 1989.

Jarass, L., Obermaier, G.M.: Raumordnungsgerechte Ausführung von Hochspannungsleitungen, Energie und Umwelt, Heft 7/8, Bundesforschungsanstalt für Landeskunde und Raumplanung (German Federal Research Institute for Regional Geography and Regional Planning (Ed)), Bonn, 1984.

Pflaum, E.: Entwicklung der Löschprinzipien von Hochspannungs-Leistungsschaltern, etz., 1988, Heft 9.

Rat von Sachverständigen für Umweltfragen: Sondergutachten März 1981, Energie und Umwelt, Verlag W. Kohlhammer GmbH, 1981.

Rauhaut, A.: PCB-Bilanz, etz., Bd. 104, Heft 23, 1983.

Sander, R.: Biologische Wirkungen magnetischer 50-Hz-Felder, Medizinisch-technischer Bericht, Elektrizitätswirtschaft, Bd. 82, Heft 26, Institut zur Erforschung elektrischer Unfälle der Berufsgenossenschaft der Feinmechanik und Elektrotechnik, Cologne, 1982.

Sauer, E. u.a.: Energietransport, -speicherung und -verteilung, Handbuchreihe Energie, Bd. 11, Cologne.

Schemmann, B.: Vakuum-Leistungsschalter in Ortsnetz-Verteilerstationen, etz., 1987, Heft 16.

Silny, J.: Der Mensch in energietechnischen Feldern, Elektrizitätswirtschaft, Jg. 84, Heft 7, 1984.

Soldner, K., Gollmer, G.: Probleme mit PCB-gefüllten Transformatoren, Elektrizitätswirtschaft, Jg. 82, Heft 17/18, 1982.

Theml, Horst: Schutz gegen gefährliche Körperströme - Anordnung von Betätigungselementen in der Nähe berührungsgefährlicher Teile, Elektrizitätswirtschaft, 1982, Heft 25.

42

Umweltbundesamt (German Federal Environmental Agency): Ersatzstoffe für in Kondensatoren, Transformatoren und als Hydraulikflüssigkeiten im Untertagebergbau verwendete Polychlorierte Biphenyle, Berlin 1986.

Umweltbundesamt: (German Federal Environmental Agency): Lärmbekämpfung 1988, Berlin, 1989.

United States Agency for International Development: Environmental Design Considerations for Rural Development Projects; Washington 1980.

VDEW: Begriffsbestimmungen in der Energiewirtschaft - Teil 4, Begriffsbestimmungen der Elektrizitätsübertragung und -verteilung, 4. Ausgabe, Frankfurt 1979.

VDEW: PCB oder Askarel, VDEW zum Thema Askarel Elektrizitätswirtschaft, Jg. 82, Heft 17/18, 1982.

VDEW: Vogelschutz an Freileitungen, 1986.

WHO: Environmental health criteria, Magnetic fields, Dec. 1985.

Zahn, B.: Weiterentwicklung SF_6-gasisolierter Schaltanlagen, etz., 1988, Heft 9.

(-): ANSI (American National Standards Institute) Standards.

(-): DIN VDE - Vorschriften zur Errichtung und Betrieb von elektrischen Anlagen (Standards relating to the erection and operation of electrical installations)

(-): Höchstzulässige Geräuschwerte für Transformatoren, Technische Angaben Trafo-Union, 1982.

(-): IEC (International Electrotechnical Commission) Publications.

Mining and Energy

Renewable Sources of Energy

43.   Renewable Sources of Energy

<u>Contents</u>

1. Scope

In addition to finite deposits of fossil and mineral fuels such as oil, gas, coal and uranium, the earth also offers various natural, auto-regenerative - or renewable - sources of energy that derive from sun **insolation, geothermal activity and gravitational forces**.

Theoretically, the **global supply of energy** from such renewable sources by far exceeds the earth's present **total energy demand**. The supply of energy is subject in part to pronounced technical and economic utility limitations, e.g., the **disparity** between the temporal/spatial demand for energy and the actually available supply of renewable energies, and the latter's modest power density compared to conventional energy vehicles.

The main renewable energy (RE) sources are:

1.   Insolation, i.e., the direct radiant energy of the sun (made useful by collectors, solar cells, etc.)

2.   Energy obtained from biomass; biochemical energy of photosynthetic products; made useful by

     -   burning (of wood, straw, etc.)
     -   gasification (of wood, etc.)
     -   anaerobic digestion (= biogas)
     -   alcoholic fermentation

3.   The kinetic energy of wind

4.   The kinetic energy of moving water:

     -   low-pressure systems
     -   high-pressure systems
     -   micro-hydropower plants
     -   tides, waves, ocean currents

5.   Miscellaneous

     -   geothermal energy
     -   thermal energy deriving from differences in seawater temperature
     -   osmotic energy deriving from concentration gradients between saltwater and freshwater.

43

With a view to the proper and adequate sizing and, hence, limitation of the environmental consequences of renewable energy systems, the energy consumers' options for the **conservation and rational use of energy** should always be given full consideration, whereas boundary conditions in the form of prices, tariffs, etc. are major factors.

The environmental impacts resulting from utilization of the following renewable sources of energy are dealt with in this brief:

- solar energy (heat and photovoltaics)
- energy from biomass
- wind energy
- hydropower
- geothermal energy.

To the extent deemed relevant, other renewable sources of energy are dealt with in other briefs.
With regard to the general environmental consequences of energy systems and to the supradisciplinary aspects to be considered in connection with the planning of energy policy and energy economics projects, the reader is referred to the environmental brief Overall Energy Planning.

2.   Environmental Impacts and Protective Measures

The utilization of energy, no matter what the source, is bound to have certain environmental consequences (land consumption, pollution, ...) that need to be identified and evaluated, preferably in advance.

2.1   Solar energy

The use of solar energy via collectors or photovoltaic systems places no immediate material burden on the environment. However, the collector system can be expected to contain a **heat transfer medium** (fluid), the escape of which could result in pollution. The acceptable media include such readily degradable substances as propylene glycols. Noxious additives serving as preservatives should be replaced by less harmful alternatives (carboxylic acid).
The use of solar cookers involves the **danger of blinding**, and solar energy collected by solar cells and stored in batteries demands proper handling and appropriate disposal of the spent batteries. The materials used for the battery case, as well as the hydrochloric acid and lead contents, can be recycled in suitable facilities.

Land consumption for small-scale systems can be avoided by installing them on roofs and facades. Well-considered integration can prevent optical/aesthetic impairment, and annoying reflections can be diminished by lumenizing and/or delustering.

With the exception of **reduced reflections,** no such measures can be applied to large-area systems. Consequently, optical/aesthetic expectations may stand in conflict with other natural surface potentials (soils for agricultural production, protection of species and biotopes; unless, of course, the site in question is located in the desert).

Depending on the local situation, the **shading and altered albedo** resulting from large-scale installations can affect the flora, fauna and microclimate (evaporation rates, airflow, temperature).

Solar cells and various collectors have a **substantial space requirement** relative to the amount of energy produced (per 100 MW: ~ 1 km² for solar cells and ~ 3 km² for solar-thermal power plants, compared to ~ 0.4 km² for hard-coal power plants).

Additional environmental impacts derive from the manufacture of **materials** used in the production of collectors and solar cells. Steel, copper and aluminum, all of which are used frequently, cause environmental problems in the form of emissions, i.e., particulates, fluorine compounds, solid and liquid waste and high levels of energy consumption, particularly for aluminum.

Some rare and **toxic metals** such as cadmium, arsenic, selenium and gallium used in solar cells are mildly pollutive at the processing stage (wastewater, exhaust gases). These substances are characterized by high chemical stability, and the environmental risk remains confined to the production site. Thus, adequate monitoring and safety measures can minimize the risk; cf. environmental brief Non-ferrous Metals.

2.2   Biomass energy

Used as a substitute for metal, cement, plastic and diverse other raw materials, biomass can help reduce the energy expenditures for processing and manufacturing such materials.

In the present context, however, our interest in biomass is limited to its being a source of energy.

Significant utilization of biomass presupposes that the biomass cycle of growth and extraction remains essentially intact, i.e., that the biomass source (a forest, perhaps), is always allowed to adequately regenerate.

43

2.2.1 Burning

The burning of biomass (wood, straw, dung, etc.) liberates **pollutants** -

— from the fuel and the combustion air
— or which form as a result of incomplete combustion [CO, tar, soot and hydrocarbons, including carcinogenic polycyclic aromatic hydrocarbons (PAH)].

The main cause of emission problems with biomass is **incomplete combustion**. The following measures can help achieve **complete combustion**:

Combustion plant

— sufficiently large incinerator
— sufficiently hot combustion chamber

Those conditions are inherently satisfied by systems equipped with prefiring chambers or for bottom firing.

Fuel conditions

— use of dry fuel (< 20 % wood moisture).

Mode of operation

— full-load operation
— uniform fuel supply.

The exhaust gases, particularly in the case of straw, contain large amounts of solid particulates; large-scale systems therefore should include appropriate cyclone separators or filters.

On a country-specific basis, **biomass can cover as much as 90 % of the overall demand for energy**. As a rule, wood, dung and straw are burned in **open fires** from which the aforementioned pollutants escape and can be inhaled by the users (primarily women and children).
This can amount to a formidable **health hazard**, particularly because of the carcinogenicity of polycyclic hydrocarbons. In addition, respiratory ailments can also result from such exposure.
The use of stoves with some form of chimney substantially reduces the indoor smoke nuisance and improves the combustion efficiency, thereby reducing fuel consumption and, hence, emission levels.

The **use of straw and dung as fuel** can lead to conflicts concerning agricultural production and the sustenance of soil fertility due to **loss of nitrogen and reduced humification**, because what has been burned cannot be returned to the soil. In some climate zones, using the ashes as fertilizer can cause a dust-evolution problem.

From an ecological standpoint, the use of **scrap wood and various forms of wood residue** calls for a somewhat sophisticated frame of reference: while tending felling can be both ecologically compatible and advisable, the safe extent of **wood removal from forests** and plantations depends on the climate, the soil conditions and the vegetation. The removal of wood residue impacts the nutrient cycle, humification, microflora and microfauna. This applies as well to large-scale stump-grubbing, which also makes the ground more **susceptible to erosion**.

Long-term natural wood production does not satisfy the "firewood criteria" of easy, short-term availability. **Agroforestry projects** involving certain harmonized plant species in certain spatial arrangements designed to make the individual species and combinations serve different functions (shading, soil amelioration, shelterbelting, improvement of water regimen, mulching, fuel, food/fodder, starting material), are able to more quickly satisfy fuel requirements by reason of brief rotation periods. Such - noncentralized - configurations facilitate the gathering of wood while abating environmental burdens in connection with road transport and helping to bridge over fuel shortages.

Intensive (energy farming) techniques based on fast-growing combustibles treated with high doses of pesticides and fertilizers can **pollute, i.e., eutrophize, surface waters** due to nutrient loading, possibly in combination with erosion, a loss of diversity, and health hazards emanating from residual pesticides. The use of machines on sensitive ground (marginal soils) can induce erosion; cf. environmental brief Forestry.

Large-scale felling of trees (= land clearing) affects the water economy and microclimate, is harmful to flora and fauna, and can cause erosion, the extent of which depends on the type of soil, the climate and the angle of slope.

If cleared land is not appropriately reafforested, or if the soil is overused for a prolonged period, both the soil and the water regimen may sustain irreversible damage.

Any attempt to substantially expand fuelwood production without **integrating the effort into the general agricultural scheme** can generate conflicts over space requirements for food production; cf. corresponding environmental briefs on agriculture, such as Plant Production, Forestry etc.

43

2.2.2 Gasification

As a rule, any gas extracted from biomass by such means as pyrolysis is used as fuel, either for heating purposes or for driving gas-fueled power generators.

While the environmental effects of fuel extraction from biomass are dealt with in section 2.2.1, additional ecological impacts can derive from:

—   carburetion (accidents, deflagration);
—   the gas itself (accidents, fire, poisoning due to leaks);
—   wastewater from gas scrubbing;
—   carbonization residue (ash, tar);
—   combustion emissions (exhaust, cooling water, lubricant).

Generator gas obtained from large plants (as opposed to small wood gasifiers, e.g., for tractors) should be cleaned and dedusted prior to use. The **wastewater** from gas scrubbing can be expected to contain ammonia, phenols, perhaps even cyanides and potentially carcinogenic polycyclic aromatic hydrocarbons (PAH). Consequently, they cannot be disposed of freely. To the extent possible, the incidental **tars and oils** should be returned to the gasification process. In addition to the mechanical extraction of solids, e.g., in a settling basin, the effluent can be put through a biological clarifying plant in which phenols are digested by suitable strains of bacteria.

Solid residue from the gasification process is usually heavily polluted and therefore **problematic** with regard to its **disposal**. The **harmful-substance contents** require case-by-case determination, because they vary according to the raw material in question and the process employed.

The **exhaust** from generator gas combustion may also require treatment, depending on the quantity involved and its pollutive load. It is likely to contain oxides of nitrogen, PAH's, carbon monoxide or soot (plus negligible amounts of sulfur dioxide). The NO_x and hydrocarbon contents can be extensively decomposed with the aid of **catalytic converters**.

2.2.3 Biogas

Biogas resulting from **anaerobic bacterial fermentation of biomass** consists primarily of **methane** (principal component), **carbon dioxide, carbon monoxide** and small amounts of **hydrogen sulfide**. Small biogas plants provide fuel for cooking, lighting, etc., while large-scale facilities can produce enough biogas for fueling gas motors.

Accidents can occur when a slurry pit or a fixed-dome digester has to be entered for cleaning (**danger of asphyxiation**).

Since hydrogen sulfide has **toxic** effects of humans, corrodes materials, and forms sulfur dioxide in the combustion process, its removal should be given due consideration. However, the precleaning process is rather complicated and generates end products with a pollutive potential. The chemicals used for cleaning biogas (e.g., iron oxide), as well as their reaction products (mixture of iron oxide and sulfur) demand proper storage, use and subsequent disposal.

Whereas biogas often requires interim storage, appropriate pertinent **safety standards** must be heeded (danger of poisoning, fire, explosion); cf. environmental brief Petroleum and Natural Gas.

The raw material may contain **toxic heavy metals** that are prejudicial to health. While such constituents (deriving from polluted soil) remain unaffected by the digestion process they nevertheless should be monitored (tested for). And while the digestion process does not kill off all pathogens and worm ova, the digested sludge nonetheless counts as safe and benign from the standpoint of epidemic control. Used improperly, its high nitrogen content can **emburden both surface water and groundwater**. Thus, the use of biosludge as a fertilizer must be properly timed (availability for plants), effected with suitable equipment, and applied in accordance with the soil's nutrient reserves.

Considering methane's relevance as a greenhouse gas, its collection and combustion is ecologically advantageous as long as it is being generated by anaerobic digestive processes.

2.2.4 Biofuels

Various technical processes are available for deriving oil and alcohol from biomass and using them as **substitutes for conventional fuels**.

The cultivation of biomass as a raw material for obtaining fuel by alcoholic fermentation (e.g., of sugar cane) or by extracting oil from soybeans stands in direct competition with foodstuff farming. Large **monocultures** involving high levels of fertilization and pesticide spraying have environmental impacts of the kind discussed in section 2.2.1; cf. environmental brief Plant Protection.

43

The following environmental loads result from the **production of ethanol and oil**:

—   **exhaust gases** deriving from the provision of process energy (e.g., distillation, burning or refining of crude oil) - cf. section 2.2.1;

—   **carbon dioxide** as a product of fermentation;

—   nontoxic but very **pollutive organic sludge and wastewater** (slops) from ethanol production, all containing large amounts of nitrogen-phosphorus and potassium components.

The **slops**, or distiller's wash, can serve as a fertilizer or fodder additive. If it contains enough residual sugar or starch, it is suitable for fermentation, i.e., biodigestion.
The biogas yield can serve as a substitute for part of the conventional process energy, while the organic substances remaining in the effluent must be decomposed in a clarifying plant.
The production of alcohol is very **energy-intensive**.
The **use of ethyl alcohol (ethanol) as a fuel additive** in internal-combustion engines produces relatively low pollution in the form of NO_x, CO, soot and simple hydrocarbons, but is accompanied by certain aldehydes, some of which are carcinogenic.
Motors fueled by alcohol alone should be specially tuned and optimized in order to minimize harmful emissions. **Catalytic converters**, for example, reduce the aldehyde emission levels to that of gasoline engines. Compared to gasoline/petrol, ethyl alcohol contains practically no carcinogenic polycyclic hydrocarbons.
Like alcohol, **biomass-base oil for diesel engines** gives off no sulfur or lead but some amounts of soot, simple hydrocarbons and particulate emissions. **Soot filters** are conditionally suitable for cleaning the exhaust gases.

2.3   Wind energy

Even large wind power plants have **modest environmental impacts**. Their material and space requirements are also relatively modest. The manufacture of some steel and plastic components, however, does involve certain environmental problems.

The following substantial environmental problems arise in connection with their operation:

—   noise;
—   landscape impairment;
—   danger of accidents due to rotor-blade detachment;
—   electromagnetic interference;
—   negative effects on fauna, birds in particular.

How much **noise** is produced depends on how fast the propellor is rotating. The faster the speed of rotation, the louder the noise.
Old aerogenerators have been known to produce sound intensities on the order of 130 dB(A). Small wind generators tend to make more "wind" noise than running noise. New facilities have aerodynamically optimized blades and encapsulated generators-cum-transmissions that minimize the noise nuisance. Nevertheless, a minimum distance of roughly 100 meters should be maintained between wind generators and residential areas. There is, of course, always the possibility that the safe clearances designated at the planning stage will eventually be transgressed by uncontrolled settlement (squatting).
Impairment of the landscape is unavoidable. The degree of impairment depends on local circumstances, including the intensity of wind-power utilization. Wind parks do more to impair the landscape than individual plants. Especially large aerogenerators with metal motors tend to **disrupt natural electromagnetic fields** and interfere with radio reception. Modern wind power plants have fiberglass rotor blades and therefore cause no such interference.
The **danger of accidents** attributable to rotor-blade detachment can be minimized, if not precluded, by routine inspections and maintenance, plus adherence to the appropriate safety clearances.

2.4   Hydropower

Hydropower is the by far the most important renewable source of energy. The incidental reservoirs often serve other, additional purposes such as irrigation and the supply of drinking water.
The harnessing of hydropower entails substantial intervention in the environment (land consumption, altered hydrological regimen, etc.). Due to the importance of hydraulic engineering with respect to the environment, and with deference to the vast experience that has been accumulated in connection with such facilities, a separate brief has been devoted to that sector.

43

2.5   Geothermal energy

Geothermal sources of energy include:

—   warm and hot water in deep-reaching joint systems of crystalline rock for-
     mations or deep-lying groundwater stories within expansive sedimentary
     basins,

—   hot-water and steam occurring deep within structurally disturbed zones or in
     regions marked by current or recent volcanic activity,

—   exploitation of geothermal energy according to the dry hot rock process
     (DHR technology presently under development).

DHR technology aims to establish artificial heat-exchange surfaces in hot rock
(with temperatures > 200°C) from which geothermal energy can then be extracted
by pumping water into and back out of the artificial hot-rock joint system. Despite
substantial research funding to date, however, the method's economic feasibility has
not yet been established.

The environmental impacts of exploiting geothermal energy depend on the concrete
situation. **Environmental burdens** can result from entrained pollutants (various
salts, sulfur compounds, arsenic, boron) and gases in the geothermal fluids. In
modern geothermal facilities the spent (cooled-down) fluids and their entrained
pollutants are pumped back into the ground, preferably to a point below the pay
zone of the occurrence, while the incidental gases are released to the atmosphere.

The extraction of geothermal fluids, particularly in dry-climate regions, can negatively influence **near-surface groundwater stories** and, hence, their utilization
(potable water, irrigation) by causing the groundwater table to recede (phreatic
decline).

Sustained use of a particular geothermal reservoir can lead to gradual and extensive
subsidence and frequent consequential damage to railroads, highways, power
transmission lines and, particularly, the pipelines through which the geothermal
fluids are pumped from the wells to the power plant/user. The local **hydrological
situation** can be substantially influenced and modified by attendant phenomena
such as the diversion of streams and rivers or even the formation of lakes in ground
depressions.

The **space requirements** of geothermal installations (wells, pipelines) are quite
modest - so much so that such facilities hardly interfere with agricultural utilization
of the surrounding land.

The drilling of wells in a geothermal field is somewhat hazardous in that unforeseen eruptions of steam can occur without notice and then take weeks or even months to get under control. In the meantime, the environment may have become substantially contaminated by impurities in the steam.

3. Notes on the Analysis and Evaluation of Environmental Impacts

The main environmental consequences of renewable energy systems are the **consumption of land area** and the **loss of plant and animal species and biotopes**. Biomass utilization also involves **solid waste, wastewater and air pollution**.

The environmental consequences of renewable energy systems can be limited in quantity, but normally require qualitative analysis with due regard for avoidance effects (e.g., CO_2 emissions) in comparison with nonrenewable energy sources. To evaluate the environmental impacts of any such system, one must begin with an analysis of the biotic (flora and fauna) and abiotic (water, soil, air) ecological factors. For the biotic domain, mapping and charting activities are necessary. For the abiotic range, water, air and soil samples should be analyzed according to standard techniques such as those described in DIN/EN and ISO standards, NIOSH standards, guidelines of the Association of German Engineers VDI, WHO recommendations, etc.).

The **evaluation of environmental consequences** is a deficitary matter in that, for example, no limit values can be quoted for the loss of animal species, biotopes, etc. Nor do any **generally recognized standards of evaluation exist** - quantitative or otherwise - for landscape impairment. The **criteria** need not always be as unequivocally quantifiable as "rarity" (e.g., as defined by **international conventions** within the pollutants' sphere of influence); it is also difficult to attach a particular value to consumed land area with allowance for alternative uses. For the abiotic domain, though, certain limit values and recommendations can be enlisted in connection with various types of pollution (wastewater, exhaust, noise).

To the extent available, **effect-specific reference/limit values** should be consulted for evaluating immissions (airborne pollutants, noise, ...) as a means of anticipating the sensitivity (reaction) of existing and planned forms of utilization (housing, farming) to the projected impairment.

For all forms of renewable energy utilization, the importance of immissions and pollutant levels increases along with the size of the project.

In connection with the extraction of energy from biomass, any solid substances that are re-utilized instead of being treated as waste count as a positive effect that must be given due consideration.

43

4.   Interaction with Other Sectors

If a planned renewable energy system will involve material emissions, the local
prior load must be determined in advance of the project's implementation (e.g.,
condition of recipient water in conjunction with wastewater-producing processes).
In addition to the effects of renewable energy utilization listed in section 2, such
secondary effects are also important. Apart from the project's consequences for the
basic needs of certain sections of the population, its possible impacts on agriculture,
water supplies, transportation and diverse aftereffects must also be accounted for
(whereas allowance must be made for the fact that improving the supply of energy
to or within a given region can have practically identical consequences for the
sectors in question):

— The loss of farmland alters the food market structure and/or necessitates the
   agricultural utilization of formerly more or less "virgin" areas. For additional
   information, the reader is referred to the environmental briefs on agriculture
   (e.g., Plant Production).

— Any more intensive use of water resources naturally involves higher rates of
   water consumption, larger volumes of wastewater and, hence, changes in the
   water regimen. That, in turn, affects the soil, the microclimate, the composition of the microsystem, and the hygienic situation (salinization, spread of
   pathogens; cf. environmental briefs **Rural Water Supply, Rural Hydraulic
   Engineering Large-scale Hydraulic Engineering, Water Framework
   Planning.**

— Increased traffic due to transportation in connection with large-scale renewable
   energy applications (or simply attributable to an improved energy supply
   situation) necessitates more and better traffic infrastructure. Its provision, in
   turn, has primary and secondary development effects; cf. environmental briefs
   Road Traffic, Transport and Traffic Planning. The general environmental
   impacts of renewable energy exploitation systems are discussed in the
   environmental brief **Overall Energy Planning**.

5.   Summary Assessment of Environmental Relevance

This environmental brief summarizes the environmental consequences of renewable energy sources. Such consequences include gaseous and liquid emissions, solid wastes, noise evolution, use of sensitive materials, land consumption and other forms of impairment.

The renewable-energy utilization options involving **little or no replacement or decomposition of material** (solar, wind) and, hence, fewer direct consequences for the environment are deserving of preferential treatment.

The fact that long-term sustained use of renewable energy sources can fit neatly into the natural biochemical and energy cycles produces a situation in which combustion and digestion processes (wood, straw, biogas, alcohol), unlike those involving fossil fuels, **add no carbon dioxide to the atmosphere**, because the amount emitted is offset by the incorporation of equal amounts into the *regeneration* of biomass. In other words, biomass enables the CO_2 - neutral generation of energy.

On the other hand, again unlike fossil fuels, the **continuous renewal process of biomass** as an energy vehicle ties up land area, i.e., soil, that otherwise could be put to some other or additional use, e.g., for agricultural production or agroforestry.

Land consumption is unavoidable. Accordingly, valuable ecosystems must be protected - *instead* of simply being exploited as a renewable source of energy.

As long as the requisite facilities are properly maintained and serviced by skilled specialists, and as long as the operating personnel is well-trained, the use of renewable energy sources poses little **danger of accidents**.

Like most finite sources of energy, the majority of renewable energy sources can be exploited both on a large, centralized scale as well as through **small, noncentralized facilities**. Some renewable sources of energy (e.g., solar cells, solar collectors, biogas, wind power) are inherently suited to noncentralized forms of energy generation, particularly in connection with energy supply and development strategies for rural, village-level and regional development projects involving little or no transport costs. Such constellations help minimize energy conveyance losses and avoid such secondary environmental problems emanating from the socioeconomic ramifications of centralized development strategies as urbanization, rural-urban drift and their consequential effects; cf. environmental briefs Spatial and Regional Planning, Overall Energy Planning, Planning of Locations for Trade and Industry.

43

6.　References

AKN Reddy: Rural Technology, Bangalore, 1980.

Albrecht, Buchholz, Deppe u.a.: Nachwachsende Rohstoffe, Bochum, 1986.

Bonnet, D. u.a.: Nutzung regenerativer Energien, Handbuchreihe Energie, Bd. 13, Cologne, 1988.

Bundesgesundheitsamt (German Federal Health Office): Vom Umgang mit Holzschutzmitteln, Berlin.

Bundesminister für Forschung und Technologie (German Federal Minister of Research and Technology) (Ed.): Expertenkolloquium "Nachwachsende Rohstoffe, Band 1 und 2, 1986.

EC-Council Directives relating to clean-air standards.

Edelmann, Fawre, Seiler, Woschitz (Ed.): Biogas-Handbuch, Aarau, 1984.

Fleischhauer, W.: Neue Technologien zum Schutz der Umwelt, Essen, 1984.

Fort, V.: Environmental Soundness, Proceedings of a Workshop on Energy, Forestry and Environment, Bureau for Africa, Agency for International Development, 1982.

Gadgil, M.: Hills, dams and forests; Some field observations from the Western Ghats, AKN Reddy, 1980.

Gieseler, G., Rauschenberger, H., Schnell, C.: Umweltauswirkungen neuer Energiesysteme, Dornier System/Bayerisches Staatsministerium für Landesentwicklung und Umwelt, March 1982.

Hartje, V.J.: Umwelt- und Ressourcenschutz in der Entwicklungshilfe: Beispiele zum Überleben? Frankfurt, New York, 1982.

Kannan, K.P.: Ecological and socio-economic consequences of water-control projects in the Kuttanad region of Kerala, AKN Reddy, 1980.

Kaupp, A. und Goss, J.R.: Small-scale Gas Producer-engine Systems, Braunschweig, Wiesbaden, 1984.

Kleemann, M.; Meliß, M.: Regenerative Energiequellen, Berlin 1988.

Lehner, G. und Honstetter, K.: Solartechnik, Ullmanns Encyklopädie der technischen Chemie, Band 21, Vienna, 1982.

Menrad, H.; König, A.: Alkoholkraftstoffe, Vienna, 1982.

Meier, P.: Energy Systems Analysis for Developing Countries, Berlin, 1984.

Montalembert, de, M.R.: The forestry/fuelwood problem in Africa and its environmental consequences, Proceedings of a Workshop on Energy, Forestry and Environment, Bureau for Africa, Agency for International Development, 1982.

Osterwind, D., Renn, O. und Voß, A.: Sanfte Energieversorgung, Jülich, 1984.

Porst, J.: Holz-Zyklus Kenia - Zusammenfassung der Studien des Beijer-Instituts, Berichte für GATE/GTZ, 1984.

Porst, J.: Überwachung eines chinesischen Reisspelzengasgenerators in Mali, Berichte für GATE/GTZ, 1986 und 1987.

Rat von Sachverständigen für Umweltfragen: Energie und Umwelt, Sondergutachten, Stuttgart, 1981.

Ripke, M. und Schmit, G.: Erschließung und Nutzung alternativer Energiequellen in Entwicklungsländern, Cologne, 1982.

Sorensen, B.: Renewable Energy, London, 1979.

UNESCO: Programme on Man and the Biosphere; Ecological effects of energy utilization in urban and industrial systems, Bad Nauheim, 1973.

Umweltbundesamt (Federal Environmental Agency (Ed.)): Lärmbekämpfung 1988, Berlin, 1989.

VDI - Richtlinien - Maximale Immissions-Werte.

Weih, H.; Engelhorn, H.: Wärme und Strom aus Sonnenenergie, Altlußheim, 1990.

WHO (World Health Organization): Environmental Health Criteria, Geneva.

43

Trade and Industry

Nitrogenous Fertilisers
(Raw Materials, Ammonia and Urea Production)

44

44. Nitrogenous fertilisers
(raw materials, ammonia and urea production)

Contents

1.   Scope

Worldwide demand for synthetic nitrogenous fertilisers currently stands at some 80 million tonnes per year.   Practically the sole **source of nitrogen** for all synthetic nitrogenous fertilisers is **ammonia** - chemical formula NH_3 - which has a characteristic pungent odour, is gaseous under ambient conditions and liquid at -33°C under atmospheric pressure.

Since 1913, **ammonia** has been produced on a large scale **from atmospheric nitrogen** and **hydrogen by catalytic synthesis**.

Naturally occurring hydrocarbons are converted with steam at high temperatures to **produce hydrogen**.

$$C_n H_m + 2nH_2 O = (m/2 + 2n) H_2 + nCO_2 \text{ (endothermic)}$$

The following **raw materials** are used in **ammonia synthesis gas production**:

- pit coal
- lignite
- peat
- non-volatile hydrocarbon residues
- light petrol
- natural gas and other gases.

For economic reasons the **electrolytic disintegration of water** to produce hydrogen can only play a **minor role** in ammonia synthesis.

The **synthesis gas** produced is in all cases converted **directly** into ammonia:

$$3 H_2 + N_2 = 2 NH_3$$

As **ammonia in liquefied gas form** is only suitable for **direct fertilisation under certain circumstances,** and only at **a considerable cost,** some or all of the **ammonia** produced is **processed** in situ to produce **urea** or other **nitrogenous fertilisers**. Only a few production plants are totally export-oriented.

In this section of the brief, only the **synthetic manufacture of urea from ammonia and carbon dioxide** (CO_2) - which occurs as a by-product of hydrocarbon reforming - will be considered.

Normal current **production capacities** ranges from approximately 400 to 2,000 t of NH_3/day and 600 to 3,000 t of urea/day.

44

Sites are **not** selected on the basis of any specific criteria; some **plants are both raw material oriented** and **consumer and transport oriented**.

The **environmental impact** of the production plants derives from **waste gases, wastewater, waste heat, dust, solid residues** and from **noise, transport routes, space requirements (pressure on space)** and **general industrialisation phenomena**.

We will not consider in this brief the impact on the environment from noise, transport, space requirement and other general industrialisation phenomena; this subject is dealt with in the **environmental brief Planning of Locations for Trade and Industry**.

We examine in the following the **process materials, intermediate products, by-products and waste products** which arise in the production processes and the measures required to **dispose of waste, to prevent any harmful impact on the environment and to keep within prescribed limits**.

2.   Environmental Impacts and Protective Measures

2.1   Ammonia synthesis gas production (ASGP)

2.1.1   ASGP from light hydrocarbons

Because it is **economical**, the **catalytic steam reforming** of **light hydrocarbons**, such as natural gas, petroleum-associated gas, LPG, light petrol and other gases containing H_2, and hydrocarbons such as coke oven and refinery gas, has become generally accepted.

Some **80% of all ammonia synthesis gas plants** use this highly endothermic process which can be illustrated - taking methane reforming as an example - by the following molecular formula:

$$CH_4 + 1.39\,H_2O + 1.45\,AIR = CO_2 + 2.26\,(H_3 + N)$$

In the **initial stage** of this process, **light hydrocarbons** are catalytically reformed with **steam** at temperatures of between 750°C and 800°C with the addition of heat (primary reforming) and, in a **second** autothermic stage, with **air** at approx. 1,000°C (secondary reforming); depending on pressure and temperature determined equilibrium conditions, this produces a mixture of H_2, CO, CO_2, N, CH_4 and traces of Ar. The nitrogen required for ammonia synthesis is introduced into the system by the air used for autothermic conversion in the secondary reformer. The **carbon monoxide** (CO) which forms is then converted catalytically into H_2 and CO_2 (usually in two stages) with steam at 300°C to 450°C.

Figure 1

Ammonia Production from Light Hydrocarbons

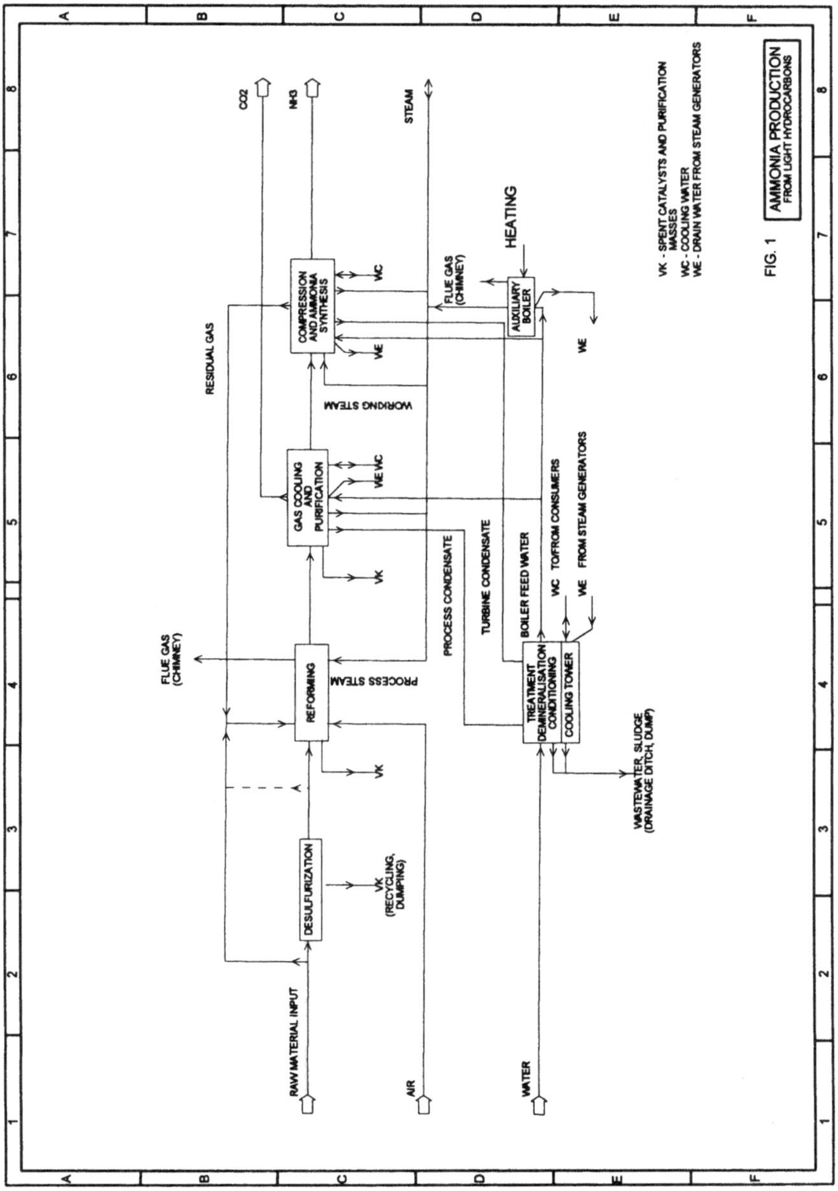

Before catalytic reforming, **sulphur, chlorine and other compounds,** which toxify the catalysts, must be removed, and this is performed in a single or multi-stage **gas purification process**.

Once the carbon monoxide from the reforming gases has been converted to hydrogen, the **carbon dioxide is separated** by **chemical or physical scrubbing,** from which a CO_2 stream can also be produced for urea synthesis.

The **purity** of the H_2/N mixture **necessary** for ammonia synthesis is obtained by a **fine purification stage** following CO_2 removal.

In most plants, the **primary reformer is heated** with the **process raw material**.

Thanks to the **intensive utilisation of waste heat,** almost all known processes involved in ammonia synthesis work **autonomously,** i.e. **steam for heating and power** from an external source is required or must be **produced by an auxiliary boiler** only at start-up. The **total energy requirement** of modern autonomous plants is less than 29 GJ/t NH_3.

- **Waste streams, pollutants and protective measures:**

(a)   **Waste gases**

-    **Carbon dioxide (CO_2):**
     It occurs at a concentration of around 98.5 % by volume, is used **in full** or **in part** as a **raw material** for urea synthesis and can be released **into the atmosphere untreated** as in practice the only impurities contained are H_2, N_2 and CH_4.

-    **Flue gases** from the primary reformer and steam boilers:
     If the heating medium contains **too much sulphur,** it may undergo a **purification process** to keep SO_2 values in the flue gases to within admissible levels. Primary measures to reduce the NO_x emission can be taken in the primary reformer. Flue gases are released **into the atmosphere** through a chimney so as to **comply with** the **values of the TA-Luft** [Technical Instructions on Air Quality Control] valid in Germany, for example.

-    **Other waste gases:**
     All other waste gases formed in the plant contain **combustible components** and are fed into the plant's **heating gas system**. If there is any **unscheduled stoppage,** process gases (H_2, CH_4, CO, CO_2, NH_3, N_2, steam) have to be burnt in a flare as a temporary measure so that only flue gases are released into the atmosphere.

(b)   **Wastewater**

-       **Process condensate:**
        is generally **reprocessed** and used as **boiler feedwater**.

-       **Blow-down water from steam generators:**
        does not contain **any toxic components** and can be **discharged untreated** or
        fed into the **cooling water circuit**.

-       **Blow-down water from cooling water circuits:**
        is to be **treated** before disposal depending on the degree of concentration and
        the content of corrosion inhibitors, hardness stabilisers and biocides.

-       Wastewater from **demineralisation plants** for boiler feedwater conditioning:
        can be drained following a **neutralisation stage**.

-       Spent lye from CO_2 **scrubbing:**
        In normal operation, no **waste streams** are produced. **Wash water** is to be
        treated in the same way as wastewater from demineralisation plant or cooling
        water circuits.

(On the general subject of wastewater, see also the environmental brief on
Wastewater Disposal).

(c)   **Solids**

-       **Sludges:**
        The purification of blow-down water from cooling circuits can produce
        sludge residues which then need to be **dumped** by a method appropriate to
        their composition.

-       **Spent catalysts** and **purification masses:**
        The **useful life** of **catalysts** used in ammonia production plants ranges from
        about **2 to 8 years** depending on the particular use and method of operation.
        When the activity of catalysts falls below a predetermined level, they are
        replaced by new active ones. Most **catalysts** contain notable quantities of
        oxides and sulphides of the heavy metals Co, Ni, Mo, Cu, Zn and Fe, which
        are insoluble in water, while spent **sulphur purification masses** consist in
        the main only of **water-soluble** oxides and sulphides of Zn or Fe, and chlo-
        rine purification masses of $NaCl/Na_2O$ on Al_2O_3. Some of these **waste
        products** are recovered by the manufacturers for **reprocessing** or are passed
        on to **smelting works** for **metal recycling**. Otherwise, they have to be
        dumped by a method appropriate to their composition; for example, the

44

water-soluble HT conversion catalyst containing Cr must be dumped so that no soil or water pollution is possible.

(On the general subject of waste, see also the Environmental Briefs Solid Waste Disposal and Disposal of Hazardous Waste).

2.1.2 ASGP from heavy residual oils

The **residual oils containing sulphur and heavy metals** produced in **crude oil processing** should today **no longer be burnt untreated** for reasons of environmental protection. They can however be successfully used for **the production of ammonia synthesis gas**.

The residues are gasified by **partial oxidation** with oxygen from an air separation plant - in which the nitrogen required for ammonia synthesis is also produced - according to the following simplified molecular formula:

$$C_n H_m + n/2\ O_2 = n\ CO + m/2\ H_2$$

The hydrogen required for ammonia synthesis is produced by further conversion with steam and **disintegration** of **contaminants** - such as H_2S, COS, CNS, HCN, soot and metal residues - **formed** due to the raw material composition and the particular process conditions.

As the process generally consumes **a large amount of energy, there is intensive waste heat utilisation** and **all combustible by-products and waste products formed are used internally** for reasons of economy.

Figure 2

Ammonia Production from Heavy Residue Oils

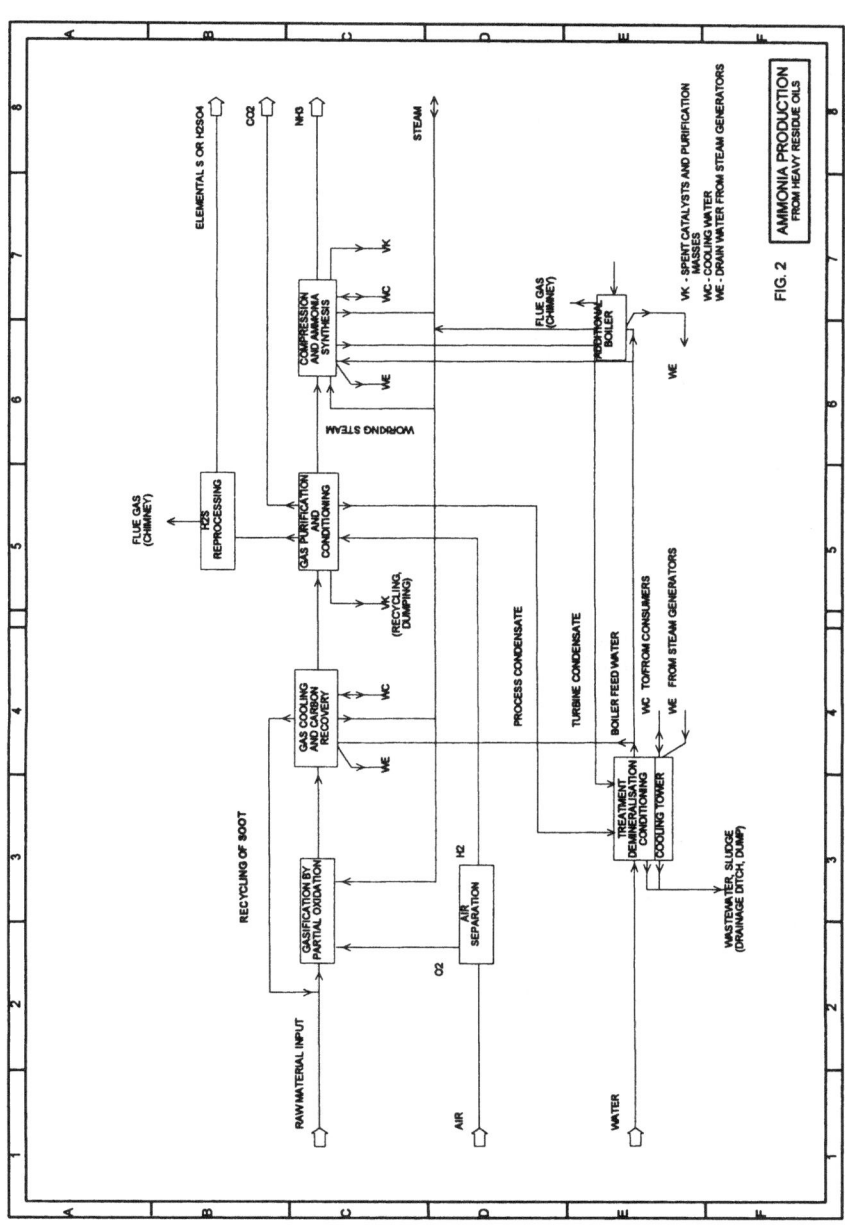

- ## Waste streams, pollutants and safety measures

Solid residues, such as ash and salts, and also **liquid and gaseous by-products and waste products** are formed during the process due to raw material composition and the gasification and purification processes.

Numerous processes are available for **waste reprocessing** and **pollutant disposal**, thus plants of this kind can even operate within the strict environmental regulations of the Federal Republic of Germany. Generally, the details given in section 2.1.1 apply to the reprocessing of the corresponding waste gases, wastewater and solid residues.

The following are **also** produced:

-    H_2S as a conversion product of the sulphur contained in the raw material. **Elementary sulphur** is produced with a 98% yield by the Claus process (a 99% yield can even be achieved by means of additional stages); alternatively a 98% yield can likewise be obtained by wet catalysis of **sulphuric acid**.

-    **Process water** contaminated with the **metals** contained in the raw material, such as Ni, V, Co etc., and the **water-soluble compounds** formed in the gasification process from other elements present in the raw material, such as H_2S, CNS, HCN, As, NH_3, Cl, MeOH etc. Before it can be discharged into drains, this wastewater must be purified by means of **appropriate purification processes** and **biodegradation**. In most cases provision must be made for a **demetallisation stage**, the **heavy metals** deriving from this being transported to **special dumps** or to special works where **the metal is recovered**.

2.1.3   ASGP from solid fuels

A **crude gas** consisting of H_2, CO, CO_2 and CH_4 is produced with steam at temperatures of **over 1200°C** and by the partial oxidation of **hard coal, lignite, coke, peat** etc., with oxygen from an air separation plant in which the nitrogen required for ammonia synthesis is also produced.

As with the partial oxidation of liquid hydrocarbons (section 2.1.2), the impurities in the crude gas are largely determined by the raw material composition and process conditions (pressure and temperature), the sulphur in the raw material being present almost exclusively in the form of H_2S. In the subsequent **purification and conditioning stages**, which in principle correspond to the operations involved in the reprocessing of heavy oil residues (section 2.1.2), **pure hydrogen** is extracted and this is used for ammonia synthesis with the oxygen from the air separation process.

On a large scale, the following methods of **solid gasification** have proved successful:

- moving bed process,
- fluidised bed process and
- entrained bed process.

Feed and storage installations for the fuel and also **conditioning stages** tailored to the particular gasification process used, are **always found upstream** of the gasification process.

As the overall process **consumes a great deal of energy, there is intensive waste heat utilisation**.

- **Waste streams, pollutants and protective measures**

In all processes, **solid residues** such as ash, slag and salts are produced, as are also **liquid and gaseous by-products and waste products**, in quantities and of compositions which are determined by the raw material composition and the gasification and gas purification processes.

A large number of **processes can be used for waste recycling** and **pollutant disposal**, thus plants of this kind can operate within the strict environmental regulations of the Federal Republic of Germany applicable in the energy supply sector.

The type and reprocessing of waste gases, wastewater and solid residues conform in principle to the provisions of sections 2.1.1 and 2.1.2.

44

Figure 3

Ammonia Production from Solid Fuels

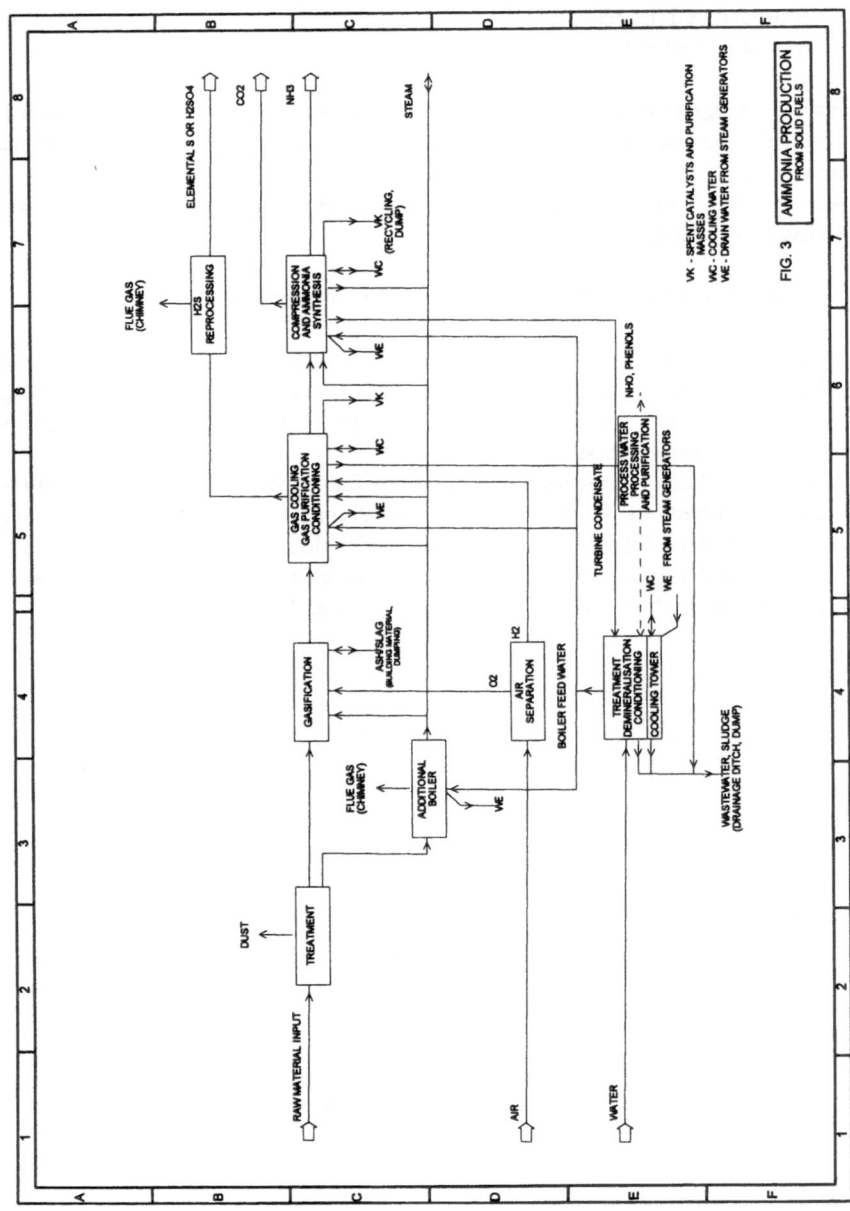

In addition the following are formed:

-   **Dust**, formed during fuel transport, storage and reprocessing. The problem of dust can however be controlled effectively by the implementation of **measures** which are **commonplace in coal power stations** and which have proved to be highly successful in overcoming the dust problem.

-   **Leakage water** from the fuel store. Any harmful effects can be avoided by **drainage** and/or by **covering** the ground water area with an impermeable layer of clay.

-   In many processes **wastewater** containing ammonia, phenol, cyanide and tar is formed, but there are also processes which can be used to separate these contaminants and **recover** them to a technically pure level.

-   **Ash** and/or **slag** from the gasifiers. It is essential to check in each individual case whether this can be **recycled**, e.g. in the construction industry, and to determine what form of **dumping** is appropriate.

2.1.4   Water electrolysis and air separation

The feed product is **fully demineralised water**; this is produced in ion exchangers and mixed bed filters. Water electrolysis **consumes a great deal of power** and is thus an option only where cheap excess energy is available or where other raw materials are in short supply. The **nitrogen** required for NH_3 synthesis is obtained by **air separation**. In electrolysis, **very pure oxygen, suitable for a large number of technical applications**, is formed, whereas in air separation only an oxygen-enriched spent air flow is generated which is normally released into the atmosphere.

•   **Waste streams**

Only **wastewater** from the demineralisation plant and **blow-down water** from the cooling water circuit are continuously formed; they must be treated as described in section 2.1.1. The **precious metal catalyst** for the removal of residual oxygen from the synthesis gas is only replaced at intervals of several years and can be returned to the manufacturer for **reprocessing**.

44

Figure 4

Ammonia Production by Water Electrolysis

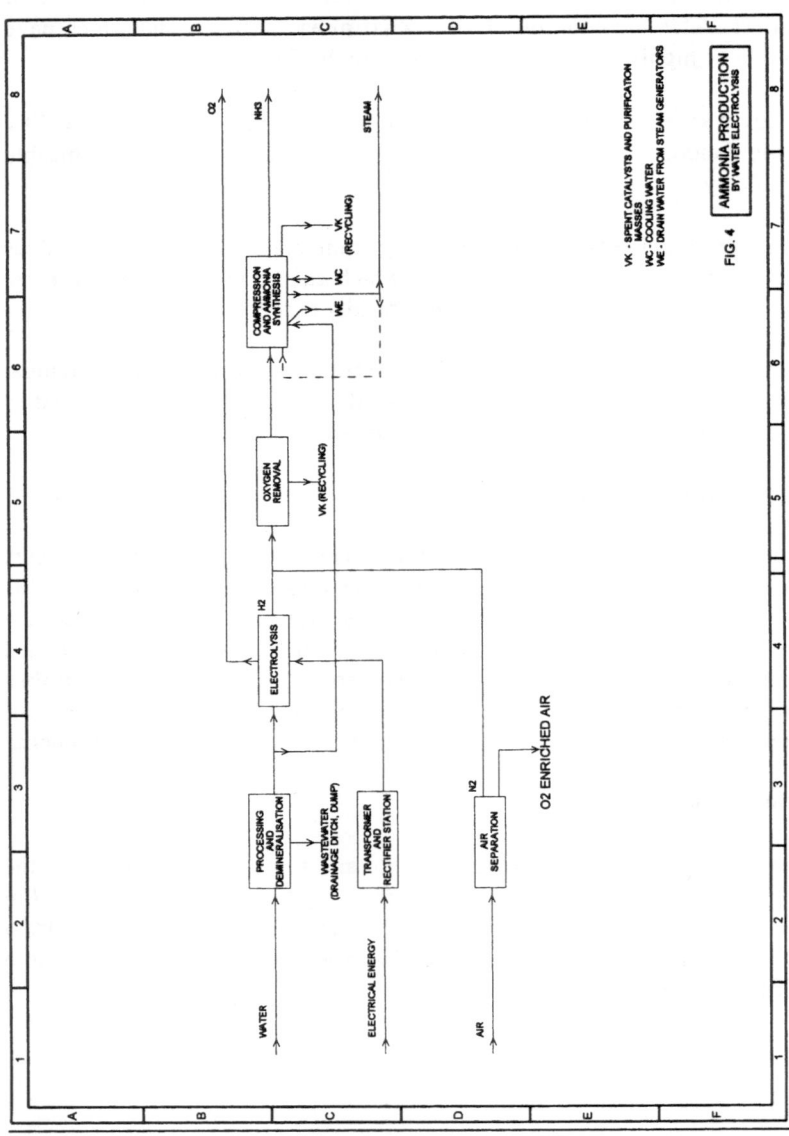

2.2   Ammonia synthesis and storage

Very pure **hydrogen** and **nitrogen** are converted catalytically in an exothermic process to **ammonia** at **pressures** of over 100 bar and **temperatures** of around 350°C - 550°C.

$$3\,H_2 + N_2 = 2\,NH_3$$

The **conversion** is **not complete** due to the equilibrium conditions. The **ammonia** formed is condensed by cooling (air, cooling water, cold) and **released from** the process **in liquid form. Any gases not converted remain in a recycle.** This results in an accumulation of **inert components** (CH_4, Ar, He) which must then be removed from the process by a continuous stream of **purge gas.** The purge gas stream, together **with the flash gases** from the ammonia produced, can be used as **heating gas** in the synthesis gas production plant, in which case NH_3, H_2, N_2 and Ar can first be separated in **recovery plants.**

The **liquid ammonia** goes either **directly into processing plants** or into **a storage tank, storage** taking place under pressure but at ambient temperature or slightly lower, or alternatively at atmospheric pressure and at a temperature of around -33°C.

* **Waste streams, pollutants and protective measures**

In normal operation, the plant does not release **any pollutants** into the environment. The continuously formed **waste gas streams** are processed **internally** or in the synthesis gas production plant.

No problems arise with the **disposal** of the **catalyst**, consisting of iron with small quantities of Al_2O_3, K_2O, MgO, CaO and SiO_2, an operation which **takes place** at intervals of around **5 to 10 years** (e.g. **smelting, road-building**).

As **ammonia fumes** are **highly irritant** and the liquid **is caustic** and causes **freezing**, appropriate **safety precautions** - particularly during storage - need to be taken, such as double-shell tanks, collecting basins and water spray curtains.

2.3   Urea synthesis and granulation

Urea is produced from **ammonia** and the **carbon dioxide** which is a by-product of ammonia synthesis gas production from hydrocarbons, in a 2-stage process at pressures of 140 to 250 bar.

44

1st stage: Ammonia carbamate synthesis (exothermic)

$$2NH_3 + CO_2 = NH_2 - CO - ONH_4$$

2nd stage: Thermal carbamate decomposition to urea (endothermic)

$$NH_2 - CO - ONH_4 - CO(NH_2)_2 + H_2O$$

The **urea** is present in the form of an **aqueous solution** in a concentration of some 70 to 80%, from which a pumpable melt is extracted for further processing by the vacuum evaporation of the solution water.

It is then processed to **granular urea fertiliser** either by **prilling in towers using a countercurrent of cold air** or by **fusion granulation on rotary plates** or other cooled installations and by the **fluidization technique.**

The **granular product** is then **poured** directly into bags and/or **stored** temporarily in warehouses as **bulk product.**

- **Waste streams, pollutants and protective measures**

(a)   **Waste gases:**
- Waste gases from synthesis contain only CO_2 and air, together with traces of the gases dissolved in the ammonia: H_2, CH_4, Ar, as all **waste gases** have to be **scrubbed** before they are released into the atmosphere.
- Waste gases from prilling towers or granulation installations always carry a certain amount of product dust with them, the **release** of which must be contained **by filtration** to prevent **"overfertilisation"** of the environment with the repercussions this has on soil and water quality.

(b)   **Wastewater:**
- Wastewater derives mainly from the gas scrubbing operations and contains NH_3, CO_2 and urea. All wastewater is **recycled in the process itself,** to keep the addition of water to the process as low as possible and to minimise raw material and product losses. The wastewater which does arise can be simply **biologically purified.**

(c)   **Solids:**
- Residue produced during waste gas dust extraction, which is practically pure product, is **returned to the process.**

Figure 5

Urea Production and Granulation

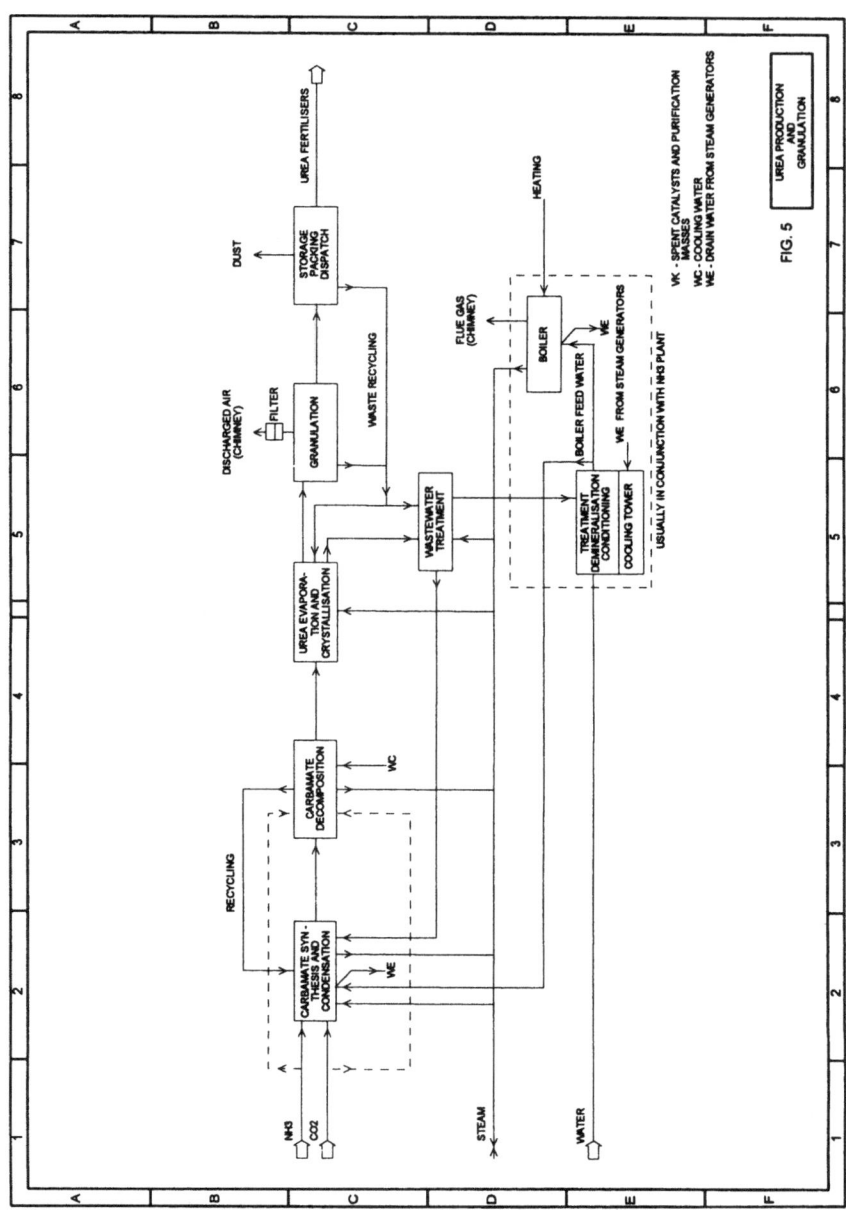

3.   Notes on Analysis and Evaluation of Environmental Impacts

In the fertiliser production plants described here, environmental impacts, in the form of **emission into the atmosphere, watercourses** and **soil,** as well as **noise emissions,** may be anticipated. However, there are **process stages** for all production plants which can be implemented to **contain this impact**.

In **Germany**, the **TA-Luft** [Technical Instructions on Air Quality Control] is the main instrument as regards air quality. Pollutant limit values relating to specific plants and substances are listed in the **Allgemeine Verwaltungsvorschrift zum Bundesimmissionsschutzgesetz** [General Administrative Regulations pertaining to the Federal Immission Control Act] of 27.02.1986. It also contains a series of **Richtlinien des Vereins Deutscher Ingenieure** (VDI-Richtlinien - guidelines of the Association of German Engineers) regarding process and gas purification techniques and emission measurement techniques, which must be complied with. There are similar provisions in other countries, e.g. the Clean Air Act in the USA or its Swiss equivalent, the Luftreinhalteverordnung.

In countries which do not have their own regulations, reference is frequently made to the TA-Luft or other foreign regulations at the planning stage.

Most **atmospheric pollution** in such plants derives from **SO_2** in the waste gas. Under TA-Luft, a sulphur emission level of 3% down to 0.5%, depending on plant size, must not be exceeded in sulphur extraction plants. **Not all purification processes** achieve this, but they are nonetheless used where **less stringent regulations** are in force.

In wet catalysis for sulphuric acid extraction, a minimum conversion level of 97.5% must be complied with. **Sulphur trioxide emissions** in the waste gas must not exceed 60 mg/m^3 under constant gas conditions, and must not exceed 120 mg/m^3 otherwise.

Limits which can also be adhered to are established in TA-Luft for **NO$_x$ emissions** in furnace flue gas streams - tube furnaces, steam generators, booster heaters.

Dust emissions from UREA fertiliser production facilities are restricted to 50 mg/m^3, while the **free ammonia content** in waste gases must not exceed 35 mg/m^3. The **dust load is measured** gravimetrically with filter head equipment and the **free ammonia** is determined by titration.

The **wastewater treatment processes** used are subject to local regulations. In **Germany**, the **Wasserhaushaltsgesetz (WHG)** [Federal Water Act] applies, with its associated Verwaltungsvorschrift [Administrative Regulation] relating to minimum requirements for the disposal of wastewater in drains. In fertiliser production plants, the associated 44. Verwaltungsvorschrift [44th Administrative Regulation] can be observed.

In the extreme case of wastewater treatment, no wastewater is produced, merely **combustion residues** which are finally disposed of on **special dumps** where no leaching can occur, or concentrated residual solutions which require disposal in deep wells, for example, may be formed.

The **catalyst and purification mass residues, most of which are formed** at intervals of two years or more, do not cause any problem in terms of quantity and, as already stated, are passed on to smelting works for **metal recycling** or must be dumped as **special waste**.

With regard to the **ash and slag** from solid-fuel ammonia production, the possibility of **recycling** or **dumping** has to be examined in each individual case.

The TA-Lärm [Technical Instructions on Noise Abatement] which is the comparable administrative regulation for **noise protection**, specifies immission values which are graded by location and time for areas, based on a variety of uses. The determining criterion is that of total impact level. Noise protection measures must be taken into account at planning stage as they are costly if implemented at a later date. In site planning, therefore, **adequate distances** from protected property, such as residential housing development, and a **shortening** of this distance must be **prevented**.

In Germany the TRgA 900[1] for limiting the maximum pollutant concentration at the workplace (MAK/TRK values[2]), the **Arbeitsstättenverordnung** [Ordinance on Workplaces] including workshop guidelines for workplace design and the accident prevention regulations **Unfallverhütungsvorschriften** of the Berufsgenossenschaften (employers' liability insurance associations), as being the body responsible for insuring accidents at work, apply to **workplace conditions** in terms of pollutant concentration, noise nuisance and industrial safety. Comparable regulations exist in other countries, e.g. in the USSR, with Health Standards for Industrial Concerns (SN 245-71).

4.   Interaction with Other Sectors

In view of the **high energy and raw material requirement**, ammonia and urea production plants are normally built **close to raw material sources or transport routes**; these include natural gas and crude oil conveying plants, refineries, pipeline terminals, LNG stores, coal mines, power stations and coking plants - or hydroelectric power stations with high excess energy (for water electrolysis).

Proximity to other fertiliser production facilities is also useful, e.g. NP or NPK fertiliser production.

Less practical, in contrast, are **purely consumption-oriented sites** if these do not also enjoy favourable conditions for the supply of raw materials or energy (e.g. port installations, power stations).

44

---

[1]   TRgA   - Technische Regeln zur Arbeitsstoffverordnung [technical regulations on the industrial substances decree]

[2]   MAK   - Maximale Arbeitsplatzkonzentration [maximum workplace concentration]

        TRK   - Technische Richtkonzentration [technical approximate concentration]

5.   Summary Assessment of Environmental Relevance

In ammonia and urea production plants, mainly **gaseous by-products and residues** are formed due to the raw materials used, together with **wastewater, waste heat** and **spent catalysts** resulting from the processes used. Moreover, **noise** and other **industrial influences** also occur.

Because of the **high energy requirement** for ammonia production, which is about 29 GJ/t of NH_3 in modern natural gas fed plants and over 70 GJ/t of NH_3 where coal is the raw material, the environmental impact is comparable to that of power stations (cf. environmental brief Thermal Power Stations).

With today's **gas and water purification methods**, even the **most stringent environmental protection regulations** can be complied with, the **lowest** costs being incurred where **natural gas** is the raw material, and the **highest** being incurred for **coal** - due to its complex composition. In the **manufacture of granular urea fertiliser**, particular emphasis must be placed on effective **dedusting techniques**. Likewise, **suitable wastewater purification plant** and **environmentally friendly dumping facilities** must be available.

In industrial **conurbations, air coolers** or **dry cooling towers** may be required to prevent the environmental pollution which can occur where cooling water is used to deal with **waste heat**.

The population affected should be **involved** at the planning phase; likewise, the population resident in the area of the project should have **access to medical care**.

In the case of new planning measures without any differentiated (state) monitoring system in the environmental field, the aim must be to choose a technique which **is best adapted to the particular circumstances**.

It is extremely important for plants of this kind to be **systematically monitored** and **maintained** to guarantee correct operation - a point which is all too easily ignored. Thus, a **works environmental protection officer** with appropriate powers must be appointed who will also be responsible for **increasing the awareness, and for the education and training** of operating personnel with regard to environmental issues.

It may generally be stated that apart from the pollutants due to waste heat and contained in the raw materials, very little environmental impact need be feared from ammonia and urea production provided that environmental protection aspects are taken into account during planning and operation.

6.   References

Allgemeine Verwaltungsvorschrift über genehmigungsbedürftige Anlagen nach 16 der Gewerbeordnung - GewO; Technische Anleitung zum Schutz gegen Lärm (TA-Lärm), 1968.

Gesetz zur Ordnung des Wasserhaushalts (Wasserhaushaltsgesetz - WHG), 1976.

Gesetz zum Schutz vor schädlichen Umwelteinwirkungen durch Luftverunreinigungen, Geräusche, Erschütterungen und ähnliche Vorgänge, BundesImmissionsschutzgesetz - BImSchG, 1985.

Katalog wassergefährdender Stoffe, Bekanntmachung des BMI, 1985.

Technische Regeln für brennbare Flüssigkeiten - TRbF
    TRbF 100 Allgem. Sicherheitsanforderungen
    TRbF 110 Läger
    TRbF 210 Läger
    TRbF 180 Betriebsvorschriften
    TRbF 280 Betriebsvorschriften.

1. Allgemeine Verwaltungsvorschrift zum BundesImmissionsschutzgesetz (Technische Anleitung zur Reinhaltung der Luft - TA-Luft), 1986.

1. Allgemeine Verwaltungsvorschrift (VwV) zur Störfall-Verordnung (1. Störfall-VwV), 1981.

2. Allgemeine Verwaltungsvorschrift zur Störfall-Verordnung (2. Störfall-VwV), 1982.

4. Verordnung zur Durchführung des BundesImmissionsschutzgesetz (Verordnung über genehmigungsbedürftige Anlagen - 4. BImSchV), 1985.

9. Verordnung der Bundesregierung zur Durchführung des BundesImmissionsschutzgesetzes, (Grundsätze des Genehmigungsverfahrens - 9. BImSchV), 1980.

12. Verordnung der Bundesregierung zur Durchführung des BundesImmissionsschutzgesetzes, (Störfall-Verordnung - 12. BImSchV), 1985.

13. Verordnung zur Durchführung des BundesImmissionsschutzgesetzes, (Verordnung über Großfeuerungsanlagen - 13. BImSchV), 1983.

44

Verordnung über Anlagen zur Lagerung, Abfüllung und Beförderung brennbarer Flüssigkeiten zu Lande (Verordnung über brennbare Flüssigkeiten - VbF), 1982.

Verordnungen der Bundesländer über Anlagen zum Lagern, Abfüllen und Umschlagen wassergefährdender Stoffe - VAwS.

Trade and Industry

Nitrogenous Fertilisers (Starting Materials and End Products)

45. Nitrogenous Fertilisers (Starting Materials and End Products)

Contents

1.   Scope

Nitrogenous fertilisers in the strict sense of the term include the following, which are considered in the context of this environmental brief:

- ammonium nitrate	(abbreviation AN)
- calcium-ammonium nitrate	(abbreviation CAN)
- ammonium sulphate	(abbreviation AS)
- calcium nitrate	(abbreviation CN)
- nitrogen solutions	(abbreviation N solutions)
- ammonium chloride	
- ammonium phosphates.	

The nitrogenous fertilisers examined here are produced for **agriculture** in a **granulated** or **prilled** form with the **exception** of **nitrogenous solutions,** the use of which requires a system of mixing and distributor stations.

The primary products required for the manufacture of these fertilisers comprise:

- ammonia, covered by the environmental brief Nitrogenous Fertilisers (raw materials, ammonia and urea production)
- nitric acid
- sulphuric acid
- urea
- limestone.

The **capacities** of individual plants vary considerably; the upper limit for nitric acid, for example, is 2000 t HNO_3/day, for sulphuric acid 3000 t H_2SO_4/day and for ammonium nitrate and calcium-ammonium nitrate 2000 t/day on one line.

2.   Environmental Impacts and Protective Measures

With the use of modern processes, **environmental impacts** can be confined to **gaseous emissions** in the overwhelming majority of cases. Any liquid emissions produced can usually be avoided by internal recycling, although in a few cases **solid waste** cannot be avoided, and **noise emissions** occur with most processes.

45

Figure 1

Nitrogenous Fertiliser Production

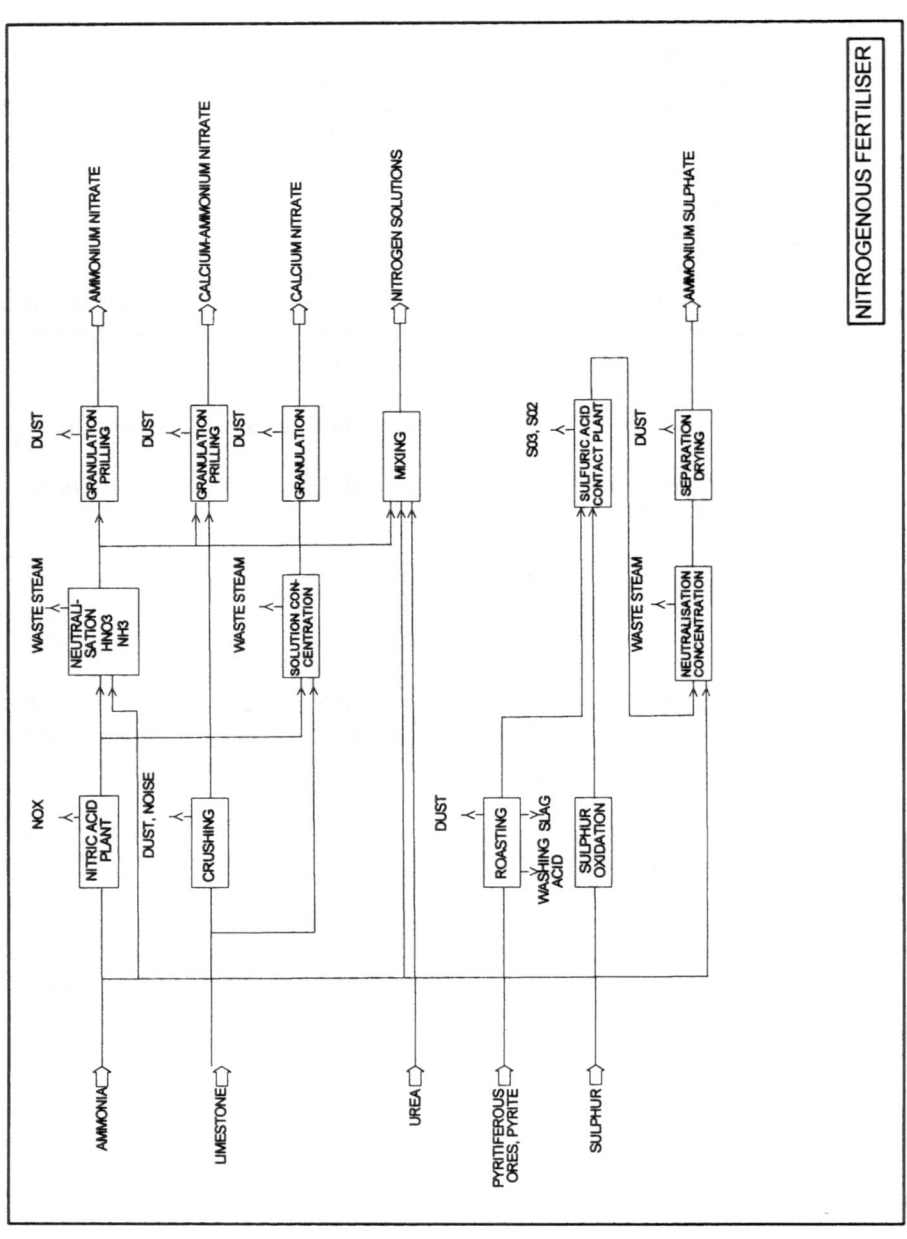

2.1 Nitric acid production

Industrial production of nitric acid is based on the **catalytic oxidation** of ammonia and subsequent absorption of the nitric oxides, formed during oxidation, in water. The various **processes** used in industrial production **differ** mainly with regard to the **pressure** used in the burning or absorption stage and the efficiency of the **heat recovery** system. The acid produced for further processing into fertilisers is an aqueous solution containing up to about 60% HNO_3.

- **Pollutants produced and counter-measures**

The **process** does not give rise to continuous liquid emission flows. Where liquid ammonia is used, an **oily waste** is produced intermittently depending on the oil content of the ammonia, which is collected and **burnt** in a suitable incineration plant. Gaseous emissions are the tail gas containing $(NO + NO_2) = NO_x$ from the absorption column.

The higher the NO_2 content, the more intense the **brown colour of the waste gas**, as is plain to see for miles around.

While the NO_x content in older plants can be several thousand mg NO_2/m^3, modern facilities are designed for around 400 mg NO_2/m^3. There are a number of ways of **removing nitrogen oxides completely**, e.g. **catalytic tail gas burning** with hydrogen, ammonia or methane.

If **neither fresh water nor seawater** can be used as **cooling water, blow-down water from the cooling water recycle** arises which, in compliance with local provisions, cannot always be discharged directly as wastewater because of its increased salt concentration and other additives. In this case, it is conditioned in the **wastewater treatment plant** together with the other wastewater flows in the works. The **residues** must then be taken to a **controlled dump** or, in the case of **biological wastewater purification**, can be **incinerated**. Where fresh water is used for cooling purposes, the heat transferred to the river or lake must be taken into account; if necessary, measures are to be taken to cool it before it is discharged.

45

2.2   Sulphuric acid production

Today sulphuric acid is **produced on an industrial scale** almost exclusively using the **contact process** in which gases containing sulphur dioxide are channelled through a vanadium catalyst. The gases containing sulphur dioxide required as the primary product for sulphuric acid production come mainly from:

- the burning of elemental sulphur,
- roaster gases from pyrite,
- roaster gases from sulphide ores of non-ferrous metals.

A **modern sulphuric acid plant** can be identified by optimum **use** of the **reaction heat** in the individual process stages. Most **surplus steam** is used for **energy production**, and in some plants, the low-temperature energy produced in the acid coolers is already being utilised.

The SO_3 formed in catalytic SO_2 oxidation is absorbed in 98% to 99% sulphuric acid, which yields H_2SO_4 in a reaction with water.

- **Pollutants produced and counter-measures**

There are **no** process-specific **liquid emissions** if sulphuric acid is produced by sulphur oxidation.

The tail gas from sulphuric acid facilities contains SO_2 and SO_3.

For sulphuric acid facilities, **emissions of sulphur trioxide in the waste gas**, at constant gas conditions, are limited to maximum 60 mg/m³. Moreover, the **emissions** can be further **reduced** by the use of the peracidox process, a fifth tray stage (5th catalyst level) or equivalent measures.

Where roasters are installed upstream, small quantities of **contaminated sulphuric acid** are produced in the form of washing acid, which, if it does **not** contain **any harmful pollutants**, can be **concentrated** and used, for example, in a **fertiliser plant**. If it contains **harmful pollutants** from the raw materials which have not been removed by the waste gas plant upstream, the acid must be **neutralised** and the residue **dumped**.

The **slag** may, depending on the feedstock analysis and possibly following an intermediate stage in which elements of any value are extracted, be passed to the **steel industry** or **dumped**. The remarks made in section 2.1 apply with regard to the **cooling water problem**.

In Germany, pure liquid sulphur is used almost exclusively. In the rare cases in which the sulphur contains arsenic or selenium, **purification** is essential and filtration **residues** must be **dumped** with care. Where the **dumps are in the open**, it must be ensured that the **sulphurous acid** formed by oxidation of the sulphur in the atmosphere does not percolate into the ground water with rainwater.

2.3   Ammonium nitrate production

Along with urea, ammonium nitrate is one of the **most frequently used nitrogenous fertilisers worldwide**. It is mainly produced by the neutralisation of 45 - 65% nitric acid with ammonia.

Ammonium nitrate is also a **by-product** of the nitrophosphate process in which NP or NPK fertilisers are made by the nitric acid decomposition of crude phosphates. The neutralisation reaction yields 95 to 97% solutions of ammonium nitrate.

The solution is **processed further** to obtain a marketable product by **granulating** or, after concentrating further to 99.5%, by **prilling**.

- **Pollutants produced and counter-measures**

Where **the prilling process is used,** the prilling tower in the dry part of the plant can give rise to serious **emission problems,** as the relatively large quantities of discharged air are **extremely costly to purify**. In time, **ammonium nitrate dust kills vegetation** in the surrounding area. Such **problems** can be **dealt with** far **more easily in granulation installations**. Thus, this aspect should be studied in depth before any new investment is made and before any decision is taken on the process to be used.

With granulation, the **process gas flows** must be **purified** in effective **wet scrubbers** before they are discharged into the atmosphere. The installation should be fitted with a **dust extraction system** to ensure the **safety of operating personnel**.

Waste fumes from neutralisation and evaporation also must be **scrubbed** if they are to be discharged into the atmosphere as vapour. The preferred solution is the **condensation of purified fumes**, which yields condensates polluted with ammonium nitrate and ammonia, some of which can be used as **process water** for an adjoining nitric acid plant. **Condensate which contains small amounts of impurities** can be fed through an ion exchanger installation and **reprocessed** to **boiler feed water**.

45

2.4   Calcium-ammonium nitrate production

While the ammonium nitrate considered in section 2.3 has an N content of 33.5 - 34.5%, the nitrogen content of calcium-ammonium nitrate is 20.5 - 28%, and **EC regulations** do not permit a nitrogen content of over 28%. The **nitrogen content** is **reduced** by the **addition of crushed limestone**. With the exception of this addition of crushed limestone and mixing with the ammonium nitrate melt immediately before the prilling or granulation process, calcium-ammonium nitrate **is made in the same way as ammonium nitrate**. For this reason, the comments made in section 2.3 regarding pollutants and counter-measures apply here too but, in addition, because of the crushing plant for the lime, increased **noise emissions** must be anticipated. An **effective dedusting unit** is to be provided for the **crushing process.** Where there is a **constant electricity supply** and the plant is **maintained** to West European standards, **continuous dust removal to less than 50 mg/m³** can be achieved.

2.5   Ammonium sulphate

In view of the popularity of more highly concentrated nitrogenous fertilisers, the consumption of **ammonium sulphate** with just 20.5% N is constantly **declining** and now, worldwide, accounts for just 6% of nitrogenous fertiliser consumption. The **strong physiologically acidic effect** of this fertiliser is also to blame for the decline in its use.

The main industrial-scale **production methods** are:

- from coke-oven or coal gasification;
- from ammonia and sulphuric acid;
- as a by-product of organic syntheses, e.g. caprolactam manufacture;
- from gypsum, either from natural deposits or as a by-product of other processes, by reaction with ammonia and carbon dioxide.

2.5.1   Production from coke-oven or coal gasification

In both **dry distillation** and **pressure gasification**, some of the nitrogen in the coal forms ammonia. This **ammonia** is also found in the **aqueous and carbon dioxide-rich condensate** produced when the **gas is cooled**. The **gas condensate** also contains **tar, phenols, pyridins, hydrogen sulphide, hydrocyanic acid** etc., which cause serious **problems** when it comes to **ammonia recovery** and **wastewater purification**. When the **tar** has been **separated** and the **phenols** removed, the **volatile components of the gas condensate** are **stripped** in a column **by steam injection**. The **fumes from the stripper** are scrubbed with sulphuric acid in coking plants, and the **acidic gases** remaining after sulphuric acid scrubbing are either **processed to sulphur** in a Claus plant or converted directly **to sulphuric acid** in a wet catalysis installation. Fume burning could well be an option for consideration where only small quantities are produced, but this must be in line with sulphur emission regulations.

The **wastewater** must undergo a **biological treatment** as it contains various sulphur compounds, phenol and other organic compounds.

- **Pollutants produced and counter-measures**

The problems arising from ammonia production have already been examined in the previous section and should be the topic of a separate study - on coal. The **dust** needs to be **removed** from **waste gases** produced by ammonium sulphate drying before they can be discharged into the atmosphere, as otherwise they lead to **over-fertilization** with the associated negative consequences for soil and water quality.

2.5.2   Production from ammonia and sulphuric acid

Neutralisation and crystallisation are carried out under vacuum or at atmospheric pressure. Crystallised ammonium sulphate is removed from the resulting mash in centrifuges and then **dried**.

- **Pollutants produced and counter-measures**

The fumes produced by the exothermic reaction of sulphuric acid and ammonia, in particular the ammonia in the waste gas which can cause caustic burns to man, animals and plants, may contain impurities depending on the process used, and should be fed though a **scrubber** before being discharged into the atmosphere.

Dedusting systems are needed to remove the dust content from **drying plant waste gases** before they are released into the atmosphere.

45

2.5.3   As a by-product

Ammonium sulphate is obtained from the **liquid waste of some organic processes**, e.g. the production of caprolactam or acrylonitrile which yields a **dilute ammonium sulphate solution**, by evaporation, crystallization, centrifuging and drying.

For information on pollutants and counter-measures, see section 2.5.2.

2.5.4   Production from gypsum, ammonia and CO_2

The **feedstock** is finely ground natural gypsum or anhydrite, or alternatively calcium sulphate - a by-product, for example, of phosphoric acid production - which is converted with ammonia and carbon dioxide. The calcium carbonate obtained from the reaction is filtered off and the **ammonium sulphate solution** evaporated, crystallised and treated as described in section 2.5.3.

- **Pollutants produced and counter-measures**

In principle, the same factors as stated in 2.5.2 need to be considered. Where natural gypsum is used, there is the added **nuisance of noise from the grinding plant**. The details given in 2.4 apply with regard to the **dust** produced in the grinding process.

2.6   Calcium nitrate production

$Ca(NO_3)_2$ is produced either **directly** via the reaction of nitric acid with limestone or, alternatively, produced as a by-product of the nitrophosphate process.

In **direct manufacture**, **limestone** is **dissolved** in dilute nitric acid and **granulated** or **prilled** after evaporation of the dilute calcium nitrate solution.

In the **nitrophosphate process**, in which crude phosphate is decomposed with nitric acid, the **calcium nitrate** is **crystallised** by cooling, **separated** and, after appropriate treatment, **granulated** or **prilled**.

- **Pollutants produced and counter-measures**

In direct manufacture, the dissolution process yields **gases which contain NO_x** and need to be **extracted** and **absorbed**, mainly to **protect** the health of **operating personnel**, although the gases are also responsible for **corrosion** of equipment and buildings.

Either appropriate precautions have to be taken at the design stage, or a **scrubber installation** is to be provided to **reduce** the **pollutant content** of the fumes produced during evaporation. Any purification stage installed after dissolving generates a **moist waste** which - depending on its composition - can be **used in other plants** or must be **dumped**.

Dust-laden gases must be **cleaned** before discharge into the atmosphere. Any **washing solutions** produced by these cleaning operations are to be **concentrated** and **recirculated**.

2.7 Production of nitrogen solutions

The following are used as **liquid nitrogenous fertilisers**:

- liquid ammonia;
- aqueous ammonia solutions (e.g. 25%);
- solutions which contain free ammonia together with either ammonium nitrate or urea, or both;
- solutions of ammonium nitrate or urea, or both.

Liquid ammonia is used **directly** as a fertiliser principally in the United States, where it is **injected** 15 - 25 cm deep **into the soil** with special equipment.

Where applied in this way, **storage, transport and transfer equipment** are the basic essentials, and the precautionary measures stated in the first section with regard to ammonia are to be observed.

These same precautionary measures are also to be taken in a somewhat diluted form for other nitrogenous solutions containing free ammonia.

The **long-term implications** - especially on soil microorganisms and the humus layer - should be **examined** for the particular soil type concerned before liquid ammonia or nitrogenous solutions containing free ammonia are used.

2.8 Ammonium chloride production

This salt, which - at 26% N - has a somewhat higher nitrogen content than ammonium sulphate, is not used alone as a nitrogenous fertiliser in Germany. Its **main areas of use** are China, Japan and India, principally in **rice paddies** as an alternative to ammonium sulphate, which decomposes into toxic sulphides where rice is attacked by fungus. The **use** of ammonium chloride is now on the **decline** as **soils** become **overchlorinated** if chloride is **used for prolonged periods**.

45

By far the largest share of ammonium chloride made for use as fertiliser is produced in solvay plants modified for **soda production**. After separating the sodium bicarbonate, **ammonium chloride** is **crystallised out** of the remaining solution by additional process stages, thus obviating the need for the usual ammonia recovery with its attendant yield of relatively useless calcium chloride, and instead ammonium chloride fertiliser is obtained as a by-product.

- **Environmental impacts and counter-measures**

As facilities of this kind yield ammonium chloride as a by-product of soda manufacture, the main measures applicable are those relating to soda works. The additional equipment required for ammonium chloride production must be fitted **with efficient dedusting systems**, especially for waste gases from driers.

2.9   Ammonium bicarbonate

To complete the picture, mention must also be made of this nitrogenous fertiliser, which is **only** produced and used **in China**. According to statistics, of the 11.1 million tonnes made in China in 1983, 6.4 million tonnes went to the fertiliser market in the form of ammonium bicarbonate. The reason for this one-off development lies in the rapid establishment of nitrogenous fertiliser production from 1960 on, with the creation of a large number of small facilities for ammonia production using carbon gasification. The CO_2 obtained as a by-product is used directly for neutralisation of the ammonia produced.

Please refer to the section on ammonia synthesis using coal gasification for information on **environmental impacts and counter-measures**.

2.10   Transport, storage and bagging of solid fertilisers

Because they are **water soluble,** and in view of their hygroscopicity, fertilisers must be **stored** in **bulk goods stores** which are **roofed** and **enclosed on all sides** and then transferred to a **bagging and transfer station in the immediate vicinity** for dispatch. **The delivery, removal and transfer points** are to be of an as **dust-tight** as possible design, and - as in production plants - at critical points, where enclosure is not feasible, **dust-laden waste gases** must be **collected** and transferred to a **dedusting installation**.

3. Notes on the Analysis and Evaluation of Environmental Impacts

The basic regulations to be considered for this environmental brief are found, in Germany, in the *1. Allgemeinen Verwaltungsvorschrift* [1st General Administrative Regulation] to the *Bundes-Immissionsschutzgesetz* [Federal Immission Control Act] (*Technische Anleitung zur Reinhaltung der Luft* [Technical Instructions on Air Quality Control] - *TA-Luft*) of 27.02.1986.

It is often the case in countries without firm regulations that the relevant German provisions are used when designing such facilities.

The **NO_x emission** for new **nitric acid facilities** is now restricted to 0.45 mg/m^3, expressed as nitrogen dioxide, and **waste gases** must be **colourless** before discharge. **NO_x is determined analytically** by titration or photometry.

In **sulphuric acid plants, sulphur trioxide emissions** in waste gas, at constant gas conditions, are restricted to 60 mg/m^3 maximum. The **sulphur dioxide content of the tail gas is determined** by the conversion level, which must be at least 99.6% in the double-contact process, with a minimum sulphur dioxide volume content of 8% in the input gas and at constant gas conditions. Furthermore, **emissions** are to be further **reduced** by the use of the peracidox process, a fifth tray stage or equivalent measures. **Sulphur dioxide** can be determined iodometrically, titrimetrically, gravimetrically or colorimetrically. For **continuous measurement, recording analyzers** are used, working on the basis of optical absorption in the infrared or ultraviolet spectral range or the electrical conductivity of the sulphur dioxide.

For **fertiliser plants, dust emissions** from granulation and drying installations for multinutrient fertilisers with an ammonium nitrate content of over 50% or a sulphate content of over 10% are **restricted** to 75 mg/m^3 maximum. This category includes, for example, the following fertilisers: ammonium nitrate, calcium-ammonium nitrate and ammonium sulphate. For other fertiliser plants, the dust emission is to be kept at no more than 50 mg/m^3. Operating licenses set values of 35 mg/m^3 maximum for the **free ammonia content** of waste gases. **Dust is analyzed** gravimetrically with filter head equipment. Compliance with **sampling technique** rules is of utmost importance for the **reliability** of analyses and thus compliance with statutory limits. **Free ammonia is determined** by titration.

45

4.   Interaction with Other Sectors

Today, it is frequently the case that **complexes are not confined solely to the production of nitrogenous fertilisers** but **make NP and NPK fertilisers, too**. In this case, the sulphuric acid obtained is used for phosphoric acid production. The phosphoric acid is then neutralised with ammonia to ammonium phosphates which are processed in granulation operations to DAP fertilisers or, after adding potassium salts and micronutrients as necessary, to NPK fertilisers. This sort of **combined economic management** is characterised by a high level of **flexibility** with regard to fertiliser type. Furthermore, individual plants, including any ammonia synthesis upstream, can have **increased capacities** and thus **manufacture their products economically**; finally, a complex of this kind is **self-sufficient in electricity** because of the extra energy provided by the sulphuric acid installation. A further possibility is that of using the Müller-Kühne process or a modern variation of it to **reconvert** into **sulphuric acid** the **gypsum** produced in the phosphoric acid plant, which in many instances represents a major **dumping problem**.

The **slag** from a roaster plant can be **a raw material for non-ferrous metal and/or steel works**.

Use of the nitrophosphate process obviates the need for sulphuric acid, in which case **calcium nitrate** is a **by-product** that can be converted to **ammonium nitrate and fertiliser lime or calcium-ammonium nitrate** where cheap carbon dioxide is available, e.g. from an adjoining ammonia synthesis plant.

The special variant of the solvay process for **soda production** practised in the **Far East**, of which ammonium chloride is a **by-product**, has already been mentioned.

For high-capacity **nitrogenous fertiliser facilities**, having ammonia synthesis close by is always worthwhile unless the plant enjoys an excellent transport infrastructure (e.g. ports and harbours, cf. environmental brief) and can also conclude favourable long-term supply contracts.

References are given in the relevant environmental briefs.

5.   Summary Assessment of Environmental Relevance

In nitrogenous fertiliser production facilities, the implications for the environment concern in the main **gaseous waste** (dust, ammonia, nitrous gases, sulphur dioxide), and noise, plus, **in the case of roaster installations**, process-specific **by-products and residues**.

Nitric acid installations can be operated such that gaseous emissions are practically **colourless**, i.e. NO_x-free, by the use of catalytic tail gas treatment where the NO_x design value is not sufficient.

In **sulphuric acid plants**, the officially prescribed **emission values** listed in section 3 are to be further **reduced** by the installation of a fifth tray stage, the use of the peracidox process or equivalent measures. Where roasters are installed upstream, the **slag**, if it cannot be further used, must be **dumped**, the **washing acid neutralised** and **residues dumped** if further utilisation is not possible in view of the impurities they contain.

In plants for the production of salt, prilled or granulated **fertilisers**, an efficient **dedusting system** is of prime importance. This requires the separate treatment of the individual waste gas flows in specific dedusting installations. As stated, liquid waste from gas scrubbers is returned to the process. With modern technology, the **harm** to the environment can be kept **low** in the processes described here.

On the process management side of such plants, **all waste gas purification installations** must be systematically **monitored** and **maintained**. In particular, regular maintenance - which includes the **cleaning** of machines, motors and plant - is a major determining factor in the operating efficiency of such systems. Another important factor is the **timely provision of the necessary spare parts**. Monitoring also includes regular **analyses** by an **efficient laboratory** so that appropriate measures can be taken promptly when values drift out of the permitted range. **Works environmental safety officers** should also be appointed; they should have the appropriate powers and should be responsible for the **training and upgrading** of personnel and **for raising their awareness** with regard to environmental matters.

Retention basins are also to be provided so that, if there should be any **process incident** resulting in an unforeseen production of **wastewater**, the plant does not have to be immediately shut down.

Although the dusts and gases produced are fertilizing substances, attention must be paid to compliance with prescribed emissions as, in the long-term, **excessive immissions** can be **harmful to plant crops** or **trees** in the surrounding area.

The affected population should be **involved** at the planning stage, and **access to medical care** must be guaranteed.

45

6.   References

Abfallbeseitigungsgesetz, 04.03.1982.

31. Abwasser VwV Wasseraufbereitung, Kühlsysteme, 13.09.1983.

American National Standards Institute Safety Requirements for storage and handling of anhydrous ammonia, ANSI K 61.1., 1972.

1.   Allgemeine   Verwaltungsvorschrift   zum   Bundes-Immissionsschutzgesetz (Technische Anleitung zur Reinhaltung der Luft - TA-Luft), 27.02.1986.

44. Allgemeine Verwaltungsvorschrift über Mindestanforderungen an das Einleiten von Abwasser in Gewässer, Herstellung von mineralischen Düngemitteln außer Kali, 44. Abwasser VwV, 05.09.1984.

Arbeitsstätten-Richtlinien (ASR).

The relevant accident prevention regulations of the employers' liability insurance associations   (Berufsgenossenschaften)   relating   to   the   handling   of hazardous materials.

Gesetz zur Ordnung des Wasserhaushalts, Wasserhaushaltsgesetz, 16.10.1976.

Gesetz zum Schutz vor schädlichen Umwelteinwirkungen durch Luftverun-reinigungen,   Geräusche,   Erschütterungen   und   ähnliche   Vorgänge, Bundes-Immissionsschutzgesetz BImSchG, 04.10.85, and the associated enforcement ordinances and general administrative provisions.

Katalog wassergefährdender Stoffe, German Federal Ministry of the Interior (BMI) publication, 01.03.1985.

Merkblätter Gefährliche Arbeitsstoffe (codes of practice for hazardous materials), e.g.:

Blatt S 24	Nitrogen dioxide (Stickstoffdioxyd)
Blatt S 33	Nitrogen oxide (Stickstoffoxyd)
Blatt S 03	Nitric acid (Salpetersäure)
Blatt A 64	Ammonium nitrate (Ammoniumnitrat)
Blatt A 59	Ammonia solution (Ammoniaklösung)
	etc.

Technische Anleitung zum Schutz gegen Lärm (TA-Lärm), 16.07.1968.

Technische Regeln zur Arbeitstoffverordnung TRgA 511, Ammoniumnitrat, September 1983.

TRgA 951 Ausnahmeempfehlung nach 12 Abs.2 in Verbindung mit Anhang II, Nr. 11 of the ArbStoffV für die Lagerung von Ammoniumnitrat und ammoniumnitrathaltigen Zubereitungen, October 1982.

Ullmanns Enzyclopädie der technischen Chemie, 4. Auflage.

VDI guidelines, e.g.:
VDI-2066 Staubmesssungen in strömenden Gasen, pages 1 (10.75), 2 (6.81), 4 (5.80)
VDI-2456 Messung gasförmiger Emissionen; Messen der Summe von Stickstoffmonoxyd und Stickstoffdioxyd, pages 1 + 2 (12.73).
Messen von Stickstoffmonoxyd, Infrarot-absorptionsgeräte URAS, UNOR, BECKMANN, Modell 315, page 3 (4.75).
Messen von Stickstoffdioxydgehalten, Ultraviolettabsorptionsgerät - LIMAS G, page 4 (5.76).
Analytische Bestimmung der Summe von Stickstoffmonoxyd und Stickstoffdioxid, Natriumsalicylatverfahren, page 8 (11.83).
VDI-2298 Emissionsminderung in Schwefelsäureanlagen.

Verordnung über gefährliche Arbeitsstoffe, Arbeitsstoffverordnung - Arbstoff V., 11.02.1982.

Verordnung über Arbeitsstätten, Arbeitsstättenverordnung, ArbStätt V, 01.08.1983.

45

Trade and Industry

Cement and lime, gypsum

46.   Cement and lime, gypsum

Contents

1. **Scope**

Companies in the cement, lime and gypsum industries produce mainly powdery products which are **mouldable** when water is added to them and **set** after a certain reaction time. The following **production stages** are required to manufacture the products:

- **Extraction:** Transport, crushing, dosing of additives, storage, dressing of the raw materials;

- **Burning;**

- **Storage and crushing** of the burnt products;

- **Addition of additives:** e.g. gypsum in the case of cement or water in the case of lime;

- **Packing and dispatch**

In the **cement industry** there are essentially two production processes which are used to dress and burn the raw material, the so-called **wet process** and the **dry process**. In most cases the raw material consists of a mixture of limestone and clay in the ratio of approximately 4:1.

- In the **wet process** the raw material is ground, with the addition of water, to form a **sludge** which contains 35-40% water. During **burning** the water evaporates. The **amount of energy** required for this is 100% greater than in the dry process. Because of the conditions of the wet process the specific **waste gas flow rate** is **higher.** New furnaces for the wet process are now only being constructed for **extreme raw material conditions**, whilst older plants are being converted increasingly to the energy-saving dry process.

- In the **dry process** the raw material is **crushed** whilst being dried, preheated by the counterflow process in a heat-exchanger by the hot kiln waste gases and in most cases **burnt** in a **rotary kiln** at the required sintering temperature of approx. 1400°C. Some of the modern plants have **capacities** of over 5000 t/day, whilst the capacity of the wet kilns rarely exceeds 1000 t/day. **Shaft kilns** are only used occasionally in special cases where market or raw material conditions dictate, and for the most part their capacity is less than 200 t/day.

46

In the **lime industry** both shaft and rotary kilns are used for **burning** the limestone, the combustion temperature being 850-1000°C. In some cases ring kiln and similar internally developed shaft kiln processes are still used. Compared with the cement industry the **capacities** of the lime kiln plants are lower, rarely above 1000 t/day. **Small producers** with simple shaft kilns having a capacity of only a few tonnes per annum are commonly found in many countries.

Gypsum is **dewatered** at temperatures of 200 - 300°C max. and converted from dihydrate to hemihydrate. Direct current rotary kilns, calcining mills or calciners and boilers are used for **burning**. The **capacities** of modern gypsum works are between 600 and 1100 t/day, but some of the plants still have relatively low capacities.

Anhydrite accompanied by gypsum is found in nearly all gypsum deposits. Anhydrite is an anhydrous form of calcium sulphate ($CaSO_4$) which, after crushing and classifying, can sometimes be used as a **quick binding agent** without prior thermal treatment.

2.   Environmental Impacts and Protective Measures

2.1   Air

2.1.1   Waste gases/flue gases

No waste gases are produced in the **extraction and crushing** of cement, lime and gypsum raw materials (principally limestone, gypsum and anhydrite), processes which are mainly carried out in the quarries.

The **cement raw materials** are frequently **dried** during dressing and crushing so that the moisture produced can be driven off as harmless **water vapour**. During the **burning** of the raw materials for cement production, calcium carbonate is converted to calcium oxide when the **carbon dioxide** (CO_2) contained in the limestone is driven off. **Sulphur compounds** (mostly in the form of SO_2) and **nitrous oxides** (NO_x) may also be contained in the waste gas. **Chlorine and fluorine gas and vapour emissions** are **prevented** in the normal process by the fact that these impurities are deposited in the burnt product.

Water vapour and CO_2 emissions are **process-related**, whilst the occurrence of **sulphur compounds** can be **greatly reduced** by the use of suitable raw materials and fuels and control of the burning process. Up to certain limits, **sulphur components** are **bound** by the cement clinker during burning. Only under **extraordinary operating conditions**, e.g. where there is an excess of sulphur in the raw material and fuel, or in the case of reducing burning, will there be occasional short-term **emissions of appreciable quantities of SO_2**.

The **flame temperature** at which cement is manufactured may be as high as 1800°C, with the result that more **nitrous oxides** are formed by oxidation of the atmospheric nitrogen than in lime burning.

The **NO_x values** of 1300 - 1800 mg/Nm^3h permitted in the waste gas in Germany (TA-Luft - Technical Instructions on Air Quality Control - Table 1) will probably **become subject to more stringent requirements** in the next few years. At the present time, possible ways of **reducing the NO_x values** are the subject of large-scale trials, and there currently appear to be **four potential methods**:

- non-catalytic combustion;
- plants with activated carbon filters;
- optimisation of the burning operation;
- conversion of plants to a two-stage calcining installation (oxidising, reducing).

These processes require **different levels of investment** and they all presuppose **continuous operational monitoring.**

In the **cement industry** oils, solvents, paint residues, old tyres or other combustible waste materials are frequently used as **additional fuels.** Some of these waste products introduce **contaminants** which are normally **bound** by the clinker and do **not reach the waste gas.** If such fuels are used, the process must be monitored by **special safety inspections** to prevent the emission of additional contaminants.

In **lime burning**, which takes place in much smaller plants than in cement production, CO_2 is also emitted with the flue gas, but the **quantity of waste gas** is much **smaller** than in cement works because of the size of the plant and because of the lower combustion temperatures in the process.

In **lime slaking** calcium carbonate is converted to calcium hydroxide with the addition of water, some of the water added being discharged again as **water vapour**, since the process is exothermic. However, this water vapour is **harmless.**

46

In **gypsum burning** water vapour and small quantities of flue gas are discharged into the atmosphere. Since the **combustion temperatures** of 300-400°C are **not very high**, and since in most cases the **mass flows** are **very low**, these burning plants only cause slight environmental pollution.

Anhydrite from **natural deposits** is only **crushed** before use, but anhydrite from **phosphoric acid production** must be dried before further use, in which case water vapour will be given off. However, this anhydrite is rarely suitable for industrial use, because it is often **toxic**.

2.1.2 Dust

During the **extraction and further processing** of cement, lime and gypsum dust is produced in various stages of the work **due to process conditions**. In the case of **cement** this dust is a mixture of limestone, calcium oxide, cement minerals, and sometimes even completely burnt cement, whilst in the case of **gypsum** the dust contains anhydrite and mainly calcium sulphate. With the exception of the **pure CaO dust**, which is produced during lime burning, the dust is harmless, but on the other hand it does give rise to considerable nuisance. In the case of the individual production units and conveying installations of a **cement works 6-12 m³ of spent air and waste gas per kg of material** have to be extracted and dedusted. The **major sources of dust** in a plant include:

-      crushing and mixing of the raw material;
-      burning of the cement;
-      crushing of the cement (clinker + gypsum);
-      slaking of the lime.

The proper use of **high-performance extraction plants and dedusting installations,** such as electrostatic separators, fabric and gravel bed filters, and often cyclones used in conjunction with these, is essential, otherwise **correct process management** cannot be guaranteed, costs due to machinery wear rise disproportionately and high dust levels impair working conditions, simultaneously causing loss of production.

The separated **dusts** are **mainly recycled,** provided no enrichment of heavy metal components such thallium is expected in the waste gas. Only under **unfavourable raw material and fuel conditions** will it perhaps be necessary to separate and eject **partial quantities of the dust** because of the excessive concentration of detrimental components in the product, e.g. alkaline chlorides. Occasionally the **use** of these dusts is **possible** in other branches of industry. If the dusts are **dumped,** the **groundwater protection** requirements must be met due to the water solubility of individual components.

In **lime production** the **quantity of dust produced is smaller** because a powdery product is only involved during the slaking, packing and loading of the lime. In the **gypsum and anhydrite industry** the **amount of dust produced** is also **small**.

High quality **filters** (electrostatic or fabric filters) now make it possible to achieve a **dust concentration** of less than 25 mg/Nm^3 in the spent air in the cement, lime and gypsum industry. At present, values of below 25 mg/Nm^3 are being discussed by the European authorities for new plants, whereas the German TA-Luft (Technical Instructions on Air Quality Control) still requires 50 mg/Nm^3.

2.2   Noise

Cement works emit far higher noise levels than lime and gypsum works, but the latter also have production areas giving off **considerable noise**.

In the **extraction of raw materials,** noise and associated vibrations may occur as a result of **blasting,** but such noise emissions can be substantially reduced by means of **suitable ignition processes.** Moreover, the **machines** used for mining can be **soundproofed** to such an extent that they meet the requirements of the German TA-Lärm (Technical Instructions on Noise Abatement).

During **dressing, noise pollution** is liable to occur e.g. through the use of rebound crushers and mills for the crushing of hard materials. These **crushing installations** and the adjoining dressing installations can be **enclosed** in such a way as to protect the environment from oppressive noise. The noise generated by the majority of rock- and cement-crushing plants is so intense that they have to be installed in **soundproofed premises** in which personnel cannot work on a permanent basis.

Burning plants require numerous large **fans** which generate extremely penetrating noise, with the result that noise protection measures, e.g. in the form of enclosures, are also necessary.

In order to **avoid nuisance**, plants in the lime, gypsum and particularly the cement industries must be erected at least 500 metres from residential areas. The immissions values for nearby residential areas should not exceed 50 to 60 dB(A) during the day, and 35-45 dB(A) at night.

46

2.3   Water

In the vicinity of pits in the German cement, lime and gypsum industry the **waste-water** may contain up to 0.05 ml/l of total suspended solids. To avoid exceeding this value the pit water produced must be discharged via **stilling basins.** Water used for washing limestone must always be discharged via **sedimentation ponds**, and the **surface water** produced in the area surrounding the pits must be **discharged separately.**

Some **cement and lime works** are major **water consumers**, but because of the process involved they cause **no water pollution.** In **cement works** approximately 0.6 m^3 of water per tonne of cement is required to **cool the machines.** Most of this water is in **circulation**, thus only the water losses need be made up. In plants involved in the drying process, water is also used for **cooling the kiln exhaust gases**, resulting in a calculated net consumption of approx. 0.4 - 0.6 m^3 of water/t of cement. In plants using the **wet process** an additional 1 m^3 of water/t of cement or so is required for the **sludge milling**. This water is discharged again by **evaporation.**

In the **lime industry** water is required for **slaking** burnt lime (approx. 0.33 m^3/t of lime). Some lime works consume an additional 1 m^3 or so of water per tonne of lime for **washing** the raw limestone when extremely pure qualities are required. After use, this washing water is fed to **settling basins or settling ponds** where the fine particles are deposited and the residual water **evaporated** or partially **re-used.**

The **gypsum industry** requires **relatively little water** because the processes take place at low temperatures, with the result that no cooling energy is required. In **plasterboard production**, water is added to the raw gypsum and remains in the product to set the gypsum (conversion of hemihydrate to dihydrate).

Water demand can be reduced by **increasing the proportion of circulating water** or by **minimising the water losses.**

In **dry areas** the **cooling water demand** can be reduced by installing special **electrostatic precipitators** which are operational at higher exhaust gas temperatures.

Any sanitary water produced must be discharged and disposed of separately.

2.4 Soils

In the area **surrounding** cement, lime and gypsum works the soils may be impaired by falling dust where the dedusting plants are **inadequately maintained.**

Although potentially **environmentally relevant trace elements** can be introduced into the **cement production** process by special raw material components such as iron ore and, more recently, by the increased use of combustible **waste materials**, these hazardous substances are almost completely **absorbed** by the cement clinker in the molten state, chemically **bonded** and therefore rendered harmless. To **rule out** the possibility of **adverse effects** when using special raw material components or waste products from other industries as fuel from the outset, **analyses** must be carried out te detect environmentally relevant trace elements such as lead (Pb), cadmium (Cd), tellurium (Tl), mercury (Hg) and zinc (Zn), which are deposited in the filter dusts. If necessary, technical measures such as dust separation must be applied to **prevent the accumulation** of hazardous substances in the process.

2.5 Workplace

Numerous machines generating **noise levels** of 90 dB(A) are still operated in cement, lime and gypsum works, even with the present state of the art. Noise levels can generally be reduced by means of static devices. **Permanent workplaces** inside the plants, e.g. control platforms, must be **soundproofed**, but if continuous noise levels of 85 dB(A) are still produced, **hearing protection** must be made available. At noise levels in excess of 90 dB(A) this protection must compulsorily be worn to avoid hearing impairment. Even where personnel remain in high-noise process areas for short periods, hearing protection is recommended.

In **exceptional cases**, e.g. during **repair work** or when **rectifying faults**, personnel may be exposed to high temperatures and higher levels of noise and dust for long periods, and **suitable protective devices** and **protective clothing** must be provided for these tasks. Moreover, work in the danger area must be restricted and supervised.

2.6 Ecosystems

Cement, lime and gypsum works require **raw materials close to the surface**, thus **interference with the surrounding landscape** cannot be avoided in the extraction of raw materials. The **environmental effects** of extraction are described in the environmental brief Surface Mining.

46

When **selecting locations** for cement, lime and gypsum works, due consideration must be given to the environmental aspects. In the case of locations in **areas** previously **used for agriculture, possibilities for alternative employment** must be examined, particularly for affected women. Besides complying with the **regulations** concerning waste gases, dust, noise and water, the conditions as regards the **building land**, integration in the **landscape**, and the **infrastructure** of the location must also be examined. Infrastructural considerations include, amongst other things, the recruitment and housing of employees, transport systems and traffic density and the existing and planned industrialisation of the area.

Since the **environmental impact** is not limited to the factory area, the **local population**, including women and children in particular, should be given access to **medical care.**

In **cement production** approximately 1.6 t of raw material per tonne of cement and additional quantities of gypsum are required, bringing the total raw material requirement to approximately 1.65 tonnes. In **lime production** the raw material requirement of approx. 1.8 t per tonne of finished product is about 10% higher than for cement production. In calculating this raw material requirement, the over-burden, which varies considerably from deposit to deposit, is not taken into account. In Germany most of the **gypsum requirement** could now be covered by the gypsum produced in **flue gas desulphurisation plants**, so that producing this raw material would no longer affect the landscape.

It is advisable to build up **financial reserves** for the subsequent **recultivation** of a quarry, even while the quarry is operational.

3.   Notes on the Analysis and Evaluation of Environmental Impacts

Limit values for exhaust gas, dust and water have been formulated for dischargers of wastewater in the provisions of TA-Luft and TA-Lärm (Technical Instructions on Air Quality Control and Technical Instructions on Noise Abatement), in the Guidelines adopted by the Association of German Engineers (VDI) and in the administrative regulations specific to the various industries. Similar values are being adopted by most European countries. The US regulations published by the Environmental Protection Agency (EPA) are frequently **more stringent** than the German regulations, particularly in California.

For **countries without their own environmental protection laws,** these values must be **examined** and **adapted** in the individual case, taking the prevailing environmental conditions into consideration. In exceptional cases, particularly for **rehabilitation of plants, special regulations** must be established, but **new plants** should conform to the European **standard values** for environmental protection.

The **Compendium of Environmental Standards** offers advice on assessing environmental relevance for individual substances.

Table 1

Limitation of hazardous substances under TA-Luft (Technical Instructions on Air Quality Control) and the 17th Administrative Regulation according to § 7a of the Federal Water Act

Cement and lime, gypsum		Air mg/Nm³	Water		
			Direct discharger g/m³	Sample type	Indirect discharger** g/m³
Dust		50			
NOx nitrous oxide grill preheater	NOx	1.500			
NOx nitrous oxide cyclone preheater and exhaust gas heat utilisation	NOx	1.300			
NOx nitrous oxide cyclone preheater without exhaust gas heat utilisation	NOx	1.800			
NOx nitrous oxide grill preheater	NOx				
SOx sulphur oxide as SO2	SOx	400			
Fluorides	F	5			50
Chlorine	Cl	30			
Filterable solids			100	1)	1
Total suspended solids	TSS		0.5	2)	1
Chemical oxygen demand	COD	80			
Antimony	Sb	5			
Arsenic	As	1			
Lead	Pb	5	0.50	2)	2
Cadmium	Cd	0.2	0.07	2)	0.5
Chromium	Cr	5	0.10	2)	2
Cobalt	Co	1	0.10	2)	
Cyanides (*)	-CN	5			0.2
Copper	Cu	5	0.10	2)	2
Manganese	Mn	5			
Nickel	Ni	1	0.10	2)	3
Palladium	Pd	5			
Platinum	Pt	5			
Mercury	Hg	5			
Rhodium	Rh	0.2			0.05
Selenium	Se	1			
Tellurium	Te	1			
Thallium	Tl	5			
Vanadium	V	0.2			
Zinc	Zn		2.00	2)	
Tin	Sn	5			

*	May be formed in reduced burning	COD	Chemical Oxygen Demand
**	Law applicable in the German state of Baden-Württemberg	TSS	Total Suspended Solids
		TA-Luft	Technical Instructions on Air Quality Control
1)	Two hour mixed sample	VwV	Administrative Regulation
2)	Random sample	WHG	Federal Water Act

In developing countries **dust emissions** of 100 mg/Nm³ of exhaust gas or spent air should on no account be exceeded. Higher dust emissions will cause both internal and external environmental burdens.

Similarly, **wastewater disposal** should meet the **minimum requirements** imposed by the regulations laying down limits for dischargers of wastewater into receiving bodies of water.

The **noise problem** is **underrated** in many countries, but constant noise can lead to **permanent damage.** Here too, therefore, the prescribed **noise limits** must be adhered to in the workplace and in the surrounding residential areas (Section 2.2), and encroachment on residential areas must be prohibited.

All parameters must be regularly **checked** by means of internal audits, for which purpose **training** must be given and **personnel generally sensitised** to environmental matters if necessary.

The use of land by the cement, lime and gypsum industry must be kept within definable limits by forward-looking and detailed planning covering the areas of mining, recultivation and water management. The high costs often mean that there is no money available for **recultivation** of pits, often resulting in direct or consequential damage that may be difficult to repair (see environmental brief Surface Mining)

3.1   Inspection and maintenance of environmental protection installations

A **control centre independent** of the production process must be established to comply with existing environmental protection regulations. The responsible personnel must be enabled to perform and monitor **all inspection functions** including measurements relating to environmental protection in the works. They should be available for consultation on investments and take charge of negotiations with environmental protection authorities. Moreover, this department is responsible for ensuring that all **environmental protection installations** are regularly **maintained** and **upgraded.** This internal environmental department is also responsible for **staff training.**

4.   Interaction with Other Sectors

Cement production may touch on other project areas, particularly where **additional raw material components are used.** For example, use is made of materials produced in lime works with inadequate lime content, other waste materials such as crystallised calcium carbonate from the chemical industry or ferrous residues from sulphuric acid production. Up to 5% gypsum per tonne of cement is required to control the rate of setting in the cement, and a major proportion of this gypsum requirement is now met in Europe by gypsum from flue gas desulphurisation plants. Up to 85% of **fly ashes** from power station dedusters and slags can also be added to the clinker to produce cement varieties with special properties.

Because of the **high temperatures** and comparatively **long holding times** of the materials in the relevant areas, cement kilns in particular are ideal for **disposing of combustible waste.** This possibility is increasingly important in countries where large quantities of **vegetable waste** with high potential energy, such as rice chaff, are produced in the region.

In the cement, lime and gypsum industry, **secondary activities** such as quarries, fuel stores, workshops etc. also exert **environmental impact.**

Table 2
Environmental impacts of adjacent project areas - cement, lime and gypsum

Interacting project areas	Nature of intensification of impact	Environmental briefs
Extraction/storage of raw materials and fuels	- Landscape impairment - Pollution of bodies of water - Waste storage in former pits	Surface Mining Planning of Locations for Trade and Industry Urban Water Supply Rural Water Supply
Disposal of solid and liquid waste	- Discharge of deposited solids e.g. filter dusts - Pollution of bodies of water by wastewaters	Solid Waste Disposal Disposal of Hazardous Waste
Maintenance of workshops and transport facilities	- Risks of handling water pollutants (e.g. solvents) - Impacts of transport and traffic (noise, link roads)	Mechanical Engineering, Workshops Road Building and Maintenance Planning of Locations for Trade and Industry

46

5.   Summary Assessment of Environmental Relevance

The environmental impacts of cement, lime and gypsum works are caused by **exhaust gas, dust, noise and water**. The following table **assigns values** to the individual process stages as regards the environmental burden which they impose.

Table 3
Environmental impact of process stages
(cement/lime/gypsum)

Process	Air		Noise	Water[1]	Soil	Work-place
	Exhaust gas /flue gas	Dust				
Extraction	1	1	2	2	3	2
Precrushing	1	1	3	1	1	2
Rough milling/mixing	2	3	4	2	2	3
Burning	3	3	3	2	2	3
Cement milling	1	3	4	2	2	2
Lime slaking	2	3	2	3	3	2
Packing	1	2	1	1	1	1
Loading	1	2	1	1	1	1

Key: 1 very slight; 2 slight; 3 moderate; 4 considerable
[1] dry process only

Proven technologies have been available for a good many years to reduce pollutant loads. In new plants for the cement industry in the industrialised countries, the **costs of environmental protection measures**, in the widest sense of the term, already account for as much as 20% of the total investment cost, and in the future this proportion will **increase still further.**

The **more sophisticated** the dedusting method, the greater the importance of **systematic monitoring and maintenance** for the continuing reliability and efficiency of the plants. Besides dedusting plants, **changes in burning technology** are becoming increasingly important for reducing NO_x values.

Catering for the **needs of the environment** when planning and erecting cement, lime and gypsum works can also **save money**. The dusts generated are mainly preliminary, intermediate or end products which can **reduce** the direct **production costs** if recycled and returned to the process. Reduced ejection of dust also reduces wear on machines, thereby increasing their availability and saving repair costs.

The cement industry is becoming increasingly important as a **recycler of waste materials** such as food, waste oil or rubber tyres, thereby reducing the need for dumping. The initial fear that this disposal might lead to an increased emission of environmentally relevant trace elements has been allayed by measurements carried out during operation. When the materials are **burnt**, particular attention must be paid to **correct firing, design and monitoring of the plants.** Therefore the regulations concerning waste gas emissions and monitoring of such plants have been made more stringent.

The designers of a new plant must consider what **environmental protection measures** are necessary and appropriate as early as the **planning phase**. Suitable guidelines must also be established during the planning phase for countries which do not have their own regulations in this area.

Early **involvement of neighbouring population groups** in the planning and decision-making processes will enable measures to be devised to deal with any problems arising.

46

6. References

Erste Allgemeine Verwaltungsvorschrift zur Reinhaltung der Luft -TA-Luft - GMBl (joint ministerial circular) Nr. 24.

Allgemeine Verwaltungsvorschriften über genehmigungsbedürftige Anlagen nach § 16 der Gewerbeordnung GeWO (technische Anleitung zum Schutz gegen Lärm - TA-Lärm) verschiedene Ausgaben.

Allnoch, G. et. al.: Umweltverträglichkeitsprüfung von Entwicklungshilfeprojekten - Erstellung eines Kataloges von Emission- und Immissionsstandards, im Auftrag der GTZ Eschborn, 1984.

Betriebswacht, Datenjahresbuch 1991: Berufsgenossenschaft der keramischen und Glasindustrie, Würzburg.

Emissionsminderung Zementwerke VDI-Richtlinie 2094, Entwurf May 1981.

Entwurf zur Abwasserverordnung, Deutscher Industrie- und Handelstag, Anhang 17, Sept. 21, 1990.

Environmental Protection Agency: New source performance standards - Clean Air Act (USA).

Environmental Assessment Sourcebook Nov. 1990, Worldbank Draft Part 9.3-1 Cement /93-101, Mining and Mineral Processing 31.10.1990.

Funke, G.: Immissionsprognosen für Genehmigungsverfahren Zement, Kalk, Gips 33, p. 15-23, 1980.

Göke H.: Grenzen des Umweltschutzes aus der Sicht der Tagebau- und Steinbruchindustrie, Zement, Kalk, Gips 31, p. 252 - 254, 1978.

Gesetz zum Schutze vor Umwelteinwirkungen durch Luftverunreinigungen, Geräusche, Erschütterungen und ähnliche Vorgänge. Bundesimmissionsschutzgesetz - BImSchG - dated 15.03.1974 - BGBl. I (Federal Law Gazette I), p. 721 - 1193.

Hinz, W.: Umweltschutz und Energiewirtschaft Zement, Kalk, Gips 31, p. 215 - 229, 1979.

Luftreinhalte-Verordnung (LVR) Switzerland, of 16.12.1985, edition of 1 July 1990.

Luftreinhalteplan bei der Basel, February 1990.

Schulze, K.-H.: Immissionsmessungen und ihre Fehlergrenzen Zement, Kalk, Gips 36: p. 7 - 11, 1980.

Technical note on best available technologies not entailing excessive cost for the manufacture of cement: Commission of the European Communities, Report EUR 13005 EN, 1990.

Umweltschutz in der Steine - und Erden-Industrie Zement, Kalk, Gips 31, p. 215 - 229, 1979.

Verein Deutscher Zementwerke: Forschungsbericht der Zementindustrie, Tätigkeitsbericht, 1978 - 1981 - VDZ Düsseldorf 1981.

Ditto Tätigkeitsbericht 1981 - 1984

Siebzehnte Verordnung zur Durchführung des Bundesimmissionsschutzgesetzes 1990, (Verordnung über Verbrennungsanlagen für Abfälle und ähnliche brennbare Stoffe, 17. BImSchV)

Zukünftige Probleme des Umweltschutzes in der Zementindustrie, Zement, Kalk, Gips 33: p. 1 - 9, 1980.

46

Trade and Industry

Ceramics
Fine, Utilitarian and Industrial

47. Ceramics - Fine, Utilitarian and Industrial

Contents

1.   Scope

Fine, industrial and utilitarian ceramics cover the following **industrial sectors:**

- **Ordinary ceramics:** tiles, roof-tiles, earthenware, expanded clay, wall tiles and floor slabs, refractory products

- **Fine ceramics**: earthenware, pottery, fine earthenware, porcelain, electrical porcelain, sanitary products, grinding discs and abrasive wheels

- **Technical ceramics**

Most ceramics companies are established in the **vicinity of clay deposits.** (This environmental brief deals only briefly with the **extraction** of raw materials; for further details refer to the environmental brief Surface Mining. Advice on processing and transportation of raw materials is also given in the relevant environmental brief. The **size** of ceramic plants and their daily **throughputs** vary from a few kilograms for technical ceramic plants, normally 10 to 50 t/day for fine ceramics, to as much as 450 t/day in the tile industry. Since many companies operate **different types of production**, the total output of the works is often higher than the typical daily output of a specific product.

The fine, industrial and utilitarian ceramics industries use all types of clays, kaolins and fireclays (burnt clay), feldspars and sands as a **raw material base.** The refractory, abrasives and technical ceramics industries **also** use numerous high-temperature-resistant and abrasion-resistant **oxides** such as corundum (Al_2O_3), zirconium oxide (ZrO_2) and silicon carbide (SiC).

Besides using their own, readily available raw materials, many companies are increasingly purchasing **ready-processed raw materials**, particularly for refractory products, abrasives and technical ceramics, as well as the raw materials required for glazes and frits.

The following **process sequence** is typical of the **production processes** in industrial, utilitarian and fine ceramics:

- extraction, processing, forming, drying, partial glazing or enamelling, firing, sorting/packing and transportation.

Execution of the individual **process stages varies** according to the selected method. Generally speaking, casting, plastic or drying processes are employed, **with smooth transitions** between the process stages.

47

Table 1

Production processes

Casting processes	Plastic processes	Dry pressing processes
- Porcelain	- Tiles	- Refractory products
- Sanitary products	- Roof tiles	- Wall tiles, floor slabs
- Electrical porcelain	- Expanded clay	- Pottery
- Refractory	- Cleaving tiles	- Earthenware tiles
-	- Electrical porcelain	- Technical ceramics
	- Pottery	- Steatite
	- Earthenware	- Abrasive wheels

-      In the **casting process** the raw materials are **dosed, wet-ground** and **poured** into plaster moulds as so-called slip. During pressure casting, the slip is shaped to produce the blank under pressure in machines.

-      In the **plastic process** the raw materials are normally **prepared in the wet state,** mixed and **shaped** with moisture content of 15 - 20% water.

-      In the **dry pressing process** used in fine ceramics, the raw materials are frequently **prepared in the wet state,** then **dried** in a spraying tower to a residual moisture content of 5-7%. In the refractory industry the raw materials are mixed dry and are often processed with pressing moisture content of less than 2%, also using organic and inorganic binding agents.

The **moulded products** are **dried** and then **fired.** They are generally fired in high-power tunnel kilns; **special products** are fired mainly in individual, hood-type or batch kilns, while fast-burning products are fired in roller hearth kilns of various designs. In many countries, tile products in particular are often fired in **self-built single-chamber and ring kilns** or in charcoal kiln systems.

Many **fine ceramic products** are glazed or enamelled before firing.

Depending on the raw materials used, the **firing temperatures** in **industrial, utilitarian and fine ceramics** begin at 950°C for some tile products, for example, whereas most **fine ceramic products** are fired at between 1100°C and 1400°C. **Refractory** and **technical ceramic products** have firing temperatures of 1280 to 1900°C. (Pure glaze baking is done at lower temperatures.) The dual firing process is sometimes used for porcelain and very rarely for wall tiles.

Energy consumption depends on the product and the process; in the **tile industry,** because of the low firing temperatures, it is between 800 and 2100 kJ/kg of manufactured product, but in almost all other areas of **industrial, utilitarian and fine ceramics** it is on average much higher per manufactured product, and may be as much as 8000 kJ/kg of product.

After firing the products must be **sorted** and sometimes **reworked,** which will involve varying labour costs depending on the product.

2. Environmental Impacts and Protective Measures

2.1 Air

2.1.1 Waste gases/flue gases

Hardly any waste gases are produced in the extraction, processing and moulding of ceramic products. Exceptions to this are the **demoisturisation** in the spraying tower, e.g. during the production of tiles, and the dry crushing plants used in clay processing, where harmless **water vapour** is given off.

During the **glazing process,** care must be taken to prevent glazing vapours, some of which contain **heavy metals** and other **toxic substances,** being discharged to the environment or being inhaled by personnel. Therefore only glazing plants which are equipped with the necessary **extraction and wastewater discharge equipment** should be licensed. Operating or maintenance personnel working in this area must be protected by **breathing filters.** When the glazed products are **dried,** mainly harmless **water vapour** is given off.

The amount of **flue gas** produced during firing depends on the emission of the fired product and on the type of fuel used. **Volatile components** are sometimes given off from the product mass and from the fuel.

The adverse environmental effects of **fluorine emissions** from the ceramics industry have come to be recognised as a **serious problem,** particularly in recent years, in view of the damage occurring in the vicinity of ceramic works (animals and plant diseases). Fluorides are present in all ceramic raw materials and are sometimes emitted in the waste gas during firing. Because of this, fluorine emissions from new plants built in Europe must be **less than 5 mg/Nm^3.**

47

Because ceramic firing plants operate continuously, **residual substances** from other sectors such as waste oils or organic components from water treatment plants are sometimes used as **fuel additives**. Plants which use such materials are subject to **special regulations** because dangerous oxides may be introduced via these **waste substances** and re-emitted with the flue gas.

German companies must conform to the following values when burning waste substances:

-	Total dust	10	mg/Nm³ max.
	Sulphur dioxide	50	mg/Nm³ max.
	Cd, Tl, Hg,	0.1	mg/Nm³ (per element)
	(cadmium, tellurium, mercury)		
	Other heavy metals	1	mg/Nm³

Because of these conditions, waste substances cannot be used in the ceramic industry without the installation of additional **water-spray separators.**

Nitrous oxide emission during firing appears **not to be a problem** in most plants which are operated at relatively low temperatures, but **special solutions** must be found for high-temperature firing plants in the **refractory industry** for denitrifying the waste gases.

No waste gases are generally produced during sorting, packing, internal conveying, processing or refining. Only in very rare cases, e.g. during subsequent **colouring** or **printing**, may environmental pollution be caused by waste gases. These problems must be solved on a **case-to-case** basis.

2.1.2 Dust

Dust presents a **latent risk** in fine, industrial and utilitarian ceramic plants, particularly for the **labour force**. Fine quartz dusts < 5 μm may cause silicosis.

Depending on the geological and meteorological conditions, **dusts** may occur in pits during **extraction of the raw materials** which can be reduced by wetting and by the use of appropriate extracting and conveying methods. (See environmental brief Surface Mining).

Whilst **hardly any dust** is produced in the **wet medium** of the plastic processes, in the **preparation, moulding** and **drying** processes a variety of **methods** can be adopted to minimise dust formation, such as continuous cleaning of the works, concreting and sealing of floors, efficient dedusting systems and wet grinding of porcelain and sanitary products.

Silicosis in the German porcelain and refractory industry, particularly in the case of silicate products, has been successfully **minimised** by **systematic dust control** in all working areas, but in many countries it is still a problem. The **statutory limits** for quartz dusts impose a maximum allowable concentration (MAK) of 0.15 mg/Nm^3 of fine dust, and the air may contain no more than 4 mg/Nm^3 of fine dust containing more than 1% by weight of quartz.

In Germany according to TA-Luft [Technical Instructions on Air Quality Control] the **total dust content** must not exceed 50 mg/Nm^3 in the waste gas at a mass flow of more than 0.5 kg/h, or 150 mg/Nm^3 at a mass flow up to and including 0.5 kg/h.

During **firing** the **dust burden** is generally very slight. Dry filters are now frequently installed in kilns, water-spray separators more rarely. Dry absorption systems may create dust, thus care must be taken to ensure that when such systems are used the maximum dust quantity of 50 mg/Nm^3 in the flue gas is not exceeded. These plants require **regular maintenance** to preserve their **efficiency** (see 3.1).

2.2   Noise

In most production processes in the ceramic industry, **noise** is emitted but rarely exceeds 85 dB(A) (see 2.5 - Workplace).

During the **extraction of raw materials,** noise and associated vibrations may occur for a short time as a result of **blasting**, sometimes causing a serious nuisance to residents living close by. However, such noise can be substantially reduced by means of **suitable detonation methods.** The **machines** used for mining can now be **soundproofed** to such an extent that they meet the noise protection requirements. (See environmental brief Surface Mining).

During **dressing, noise pollution** is liable to occur e.g. through the use of rebound crushers and mills for the crushing of hard materials. These crushing installations and the adjoining dressing installations can be **encapsulated** or soundproofed in such a way as to protect the environment from oppressive noise.

During the drying and firing phases, **fans** are used which may generate noise levels in excess of 85 dB(A). These noise sources must be installed **outside permanent workplaces.** During special ceramic **production processes,** e.g. when splitting cleaving tiles and when using sheet metal plates, frames or pallets for internal conveying systems, typical noise problems arise. However, such **noise levels** can be **reduced** by taking appropriate measures, e.g. **encapsulating** permanent workplaces and **buffering** mobile conveying systems with rubber.

47

To avoid noise nuisance the **immission values** for the **residential areas located close** to the ceramic production centres should not exceed 50 - 60 dB(A) during the day and 35 - 45 dB(A) at night. **Housing developments** should be sited at least 500 m from a ceramic factory.

2.3 Water

In Germany, ceramic works must comply with the administrative regulations regarding permitted **substances in the wastewater.**

Works laboratories must be established to **monitor** the works in question.

Table 2
**Maximum permissible values for direct dischargers
according to the 17. VwV of the WHG
[17th Administrative Regulation of the Federal Water Act]**

Parameters	Maximum value
Filterable solids from the 2-hour mixed sample	100 ml
Total suspended solids from the random sample	0.5 mg/l
Chemical oxygen demand (COD) from the 2-hour mixed sample	80 mg/l
Lead content from the 2-hour mixed sample	0.5 mg/l
Cadmium content from the 2-hour mixed sample	0.07 mg/l

To avoid exceeding the applicable values, the **water produced** in the area of the pit must be fed through **stilling basins,** with the addition of **sedimenting agents** if necessary. The **surface water** occurring in the area surrounding the pit **must be discharged separately.**

Fresh water consumption in modern ceramic plants is **low** because the water required for the process is **circulated** internally. Some of the water used is driven off again as **water vapour** in the production of granulates in the spray tower and in the drying of the products. **Wastewaters** produced contain clay, flux and other ceramic **raw materials** which are precipitated and returned to the process by internal **circulation.**

Sanitary water produced in fine, industrial and utilitarian ceramic works must be **discharged and disposed of separately.**

2.4   Soils

Nowadays old **clay pits** are frequently used for **storing waste products** of all kinds, because of their relatively low water permeability. **Soil damage** may occur due to elutriation and water accumulation in **old pits,** because when the pit was worked, water management was not normally up to present-day environmental standards.

Soil is rarely impaired by **spoil from ceramic works** because the waste generated during production is **reused** in the plant's own production or in other ceramic works, so that spoil dumps are only formed where the plant is operated inefficiently. Exceptions to this are the small quantities of **gypsum** produced during porcelain, sanitary and roof tile production, which have to be properly **disposed of.**

2.5   Workplace

Personnel working in ceramic plants may be endangered or oppressed by **noise, dust** and **heat** in certain work areas.

Permanent workplaces near **sources of loud noise** must be **soundproofed.** If the noise level is still not less than 85 dB(A) despite soundproofing measures, **hearing protection** must be made available, and from 90 dB(A) upwards it must be compulsorily worn to prevent resulting hearing impairment. Hearing protection must also be worn by personnel working in high-noise production areas for short periods.

During **firing** in tunnel, reciprocating, roll-over or bogie hearth kilns, **temperature stress** on personnel is relatively **low** in modern plants, but in plants with **old** single chamber and ring kilns, there may be **considerable exposure to heat** when the product is inserted and removed. In **special cases,** e.g. if a tunnel kiln car caves in, work must be carried out for a short time under **conditions of extremely high temperature.** In this case, strict protective measures, e.g. the wearing of **thermal suits,** must be complied with. Moreover, such work must only be carried out under appropriate **supervision.**

47

In fine ceramic works, particularly in the porcelain and silicate industry (refractory products), personnel may be at risk from continuous exposure to **quartz dust**. In addition to **technical precautions, regular medical check-ups** are essential here to ensure that fibrotic changes (changes in the pulmonary alveoli) are detected early, so that the employee in question can be **protected from permanent injury** through redeployment.

2.6   Ecosystems

When raw materials are extracted **the landscape is impaired** and there is an **alteration to the surface** (see environmental brief Surface Mining). Since the raw material requirement per plant is not very high, the individual **mining areas are** generally also **relatively small. Many different types of clay** are present in each clay pit, and with the introduction of suitable processing methods even low quality clays have been successfully processed in recent years, thereby **reducing the amount of spoil** in the vicinity of clay pits.

When selecting a site for a ceramic plant, due consideration must be given to the environmental aspects. In the case of locations in **areas** previously **used for agriculture, possibilities for alternative employment** must be examined, particularly for affected women. Besides complying with the **regulations** concerning waste gases, dust, noise and water, the conditions as regards the **building land**, integration in the **landscape**, and the **infrastructure** of the location must also be examined.

Infrastructural considerations include, amongst other things, the recruitment and housing of employees, transport systems and traffic density and the existing and planned industrialisation of the area.

Since the **environmental impact** is not limited to the factory area, the **local population,** including women and children in particular, should be given access to **medical care.**

Recycling of fine ceramic consumer goods, after use on or in buildings or in the home, is **hardly feasible** because of the **variety of materials** and **small quantities involved** at the points of consumption. On the other hand, in the **refractory industry**, particularly in steel works, over 30% of the refractory products are **recycled.**

3.   Notes on the Analysis and Evaluation of Environmental Impacts

Emission limits for waste gas, dust and water have been formulated in the provisions of the German TA-Luft and TA-Lärm [Technical Instructions on Air Quality Control and Technical Instructions on Noise Abatement], in the guidelines adopted by the Association of German Engineers (VDI) and in the regulations specific to the various industries for dischargers (under the WHG - German Federal Water Act) and MAK (maximum allowable concentration) values have been established by the *Berufsgenossenschaft* (employers' liability insurance association) of the ceramic and glass industry for avoiding silicosis. These emission limits are being adopted in similar form by most European countries. The **US** regulations published by the Environmental Protection Agency (EPA) are frequently **even more stringent** than the German regulations, particularly in California.

For **countries without their own environmental protection laws,** these values must be **examined** taking into consideration the prevailing environmental conditions in the individual case and **adapted** to the particular circumstances. In exceptional cases, particularly for **rehabilitation of plants, special regulations** must be established, but **new plants** should conform to the **standard values** of environmental protection.

The Compendium of Environmental Standards offers advice on assessing environmental relevance for individual substances.

47

Table 3

Limitation of hazardous substances under TA-Luft (Technical Instructions on Air Quality Control) and the 17th Administrative Regulation according to § 7a of the German Federal Water Act

Ceramics		Air mg/Nm³	Direct discharger g/m³	Sample type	Indirect discharger** g/m³
Dust		50			
Sulphur dioxide as SO2 at a mass flow < 10 kg/h	SO2	500			
Sulphur dioxide as SO2 at a mass flow > 10 kg/h	SO2	1,500			
Nitrous oxide NOx	NOX	500			
Fluorides	F	5			50
Chlorine	Cl	30			
Filterable solids			100	1)	1
Total suspended solids	TSS		0.5	2)	1
Chemical oxygen demand	COD		80		
Antimony	Sb	5			
Arsenic	As	1			
Lead	Pb	5	0.50	2)	2
Cadmium	Cd	0.2	0.07	2)	0.5
Chromium	Cr	5	0.10	2)	2
Cobalt	Co	1	0.10	2)	
Cyanides (*)	-CN	5			0.2
Copper	Cu	5	0.10	2)	2
Manganese	Mn	5			
Nickel	Ni	1	0.10	2)	3
Palladium	Pd	5			
Platinum	Pt	5			
Mercury	Hg	5			
Rhodium	Rh	0.2			0.05
Selenium	Se	1			
Tellurium	Te	1			
Thallium	Tl	5			
Vanadium	V	0.2			
Zinc	Zn		2.00	2)	
Tin	Sn	5			

*	May be formed in reduced burning	COD Chemical Oxygen Demand
**	Law applicable in the German state of Baden-Württemberg	TSS Total Suspended Solids TA-Luft Technical Instructions on Air Quality Control
1)	Two hour mixed sample	VwV Administrative Regulation
2)	Random sample	WHG Federal Water Act

When **waste materials are used as fuel**, the above **emission limit values** must on no account be exceeded, and **regular inspection** of the charge material, firing system and process, as well as of the waste gases and dusts, is **essential** (see 3.1).

It is **vital** that the **dust regulations**, based on the **maximum allowable concentrations in the workplace,** are **adhered to,** particularly in the porcelain and silicate industry. Non-compliance with these regulations leads to **diseases** with **long-term consequential damage. Intensive dust abatement** in all plants and in all sections of plants is **imperative** in this regard also.

The **noise problem** is **underrated** in many countries, but constant noise can lead to **permanent damage.** Here too, therefore, the prescribed **noise limits** must be adhered to in the workplace and in the surrounding residential areas, and encroachment on residential areas must be prohibited (see 2.2 and 2.5).

Managers of ceramic production plants must be alerted to the **specific risks to employees** and must be trained in the **use of protective measures** so that employees are not exposed to health hazards through ignorance (see 3.1). Suitable **training** must be given and **personnel generally made aware** of environmental concerns.

In all plants an **internal water circuit** must be carefully planned. **Treated wastewaters** which are discharged into receiving bodies of water are subject to **minimum requirements** which must be met to avoid damage to the ecosystem in areas close to the works.

All the parameters must be regularly **checked** by internal audits (see 3.1), and **works laboratories** must be set up to monitor adherence to the specified values.

3.1 Inspection and maintenance of environmental protection installations

A **control centre independent** of the production process must be established to comply with existing environmental protection regulations. The responsible personnel must be enabled to perform and monitor **all inspection functions** including measurements relating to environmental protection in the works. They should be available for consultation on investments and take charge of negotiations with environmental protection authorities. Moreover, this department is responsible for ensuring that all **environmental protection installations** are regularly **maintained** and **upgraded.** This internal environmental department is also responsible for **staff training.**

47

4.   Interaction with Other Sectors

In the ceramic industry, **interaction** between **different branches of production** is common and is often **necessary** for a smooth production process. Fine, industrial and utilitarian ceramic works rely on **numerous secondary operations,** such as extraction plants, fuel stores, workshops and transport systems involving a number of other sectors.

Table 4
**Environmental impacts of adjacent sectors
- fine, industrial and utilitarian ceramics -**

Interacting sectors	Nature of intensification of impact	Environmental briefs
Extraction/storage of raw materials and fuels	- Landscape impairment - Pollution of bodies of water - Waste storage in former pits	Surface Mining Planning of Locations for Trade and Industry Urban Water Supply Rural Water Supply
Disposal of solid and liquid waste	- Discharge of deposited solids e.g. filter dusts - Pollution of bodies of water by wastewaters	Solid Waste Disposal Disposal of Hazardous Waste
Maintenance of workshops and transport facilities	- Risks of handling water pollutants (e.g. solvents) - Impacts of transport and traffic (noise, link roads)	Mechanical Engineering, Workshops Road Building and Maintenance Planning of Locations for Trade and Industry

All ceramic products must be packed, and the **packing materials** required for this purpose must be disposed of or recycled after use. Environmental impacts can be avoided in this area by making use of modern processes employed in the packaging industry. Moreover, the ceramic industry is highly **transport intensive**, since tiles, roof tiles, cleaving tiles and refractory products have **high bulk weights** and therefore require suitable means of transport.

5.   Summary Assessment of Environmental Relevance

The individual **process stages** in the industrial, utilitarian and fine ceramic industry do **not** generally give rise to **severe environmental burdens**.

Table 5
Environmental impact of process stages
(ceramics)

Process	Air		Noise	Water[1]	Soil	Work-place
	Exhaust gas /flue gas	Dust				
Extraction	1	2	2	3	3	1
Preparation	1	3	3	2	1	2
Moulding	2	2	2	1	1	2
Glazing	3	3	2	3	2	3
Drying	2	1	2	1	2	1
Firing	3	1	3	1	2	1
Sorting	1	1	3	1	1	2
Packing	1	1	1	1	1	1
Internal transport	1	1	1	1	1	2
Processing/ Refining	1	2	2	2	2	2

Key: 1 very slight; 2 slight; 3 moderate; 4 considerable
[1]   Depending on composition
Particularly dangerous in the case of free quartz with grain sizes smaller than 5 μm

47

Moreover, numerous measures to protect employees and the environment have been introduced through **modernisation of the technologies applied** and by **installing protective equipment**, e.g.:

-       Surface mining: **pit problems** can be overcome by suitable mining planning, water management and recultivation.

-       Internal water circuits and downstream stilling basins minimise the **wastewater** burden.

-       **Soundproofing** of systems and processes prevents long-term hearing impairment.

-       **Fluorine and sulphur dioxide emissions** are reduced to the required levels in the waste gas by controlling the firing processes or by means of downstream separation systems.

-       The **risk of silicosis** is eliminated in relevant plants by technological improvements and dedusting systems, and is monitored by staff conducting routine preventive checks.

The **environmental protection installations** required **in ceramic works** may account for as much as 20% of the total investment costs. To achieve the desired results from the equipment in the long term, its efficiency must be guaranteed by proper maintenance. Improvements in the area of personnel and environmental protection can only be achieved by providing proper **information** and **training.**

Early **involvement of neighbouring population groups** in the planning and decision-making processes will enable measures to be devised to deal with any problems arising.

In countries which have no legal guidelines it should be ascertained as early as the **planning stage,** based on the raw materials to be used and the process technology applied, what **environmental protection measures** are necessary and appropriate. Environmental protection equipment provided should be of **robust** design so that the life of this equipment is appropriate to the overall project and so that **simple, low-cost maintenance** can be guaranteed.

6.   References

Allgemeine Verwaltungsvorschrift über genehmigungsbedürftige Anlagen nach §16 der Gewerbeordnung - GewO.: Technische Anleitung zum Schutz gegen Lärm (TA-Lärm), 1985.

Siebzehnte Allgemeine Verwaltungsvorschrift über Mindestanforderungen an das Einleiten von Abwasser in Gewässer - 17. Abwasser VwV-, GMBL (joint ministerial circular) 1982.

Bauer, H.D., Mayer, P.: Zusammenführung staubtechnischer Daten und arbeitsmedizinischer Befunde am Beispiel von Asbesteinwirkungen, Sonderdruck aus "der Kompaß"91, Nr./1981.

Betriebswacht, Datenjahresbuch 1991: Berufsgenossenschaft der keramischen und Glas- Industrie, Würzburg.

1. Bundesimmissionsschutzgesetz (BImSchG), 1985.

Entwurf zur Abwasserverordnung Deutscher Industrie- und Handelstag, Anhang 17, Sept. 21, 1990.

Environmental Assessment Sourcebook: Environmental Department, November 1990, Draft, World Bank.

Industrial Minerals and Rocks: 5th Edition 1983.

Mayer, P.: Grenzwerte für Asbest am Arbeitsplatz und in der Umwelt unter besonderer Berücksichtigung der keramischen und Glas-Industrie "Sprechsaal" 1/80, 1980.

Mining and Mineral Processing: Environmental Department World Bank, October 1990, Draft.

Mineral Commodity Summaries U.S.: Department of the Interior, Bureau of Mines, 1991.

Guidelines of the German Federal Ministry of the Interior (Bundesdesministerium des Inneren) regarding BImSchG -Zugelassene Stellen zur Ermittlung von Luftverunreinigungen im Emissions- und Immissionsbereich nach BImSchG - Guidelines of the Council of the European Community

47

Schaller, K.H.; Weltle, D.; Schile, R.; Weissflog, S.; Mayer P. und Valentin, H.: Pilotstudie zur Quantifizierung der Bleieinwirkung in der keramischen und Glas-Industrie, Sonderdruck aus "Zentralblatt" Zbl: Arbeitsmed. Bd.31, Nr.11, 1981.

Schlandt, W.: Umweltschutz in der Keramischen Industrie, Beilage zur Keramischen Zeitschrift 36, Nr.10, 1984.

TA-Luft (Technische Anleitung Luft): Erste Allgemeine Verwaltungsvorschrift zum Bundes-Immissionsschutzgesetz (Technische Anleitung zur Reinhaltung der Luft - TA-Luft-), 1986.

Siebzehnte Verordnung zur Durchführung des Bundes-Immissionsschutzgesetzes 1990 (Verordnung über Verbrennungsanlagen für Abfälle und ähnliche brennbare Stoffe, 17. BImSchV).

Trade and Industry

Glass

48. Glass

Contents

1.   Scope

The main raw materials used by the glass industry are sand, lime, dolomite, feldspar, as well as soda, borosilicates and numerous **additives** which embrace practically the entire periodic system of elements. Its products are a **large number of glasses** with different properties, many of them **further processed** after manufacture (Table 1).

Table 1
Glass products

Container glass	Sheet glass	Utility glass and special glass
- Tall jar - Preserve jar - Medical glass - Packing glass	- Sheet glass (float glass) - Casting glass - Moulding glass - Wire (reinforced) glass	- Optical glasses - Lighting glass - Glass hardware - Laboratory glass - Flasks
Lead crystal and	Mineral fibres	
- Bleaching glass - Goblet glass - Television tubes - Glass fibres for optical transmission	- Glass fibres - Mineral fibres - Borosilicate fibres - Ceramic fibres (high-temperature resistant)	

In the modern glass industry the **raw materials** are no longer generally extracted by the companies themselves but are **purchased** in the desired chemical and physical composition, e.g. in terms of granulation, moisture content, impurity (for the environmental relevance of the extraction of raw materials refer to the environmental brief Surface Mining). The substantial differences between the materials to be dosed and mixed necessitate the use of **mixing and processing plants** where the **mixtures** are **melted** in tank furnaces, more rarely in pot furnaces, or special furnaces. Cupola furnaces are still sometimes used for mineral fibres, and electric melting systems are used for manufacturing ceramic fibres. The **flue gases** formed during melting are nowadays **cooled** by regenerative or recuperative plants, thereby **reducing** the specific **fuel consumption.**

After melting, the glasses are **moulded**. Most glasses must then be **cooled** according to the subsequent application, to avoid glass stresses. Glasses are frequently further **processed** by thermal, chemical and physical **post-treatments**, such as clamping, pouring, bending, gluing, welding and grinding. **Hollow glassware** is frequently decorated. **Fibres** are drawn, centrifuged, blown or extruded after melting, using a variety of technologies.

The **capacities** of the individual glass-producing companies vary considerably, and it is often the case that **several melting systems** with different production programmes are combined in one works. **Pot furnaces** have a capacity of 3-8 t/day, whilst the **tank capacities for special glasses** range from 8 to 15 t/day in most cases. In specialist fields, however, the outputs are much higher, e.g. **tanks for container glass** melt between 180 and 400 t/day, **float glass tanks** attain melting capacities of between 600 and 1000 t/day.

The **melting temperatures** of the glass generally range between 1200 and 1500°C, the temperature depending for the most part on the mixture and the product to be manufactured. The **amount of energy** required to melt 1 kg of glass is between 3700 and 6000 kJ. The capacities and energy consumptions indicated above are **average values** which depend on the design and operating time of the tank, the production programme and the actual tank load. The specific energy consumption should be reduced by the **use of waste fragments** wherever possible.

2.   Environmental Impacts and Protective Measures

2.1   Air

2.2.1 Waste gases/flue gases

In a glass works **waste gases** are formed during **melting** of the glass as a result of **combustion of the fuels used**. In addition to the **combustion residues**, such as sulphur dioxide (SO_2) and nitrous oxides (NO_x), flue gases also contain **compound components** such as alkalis (Na, K), chlorides (-Cl), fluorides (-F) and sulphates (-SO_4).

- **Sulphur dioxide (SO$_2$)**

Sulphur dioxide or SO$_x$ emissions, made up of SO$_2$ + SO$_3$, lie within the range of 1100 to 3500 mg/Nm^3 of waste gas in the case of **regeneratively heated glass tanks** within one firing period. Where the chambers are insufficiently scrubbed **much higher peak values**, as high as 5800 mg/Nm^3 of waste gas, are found at the start of the firing change.

Electrically heated or electrically booster-heated tanks can be operated continuously at a **lower SO$_x$ load** (< 500 mg/Nm^3). On the other hand, the **use of heavy oil** with a very high sulphur content (up to 3.7%) gives rise to extremely **high emission values. Natural gas**, which does not normally contain any sulphur, does not affect the formation of SO$_x$. Some of the sulphur emission is also caused by the **addition of sulphate** to the mixture.

The currently applicable **Technical Instructions on Air Quality Control** (TA-Luft 1986) indicates a **maximum value** for sulphur dioxide of 1800 mg/Nm^3 of waste gas, thus in normal glass tanks **absorption** of the excess sulphur dioxide is required. The **sulphur dioxide content** can be **reduced** by feeding magnesium, calcium carbonate and soda into the flue gas. The dusts forming during this process must also be filtered out again.

- **Nitrous oxides (NO$_x$)**

A further environmental problem in glass manufacture is posed by the **NO$_x$ loads** occurring, which can range from 400 to 4000 mg/Nm^3 of waste gas. During nitrate refining, i.e. the reduction of the proportion of bubbles or nodules in the glass mass by nitrates, these values are **considerably increased**. The No$_x$ content depends on the air preheating temperature, the air coefficient (excess air) and the process and type of tank used. NO$_x$ content can be reduced **using catalysts** with ammonia (NH$_4$). This process, which is currently undergoing **large-scale trials**, promises to reduce NO$_x$ content to below 500 mg/Nm^3 NO$_x$ load.

The **NO$_x$ limits applicable in Germany** (1991) for the different tanks are summarised in Table 2.

48

Table 2
Nitrous oxide emissions
under applicable version of TA-Luft
[Technical Instructions on Air Quality Control]

Plant	Oil-fired mg/Nm³	Gas-fired mg/Nm³
Pot furnaces	1200	1200
Tanks with recuperative heat recovery	1200	1200
Day tanks	1600	1600
Horseshoe flame tanks with regenerative heat recovery	1800	1800
Cross-burner tanks with regenerative heat recovery	3000	3000
Values attainable for electrically heated tanks	500	

The emission values of nitrate-refined tanks must not exceed twice the above-mentioned values.

- **Fluorine/chlorine**

The **fluorine contents of the waste gas** (calculated as HF) must not exceed certain values since **plants and animals** can be **harmed** by fluorine. Fluorides are contained in almost all raw materials used in glass manufacture. Through the **addition of waste fragments** originally melted with fluorspar to the melting process, the fluorine concentration in the waste gas may exceed 30 mg/Nm³.

The low **fluorine limit value** prescribed in Germany under TA-Luft 1986 of < 5 mg/Nm³ can only be achieved through systematic **selection of raw materials** or through **additive reactions** with calcium and alkali compounds.

Chlorine compounds, which are introduced into the mixture primarily through soda or salt-contaminated raw materials, also cause problems. Measurements have indicated gaseous **chloride concentrations** of between 40 and 120 mg/Nm³ of waste gas. Problems with gaseous chlorine emissions (HCl) arise mainly in **heavy-oil-fired plants.** Like sulphur dioxide, chlorides must also be **absorbed** by calcium or sodium compounds in the mixture.

2.1.2   Dust

One problem area in the glass industry is the **dust emission of the glass melting furnaces** caused by the high temperatures, and the evaporation of mixture components which sublimate as **fine dusts.** The dust concentration of different melting tanks without filters is indicated in Table 3.

Table 3
Dust concentration in the waste gas of glass tanks
- Measured values -

Glass type	Firing	Dust in the waste gas[1] mg/mg³
Soda-lime glass	Natural gas	68 - 280
Soda-lime glass	Fuel oil S	103 - 356
Potassium crystal glass	Natural gas/Fuel oil EL	45 - 402
Lead glass	Natural gas/Fuel oil EL	272 - 1000
Borate glass	Natural gas/Fuel oil EL	120 - 975
Borosilicate glass fibres	Natural gas/Fuel oil EL	1425 - 2425

[1]   Waste gas in the normal condition, 8% oxygen in the waste gas

The values indicated in the table show that glass furnaces without filter systems have high dust concentrations in the waste gas. The prescribed **limits** of 50 mg/Nm³ of dust in Germany (TA-Luft 1986) **are difficult to achieve without dedusting plants.** Electrostatic dust precipitation, fabric dust filters with sorption or wet scrubbing may be used, depending on the type and capacity of the furnace. However, the dedusting systems must also help to **reduce fluoride, sulphate and chloride emissions**, as well as **toxic heavy metals.**

Emissions of lead, cadmium, selenium, arsenic, antimony, vanadium and nickel are particularly critical. These environmentally harmful dusts, which are formed primarily during the manufacture of special glasses in the waste gas, can only be separated by **dust filters.**

2.2   Noise

The **noise** generated is particularly **significant** in the glass industry during **melting, moulding and cooling** and in the chambers of the **compressors,** whilst hardly any problematic noise loads are generated in the areas of **extraction, processing, packing and finishing.**

In the furnaces **noise levels of up to 110 dB(A)** may be reached during melting and in the feeder. The large fans which produce the quantities of air required and the compressors also generate relatively high levels of noise. However, few work-places are situated **in the vicinity of these noise sources.** In modern works these workplaces are provided with **static noise protection devices.** The control systems of the plants can be **soundproofed** or can be installed outside the noise zone. **Hearing protection** must be worn for short-term working in these zones.

An extremely critical area in terms of **noise emission,** which is also affected by **high temperatures and oil vapours,** is the container glass **moulding area** with compressed-air-controlled machines; here the noise load generally exceeds 90 dB(A). In recent years improvements have been made with modified air guides. **So far,** attempts to **enclose** the machines for soundproofing purposes have been **unsuccessful** because of the need for regular oil lubrication of the units and cleaning of the moulds. When the glasses are **cooled,** noise is generated by fans but can be reduced by suitable **designs** and **enclosures.**

To avoid noise nuisance, glass works must be erected at least 500 m away from **areas of habitation.** The distance from residential areas should be such that no more than 50 to 60 dB(A) is immitted during the day, and no more than 35 - 45 dB(A) during the night.

2.3   Water

The total water consumption per tonne of glass produced varies considerably. **Circulating systems** should be installed so that **only small quantities of additional fresh water** are required. The **main water-consuming areas** of a glass works are:

-       **cooling** of the compressors required for generating compressed air
-       **cooling** of the diesel units sometimes used for power generation

- **quenching basins** for excess glass
- **finishing** and refining of glass by grinding, drilling etc.

The wastewater produced in these sectors is cooled and **reused**, but part is also **tapped for other functions**, such as:

- **moistening** the mixture for dust prevention
- **cooling** of flue gases, particularly in EGR dedusting plants
- **moistening** of lime products for dry sorption filter plants.

The average **water consumption** in a glass works should be less than 1 m³/t of glass produced. The cooling water of the cutting devices and moulding machines, the compressors, any emergency power diesel generators used and also the water from the quenching basins underneath the production machines may be **contaminated** by oil. This effluent must be cleaned by oil separators. In Germany, if water is discharged it must meet the **minimum requirements** regarding discharge of effluent into watercourses (direct dischargers). By virtue of these regulations no more than 0.5 mg/Nm³ of **depositable substances** may reach the effluent in glass production.

Special disposal arrangements are required for the **sewage** produced (see the environmental brief Wastewater Disposal).

2.4   Soils

In the **area surrounding modern glass works** which meet the existing environmental regulations regarding waste gas and dust, are equipped with the necessary cleaning systems and have a suitable internal wastewater circuit and water separator, there is **unlikely to be any contamination of the soil** or consequent damage to plants or animals.

2.5   Workplace

Employees of glass works may be endangered or oppressed particularly by **noise** and in certain workplaces by **heat**. **Hardly any dust problems** arise in well-maintained glass works, but in **special cases**, e.g. in the manufacture of special glasses, **toxic dusts** may pose a health hazard.

In principle no workplace within a plant should be exposed to a **continuous noise level** in excess of 85 dB(A); at this level **hearing protection** should be provided, and from 90 dB(A) protection must be worn in all cases. Hearing protection is compulsory in noise-intensive process areas, even when employees remain there only for a short time.

So far it has not been possible, for technical reasons, to enclose glass moulding machines, particularly the noisy **container glass machines**, or to automate them completely, so that employees must wear **hearing protection** in these areas. Noise from burner systems, fans and compressors can easily be avoided; firstly there are hardly any workplaces in the vicinity of these machines, and secondly the control units of the machines can be screened against dust, heat and noise. When carrying out **maintenance and repair work**, employees must wear the prescribed **hearing protection** and, if necessary, **protective clothing.**

In the event of **stoppage** or **unexpected breakdown** of tanks or faults in the preheating system **very high temperatures** may occur, since some tanks are operated at temperatures in excess of 1500°C. Work in such emergency situations must be carried out under **supervision,** and protective devices to facilitate the work, such as thermal protective suits, must be available in all works in case of emergency. **Contingency plans** must be drawn up and **regular drills** carried out to ensure rapid, targeted intervention in emergency situations.

According to recent studies, **glass and mineral fibres are suspected** of having carcinogenic effects. **Regular medical examinations** should therefore be carried out in glass works to identify any problems arising at an early stage and forestall adverse consequences.

2.6   Ecosystems

Glass works process 70 - 80% **natural raw materials** (sand, feldspar, dolomite, lime), but these are not generally extracted in the vicinity of the works. About 75% of the natural raw material is **quartz sand** which nowadays is rarely extracted by the glass works themselves. The **soda** required is manufactured in Germany synthetically from salt (NaCl) and carbon dioxide, the latter being extracted from limestone. Soda may also be extracted from natural deposits occurring mainly in the USA. Certain of the **other raw materials** are **synthetic** or **cleaned raw materials** such as sodium and boron compounds.

Approximately 1.2 - 1.3 tonnes of raw materials are required to melt one tonne of glass, but the **area required** for extracting the glass raw materials cannot be determined accurately because the deposits in question are **not** used **exclusively** for the glass industry and the extraction levels vary considerably.

If a works carries out its **own extraction**, the environmental protection aspects must be considered as early as the extraction **planning phase,** particularly as regards **water management** and the constant need for **recultivation**. The extraction and recultivation costs must be added to the raw material costs (see the environmental brief Surface Mining).

When **selecting the site** of a glass production centre, the environmental factors must also be taken into account. In the case of sites in **areas which have so far been used for agricultural purposes alternative sources of income** must be examined, particularly for affected women. Besides complying with the applicable regulations regarding waste gas, dust, noise and water, the **subsoil** conditions, **landscaping** and **infrastructure** must also be examined. The infrastructure includes, among other things, recruitment and housing of employees, traffic and transport systems and the existing and planned industrialisation of the area.

Since the **environmental impacts** are not limited to the works area, the **population groups concerned**, particularly women and children, should be provided with access to **medical care.**

The addition of a **recycling system for waste glass** may on the one hand **reduce** the energy requirement for glass manufacture and on the other hand substantially **relieve pressure** on **refuse tips.** In a similar vein, **disposable packaging systems** should be **replaced** by reusable packaging systems.

3.   Notes on the Analysis and Evaluation of Environmental Impacts

The **limits** - based on TA-Luft (Technical Instructions on Air Quality Control) and TA-Lärm (Technical Instructions on Noise Abatement) and other regulations - summarised in Table 4 for waste gas, dust and noise are now applicable in Germany and are being adopted in similar form by most European countries. The **minimum requirements in Germany** regarding treated wastewater discharged into receiving bodies of water are also indicated.

Table 4

Limitation of hazardous substances under TA-Luft (Technical Instructions on Air Quality Control) and the 17th Administrative Regulation (VwV) according to § 7a of the Federal Water Act (WHG)

Glass industry		Air	Water		
			Direct discharger g/m^3	Sample type	Indirect discharger3) g/m^3
		mg/Nm^3			
Dust		50			
Sulphur dioxide as SO_2					
Glass melting furnaces	SO_2	1800			
Pot furnaces and day tanks		1100			
NO_x nitrous oxide as NO_2	NO_x	400-3500			
Fluorides	F	5			50
Chlorine	CI	30			
Filterable solids			100	1)	1
Total suspended solids	TSS		0.50	2)	1
Chemical oxygen demand	COD		80		
Antimony	Sb	5			
Arsenic	As	1			
Lead	Pb	5	0.50	2)	2
Cadmium	Cd	0.20	0.07	2)	0.50
Chromium	Cr	5	0.10	2)	2
Cobalt	Co	1	0.10	2)	
Cyanides2)	-CN	5			0.20
Copper	Cu	5	0.10	2)	2
Manganese	Mn	5			
Nickel	Ni	1	0.10	2)	3
Palladium	Pd	5			
Platinum	Pt	5			
Mercury	Hg	5			
Rhodium	Rh	0.20			0.05
Selenium	Se	1			
Tellurium	Te	1			
Thallium	TI	5			
Vanadium	V	0.20			
Zinc	Zn		2.00	2)	
Tin	Sn	5			

*	May be formed in reduced burning		COD	Chemical Oxygen Demand
**	Law applicable in the German state of Baden-Württemberg		TSS	Total Suspended Solids
			TA-Luft	Technical Instructions on Air Quality Control
1)	Two hour mixed sample		VwV	Administrative Regulation
2)	Random sample		WHG	Federal Water Act

Glass works, which are generally large-scale plants, produce **considerable emissions**. In principle a maximum of 1800 mg SO_2Nm^3 should be established as the mean **guideline value** for avoiding serious environmental pollution. The NO_x emissions must not exceed the currently applicable values, and **nitrate refining** should be **dispensed with** because of the high NO_x levels generated.

No separate wet or dry sorption plants are required to comply with these relatively high mean values. **Accurate control of the tank heating** is vital in order to attain the required values.

Fluorine and chlorine emissions which may give rise to direct damage must be kept **as low as possible**. The values indicated above can be achieved by **suitable selection of raw materials and fuels** and **systematic monitoring of burner operation**. A further **benefit** is that **energy consumption** can be further reduced by conforming to these guideline values, resulting in greater economy.

The **dust** emission from glass furnaces should not exceed 50 mg/Nm^3. A **dedusting plant** should always be installed in order to comply with this limit.

It is vital to adhere to the emission limits for **toxic dusts** (heavy metals) such as cadmium, lead, fluorine, selenium and arsenic; the **maximum values** specified in TA-Luft must **not be exceeded.**

For individual substances, the **Compendium of Environmental Standards** contains notes on evaluating environmental relevance.

It is absolutely essential to comply with the regulations on permissible **noise levels**, since failure to prevent or protect against noise can result in **permanent injury** of employees.

To avoid environmental pollution, the limits laid down for **direct water dischargers** must be observed, particularly regarding heavy metal concentrations in the effluent.

If no national regulations exist, values in line with **German or European standards** should be established for the erection of **new glass works**, particularly in areas already suffering from serious environmental pollution. **Special regulations** must be introduced for **plants already in operation**. The parameters defined for the principal hazardous substances must in future be **regularly monitored** and disclosed by the glass works, so that appropriate steps can be taken immediately in the event of nonconformance (see 3.1).

For all practical purposes it may be assumed that in order to comply with the limits indicated all **alkali borosilicate, borate, lead** and most **special glass furnaces** must be equipped with **dedusting systems.** Allowance must be made for these dedusting and sorption systems as early as the planning phase.

48

In countries with **low-cost electricity** it is possible to construct glass furnaces of special design which produce far **lower emissions** and do not require expensive environmental protection equipment. The energy requirement per kg of glass can also be reduced by introducing such melting methods.

3.1   Inspection and maintenance of environmental protection installations

A **control centre independent** of the production process must be established to comply with existing environmental protection regulations. The responsible personnel must be enabled to perform and monitor **all inspection functions** including measurements relating to environmental protection in the works. They should be available for consultation on investments and take charge of negotiations with environmental protection authorities. Moreover, this department is responsible for ensuring that all **environmental protection installations** are regularly **maintained** and **upgraded.** This internal environmental department is also responsible for **staff training.**

4.   Interaction with Other Sectors

Glass works which rely on numerous **secondary operations,** such as workshops, compressed air generation, fuel stores, galvanisation shops, refining shops, transport and packing departments etc. are also affected by **regulations applicable in other sectors.**

Because of the relatively **high transport costs, container glass factories** must be located **near their main customers.** Modern sheet glass works, on the other hand, can only operate economically with capacities upwards of 600 t/day, thus they supply their products **to more distant sales areas** and are reliant on good transport facilities.

Table 5
Environmental impacts of adjacent sectors - Glass -

Interacting sectors	Nature of intensification of impact	Environmental briefs
Extraction/storage of raw materials and fuels	- Landscape impairment - Pollution of bodies of water - Waste storage in former pits	Planning of Locations for Trade and Industry Urban Water Supply Rural Water Supply
Disposal of solid and liquid waste	- Discharge of deposited solids e.g. filter dusts - Pollution of bodies of water by wastewaters	Solid Waste Disposal Disposal of Hazardous Waste
Maintenance of work-shops and transport fa-cilities	- Risks of handling water pollutants (e.g. solvents) - Impacts of transport and traffic (noise, link roads)	Mechanical Engineering, Workshops Road Building and Main-tenance Planning of Locations for Trade and Industry

5.   Summary Assessment of Environmental Relevance

The effects of glass works on the environment and workplace are caused by noise, dust, effluent and flue gases.

Table 6
Environmental impact of process stages (glass)

Process	Air		Noise	Water	Soil	Work-place
	Waste gas/ Flue gas	Dust[1)				
Dressing	1	2	2	1	2	2
Melting	3	3	3	3	3	3
Moulding	2	1	4	2	3	4
Cooling	2	1	3	1	2	2
Sorting	1	1	2	1	1	1
Packing	1	1	2	1	1	1
Machining/ Refining	1	2	2	3	1	2

Key: 1 very slight; 2 slight; 3 moderate; 4 considerable

In some cases **technological and processing developments and improvements** have already been implemented, e.g.:

-        **Arsenic and tellurium** are now only used as refining agents in exceptional cases.
-        **Fluorspar** is no longer used as a flux.
-        The specific **outputs of the tanks** have been increased with a simultaneous reduction in energy consumption.
-        **Wastewater circuits** have been introduced.
-        Numerous **noise protection devices** have been installed.
-        Wet, electric and dry sorption plants have been installed for **dust extraction.**
-        Tank designs and fire management systems have been improved.

Many of the **processes** so far tested in individual cases are capable of **further technical improvement** and **more economic design**, paying particular attention to environmental regulations. The expected **costs of environmental protection devices and measures** may be as much as 20% of the total investment costs of a glass works.

Proper maintenance is essential to environmentally acceptable operation of the plants. Suitable training must be given and personnel generally made aware of environmental concerns.

Early **involvement of neighbouring population groups** in the planning and decision-making processes will enable measures to be devised to deal with any problems arising.

In countries which have **no legal guidelines** it should be ascertained as early as the **planning stage,** based on the raw materials to be used and the process technology applied, what **environmental protection measures** are necessary and appropriate. Environmental protection equipment provided should be of **robust** design so that the life of this equipment is appropriate to the overall project and so that **simple, low-cost maintenance** can be guaranteed.

6. References

Allgemeine Vewaltungsvorschriften über genehmigungsbedürftige Anlagen nach § 16 der Gewerbeordnung 1985.

Allgemeine Verwaltungsvorschrift über Mindestanforderungen an das Einleiten von Abwasser in Gewässer - 41. AB-Wasser VwV, 1984.

Betriebswacht, Datenjahresbuch 1991, Berufsgenossenschaft der keramischen und Glas-Industrie, Würzburg.

1. Bundesimmissionsgesetz (BImSchG), 1985.

Entwurf zur Abwasserverordnung: Deutscher Industrie- und Handelstag, Anhang 17, Sept. 21, 1990.

Glass Manufacturing, Effluent Guidelines, World Bank, August 1983.

Guidelines of the Bundesministerium des Inneren [German Federal Ministry of the Interior] regarding the BImSchG, directives of the Council of the European Community.

TA-Luft: Erste Allgemeine Verwaltungsvorschrift zum Bundes-Immissionsgesetz, GMBR 1986 (A).

Siebzehnte Verordnung zur Durchführung des Bundes-Immissionsschutzgesetzes 1990 (Verordnung über Verbrennungsanlagen für Abfälle und ähnliche brennbare Stoffe, 17. BImSchV).

Barklage-Hilgefort H.J.: Minderung der NO_x-Emission durch feuerungstechnische Maßnahmen, Glastechnische Berichte 58, Nr. 12, 1985.

Bauer H.D., Mayer Dr. P.: Zusammenführung staubmeßtechnischer Daten und arbeitsmedizinischer Befunde am Beispiel von Asbesteinwirkungen, Sonderdruck aus "Der Kompaß" 91, Nr. 7 1981.

Bundesverband des Deutschen Flachglashandels e.V., Glasfibel, Vertrieb Kelasa GmbH Cologne, 1983.

Doyle T.J.: Glassmaking Today, An Introduction to Current Practice in Glass Manufacture, Portcullis Press, Redhill, 1979.

Förster H., Feck G.: In Vitro - Studien an künstlichen Mineralfasern, Sonderdruck, Zbl. Arbeitsmed. Bd. 35, Nr. 5, 1985.

48

Gebhardt F., Carduck E. und Arnolds J.: Chloremissionen von Glasschmelz-wannen, Glastechnische Berichte 51, Aachen 1978.

Gilbert G.: Zur Ausbreitung von Schadstoffen, insbesondere von Stickoxiden in der Atmosphäre, Glastechnische Berichte 51, Aachen 1978.

Kircher U.: Emissionen von Glasschmelzöfen - Heutiger Stand, Glastechnische Berichte 58, Frankfurt 1985.

Markgraf A.: Abgasentstaubung hinter Glasschmelzöfen mit filternden Abschei-dern und vorgeschalteter Sorptionsstufe zur Beseitigung von HF und HCI, Glastechnische Berichte 58, Nr. 12, Stadthaben, 1985.

Mayer P., Bergass: Grenzwerte für Asbest am Arbeitsplatz und in der Umwelt unter besonderer Berücksichtigung der keramischen und Glas-Industrie, Sprechsaal 2/80, 1980.

Mayer P., Bergass: Glasfaserstäube und ihr gesundheitlicher Einfluß auf den Men-schen, Sonderdruck der Zeitschrift, Die Berufsgenossenschaften e.V., Bonn.

Schaller K.H., Weltle D., Schile R., Weissflog S., Mayer P. und Valentin H.: Pilotstudie zur Quantifizierung der Bleieinwirkung in der keramischen und Glas-Industrie, Sonderdruck Zbl. Arbeitmed. Bd 31, Nr. 11, 1981.

Tiessler H.: Zum Einsatz eines Elektro-Entstaubers an einer Spezialglaswanne für Alkali-Borosilikatglas, Glastechnische Berichte 51, Nr. 7, 1978.

Winterhoff G.: Abgasentstaubung periodisch arbeitender Glasschmelzöfen, Glas-technische Berichte 58, Nr. 12, 1985.

Trade and Industry

Iron and Steel

49. Iron and steel

Contents

1. Scope

This environmental brief covers iron and steel production and processing with the following activities:

- sinter, pellet and sponge-iron production
- pig iron, cast iron and crude steel production (including continuous or strand casting)
- steel forming (hot and cold)
- foundry and forging operations.

The above activities are carried out in an **integrated** ironworks or sometimes in **separate locations**.

After delivery and pretreatment of the ore in the ore preparation, sintering and where applicable pelletising plant, **pig iron** is smelted in the **blast furnace** with the addition of coke and admixtures; coke supplies the energy and reduces the ore to pig iron. In the **converting mill** the molten pig iron is refined to form **crude steel** by top blowing or purging with oxygen and the addition of scrap. Crude steel is also produced from scrap in electric furnaces, sometimes with the addition of pig iron, ore and lime. The **crude steel** is either continuously cast as **blanks** or, after casting as slab ingots or blocks in permanent moulds, rolled in the hot rolling mill to form **sheets, billets** or **profiles.** Further processing takes place in the cold rolling mills and forges. **Continuous casting** which already represents 90% of German and 60% of worldwide steel production **improves crude steel utilisation** by some 10%, **saves energy** by rolling operations and reduces the **production scrap** yield in steel and rolling mills per tonne of finished steel by more than 50%.

The **direct-reduction** process represents an **alternative** to traditional steel production. With the addition of reduction gas, e.g. from natural gas or coal, sponge iron is produced as a solid, porous product from which crude steel is then refined in the electric furnace, often with the addition of scrap. 90% of sponge iron is produced by the gas-reduction process.

Cast iron smelting takes place in the cupola furnace, with increasing use of induction furnaces.

Moulds and cores are required for the shaped casting of cast iron; these are mostly of sand but frequently contain an organic binding agent.

49

The following are classified as **major units**:

sintering plants	20,000	t/day
blast furnaces	12,500	t/day
steel converters	400	t holding capacity
electric furnaces (arc)	250	t holding capacity
cupola furnaces	70	t/h
induction furnaces	30	t/h

In many countries steel is extensively produced from scrap in electric furnaces.

Since iron and steel production is predominantly based on pyrometallurgical processes, **air pollution** is a primary consideration. In addition to a multitude of **gaseous** air impurities, **dusts** play a special role, not only because they occur in large quantities but also due to the fact that the dusts contain some **hazardous substances** affecting both man and the environment, e.g. heavy metals. Due to the use of coolant water and wet separation methods, problems of **maintaining water purity** also occur. Continuous casting plants require high specific **water quantities** from which the wastewater is considerably contaminated with oil. Casting without spray-water cooling relieves the load on water resources.

Metallurgical processes also produce **slags** which should be recycled wherever possible. Where no effective recycling and final dumping facilities exist, **dusts and sludges** separated from the waste gas cleaning systems represent potential pollutants of the ground and water environments.

In blast furnace plants and converting mills, also in rolling mills and forging works, **noise and vibration protection** is of fundamental importance. Foundries produce large amounts of **waste** from used sand, broken cores and cupola slag.

For reasons of ecology and economy, work is taking place worldwide on **process methods** which permit the use of **coal instead of coke** and the extensive use of **lump ore instead of sinter or pellets**. This would enable coking and sintering plants to be **dispensed with** as **emission sources** in a metallurgical plant.
Other developments concern the **casting** of rolling feed stock in **approximately final dimension form**. Shortening the process chain permits reductions in energy requirements, residual substances, waste and emissions.

2. Environmental Impacts and Protective Measures

2.1 Sintering / pelletising plants

Sintering plants form **lumps of fine ore** prior to introduction into the blast furnace and **recycling** of ferriferous residues (waste materials). Sintering is the traditional method of treating residual and waste materials from the smelting plant. Factors determining the **limits** include the zinc concentration, because zinc in the sinter contributes in the blast furnace to the formation of scaffolding with impaired gas distribution.

Sintering plants produce the following emissions:

Waste gases and dust containing components with potential environmental relevance:

$$SO_2, NO_x, CO_2, HF, HCl, As, Pb, Cd, Cu, Hg, Tl, Zn$$

Of **dust components**, the **heavy metals** lead, cadmium, mercury, arsenic and thallium have the greatest environmental relevance where these are present in the charge materials. The relevance of **anthropogenic** heavy metal emissions is based less on their overall emission rate than in high localised mass flow densities or concentrations. The **iron and steel industries** are among those industries in whose vicinity the **highest immission rates** of heavy metals occur in the air and ground.

Dust is separated and returned to the sinter process in **gas cleaning systems**, normally **electrostatic precipitators**. In continuous operation the **dust content of clean gas** is between 75 and 100 mg/m^3. Heavy metal, e.g. lead, enrichment in the sinter plant dust is possible with continuous recycling. Dust with heavy concentrations of lead and zinc should be conducted to a zinc and lead **recovery system**. In the case of stoppages of the sintering belt due to faults, care must be taken to ensure that the **gas cleaning system** continues to operate at maximum possible separation capacity. In addition to sintering belt dedusting, modern sintering plants also have **room dedusting** whereby dust-laden waste air from transfer stations, chutes, crushers etc. is cleaned by a hot sieve system.

Depending on the composition of charge materials, inorganic gaseous fluoride and chloride compounds as well as sulphur dioxide and nitrous oxides are emitted. Sulphur dioxide emission can be significantly reduced by using **coke with a low sulphur content**. The emission of gaseous pollutants can also be reduced by **increased lime dosing**. This results in **problem substances** being transferred to the separated dust. Where regional conditions and process engineering do not permit these measures, **wet-process desulphurisation systems** offer a means of reduction; in this case some **problem substances** are transferred to the **wastewater**. On account of the large gas volumes - up to 10 E 6 m^3/h - only partial waste gas desulphurisation can take place. For this reason **preference should be given to primary**

measures. Concentrations in cleaned waste gas are around 500 mg/m^3 sulphur dioxide.

With respect to **noise** impact, a distinction is made between the noise immissions of operations to the neighbourhood and the effect on the staff at their work-places. Principal noise sources of the sintering plant include the **large fans** for drawing air through the sinter cakes, cooling the sinter and dedusting. Crushing and screening stations should be housed in **solidly constructed buildings** whose walls restrict the propagation of sound. Possible noise reduction measures are **silencers** in the air supply and discharge pipelines, also the **encapsulation** of individual units. The acoustic power immission level is used to evaluate the noise radiated to the open air by the plant. The acoustic power level of a noise source is a distance-dependent parameter; for sintering plants without silencers on supply and discharge air pipelines it can be as high as 133 dB(A) and for those with silencers 124 dB(A). With very good **acoustic planning and execution** an immission level of around 40 dB(A) can be achieved at a distance of 1,000 m from the individual noise sources. If this target cannot be achieved, protection of the residential area adjacent to the sintering plant is only possible by **noise protection measures on the propagation path**, e.g. a noise abatement wall. Measures for optimising noise protection are to be considered in parallel with the planning of the production unit.

By encapsulation and the separate installation of principal noise sources it is also possible to protect the **work places.** The typical noise level in the sintering hall is between 83 and 90 dB(A); attention must be paid to the use of **personal noise protection** because long-term exposure to an acoustic power level in excess of 85 dB(A) results in serious **hearing impairment.** The wearing of safety helmets and shoes also helps reduce industrial accidents. Staff in work-places particularly exposed to dust, gases, noise and heat are to have regular **preventive medical examinations by works doctors.**

In **pelletising plants**, fine ores are mixed with additives and water to form green pellets which are burned in pellet incinerators on travelling grates. The dust-laden waste gases are cleaned in dedusting plants, usually electrostatic precipitators. The filter dust is re-used. Pelletising plants are associated with **lower dust and gas emissions** than sintering plants. In contrast to sintering, pelletising is mainly performed at the ore mine.

2.2   Blast furnaces

The blast furnace is a **countercurrent reactor** loaded or charged from the **top** with layers of feed and coke, the molten pig iron and slag being drawn off from **below**. Hot air is injected in the **opposite direction** from the **bottom** of the furnace. Residual materials (waste) such as oily metal chips and oily rolling scale can be introduced after sintering.

The principal emissions, residues and waste materials are:

-   **top gas**, with the following potentially environmentally relevant components:

$$CO, CO_2, SO_2, NO_x, H_2S, HCN, CH_4, As, Cd, Hg, Pb, Ti, Zn$$

-   **top gas dust** (dry) from the gas cleaning plant with high iron contents (35 - 50%)

-   **slag** with the following major components :

$$SiO_2, Al_2O_3, CaO, MgO$$

-   **sludge** from the waste gas cleaning system

-   **wastewater** from the waste gas cleaning system, with the pollutants: cyanides, phenols, ammonia

-   **dust** from the casting house dedusting system.

The **waste gases** from the blast furnace are **pretreated** in mass force separators (dust catchers or cyclones) and, in a second stage, **finally cleaned** with a high pressure scrubber or wet electrostatic precipitators. Clean gas dust concentrations from 1 to 10 mg/m^3 are achieved.

Other dust emissions in the blast furnace area, particular from the burdening process, pig iron desulphurisation and the casting house must also be identified and cleaned.

Dust formation ("brown fume") in the casting house affects not only the **neighbourhood** but also, to a considerable extent, the **workplaces**. Efficient **casting house dedusting systems** which intercept process waste gases and peripheral emissions at the taphole, runners and cut-off points and separate dusts in horizontal electrostatic precipitators can achieve clean gas dust concentrations significantly under 50 mg/m^3 (best values 7 and 12 mg/m^3 and dust emission factors between 0.020 and 0.028 kg/t pig iron in blast furnace plants with a capacity of 4,000 to 6,000 t/day). As a replacement for the standard collection and cleaning methods, trials are currently in progress with the **suppression** of "brown fume" through inertisation with nitrogen.

In the **dedusting of pig iron desulphurisation**, clean gas dust concentrations of 50 mg/m³ are adhered to in both calcium carbide and soda desulphurisation, using radial flow scrubbers or electrostatic precipitators.

The **top gas** contains between 10 and 30, though possibly as much as 60 g/m³ **dust** with 35 to 50% iron, i.e. 30 to 80 kg/t pig iron, in older plants 50 to 130 kg/t pig iron. The dust is separated in the dry state in mostly multistage separators, from where it goes to the **sintering plant** and from there **back to the blast furnace.**

In view of the zinc and lead content and other factors, the **top gas scrubbing water sludge** must be disposed of by **dumping**, unless there is a special hydro-cyclone separation system. With higher concentrations, it should be transferred to a **non-ferrous metal works**. Recycling in this way would leave the **blast furnace process practically free of residues. Dumping** involves the risk of **leaching** and hence penetration of the **soil** and **groundwater** by compounds of zinc, lead and other heavy metals. The dump must be permanently and verifiably **sealed** and the **seepage water** must be **collected** and chemically **processed**. The special requirements imposed on such a dump must be laid down in the project planning stage.

Slag produced by the blast furnace process accounts for roughly 50% of the overall waste materials from pig iron and steel production. This slag is mostly used in **road-building**. Part of the molten slag is granulated by quenching in water. This so-called **slag sand** is also used in **road-building**. Part is used to produce **iron slag Portland cement and blast furnace cement**. Quenching and granulating releases **carbon monoxide and hydrogen sulphide**. The **wastewater** has an alkaline reaction and contains small quantities of sulphide.

Slag heaps sometimes produce **seepage water** with high levels of dissolved sulphides and strong alkaline reaction, posing a hazard for the **groundwater**. Slag heaps must be sealed and any seepage water must be treated.

Wastewater is generated by top gas scrubbing and simultaneous **wet dedusting**. The wastewater is normally clarified in **settling tanks** and, where necessary, **gravel bed filters** and **recirculated**. The wastewater contains suspended matter (dust) and sulphides, cyanides, phenols, ammonia and other substances in dissolved form. The last three substances must be removed from the wastewater using appropriate **physical and chemical treatment processes**.

The **top gas** can be used as a fuel for heating purposes within the works, in view of its high carbon monoxide content due to the reducing atmosphere in the blast furnace, though this will inevitably result in the formation of carbon dioxide, with its climatic implications.

Excessive levels of sulphur dioxide and nitrous oxide gases can be reduced by **flue gas desulphurisation and denitrification.**

Carbon monoxide concentrations in the workplace pose a particular problem. Where top gas pipes are not perfectly leakproof there is a danger of poisoning with possible fatal consequences for workers present at the furnace throat. Close attention must also be paid to CO concentrations by carrying out **measurements** and ensuring that **protective breathing equipment** is worn during repair and maintenance work on shut-down blast furnaces or gas cleaning systems.

Protective equipment for blast furnace workers includes fireproof clothing, breathing equipment and ear protectors, depending on where they are working; protective helmets and safety footwear must be worn in all areas.

Noise in blast furnace plants comes mainly from the combustion air fans and the charging process; also there is the noise generated upon changeover from blast to heating operation. Suitable **abatement measures** include silencers, enclosure of the furnace throat or encapsulation of all valves and shields. The noise level from the blast furnace plant is in the range of 110 to 125 dB(A); the level of background noise in the immediate vicinity may be 75 to 80 dB(A). Possible noise reduction measures should be selected as early as the blast furnace **planning phase.** Their effect can be determined by advance calculation, taking care to ascertain the significance of the emission sources (plant sections and operating processes). One should preferably begin by damping or eliminating occurrences and noise sources which arise only periodically.

2.3   Direct-reduction plants

Direct reduction plants function according to a variety of methods, e.g. with **shaft furnaces** or **rotary tube furnaces** which are similar to blast furnaces. In the former, the top gas is scrubbed and then enriched with natural gas and used for heating; in the latter, the gas is not used unless steel and rolling mills are available for this purpose. If this is the case, the gas should be burnt provided the CO content is sufficiently high. The **waste gas flow** is cleaned by mass force separators (dust chambers) for preliminary separation and then by fabric filters. Sulphur dioxide emissions may occur in the solids reduction process, depending on the sulphur content of the coal used.

2.4   Crude steel production

Excessive carbon content impeding **further processing of the pig iron** and substances influencing the quality of crude steel, such as silicon, phosphorus or sulphur, are either expelled in gaseous form or slagged during the steel production.
The following emissions occur in the steel works:

-       waste gases and dust containing components with potential environmental
        relevance:

        CO, NO_x, SO_2, F, Cd, Cr, Cu, Hg, Mn, Ni, Pb, Si, Tl, V, Zn

        ammonia, phenol, hydrogen sulphide and cyanide compounds may occur,
        depending on the process.

-       dust from waste gas cleaning

-       slag

In the steel works **dust is formed** mainly due to the top-blowing or through-blow-
ing with oxygen necessary for oxidation. The **solids content of the waste gases**
from the oxygen converter is between 5 and 50 g/m^3. They contain finely dispersed
evaporation products of iron oxides and primary anoxide ("brown fume"); also
sulphur and phosphorous compounds, fluorine compounds and, where fluxing
agent is used, silicon tetrafluoride.

Specific **dust masses** are approximately as follows:

-       electric furnace: 2 - 5 kg dust per tonne crude steel
-       bottom blowing converter
        oxygen bottom metallurgy (OBM) 5 - 10 kg dust per tonne crude steel
-       top blowing converter
        (LD and LDAC process)   15 - 20 kg dust per tonne crude steel

Gases occurring in addition to carbon monoxide include inorganic fluorine compounds with the addition of fluorspar, also small quantities of sulphur dioxide and
nitrous oxides, nitrous oxide formation being significantly higher in electric furnaces than in the blowing converter.
    A **technical solution exists** for the **collection and cleaning of the process
gases** from the converter. A fixed or lowerable hood over the converter prevents the
intake of large quantities of infiltrated air or the escape of converter gases. The gas
is subsequently dedusted by a wet or dry process. **Wet dedusting** takes place in a
two-stage operation by a combined wet scrubber and wet electrostatic precipitator.
For dry dedusting, dry electrostatic precipitators are used, designed to resist internal

pressures up to 2 bar (due to risk of deflagration). The **clean gas concentrations** are under 50 mg/m^3 dust and under 500 mg/m^3 sulphur dioxide. A value of under 400 mg/m^3 nitrous oxide cannot be continuously maintained. **Maintenance** of the separation equipment is important in order to achieve an adequate continuous level of separation. **Dry dedusting** is advantageous as the yielded dust can be returned to the converter after hot briquetting.

Transfer, charging and mixing processes produce **random dust emissions** which may pose a considerable nuisance for the neighbourhood. Clean gas contents of 10 mg/m^3 can be maintained by a **waste gas collection system** with a collection rate of 90% and a downstream separator using fabric filters or horizontal electro-static precipitators.

Proposals for the use of a **process-dependent control and instrumentation system** for reducing specific waste gas quantities must be examined with respect to system requirements such as robustness, error detection ability and ease of mainte-nance.

Since **waste gas collection** is **difficult** with Siemens-Martin furnaces while the furnace is in operation, the solution is to **convert to electric furnaces**. In addition to lead and zinc, chromium, nickel and vanadium occur in the dust if electric fur-naces are used to produce fine steels. Certain chromium compounds in the form of breathable dusts have proved to be carcinogenic.

A full **doghouse enclosure** is necessary to achieve 95% collection of the **waste gases** occurring with electric furnaces during charging, smelting and casting. Fab-ric filters permitting clean gas dust concentrations of under 20 mg/m^3 are used for dust separation.

When the converter is in operation, large amounts of carbon monoxide are produced which should be transferred for controlled burning in a torch or in a **boiler with energy conversion,** so as to avoid excessive air burdens (immissions). A potential source of polyhalogenated dibenzodioxin and furan emissions (though not currently thought to pose a major risk) is the **recycling of iron scrap in electric steelmaking plants.** Large quantities of iron scrap contaminated with halogen compounds and the operating conditions give rise to the formation of these substances. Initial random sample checks yielded emission concentrations of the order of a few nanograms. A comprehensive measuring programme is being prepared. Careful **selection and preliminary sorting of iron scrap** is currently a practicable way of minimising carcinogenic emissions. **Processes** for **separating** health-endangering **dioxins and furans** are currently being **developed.** Current trials of activated charcoal adsorption filters and their separation capabilities are being followed with close interest.

The **wastewater from wet dedusting** is clarified in a hydrocyclone or settling tank and recirculated. The separated **sludge** is dewatered by a vacuum drum filter and returned to the blast furnace via the sintering plant. Attention must be paid to the zinc content of the sludge upon recycling. Slag produced in steel works is used in road construction or processed into fertilisers.

Loud **noise** is generated in converting steel works by high-powered fans and dedusting systems and in electric furnaces by the arcing and transformer. Noise levels in electric steelmaking plants without noise reduction measures is between 117 and 132 dB(A), and around 100 dB(A) with noise reduction.

Noise reduction measures can include:

- arc soundproofing
- smaller apertures in the furnace shell
- encapsulation of the furnace
- acoustic separation of the furnace bay from adjacent bays
- increasing the soundproofing of bay walls
- silencers on air intakes and outlets
- slow-running cooling air fans
- enclosure of individual systems
- avoiding free-fall of scrap upon loading and charging.

Very high **peak noise levels** can occur during smelting, especially with wet scrap. Highly automated modern plants have **control rooms** which provide effective protection against noise at the workplace. The protective measures mentioned under 2.2 also apply to workplaces in steelworks.

2.5   Steel forming

The following emissions and residues occur with **forming (shaping) of crude steel into rolled steel**:

- oily rolling scale
- waste gases from the furnace
- oily wastewater
- wastewater from the waste gas cleaning

During the production of steel plate, the following are produced:

- oily wastewater
- waste air from the pickling baths
- spent pickling solutions
- sulphuric and hydrochloric acid
- or nitric and hydrofluoric acids
- mixtures

The most prolific residue produced in hot rolling mills is **rolling scale.** The specific mass is 20 to 70 kg/t finished steel. Scale comprises mainly iron oxides (70 - 75) and can therefore be **utilised in the blast furnace.** Finer components must first be sintered or pelletised. Oily scale with a small percentage of oil from the machinery lubricants can be freed of oil by combustion or by alkaline wet scrubbing. To avoid polluting the subsoil with oil, **oily scale** should not be **dumped.**

Wastewater is produced in the hot rolling mill by

- transport of the scale to the wastewater treatment system
- alkaline washing of the oily scale.

The **scale-water mixture** is **separated** in settling tanks and gravel filters (sometimes with the addition of flocculation agents). Floating rolling oil and grease is skimmed off and the settled or filtered scale is dewatered and transferred to the sintering plant. The clarified **wastewater is recirculated.**

The alkaline **scrubbing water** from the **scale scrubber** contains an **oil emulsion** which must be broken down with chemicals. The water contains oil and chemical residues. It should be transferred to a **biological filter plant**. The recovered **oil** can be processed and in certain cases re-utilised in the **rolling mill.**

In the **cold rolling mill** the steel plate is descaled in a pickling bath before further processing. Hence, no solid waste (scale) is produced in the actual cold rolling process.

With cold rolling, **wastewater** occurs due to contamination of water with rolling oils (mineral oils, palm oil) and from the pickling. The rolled down plates are once more pickled with acid and electrolytically degreased prior to tinning or galvanising.

49

Wastewater treatment requirements in rolling mills depend on the **type and extent of recycling** and the quality of the receiving body of water. Regular **monitoring of wastewater values** is necessary.

The **oil-water emulsions** produced by the cold rolling process must be **chemically treated** (flocculation with ferrous salt and lime). The **oily sludge** must be incinerated and the **ashes** transferred to the sintering plant. **Oil** separated from the emulsion can be used for **secondary lubrication purposes.**

To protect soil and groundwater from unwanted discharges, a **waste disposal and re-utilisation record** should always be kept for emulsions, mixtures of mineral oil products and mineral oil sludges.

Spent steel pickling agents contain mainly **ferrous salts.** These can be separated and **sold** (for production of pigments, precipitation agents for clarification processes, sulphuric acid). The remaining pickling agent must be **neutralised with lime milk.** The resulting **hydroxide sludges** are placed in drying beds or preferably dewatered with filter presses. Before **dumping,** the leachability and stability of these residues must be checked to ensure they are suitable for final dumping. If the solids content exceeds 40%, the residues should be taken to the sintering plant.

The **acid pickling water** must be **neutralised** and the coagulated hydroxide sludges separated in clarifying tanks. The clarified **wastewater** can be **re-used** (must be neutralised with acid); **sludges** must be placed in a suitable, sealed **dump.**

Special hoods are used to **eliminate oil mist** in rolling mills; it is separated by a mechanical preliminary separator combined with a downstream electrostatic precipitator.

The effective noise level generated by hot and cold rolling mills is 95 - 110 dB(A). In a rolling mill the noise level, e.g. 5 metres from the open bar steel train, is 106 dB(A) and in a pipe steel rolling mill, near the tube straightening machine, as much as 124 dB(A).

To protect workplaces from noise, the plant is extensively automated and provided with appropriate **control rooms.** These can be well insulated against noise. Ear protection should be worn at workplaces with high levels of noise.

2.6   Foundry and forging operations

Smelting takes place in cupola furnaces (shaft furnaces) and electric melting furnaces. **Gaseous emissions** from smelting are: carbon monoxide, sulphur dioxide, fluorine compounds and nitrous oxides; those from casting are: phenol (briefly), ammonia, amines, cyanide compounds and aromatic hydrocarbons (traces).

Dust occurs in foundries during e.g. preparation of the moulding sand and core sand, manufacture of sand moulds and cores, in casting, cooling of castings, knocking out moulds and with the surface treatment of parts of moulds, known as fettling. **Fabric filters** have proved effective for reducing dust emissions. These have permitted the achievement of concentrations of under 10 mg/m^3 in the clean gas from sand preparation dedusting systems. Optimum fine dust separation with **fabric filters** can help reduce toxic emissions, e.g. nickel, during fettling.

Dust occurring in **cupola furnaces** during smelting is intercepted by wet type dedusters or filtering separators. With **cold blast cupola furnaces** with smelting capacities below 10 t/h, wet dedusters are increasingly being replaced by **fabric filters with preliminary separators.** Clean gas dust concentrations of under 20 mg/m^3 are being adhered to. Fluorine emissions can also be reduced by dry absorption using hydrated lime.

It is essential to **intercept emissions** in all **operating phases,** including blowing and melting-down.

With **hot blast cupola furnaces** with smelting capacities exceeding 10 t/h, operators have managed to obtain clean gas dust concentrations of 20 mg/m^3, with blowing and melting-down as well, using optimised **wet type dedusters** in combination with **primary measures** on the cupola furnace. An enclosed forehearth feed bay also contributes to low-emission operation.

The use of **induction crucible furnaces** is increasing; with these, emissions from the crucible opening are intercepted by an extraction system.

When using **electric furnaces,** which produce significantly **lower dust emissions** than cupola furnaces, values of 20 mg/m^3 are possible using filtering separators. Additional emissions of hydrochloric acids, soot and traces of organic compounds (possibly dioxins) occur when smelting large amounts of scrap mixed with oil, paints and plastics. A **high-performance wet scrubber** must be used under these operating conditions.

Highly odorous substances such as formaldehyde, phenols and ammonia occur in **foundries for small castings** for which moulds are produced according to the cold-box, hot-box or Croning process. In addition to the odour nuisance, these substances are also health hazards. As formaldehyde and high ammonia concentrations are **suspected carcinogens,** steps must be taken to reduce these. Emissions can be reduced by a counter-current scrubber with a phosphoric acid solution. The scrubbing fluid is recirculated and continuously treated.

Waste gases with inorganic compounds occur during **core production,** including core sand mixing. The waste gases must be cleaned with a wet scrubber and in particular the amount of amines in the waste gas must be under 5 mg/m^3.

The **sludge-water mixture** resulting from **wet dedusting,** which may contain substances hazardous to health and the environment such as cadmium, lead and zinc, is **neutralised.** The precipitated solids are separated from the water by sedimentation. The scrubbing water is recirculated. Before **dumping the sediment,** which may contain phenols from the moulding sand binders, it must be tested for **leachability** and **treated** if necessary. In a suitably **modified process,** part of the wastewater flow can be evaporated and the circuit largely, closed, thereby **considerably reducing the scrubbing water requirement.**

The **moulds** are made of moulding sands with approx. 4 to 10% binder (clays, cement, organic materials, hardenable plastics, soda, water glass etc.). They are usually **used once** and then broken up. The used sands can be **treated** and re-used as components in clay-bonded mould production.

The ambient **noise levels** in foundries can reach 120 dB(A). **Noise sources** include loading operations, mixing, dedusting systems, fettling bays, sand preparation, conveyors and fans. **Noise reduction measures** include enclosed hall designs, installation of fans in enclosed rooms and silencers on air intakes and outlets. Machine soundproofing measures are especially necessary in the moulding, core and fettling shops. Measurements made over an 8 hour shift have yielded **workplace noise levels** of 106 dB(A) in the moulding shop, 99 dB(A) in the core shop and 103 dB(A) in the fettling shop. **Principal noise sources** affecting workplaces are: jolt moulding machines, vibratory grates, swing conveyors, fettling machines, impact pneumatic tools, grinders, fans, compressors and conveyors.

Appropriate noise protection measures in the workplace include **encapsulation** of noisy machines, **separation** of noisy machines from other parts of the shop and avoidance of manually operated machines. **Personal ear protection** must of course be worn. **Monitoring** is imperative.

Waste gases are expelled from the furnace in **forges.** Emissions can be controlled by using gas as a fuel. A forge must be regarded as an industrial installation as regards production of wastewater and waste materials.

The ambient **noise level** in a forging shop with e.g. 6 hammers (impact energy 0.6 to 1.3 Mpm) is 112 dB(A). The **background noise level** due to heating furnaces, fans etc. is already 90 to 100 dB(A); to this must be added the **pulsating noise** of the forging machinery. Forging hammers are louder than mechanical and hydraulic presses. It is important to maintain a **safe distance** between the forge and purely residential areas. This distance must be calculated and allowed for in the planning where a reasonable noise level cannot be achieved in the near vicinity through **noise reduction measures in the works.** The **maximum noise level** in the workplace of a drop hammer (1,500 kg tup weight) is 120 dB(A). That of an electric forging hammer in the workplace (tup weight 275 kg) is 97 dB(A). The **interior noise level in forging shops** is normally above 90 dB(A).

Possible **noise reduction measures** include reducing **structure-borne noise** by modifying the forging force curve, **reducing the propagation** of the structure-borne noise, **encapsulating work room openings, reducing** the noise from pneumatic control systems, placing **silencers** on air relief pipelines and using multiple tube nozzles for descaling. The wearing of **personal ear defenders** should be obligatory and should be monitored.

Besides noise, forging also produces **vibrations.** Measures to reduce vibration include the definition, at the planning stage, of suitable **foundation designs,** with appropriate **vibration insulation** at the time of installation. Vibrations in the neighbourhood must be below the threshold of perceptibility.

3. Notes on the Analysis and Evaluation of Environmental Impacts

Emissions produced by the iron and steel industry require particularly extensive measures and systems for **air protection.** Above all, **dusts** containing substances hazardous to health and the environment, such as lead, cadmium, mercury, arsenic and thallium, must be cleaned by high-performance separation systems. Nowadays, not only the primary emission sources, such as sintering plants, but also secondary sources such as blast furnace casting bays can be intercepted and dedusted. In the case of **gaseous emissions,** attention must be paid primarily to reducing carbon monoxide and sulphur dioxide, as well as nitrous oxides and fluorine compounds.

Monitoring of permissible emissions and the **effectiveness** of waste gas cleaning systems must be guaranteed by **measurements**. The dust must also be periodically analysed to detect heavy metals. Emissions must be measured after commissioning the plant to see whether the values assumed in the planning correspond to the reality. If there are discrepancies, new forecasts must be made and **further reduction measures** implemented if necessary.

Emission and immission standards applicable in Germany are detailed in TA-Luft (Technical Instructions on Air Quality Control) and in the Großfeuerungsanlagenverordnung (Ordinance on Large Firing Installations). In the USA, guidelines and standards for the iron and steel industry have been published by the Environmental Protection Agency.

The guidelines adopted by the Association of German Engineers (VDI) contain detailed descriptions for performing emission and immission measurements. Measuring equipment designed for continuous operation must be rigorously examined for robustness, error detection ability and ease of maintenance. Maintenance contracts should be concluded with suppliers. Continuously operating measuring instruments should be employed for measuring dust, sulphur dioxide, fluorine compounds and nitrous oxides (e.g. in the sintering plant and the steel works).

Recycling of water and **the use of closed-circuit cooling water** systems deliver cost-savings and a high rate of re-use in iron and steel works. **Effective water treatment systems** are needed for this purpose.

General **minimum requirements** are laid down in Germany for treated **waste-water** discharged into receiving bodies of water and for special plants. These parameters must be monitored by measuring equipment at the point of transfer of the cleaned water to the receiving body of water. **Cleaning systems for waste gases and water** can only satisfy their intended purpose when they are **correctly operated, serviced and repaired**. The provision of detailed operating, maintenance and repair manuals is imperative.

Practically all processes involve greater or lesser levels of noise. High noise emissions can cause annoyance in the vicinity if reduction and protection measures are inadequate. In Germany, TA-Lärm (Technical Instructions on Noise Abatement) and the guidelines adopted by the Association of German Engineers (VDI) are used for calculating and assessing noise immissions in the **neighbourhood**. Noise immissions are assessed against immission reference values which are graded according to the type of area to be protected and the periods of noise. Guidelines are also available for assessing noise emissions at the workplace.

As in Germany, **works environmental protection officers** should be deployed in iron and steel works who are totally independent of the production side. Their task is essentially to work towards the development and introduction of **environment-friendly processes**. In addition they are entitled and obliged to monitor **adherence to statutory regulations** and **compliance with official directives and conditions** in so far as these relate to environmental protection.

The scope and monitoring of **working conditions and health protection measures**, which vary from one workplace to another, should be set down in a manual. Proposals are detailed in the regulations of the **employers' liability insurance association (Berufsgenossenschaft) of the iron & steel industry.** Suitably qualified **safety officers** and a **works doctor** should be appointed.

4.   Interactions with Other Sectors

The erecting of iron and steel production plants involves **land-use** which is measured in terms of the works site with adjoining areas and connecting roads. Before erecting production plants, impacts on the local natural order and the geogenic and anthropogenic burdens on the soil and groundwater and on any bodies of surface water must be investigated in the context of the **location planning**. An adequate distance from the nearest residential zones must also be guaranteed. Details are contained in the environmental brief Planning of Locations for Trade & Industry.

Iron and steel works involve large-scale production and require large amounts of **raw materials**. These include primarily ores, coke and limestone. Generally speaking, to produce 1 tonne of crude steel requires 450 to 500 kg coke and fuel oil, 250 kg lime and 5 m^3 water.

In an integrated iron works for example, the specific **total energy consumption** is some 20 GJ/t crude steel. In an integrated iron works, the sintering plant, blast furnace, coking plant, steel works, rolling mill and power station areas are interconnected as a **combined energy system**. Thus, the top gas is utilised in all areas, its calorific value enriched with converter gas, coking oven gas or natural gas. Power and steam are supplied by the power station. Boilers are usually operated with gas, e.g. top gas. The burners can be fired with top gas, coking oven gas or fuel oil. **External power supplies** are used in addition to **internally generated power.** Waste heat boilers from the steel works contribute to steam production.

A mixed iron works is **linked** to the following **other sectors**:

- The raw materials (ores, coal, limestone) must be mined in large quantities in open cast or deep mines. (See environmental briefs Surface Mining and Underground Mining).

- Ores must be dressed (see environmental brief Minerals - Handling and Processing).

- Efficient transport routes (canals, railways or roads) are required for **transporting raw materials** and products. For environmental protection reasons, transport should mainly be via inland waterways and railways. Whether the location of the iron works is chosen because of where the ore, coal or sales market is situated, high-capacity transport facilities must always be provided.

- **Coke** of specified quality must be supplied for the blast furnace by a coking plant. Reference should be made to the environmental brief Coking Plants, Coal to-gas Plants, Gas Production and Distribution to assess the environmental impacts associated with coke production.

- In view of the quantities of cooling water needed, an adequate **water supply** must be available. To avoid the adverse consequences of drawing excessive quantities of water from groundwater or surface water resources, extensive recirculation systems must be provided, internal treatment of wastewater and cooling water. Water consumption must be in harmony with the **general water framwork planning**.

- The large workforce of a mixed iron works may result in the disorganised development of housing at an insufficient distance from the plant. This can lead to **water shortages,** unsatisfactory wastewater treatment and disorganised dumps, plus immission burdens affecting the **areas of habitation.**

- Other sectors directly or indirectly linked to the iron and steel industry are: **lime kiln plants, cement works, ferroalloy production plants, power generating plants** and **slag and dust recycling plants.** The above plants and establishments are associated with considerable potential atmospheric burdens. Reference is made to the relevant briefs.

A **general or single-purpose dump** is to be provided for non-recyclable **residual and waste materials** including furnace debris from the metallurgical processes with hazardous pollutants. These should be classified according to criteria of environmentally acceptable final storage (see environmental brief Disposal of Hazardous Waste).

5. Summary Assessment of Environment Relevance

The **establishment** of iron and steel production plants in areas not previously used for industry will have an impact on the landscape. **Environmental damage can be reduced** by **selecting locations** with relatively insensitive landscapes where there is unlikely to be any great effect on the regional productiveness of the natural environment.

The **environmental burdens** imposed by an iron and steel making plant and related technologies relate to the air, water, soil, flora and fauna, waste, noise and vibration.

Efficient separators are available for reducing **dust emissions.** Important in this regard is the **continuous monitoring** of the operation of these separators using suitable measuring equipment. Since a large proportion of the separated dust can be re-utilised in the process, **high-performance gas cleaning systems** are desirable, not only for environmental protection reasons but also in the interests of **economy.** Increasing attention is being paid to **random dust sources**, e.g. from working bays. Tried and tested **collection systems** are available for this purpose. High dust immissions occur in the vicinity of iron works. Although high grade cleaning of waste gases reduces dust emissions, **dust emissions** for the iron works as a whole are between 1 and 3 kg/t, depending on the number of installed process stages and the extent of dust reduction from diffuse sources. 1 kg/t should be regarded as an optimum value. Studies should be carried out in all cases to determine whether **agriculture in the vicinity of the works** is being impaired by contamination over wide areas with phytotoxic and zootoxic heavy metals, especially zinc, copper, chromium, nickel and lead, taking into account long-term deposition and accumulation in the soil. Heavy metals, especially cadmium and mercury, can be injurious to human health through accumulation in the soil and in plants, with increased absorption through the food chain. **Conflicts** can be **avoided** or **diminished** by **consulting** the affected population groups at an early stage, possibly developing and planning new sources of employment (see also Vol. III, Compendium of Environmental Standards).

49

As the **increased environmental burden** poses **additional health risks** and hazards, e.g. for women and children (during pregnancy etc.), **adequate medical care** should be provided in the project region.

In some respects **air protection measures** lead to a **shifting of problems,** e.g. where separated residues cannot be recycled. A high degree of **recycling of the material and energy** present in dusts, sludges and gases is a basic requirement for environmental compatibility, and one that can be met. For materials which cannot be recycled, a **dumping system** must be selected which will enable environmentally acceptable final dumping.

Although technological development in iron works has led to high water consumption, use of water in the plants can be minimised by recycling as much as 80% and through the use of **closed cooling circuits.** The standards applicable to the cleaning of wastewater contaminated with heavy metals must be raised from the level of the general rules previously applicable in Germany to a truly "state-of-the-art" level.

Noise levels can be minimised by extensive **noise reduction measures**. However it is also important to ensure an **adequate distance** between the works site and neighbouring areas of habitation.

Possible ways of preventing adverse environmental impacts through state-of-the-art emission reductions in **old plants** include (in the process engineering area) replacing old converter techniques for steel production with low-emission converters and electric furnaces and introducing continuous casting in approximately final dimension form. In the waste gas and air purification area, the use of multistage separators, fine dust separators and the interception of diffuse emission sources is also possible in old plants. Increased recycling of residues and water will help reduce the environmental burdens imposed by old plants. Secondary noise reduction measures are more difficult to implement than primary measures.

6. References

Statutory provisions, regulations

Erste Allgemeine Verwaltungsvorschrift zum Bundes-Immissionsschutzgesetz (Technische Anleitung zur Reinhaltung der Luft - TA-Luft) dated 27.02.1986, GMBl (joint ministerial circular). 1986, Ausgabe p, S.95.

Zweite Allgemeine Verwaltungsvorschrift zum Abfallgesetz (TA-Abfall) Teil 1: Technische Anleitung zur Lagerung, chemisch-physikalischen, biologischen Behandlung, Verbrennung und Ablagerung von besonders überwachungsbedürftigen Abfällen, Gemeinsames Ministerialblatt (joint ministerial circular) Nr. 8, p. 139 - 214 dated March 12, 1991.

24. Allgemeine Verwaltungsvorschrift über Mindestanforderungen an das Einleiten von Abwasser in Gewässer (Eisen- und Stahlerzeugung), GMBl (joint ministerial circular). 1982, p. 297.

Deutsche Forschungsgemeinschaft: Liste maximaler Arbeitsplatzkonzentrationen (MAK-Wert-Liste), 1990, Mitteilung XXVI, Bundesarbeitsblatt 12.1990, p. 35.

DIN 4301 (April 1981): Eisenhüttenschlacke und Metallhüttenschlacke im Bauwesen.

EC Council Directives of 12 May 1986 on the protection of workers from the risks related to exposure to noise at work - 86/188/EEC and of June 14 1989 - 89/392/EEC on the approximation of the laws of the Member States relating to machinery.

Environmental Protection Agency (EPA): Effluent Guidelines for Iron and Steel Manufacturing (CFR 420); Iron and Steel Development Document (Volumens I - VIII); Regulations on Standards of Performance for New Stationary Sources (40 CFR 60).

Hinweise für das Einleiten von Abwasser in eine öffentliche Kläranlage, Arbeitsblatt A115 (January 1983) der Abwassertechnischen Vereinigung e.V., St. Augustin.

Lärmschutz an Hochofen- und Sinteranlagen, herausgegeben vom Minister für Arbeit, Gesundheit und Soziales des Landes Nordrhein-Westfalen, Düsseldorf, 1982.

49

Lärmschutz an Elektrostahlwerken, herausgegeben vom Minister für Arbeit, Gesundheit und Soziales des Landes Nordrhein-Westfalen, Düsseldorf, 1982.

Technische Anleitung zum Schutz gegen Lärm (TA-Lärm) dated July 16, 1968, zur Allgemeinen Verwaltungsvorschrift über genehmigungsbedürftige Anlagen nach § 16 der Gewerbeordnung, übergeleitet nach § 66 des Bundes-Immissionsschutzgesetzes dated 15.03.1974, Beilage BAnz. No. 137.

Unfallverhütungsvorschriften, Hauptverband der gewerblichen Berufsgenossenschaften, Bonn u.a. UVV-Lärm, VBG 121 dated 01.01.1990.

VDI-Richtlinie 2288, Blatt 1: Auswurfbegrenzung, Kupolofen-Betrieb, September 1971.

VDI-Richtlinie 2288, Blatt 2: Anleitung für Staubauswurfmessungen an Kupolöfen, August 1971.

VDI-Richtlinie 3465: Auswurfbegrenzung, Stahlwerksbetrieb, Elektrolichtbogenöfen, January 1978.

VDI-Richtlinie 3887: Emissionsminderung, Gießereien, in Vorbereitung.

VDI-Richtlinie 2058, Blatt 1: Beurteilung von Arbeitslärm in der Nachbarschaft, September 1985.

VDI-Richtlinie 2561: Die Gesamtemission von Gesenk- und Freiformschmieden und Maßnahmen zu ihrer Minderung (Lärm), July 1968.

VDI-Richtlinie 2560: Persönlicher Schallschutz, December 1983.

VDI-Richtlinie 3572, Blatt 2: Emissionskennwerte technischer Schallquellen; Umformmaschinen, Schmiedepressen, October 1986.

VDI-Richtlinie 2262: Staubbekämpfung am Arbeitsplatz, December 1973.

VDI-Richtlinie 3929: Erfassen luftfremder Stoffe (Entwurf), March 1990.

VDI-Richtlinie 2058, Blatt 3: Beurteilung von Lärm am Arbeitsplatz unter Berücksichtigung unterschiedlicher Tätigkeiten, April 1981.

Verordnung über Arbeitsstätten (Arbeitsstättenverordnung ArbStättV) dated 20.03.75, BGBl I (Federal Law Gazette I), p. 729, 15: Schutz gegen Lärm.

Dreizehnte Verordnung zur Durchführung des Bundes-Immissionsschutzgesetzes (Verordnung über Großfeuerungsanlagen - 13. BImSchV) dated 22.06.83, BGBI (Federal Law Gazette), Teil I, p. 719.

Verordnung zur Bestimmung von Abfällen nach § 2 Abs. 2 des Abfallgesetzes dated April 3, 1990, BGBl I (Federal Law Gazette I), p. 614.

Verordnung zur Bestimmung von Reststoffen nach § 2 Abs. 3 des Abfallgesetzes dated April 3, 1990, BGBl I (Federal Law Gazette I), p. 631.

Verordnung über das Einsammeln und Befördern sowie über die Überwachung von Abfällen und Reststoffen dated April 3, 1990, BGBl I (Federal Law Gazette I), p. 648.

Verordnung über gefährliche Stoffe (Gefahrstoffverordnung GefStoffV) dated 26. August 1986, BGBl I (Federal Law Gazette I), p. 1470 in the version dated. August 23, 1990, BGBl I, p. 790.

Scientific / technical papers

Abwassertechnische Vereinigung: Lehr- und Handbuch der Abwassertechnik, Band VI, Industrieabwässer mit anorganischen Inhaltsstoffen, Verlag für Architektur und technische Wissenschaften, Berlin 1985.

Aichinger, H.M., Borgsschulte, B., Britz, H., Held, B., Meyer, O., Strohschein, H.: Stand des primärenergiesparenden Konvertereinsatzes in der Bundesrepublik Deutschland, Stahl u. Eisen 108, 1988, No. 13, p. 645 - 654.

Anonym: Die neue Entstaubungsanlage im Oxygenstahlwerk Beeckerwerth der Thyssen Stahl AG, Stahl u. Eisen 110, 1990, No. 4, p. 137.

Baum, J.P., Gerhardt, J.W.: Abgasreinigungsanlagen in der Eisen- und Stahlindustrie und ihre Kosten. in: Stand und Entwicklung der Anlagentechnik im Eisenhüttenwesen, Haus der Technik-Veröffentlichung No. 369, Essen.

Bogdandy, L., Nieder, W., Schmidt, G., Schroer, U.: Die Schmelzreduktion von Eisenerz nach dem Corex-Verfahren im kraftwirtschaftlichen Verbund, Stahl u. Eisen 109, 1989, No. 9, p. 445 - 452.

Buckel, M., Kersting, K., Kister, H., Lüngen, H.: Neue Entwicklungen bei der Sinterherstellung, Stahl u. Eisen 110, 1990, No. 2, p. 43 - 51.

Direktreduktion von Eisenerz, Verlag Stahleisen mbH, Düsseldorf, 1976.

Dreyhaupt, F.J.: Handbuch für Immissionsschutzbeauftragte, Verlag TÜV Rheinland, Cologne, 1981.

Fischer, B., Rüffer, H., Düppers, W., Nagels, Gl., Knorre, H.: Entgiftung cyanidhaltigen Gichtgaswaschwassers von Hochöfen, Zeitschrift für Wasser- und Abwasser-Forschung 14, 1981, No. 5/6, p. 210 - 217.

Fleischer, G.: Abfallvermeidung in der Metallindustrie, EF-Verlag für Energie- u. Umwelttechnik GmbH, Berlin, 1989.

Geiseler, J., Drissen, P., Treppenschuh, H.: Metallurgische Verwertung von Stauben und Schlämmen der Stahlindustrie, Stahl u. Eisen 109, 1989, No. 7, p. 359 - 365.

Gemeinfaßliche Darstellung des Eisenhüttenwesens, Verlag Stahleisen mbH, Düsseldorf, 1971.

Grebe, K., Grützner, G., Lehmkühler, H.J., Schmauch, H.: Die Metallurgie der Direktreduktion von Hüttenreststoffen nach dem Inmetco-Verfahren, Stahl u. Eisen 110, 1990, No. 7, p. 99 - 106.

Grützmacher, K., de Haas, H., Mohnkern, H., Ulrich, K., Kahnwald, H.: Staubunterdrückung in Hochofengießhallen, Stahl u. Eisen 111, 1991, No. 3, p. 51 - 56.

Haering, H.U.; Polthier, K.: Geräuschemission und Lärmminderung von Gesenkschmieden, Stahl u. Eisen 108, 1988, No. 4, p. 179 - 184.

Haering, H.U.; Möllers, K.H.; Neugebauer, G.; Polthier, K.: Lärmminderung durch Einhausung von Lichtbogenöfen, Stahl u. Eisen 109, 1989, No. 7, p. 343 - 349.

Haucke, M., Theobold, W.: Behandlung und Aufbereitung von Stäuben und Schlämmen in der Stahlindustrie, Gewässerschutz-Wasser-Abwasser, Aachen, Bd. 21, 1976, p. 511 - 54.

Kaas, W.: Handhabung von Walzunderschlamm, Stahl u. Eisen 101, 1981, p. 963 - 965.

Krumm, W., Fett, N., Pöttken, H,. Strohschein, H.: Optimierung der Energieverteilung im integrierten Hüttenwerk, Stahl u. Eisen 108, 1988, No. 22, p. 1097 - 1106.

Kühn, M., Haucke, M.: Erfahrungen bei der Behandlung und Verwertung von Stahlwerksstäuben and -schlämmen, Stahl u. Eisen 101, 1981, p. 701 - 705.

Lange, M., Minimierung der Dioxin- und Furanemissionen aus Abfall -verbrennungsanlagen, TÜ 32, 1991, No. 3, p. E35 - E40.

Lärmemission und Lärmminderung an Elektrolichtbogenöfen, Verbesserung des Gesundheitsschutzes für die Belegschaft. Bericht No. 809 des Betriebs -forschungsinstituts Düsseldorf.

Lärmquellen der Eisen- und Metallindustrie, Ed.: Berufsgenossenschaftliches Institut für Lärmbekämpfung, Mainz, 1973.

Meinck, F.; Stooff, H.; Kohlschütter, H.: Industrie-Abwässer, Stuttgart, Gustav Fischer Verlag, 1968, 4. Aufl.

Ministerium für Arbeit, Gesundheit und Soziales des Landes NW: Luftreinhalteplan Ruhrgebiet West, 1. Fortschreibung, 1984 - 1988, Düsseldorf, 1985.

Philipp, J.A. et al: Umweltschutz in der Stahlindustrie, Entwicklungsstand -Anforderungen - Grenzen, Stahl u. Eisen 107, 1987, No. 11, p. 507 - 514.

Philipp, J.A., Maas, H.: Abfallwirtschaft in einem Hüttenwerk, Stahl u. Eisen 104, 1984, p. 403 - 407.

Rat von Sachverständigen für Umweltfragen: Umweltgutachten 1987, Verlag W. Kohlhammer GmbH, Stuttgart.

Rat von Sachverständigen für Umweltfragen: Sondergutachten Altlasten 1989, Sondergutachten Abfallwirtschaft 1990, Verlag Metzler-Poeschel, Stuttgart.

Reichelt, W.; Kapellner, W.; Steffen, R.: Endabmessungsnahe Herstellung von Flachprodukten, Stahl u. Eisen 108, 1988, No. 9, p. 409 - 417.

Schallschutz in Gießereien, Teil 1: Beschreibung von Gießereien und Zusammenstellung von vorhandenen Erkenntnissen über das Geräuschemissions- und Immissionsverhalten, Studie des TÜV Rheinland im Auftrag des Ministers für Arbeit, Gesundheit und Soziales des Landes Nordrhein-Westfalen, Düsseldorf, 1983.

Schmidt, H.: Schalltechnisches Taschenbuch, VDI-Verlag, Düsseldorf, 1984.

Steffen, R., Lüngen, H.: Stand der Direktreduktion von Eisenerzen zu Eisenschwamm, Stahl u. Eisen 108, 1988. No. 7, p. 339 - 343.

Umweltbundesamt [German Federal Environmental Agency]: Altanlagereport 1986, Umweltbundesamt Berlin, 1986.

Umweltbundesamt [German Federal Environmental Agency] : Jahresbericht 1988, 1989 und 1990, Umweltbundesamt Berlin.

Umweltbundesamt [German Federal Environmental Agency]: Checklisten zur Prüfung der Umwelterheblichkeit raumbedeutsamer Vorhaben "Metallverarbeitende Industrie". UBA-FB 87-039, Werbung und Vertrieb Verlag, Berlin 1988.

VDI-Kommission Reinhaltung der Luft: Schwermetalle in der Umwelt, Düsseldorf, 1984.

Vigder, I.: Wasserkreisläufe für die Stahlindustrie, Stahl u. Eisen 103, 1983, p. 1195 -1197.

Wischmann, G.: Geräuschemission von Schmiedepressen und Möglichkeiten zur Lärmminderung, Schriftenreihe der Bundesanstalt für Arbeitsschutz [German Federal Institute for Occupational Health and Safety], Dortmund, 1984, Heft Fb 393.

Trade and industry

Non-ferrous Metals

50. Non-ferrous Metals

Contents

1.   Scope

Since the **non-ferrous metals** sector covers a multitude of individual **products, charge materials, fuels** and **processes**, this brief can only deal with a few **examples** of the **principal industrial non-ferrous metals.** Environmental impacts and protection measures in the production and processing of **aluminium, copper, lead** and **zinc** are dealt with as representative of the large number of other non-ferrous metals as well.

The non-ferrous metals sector comprises **the subdivisions:**

-       smelting of appropriately pretreated primary raw metals to produce metals
-       processing recycling material in secondary smelting plants, and
-       processing of metals to produce standard commercial billets and blanks.

2.   Environmental Impacts and Protective Measures

The following deals primarily with the environmental factors arising in the **application of current standard processes.** For projects using **pyrometallurgical processes** these are primarily **air protection measures; slags** are also produced which, depending on their composition, can be a danger to soil, water and living things. In **hydrometallurgical treatment processes,** measures to **protect water and soil** predominate.

Since most processes generate **noise,** the possibility of noise pollution occurring both in the work-place and in the neighbourhood must be taken into account.

Non-ferrous metal production plants occupy a considerable amount of **space** to accommodate the works site with adjoining areas and connecting roads.

Different quantities of energy are required depending on the production process. The **choice of location** partly depends on the availability of sufficient low-cost electricity, e.g. in the case of aluminium production. An encapsulated furnace producing aluminium by igneous electrolysis, with a current load of 200 kA and d.c. voltage of 4.2 V requires approximately 13 kWh/kg aluminium. Zinc production with the stages of roasting, leaching, neutralization, leachate cleaning and electrolysis requires 4 kWh/kg zinc. Values for copper production are somewhat higher. Energy requirements of **secondary smelting plants** are considerably lower: 20% of the primary smelting energy requirement with 100% scrap copper, around 40% with 100% scrap zinc and 10% with 100% scrap aluminium.

2.1 Aluminium extraction

The **Bayer process** is used almost exclusively for producing aluminium oxide, the charge material for primary aluminium smelting plants. Bauxite is treated with soda lye under pressure and heat in autoclaves to produce aluminium hydroxide and **red mud**. The latter is separated, washed and filtered, and may be recyclable or may have to be dumped. After sedimentation and filtration, the aluminium hydroxide is converted to aluminium oxide (alumina) by fluidized bed calcination at around 1100°C.

Large quantities of **red mud** (1 - 2 t/d Al_2O_3) are produced. Depending on its composition and the situation in the country in question, it should be used for **extracting** aluminium oxide and iron, producing flocculation agents for wastewater cleaning or the manufacture of building materials. Red mud which cannot be further processed must be **dumped.** Where it is stored on a dump, **special requirements** must be met in respect of sealing and treatment of percolation water. Dumping should be on a single-purpose dump subject to continuous supervision.

A considerable amount of fine dust may be produced upon **loading, unloading and transport of fine-grained materials** (bauxite, alumina) unless **enclosed conveyor systems** and **suitable storage facilities** are provided. Waste gas from the calcination furnaces contains dust with an aluminium oxide content which is deposited in dry filters and recirculated. Dust emissions in cleaned waste gas are under 50 mg/m³.

The process most commonly used for **extracting pure aluminium** is **igneous electrolysis**. Aluminium oxide at approximately 950°C is dissolved in a molten mixture of aluminium fluoride and cryolite and separated by direct current into pure aluminium and oxygen. The liquid aluminium is periodically drawn off and cast.

The following **emissions and raw materials** occur with the **extraction of pure aluminium:**

-       primary alumina dust during storage, transport and charging;
-       primary dust during anode production (petroleum coke etc.);
-       volatile binding agents, fluorine from anode residues in the waste gas from the anode burning kilns;
-       fluorides (dust and gaseous form) in the pot waste gas containing CO/CO2; hydrogen fluoride gas is highly corrosive, harmful to health and the environment (also affects plant growth);
-       used cathodes, containing fluoride;
-       furnace breakage materials with fluoride components;
-       wastewater.

The following individual protection measures are necessary:

Fine dust:	Use of enclosed conveying systems (e.g. pneumatic conveyors).
Anode production:	Extraction of dust and gaseous emissions, electrostatic waste gas cleaning, wet-chemical fluorine separation. Use of fabric filters permits clean gas dust concentrations of under 20 mg/m^3 and fluorine contents of under 1 mg/m^3.
Pots:	Pot encapsulation with anode gas extraction and waste gas cleaning, wet chemical fluorine recycling or combined dedusting and dry absorption in the Al_2O_3 fluidized bed with direct recirculation. Wet chemical separation with water recirculation produces a sludge which, after drying, can only be partly returned to the process. Dry absorption and return of the filter dust to the process is preferable as this relieves the burden on the water circuit. Clean gas dust contents of under 30 mg/m^3 and fluorine compounds of under 1 mg/m^3 are obtained with encased, centrally controlled, large-capacity furnaces with computerised waste gas regulation and dry absorption with fabric filter.
Cell house:	Shop air extraction and cleaning is compulsory with non-encased furnaces. Can be retrofitted.
Cathode and furnace breakage:	Dumping only on specially protected, single-purpose dumps. Cryolite, used as a fluxing agent for the electrolysis, can be obtained by processing (fluorine recycling).
Wastewater:	The discharge of wastewater from aluminium oxide production and aluminium smelting must satisfy the requirements laid down under the generally recognised standards regarding chemical oxygen demand for aluminium and fluorides.

With respect to **noise**, a distinction is made between noise emissions affecting the **neighbourhood** and those affecting the **workplace.** Emission from main noise sources can be restricted by encapsulation and by means of silencers on air intakes and outlets. A **noise reduction plan** should be prepared during the planning phase.

2.2   Heavy metal ore smelting

The **composition** of the concentrates or raw materials is crucial for the applicable **smelting process** and thus also for the nature and quantity of the **environmental pollutants** arising. Sulphidic ore concentrates are thus mostly smelted by **pyrometallurgical processes,** whilst **hydrometallurgical processes are employed for** oxidic, sulphidic-oxidic and complex ores.

Combined processes are also used in which, for instance, material roasted by a pyrometallurgical process undergoes further treatment by hydrometallurgy. The charge material is ore enriched by beneficiation.

- **Pyrometallurgical process stages**

Roasting: Partial or total desulphurization (dead roasting) of the charge material;

Sinter roasting: Roasting of sulphur with admission of air (conversion of sulphides to metal oxides and SO2 gas) with simultaneous agglomeration of the roasted material for use in shaft furnaces;

Rolling: Metal oxide enrichment by controlled volatilization (Zn);

Smelting: Separation of gangue (slags): production of high grade metal sulphides (Cu_2S) by partial combustion of the sulphur content and reduction of metal oxides (PbO, ZnO) under coke combustion with air admission;

Fuming: Conversion of metal sulphide to metal in a converter;

Pyrometallurgical refining:	Cleaning molten metal of oxygen, sulphur, impurities and tramp metals by intermetallic precipitation, slagging and/or volatilization;
Slag cleaning:	Thermal processing of slags to extract metal components.

Numerous **emissions and residual materials** occur with the above processes:

-   Waste gases of various origins
    -   Primary dust from the charge material,
    -   Dusts from volatilized metals, including lead, zinc, arsenic, tin, cadmium, mercury, selenium, tellurium and their compounds (condensed after cooling),
    -   gaseous materials including SO_2, HCl, HF, CO, CO_2;
-   Wastewater from coolant circuits and waste gas scrubbing;
-   Final slags with residual metal contents, sulphates, sulphides; possibility of polychlorinated dibenzo-dioxins and -furans with chlorinating methods (e.g. copper roast leaching process);
-   Furnace breakage materials, containing arsenic, lead, cadmium, mercury and cyanide.

For protective measures to be effective, it is **essential** that all emissions, including diffuse emissions of gas and dust, be **efficiently intercepted** at their points of origin. Diffuse emissions can be intercepted by **hoods, covers** or **encapsulation**, also by **constructional measures** such as encasement of conveyor belts or enclosed bays. Roasting furnaces should not be outdoor installations.

Dust:	Waste gases are normally dedusted in dry filter systems (cyclones, electrostatic precipitators, fabric filters). Dedusting efficiency of up to 99.9% is possible, but depends on the permissible solid or pollutant content. Dusts can also be separated with fabric filters in lead smelting plants. Good separation efficiency is particularly important for the environment because waste gas from smelting contains toxic substances such as arsenic, antimony and lead in the form of fine dust. High-performance filtering separators have proved effective for fine dust separation.

Dust recycling for enriching and recovering metals. Separate pyrometallurgical or hydrometallurgical processing of tramp metals, for example As, Cd. Fabric filters are the principal method of dust separation. Clean gas dust contents of 10 mg/m³ can be obtained. Best values are around 1 mg/m³, e.g. in lead smelting plants.

SO₂ gas:

Removed by waste gas scrubbing followed by neutralization. SO_2 concentrations in waste gas of over 3.5% are suitable for sulphuric acid production. In certain circumstances liquid SO_2, gypsum or elementary sulphur can be produced as a possible preliminary stage for industrial usage. Wet chemical waste gas cleaning processes are used for lower SO_2 concentrations. Only limited SO_2 concentrations and overall quantities may be discharged via chimneys.

Oil mists:

If oil mists are present in the waste gases from shaft furnaces on account of the charge material, waste gases must undergo thermal afterburning.

Final slags/
furnace breakage:

Slags and furnace breakage material should be stored in a specially protected single-purpose dump, since toxic and water-polluting substances such as heavy metals may be released through leaching and weathering. Depending on residual metal content and concentrations of other substances such as sulphides, sulphates, dioxins and furans, may possibly be used for road construction or reprocessing, or may have to be discarded.

Wastewater:

Wastewater from waste gas scrubbing and slag granulation is polluted with heavy metals. Dissolved and undissolved metallic compounds in communal treatment plants lead to excessive metal concentrations in sewage sludges, restricting or preventing agricultural use.

Measures for reducing pollutant loads include minimizing wastewater volumetric flow by recirculation, recycling of treated wastewater and separating wastewater requiring treatment from that not requiring treatment. Extremely high standards must be applied to the discharge of wastewater with metal compounds toxic to humans and the ecosystem. State-of-the-art wastewater treatment systems include selective ion exchangers, microfiltration systems, reversal osmosis and thermal concentration processes. Production-specific concentrations of cadmium, mercury, lead, zinc, arsenic, copper, nickel and chromium should be limited.

Significant waste gas and emission reductions are achieved by combining several process steps in **modern processes** such as the flash cyclone reactor and the flash smelting method. Trials in a copper smelting plant and a lead smelting plant yielded reductions of 75%.

- **Hydrometallurgical processes**

Charge materials are oxidic ores, pretreated sulphidic ore concentrates which can be hydrometallurgically treated, or sulphidic concentrates which undergo oxidizing leaching. Hydrometallurgy processes also include extraction and refinement electrolysis.

Leaching: Treatment and lixiviation of the metals to be recovered, e.g. with dilute sulphuric acid for zinc production. For dump leaching in the case of very low-grade ores (bottom sealing necessary for soil and ground water protection);

Enrichment: Concentration of weak solutions by fluid extraction, using an organic solvent with simultaneous leachate cleaning.

Cleaning: Separation of accompanying substances and impurities by solids-fluid extraction and/or precipitation (hydroxide or sulphide precipitation, cementation);

Extraction: Electrolytic metal deposition with insoluble anodes (e.g. with Zn, Cu);

Refining: Electrolytic metal deposition with soluble anodes (e.g. with Cu, Pb).

The following environmentally relevant **emissions and substances** may be produced with the above processes:

Wastewater: Greater or lesser quantities of zootoxic and phytotoxic heavy metal components may be present in the wastewater, depending on the charge materials.

Leachate residues: Leachate residues contain metallic compounds harmful to the environment.

Waste gases: Sulphuric acid mists are produced in the extraction electrolysis; metal-containing vapours, e.g. in crude copper anode furnaces; organic solvents, e.g. xerosin, during liquid extraction in the enrichment process.

Anode sludge: This sludge contains metals and metal compounds, e.g. gold, silver, lead, tin, arsenic, antimony.

Spent electrolyte: The electrolyte contains dissolved metallic compounds of iron, nickel, zinc, arsenic and cobalt.

The following **individual protection measures** are necessary:

Wastewater: The wastewater volume must be reduced by appropriate measures, e.g. recirculation, recycling. Wastewater containing heavy metal pollutants must be treated by state-of-the-art methods. Wastewater contaminated with e.g. cadmium and mercury must be channelled and treated separately.

For wastewater treatment, especially low production-specific concentrations are to be stipulated, with residual concentrations of under 1 mg/l Cd and under 0.1 mg/l Hg to be achieved. Suitable processes include ion exchange, ultrafiltration and electrolysis.

Leachate residues:

Residues must be converted by washing and neutralization processes to form compounds suitable for final dumping. Where technically possible, solvent residues are to be eliminated.

Waste gases:

Permissible work-place concentrations for sulphuric acid mist can be achieved by appropriate room air ducting and, where necessary, waste air scrubbing.
By equipping a crude copper anode furnace with fabric filters, it was possible to separate gaseous metallic compounds to clean gas concentrations of 0.001 mg cadmium/m^3, 0.05 mg lead/m^3 and 1.9 mg/arsenic/m^3. With liquid extraction using organic solvents, precautions must be taken against combustion and explosion and for fire fighting.

Anode sludge/ electrolyte:

Special hydrometallurgical or pyrometallurgical measures are to be employed for the phased recovery of useful materials and the extraction of tramp metals; e.g. electrolytic deposition of arsenic and antimony or precipitation of nickel, iron or cobalt.

The **extraction of zinc** from zinc blende or galmei inevitably yields 3 to 4 kg **cadmium** per tonne of zinc as an alloy element in crude zinc or in the form of residues. Cadmium is extracted in primary zinc smelting plants by dry and wet absorption processes. The generally preferred **wet processes** and **electrolytic cadmium extraction** result in **no direct production of cadmium dusts**. The **waste gases** resulting from the smelting of cadmium to produce commercial formats can be introduced to the air for roasting, in order to achieve total waste gas cleaning.

Due to the **toxic effects** of cadmium, **strict requirements** must be imposed on **work-place hygiene** and **waste air and water cleaning**. In heavy metal ore smelting operations, **main noise sources** are wherever possible to be restricted by encapsulation and by means of silencers on air intakes and outlets. A **noise reduction plan** should be prepared at the project planning stage. In the case of operations generating high levels of noise, one should preferably begin by damping or eliminating occurrences and noise sources which arise only periodically.

To protect **work-places** from noise, installations should be extensively **automated** and equipped with appropriate control rooms. **Protective equipment** includes fireproof clothing, breathing equipment and ear protectors, depending on where they are working; protective helmets and safety footwear must be worn in all areas.

Measures for safety in the work-place and to protect the soil of the works site include all precautions to prevent the discharge of water-polluting substances. **Special attention** is to be paid to installations for producing, handling and using water-polluting substances. Relevant precautions include storage tanks with leak-proof drip trays, overfilling safeguards, sealed and impermeable floor surfaces and leak testing, and these should be set forth in a **manual**.

2.3   Secondary smelting plants

Secondary smelting plants process mainly **recycling material** (shredder scrap, cables, batteries etc.), heavily contaminated mixed scrap, production scrap with alloy constituents that are difficult to remove, also slags, dross and other metalliferous residues. Predominantly **pyrometallurgical processes** are employed for metal recovery.

Environmental burdens stem mainly from **impurities and pollutants present in the charge material**, e.g. oil, paint, plastics, solvents or salts.

Special characteristics of the **emissions and substances** and requisite safeguards are as follows:

* **Aluminium scrap melting plants**

Salt slags: Aluminium scrap is usually melted down in rotary or hearth type furnaces under a layer of liquid salt to prevent ingress of air. The salt absorbs impurities present in the scrap and occurring during the melt-down process and produces salt slag (0.5 t/t Al).

Dumping these salt slags seriously pollutes the dump percolation water, therefore salt slag should be processed and returned to the melting process.

Waste gases: The molten aluminium is refined in converters using chlorine gas. The waste gases contain dusts, gaseous chlorine and fluorine compounds and chlorine gas; they may also contain organic substances which, depending on the operating conditions, may include traces of especially environmentally hazardous materials such as polychlorinated dibenzo-dioxins and -furans. Adequate separation of the dusts and inorganic compounds is achieved by dry absorption and fabric filters. Emissions of organic substances can be minimized by scrap sorting and cleaning or by special thermal afterburning of the waste gases.

• Copper scrap melting plants

Dust: When melting down copper-bearing residues, the interception and dry separation of emissions produced on charging and running-off are particularly important. Where oil mist occurs due to the impurity of the copper scrap, waste gases must undergo thermal afterburning before dust separation. For ecological and economic reasons, melting down should take place in a converter with top lances in a shop with waste air collection and cleaning rather than in shaft furnaces.

• Lead scrap melting plants

Waste gases: When recycling scrap batteries, PVC residues may give rise to gaseous inorganic chlorine compounds which are absorbed in the dust and in the slag.

Depending on the operating conditions, small quantities of polychlorinated dibenzo-dioxins and -furans may be present in the waste gases when recycling scrap cables. Emissions of health-endangering dioxins and furans can be restricted by careful sorting of scrap lead, scrap batteries and cables. Trials are in progress on activated-charcoal-based equipment for separation of these substances. Cleaning of scrap batteries results in varying quantities of battery acid (sulphuric acid) entering the washing water. The washing water is contaminated with lead, antimony, cadmium, arsenic and zinc. Separate interception and treatment is necessary.

2.4 Non-ferrous metal semifinishing works

In semifinishing works, the **main problems of maintaining clean air** stem from the upstream format foundries. These use large amounts of **defined scrap** in addition to **primary metal** which may call for pyrometallurgical smelting refining (e.g. with chlorine gas compounds in the case of Al).

Oily and plastic-coated scrap produces soot, oil mist, chlorine- and fluorine-bearing acid mist and similar substances on being melted down. Formation of polyhalogenated dibenzo-dioxins and -furans cannot be ruled out. For this reason scrap should be precleaned in fuming furnaces with afterburning chambers; depending on the permissible level of purity, waste gases are to be cleaned in electrostatic precipitators and/or gas scrubbers.

Waste gas from melting furnaces can contain metal oxides, volatile metalliferous vapours and halogen compounds which must be separated in dust filters or waste gas scrubbers. Through **process automation** and the use of **additional reactors**, even low-capacity secondary smelting plants (2,400 t/a) can achieve **low clean gas emission values**, e.g. 5 mg/m³ dust, less than 1 mg/m³ fluorine compounds, by chemisorption combined with cyclone and fabric filter. Separation efficiency for chlorine compounds can be as high as 98%.

Cooling bays for gas-emitting dross and slag are also to be connected to **centralized waste air extraction systems.**

Alkaline or acid solutions should be used for **degreasing, cleaning and pickling metal surfaces.** Organic solvents containing halogens should be avoided. **Flushing water** and used pickling and washing liquids are to be **treated** in neutralization plants.

Sludge residues are either pyrometallurgically **processed** in a smelting plant or, if they contain no pollutants, **dumped. Vapours** from heated pickling and rinsing baths must be extracted, precipitated by gas scrubbers and **neutralized.** Polluted waste must be placed on protected **dumps** with collection of percolation water.

As non-ferrous metal semifinishing plants are frequently situated **close to residential zones,** consideration must be given to **noise reduction measures** and the necessary **distance.**

3. Notes on the Analysis and Evaluation of Environmental Impacts

Non-ferrous metal industrial operations using thermochemical or pyrometallurgical processes produce **considerable quantities of waste gases** laden with environmentally harmful substances. **Air protection measures** must therefore be a priority.

The following **examples** illustrate the possible **pollutant content** of the waste gases:

- aluminium smelting plant, toxic fluorine components in the anode gas

    raw gas approx. 10 kg F/t Al

- copper smelting plant, sulphur dioxide in the waste gas

    raw gas approx. 2.6 t SO2/t Cu

The values indicate that even in regions with low levels of existing pollution, waste gases from metal smelting plants must **on no account** be discharged **uncleaned. Wet** and **dry processes** are available for cleaning, dry processes being preferred for ecological and economic reasons.

Continuous monitoring involving measurements to verify the effectiveness of the separation systems is necessary both after erection of the plant and during its operation. Detailed descriptions for carrying out **emission and immission measurements** are contained in the guidelines of the German Association of Engineers *VDI*. In Germany the obligatory **emission and immission values** are detailed in TA-Luft (Technical Instructions on Air Quality Control).

In plants using **hydrometallurgical processes**, to reduce environmentally harmful substances to a minimum, **intermediate products and residues** must undergo repeated chemical treatment, filtration, electrolytic precipitation or scrubbing with subsequent neutralization. **Wastewater** from gas scrubbing or pickling plants may only be returned to receiving bodies of water once it has been chemically neutralized and freed of solids. **Guideline values** for permitted pollutant concentrations must be established for discharging wastewater in accordance with the state of art. **Reference values** may be obtained from the regulations in force in Germany. In every case care must be taken to ensure that **drinking water** and other water resources are not impaired. **Analytical processes** have been defined under German DIN standards to determine pollutant concentrations in wastewater; in Germany these are detailed in *Allgemeine Verwaltungsvorschriften* [General Administrative Regulations]. **Routine measurements** are also to be carried out to monitor the efficiency of water treatment and clarification plants. The scope of measurements and the inspection and maintenance intervals of wastewater - and waste gas - cleaning systems must be defined in an **operating manual**.

Contaminated material is to be **stored** in such a way as to prevent soil and groundwater contamination. Where possible, **single-purpose dumps** should be established, with sealing and percolation water collection and treatment systems which satisfy stringent requirements.

As in Germany, **works environmental protection officers** should be deployed in non-ferrous metal works who are totally independent of the production side. They are obliged to monitor adherence to the regulations.

In addition to monitoring external pollutant discharge, internal **work-places** must also be inspected for pollutant concentrations, noise and safety. Suitably qualified **safety officers** and a **works doctor** should be appointed for these purposes.

4.   Interaction with Other Sectors

Normal **annual production capacities** of newer non-ferrous metal smelting plants are between 50,000 and 100,000 t. Allowance must be made for future **capacity expansions**. Due to the quantity of **land occupied** and the **environmental pollution** involved, projects cannot be considered in isolation. As early as the initial location selection phase, existing **prior pollution** of air, water and soil must be taken into account, making adequate allowance for the **additional burdens** imposed by such an industrial complex. As early as the planning phase, and when defining permissible immissions, effects on the environment must be considered from the point of view of community development. Adequate distancing from the nearest residential zones is to be guaranteed. Further details are contained in the **environmental brief Planning of Locations for Trade and Industry**.

Raw materials for smelting plants have to be extracted in large quantities from underground or surface mines. The environmental briefs on **mining** provide information on the environmental impacts. Efficient transport routes are necessary for transporting charge materials and products. Details are contained in the briefs **Road Traffic, Railways and Railway Operation** and **Shipping**.

A special **secondary effect** of the use of electrolytic processes is that their profitability, and particularly that of an aluminium smelting plant depends on the availability of cheap electricity. Additional pollution results from the erection or extension of power stations and the associated construction, particularly of hydraulic engineering works (see environmental briefs **Thermal Power Stations** and **Power Transmission and Distribution**).

A single-purpose dump must be established for non-recyclable products and waste, including slag and furnace breakage material (see environmental brief on **Disposal of Hazardous Waste** and **Volume III, Compendium of Environmental Standards**).

5. Summary Assessment of Environmental Relevance

Processes and raw materials utilized in non-ferrous metal smelting plants for extracting aluminium, copper, lead and zinc, and also refining and smelting plants for further processing, produce **emissions and raw materials** which can pollute the environment. Of special significance are **heavy metals** which endanger health and in some cases are carcinogenic. In many countries this concerns especially the poorer sections of the population who are particularly at risk due to malnutrition and illness. The same of course applies to metal smelting plants other than those mentioned here.

Environmental damage can be reduced by **selecting locations** with relatively insensitive landscapes where there is unlikely to be any great effect on the regional productiveness of the natural environment. It is also necessary to **exclude regions** that are already heavily burdened with high existing or background levels of fluorine compounds and heavy metals. In this regard it should be noted that anthropogenic heavy metals are often more readily plant-available than lithogenic or pedogenic heavy metals.

Pyrometallurgical processes cause mainly **air pollution** in the form of gases, mists and dusts which must be minimized in gas scrubbers or returned for further processing. Apart from the ecological benefit, this form of **emission reduction** has the **economic benefit** of **recovering** valuable metals or producing sulphuric acids. Similar conditions exist for **secondary smelting plants** but with the additional problem of polluted charge materials. Depending on the operating conditions, halogen-bearing pollutants combined with organic materials are a particular potential source of polyhalogenated dioxin and furan emissions (waste gas emission concentrations of the order of nanograms).

Emissions and residues from **hydrometallurgical processes** on the other hand can pollute **wastewater** and **dumps. Recycling of water** in the circulation system is very important. Though it is state-of-the-art practice to recirculate **liquid process materials** such as acids, alkalies or solvents by regeneration, thereby **reducing residues**, these must be subsequently **processed** and the waste products **neutralized** in more or less costly stages to recover valuable metals and/or extract pollutants. Checks must be made in every case to determine whether **pollution of groundwater or surface water** is possible due to the storage or emission of primary, intermediate or end products. Pollutant yield and hence the necessary outlay for pollutant reduction is significantly lower in the case of semifinishing works.

A survey should be conducted in every case to determine whether **agricultural use of the land in the vicinity of the works** will be impaired by large-area pollution with phytotoxic and zootoxic heavy metals, especially zinc, copper, chromium, nickel and lead, taking into account long-term deposition, accumulation and reactivity in the soil. The environmental risk resulting from heavy metals in the soil must be distinguished according to the form of bonding of their elements which in turn depends on their origin.

Heavy metals, especially **cadmium**, can be injurious to human health through accumulation in the soil and in plants, with increased absorption through the food chain, leading in particular to kidney damage. **Preliminary calculations** of the expected additional environmental burdens are necessary for assessing these **indirect effects** via the air - soil - food chain. As a precaution, it is advisable to restrict agriculture in the immediate vicinity. **Conflicts** can be **avoided** or **diminished** by **consulting** the affected population groups at an early stage, possibly developing and planning new sources of employment. The question as to whether the **increased environmental pollution** poses **additional health risks** and hazards, e.g. for women and children (during pregnancy etc.) should be investigated, and **adequate medical care** provided. In addition to the pollution burdens, attention must also be paid to the **noise** emitted by the plant machinery. Depending on the plant design, noise levels as high as 125 dB(A) may be emitted. Noise levels can be minimized by **noise reduction measures** which are to be specified in a noise reduction plan. The wearing of **personal ear protection** must be obligatory in **workplaces** with noise levels in excess of 85 dB(A) and must be monitored.

For environmental protection measures to be effective, it is vital that personnel should be made **sensitive to the issues** and **receive appropriate training**. Although the smelting industry already has a range of **proven methods and processes** at its disposal for effective pollution control, their application can be **excessively cost-intensive** where pollutant emissions are too low for improvements to be economic but too high to be ecologically harmless. In these cases, bearing in mind the long-term effects of heavy metal pollution, one must give **considerable weight to the needs of environmental protection,** even putting this before the profitability of the individual plant.

The emphasis of current **development work** is towards totally enclosed **circuits** in the production system. The aim is to enclose the circuit to prevent harmful effects on the biosphere through ever better utilization of charge material, production of pure intermediate and end products without recourse to dumping, with improved emission protection and recycling of separated dusts and solids.

6.   References

Statutory provisions, regulations

Abwassertechnische Vereinigung (ATV): Arbeitsblatt R 115, Hinweise für das Einleiten von Abwasser in eine öffentliche Abwasseranlage, January 1983.

Allgemeine Verwaltungsvorschrift zur Änderung der allgemeinen Rahmenverwaltungsvorschrift über Mindestanforderungen an das Einleiten von Abwasser in Gewässer. GMBl (joint ministerial circular). No. 37, 1989, p. 798.

Erste Allgemeine Verwaltungsvorschrift zum Bundes-Immissionsschutzgesetz (Technische Anleitung zur Reinhaltung der Luft - TA-Luft) dated 27.02.1986, GMBl (joint ministerial circular). 1986, Ausgabe A, p. 95.

Zweite Allgemeine Verwaltungsvorschrift zum Abfallgesetz (TA-Abfall) Teil 1: Technische Anleitung zur Lagerung, chemisch-physikalischen, biologischen Behandlung, Verbrennung und Ablagerung von besonders überwachungsbedürftigen Abfällen, vom März 1991, GMBl (joint ministerial circular). No. 8, p. 139.

39. Allgemeine Verwaltungsvorschrift über Mindestanforderungen an das Einleiten von Abwässer in Gewässer (Nichteisenmetallherstellung). GMBl (joint ministerial circular). No. 22, 1984, p. 350 - 351.

Deutsche Forschungsgemeinschaft: Liste maximaler Arbeitsplatzkonzentrationen (MAK-Wert-Liste), 1990, Mitteilung XXVI, Bundesarbeitsblatt 12, 1990, p. 35.

Environmental Protection Agency (EPA): Effluent Guidelines and Standards for Non-Ferrous-Metals, 40 CfR 421.

GVBl. des Landes Hessen, Teil 1, 31.03.1982.

Ministerium für Arbeit, Gesundheit und Soziales des Landes Nordrhein-Westfalen: Umweltprobleme durch Schwermetalle im Raum Stollberg, 1975, Düsseldorf.

5. Novelle zum Wasserhaushaltsgesetz: Mindestanforderungen nach § 7a, BGBl. I (Federal Law Gazette I), p. 1529.

Technische Anleitung zum Schutz gegen Lärm (TA-Lärm) vom 16.07.1968, Beilage BAnz. (Supplement to the Federal Law Gazette) No. 137.

Unfallverhütungsvorschriften: Hauptverband der gewerblichen Berufsgenossenschaften, Bonn u.a. UVV-Lärm, VBG 121 of 01.01.1990.

VDI-Richtlinie 2262: Staubbekämpfung am Arbeitsplatz, December 1973.

VDI-Richtlinie 2285: Auswurfbegrenzung, Bleihütten, December 1975.

VDI-Richtlinie 2058, Blatt 3: Beurteilung von Lärm am Arbeitsplatz unter Berücksichtigung unterschiedlicher Tätigkeiten, April 1981.

VDI-Richtlinie 2560: Persönlicher Schallschutz, December 1983.

VDI-Richtlinie 2058, Blatt 1: Beurteilung von Arbeitslärm in der Nachbarschaft, September 1985.

VDI-Richtlinie 2102: Emissionsminderung, Kupferschrotthütten und Kupferraffinierien, Entwurf February 1985.

VDI-Richtlinie 2286: Emissionsminderung, Aluminiumschmelzflußelektrolyse, Entwurf January 1987.

VDI-Richtlinie 3929: Erfassen luftfremder Stoffe, Entwurf March 1990.

VDI-Richtlinie 2310: Blätter 30 und 31: Maximale Immissionswerte für Blei (Blatt 30) und Zink (Blatt 31) zum Schutze der landwirtschaftlichen Nutztiere, July 1991.

VDI-Richtlinie 3792, Blatt 3: Messen der Immissions-Wirkdosis von Blei in Pflanzen, April 1991.

Verordnung über Arbeitsstätten (Arbeitsstättenverordnung - ArbStättV) of 20.03.75, BGBl. I (Federal Law Gazette I), p. 729, 15 Schutz gegen Lärm.

Verordnung über gefährliche Stoffe, Gefahrstoffverordnung (GefStoffV) of 26. August 1986, BGBl. I (Federal Law Gazette I), p. 1470, in the version dated 23. August 1990, BGBl. I, p. 790.

Verordnung zur Bestimmung von Abfällen nach § 2 Abs. 2 des Abfallgesetzes of April 3 1990, BGBl. I (Federal Law Gazette I), p. 614.

Verordnung zur Bestimmung von Reststoffen nach § 2 Abs. 3 des Abfallgesetzes of April 3, 1990, BGBl. I (Federal Law Gazette I), p. 631.

Verordnung über das Einsammeln und Befördern sowie über die Überwachung von Abfällen und Reststoffen of April 3, 1990, BGBl. I (Federal Law Gazette I), p. 648.

Verordnung über Anlagen zum Lagern, Abfüllen und Umschlagen wassergefährdender Stoffe und die Zulassung von Fachbetrieben.

Scientific/technical papers

Bureau of Mines, Washington 1973, Control of Sulfur Oxide Emissions in Copper, Lead and Zinc Smelting.

Bußmann, H.: Stand und Entwicklung des Kupferrecyclings in: Fleischer, G., Abfallvermeidung in der Metallindustrie, p. 159 - 166, Ef Verlag für Energie und Umwelttechnik, Berlin 1989.

Corwin, T.K. et al: International Technology for the Nonferrous Smelting Industry, Noyes Data Corporation, Park Ridge NJ, 1982.

Dengler, H.: Behandlung schwermetallhaltiger Abwässer in: UTZ Materialien, 1989; Zentrum für Umwelttechnik beim Battelle-Institut Frankfurt am Main.

Deutsche Gesellschaft für Technische Zusammenarbeit (GTZ), Dornier-Studie: Erstellung eines Kataloges von Emissions- und Immissionsstandards, October 1984.

Gesellschaft Deutscher Metallhütten- und Bergleute, Hauptversammlungsvorträge, Stuttgart 1972, Umweltschutz in der Metallhüttenindustrie.

Grün, M., Machelet, B., Podlesak, W.: Kontrolle der Schwermetallbelastung landwirtschaftlich genutzer Böden in der DDR.

Hartinger, L.: Taschenbuch der Abwasserbehandlung für die metallverarbeitende Industrie, Carl Hauser Verlag, Munich 1976.

Kirchner, G.: Die Bedeutung von Sekundäraluminium für die Aluminium-Versorgung in: Fleischer, G., Abfallvermeidung in der Metallindustrie, p. 173-179, Ef Verlag für Energie und Umwelttechnik, Berlin 1989.

Kloke, A.: Orientierungsdaten für tolerierbare Gesamtgehalte einiger Elemente in Kulturböden, Mitt. VDULFA 1980, p. 1 - 3 and 9 - 11.

Koch, C.T., Seeberger, J.: Ökologische Müllverwertung, Verlag C.F. Müller, Karlsruhe, 1984.

Landtag Nordrhein-Westfalen: Plenarprotokoll 11/28 of 03.05.1991.

Lärmquellen der Eisen- und Metallindustrie: Berufsgenossenschaftliches Institut für Lärmbekämpfung, Mainz 1973.

Merz, E.: Minimierung der Belastung durch Metalle und Metalloide, Vortrag im VDI-Kolloquium "Krebserzeugende Stoffe in der Umwelt", 23.04.1991. Mannheim, VDI-Bericht in Vorbereitung.

Miehlich, G., Lux, W.: Eintrag und Verfügbarkeit luftbürtiger Schwermetalle und Metalloide in Böden, VDI-Berichte No. 837, 1990, p. 27 - 51.

Persönliche Mitteilungen: Wirtschaftsvereinigung Metall e.V., Düsseldorf, 1991.

Rademacher, K.D., Koß, K.D.: Wassergefährdende Stoffe, Springer Verlag, Berlin 1986.

Riss, A. et. al: Schwermetalle in Böden and Grünlandaufwuchs in der Umgebung einer Kupferhütte in Brixlegg/Tirol, VDI-Berichte 837, 1990, p. 209-223.

Röpenack, von A.: Integrierter Umweltschutz - die Aufgabe der Zukunft, Erzmetall, 44 (1991), No. 2, p. 67 - 74.

Spona, K., Radtke, U.: Blei-, Cadmium- und Zinkbelastung von Böden im Emissionsgebiet einer Zinkhütte in Duisburg, VDI-Berichte 837, 1990, p. 165 - 183.

Ullmanns Enzyklopädie der technischen Chemie, 4. Auflage, Band 6 (Umweltschutz), Band 7 (Aluminium), Band 8 (Blei), Band 15 (Kupfer), Band 24 (Zink) - 1974/1983.

Umweltbundesamt [German Federal Environmental Agency] Berlin, April 1978: Stand der Technik bei Primär-Aluminiumhütten.

Umweltbundesamt [German Federal Environmental Agency] Berlin, March 1980: Richtlinien für Emissionsminderung in NE-Metallindustrien, incl. ausführliche Bibliographie.

Umweltbundesamt [German Federal Environmental Agency], Berlin, March 1983, R. Fischer: Maßnahmen und Einrichtungen zur Reinhaltung der Luft bei NE-Metallhütten und Umschmelzwerken.

Umweltbundesamt [German Federal Environmental Agency] Berlin, 1986: Altanlagenreport 1986, p. 59 - 73.

Umweltbundesamt [German Federal Environmental Agency] Berlin: Jahresberichte 1986, 1987, 1990.

Umweltbundesamt [German Federal Environmental Agency] Berlin, 1989: Luftreinhaltung '88, Tendenzen - Probleme - Lösungen, Erich Schmidt Verlag.

Umweltbundesamt Vienna: Montanwerk Brixlegg - Wirkungen auf die Umwelt, 1990.

VDI-Kommission Reinhaltung der Luft: Schwermetalle in der Umwelt, Düsseldorf, 1984.

VDI-Berichte 837, 1990, p. 593 - 612.

Verein Deutscher Ingenieure, Bericht 203, 1979, Schwermetalle als Luftverunreinigung - Blei, Zink, Cadmium.

Williams, Roy E.: Waste Production and Disposal in Mining, Milling and Metallurgical Industries,

Miller Freeman Publ., San Francisco, 1975.

Trade and Industry

Mechanical Engineering, Workshops, Shipyards

51. Mechanical Engineering, Workshops, Shipyards

Contents

1.   Scope

The different branches of mechanical engineering are concerned with the **machining and processing of ferrous and non-ferrous metals.** This covers the whole range of **production processes,** which can be **subdivided** as follows:

A: <u>Metal cutting</u>

* drilling	* milling	* turning
* planing	* broaching	* sawing
* filing	* honing	* grinding
* lapping	* sandblasting	* chiselling

B: <u>Non-cutting processes</u>

Thermal bonding
* oxy-acetylene welding   * electric welding
* inert-gas-shielded welding   * submerged arc welding
* build-up welding

Thermal cutting
* oxy-acetylene cutting   * plasma cutting

Forming
* forging   * deep drawing   * bending

Dividing
* punching   * cutting   * shearing
* nibbling

Jointing
* riveting   * adhesive bonding   * soldering

<u>Surface treatment</u>
* surface cleaning
   * degreasing   * pickling
* surface coating
   * electroplating   * phosphating   * chromatizing
   * anodizing   * enamelling   *hot-dip galvanizing
   * painting, lacquering
* surface annealing

The **raw materials** used in these processes may have **high environmental pollution potential** (e.g. heavy metals), and **hazardous production materials** may be used (e.g. cleaning agents containing chlorinated hydrocarbons). At the same time, **vapours, heat** and **noise** are generated, together with various **waste products** and **wastewater,** leading to adverse effects on the environment and on man, especially in enclosed areas.

In **shipyards,** the main process is **welding.** This is made additionally hazardous by the fact that welders working on bulkheads often have to work in enclosed areas, which further aggravates the **health risks** discussed below.

2.   Environmental Impacts and Protective Measures

A **product** undergoes **numerous production stages** in the course of the **metalworking process.** The **environmental impacts** of these stages affect the **workplace** and hence the **people** working there. They also affect the **air, water** and **soil.**

Due to their proximity to the point of origin, it is the workforce who are most seriously exposed to the production hazards. In highly industrialised countries this is the subject of comprehensive **worker protection** rules. The **workplace hazards** are listed below, taking as examples the most important and environmentally relevant **machining processes.** This is followed by a description of the wider **environmental effects** including the problems of waste disposal.

2.1   Potential hazards of selected operations

2.1.1   Metal cutting

Machining processes such as drilling, milling, turning, cutting, honing, grinding etc. make use of **oils and oil preparations** for **lubricating** and **cooling** tools and workpieces, to prevent **overheating** and possible **melting** of the workpiece and tool. Oils are dosed by spraying or pouring systems at rates of up to 100 litre/min. in order to dissipate heat. The spraying of moving and sometimes very hot tools and workpieces produces **vapours containing droplets known as aerosols.**

Metalworking techniques require **appropriate coolants** which must combine **several different properties** (non-foaming, corrosion-inhibiting, non-decomposing etc.).

Such a wide range of properties can only be achieved through the addition of varying quantities of **chemical additives**. These are added to the coolants in the form of non-water- miscible cutting oils or water-miscible concentrates.

More than 300 individual substances are used as coolant components. The following table divides these into substance groups by areas of application.

Substance group	Reason for use	Examples
Mineral oil	Lubrication effect	Hydrocarbons with different boiling ranges; fatty oil; esters
Polar additives	Enhanced lubrication properties	Natural fats and oils of synthetic esters
EP additives	To prevent micro-welds between metal surfaces at high pressures and temperatures	Sulphurized fats and oils, compounds containing phosphorous, compounds containing chlorine
Anti-corrosion additives	To prevent rusting of metal surfaces	Alkano-amines, sulphonates, organic boron compounds, sodium nitrite
Anti-misting additives	To prevent breakdown of the oil and thus generate less mist	High molecular substances
Anti-ageing additives	To prevent reactions within the coolant	Organic sulphides, zinc dithiophosphates, aromatic amines
Solid lubricants	To improve lubrication	Graphites, molybdenum sulphides, ammonium molybdenum
Emulsifiers	To combine oils with water	Surfactants, petroleum sulphonates, alkali soaps, amine soaps
Foam-inhibitors	To prevent foaming	Silicon polymers, tributyl phosphate
Biocides	To prevent formation of bacteria/fungus	Formaldehyde, phenol, formaldehyde derivatives, cathon MW

A significant **increase in certain occupational diseases** has occurred parallel with the **introduction** of the coolants which are now commonplace. According to scientific findings, diseases of the skin and respiratory tracts and cancer may occur.

Where coolant use is unavoidable, **mist extraction** as close as possible to the point of origin or **encapsulation** is necessary. Consistent use must be made of personal protection measures such as **the wearing of protective clothing** and the use of special **skin protection substances**. Factories should produce **skin protection plans.**

Bacteria which can have severe **effects on health** can occur due to the organic nature of coolants. Bacteria formation is promoted by warm/hot ambient temperatures. **Anti-bacterial additives** are introduced to counter this. **Timely replacement** of coolants **avoids** the need for **high doses** of anti-bacterial additives, which also represent a health hazard. However, this increases the **total quantity of waste** to be disposed of. **Proper storage** of "spent" coolants and subsequent **separation** of emulsified oils and greases, and also of metal compounds and other components, is **imperative.**

Safety data sheets informing of the danger of coolants and **instructions for use** should be **displayed** in the national language(s). It is important that **staff** are **aware** of the long-term dangers of coolants; a particular difficulty here being the often creamy, pleasant smelling and seemingly harmless nature of coolants.

No generally applicable **limit values** exist for coolants in the breathing air. The only guide is the relevant **MAK values**[1] for the individual substances. The management should find out which are the most environment-friendly coolants and ensure that these are procured.

2.1.2   Cleaning and degreasing of workpieces

For subsequent surface treatment, adhesive or thermal bonding etc., workpieces have to be **freed** of substances such as oils, fats, resin, wax, cellulose, rubber or plastics. **Solvents** are widely used for this purpose. Workpieces can be **degreased and cleaned** by various methods, for instance by **cold, hot and/or vapour degreasing** or combined processes.

---

[1]   The term MAK (maximum allowable concentration) in Germany refers to the maximum possible concentration of a substance in the air of the workplace, in the form of gas, vapour or suspended matter.

Cold cleaning frequently involves the use at room temperature in open baths of **solvent mixtures** whose precise composition is not known to the user. Mixed with air, the vapours of these solvents or solvent mixtures can be **explosive**. Most solvents represent a **health hazard** for man.

Solvents are classified as **organic compounds** such as hydrocarbons, halogenated hydrocarbons, ethers (diethyl-ether, tetrohydrofuran, dioxan), ketones (acetones, methylethylketone) and **organic alkalis** (sodium hydroxide solution, ammonia) and acids (hydrochloric acid, nitric acid, sulphuric acid).

The most important halogenated hydrocarbons are **chlorinated hydrocarbons (CHCs)**, such as tri-, tetra-, perchloroethylene, dichloromethane, tetrachloroethane etc.[2] On account of their **grease-dissolving properties** and **high volatility**, CHCs are used in almost every branch of metal working as **cleaning agents** in cold cleaners and in hot degreasing. The **high volatility** ensures **quick drying** after cleaning, but also means it is necessary to **monitor solvent concentrations in the workplace**. Through skin contact and inhalation, CHCs can damage **mucous membranes, central nervous system, liver, kidneys and lungs**.

In addition, most **solvents** are inflammable and represent a particular pollution hazard for water.

Alternative processes use **alkaline aqueous solutions** (with surfactants and other washing components in varying concentrations) or **water** (high-pressure cleaning).

Apart from the need for worker protection, it should be remembered that practically all solvents seriously pollute the environment. Particular problems in this regard include damage due to solvent evaporation, **soil and groundwater pollution** and the **difficulties of disposing** of used solvents and solvent sludges.

Foremost among modern methods of **alleviating the problems of disposal** are primary measures to **prevent wastewater** occurring in the first place, rather than subsequently treating highly contaminated bath and flushing water before it enters the drainage system. Membrane filtration and ion exchange processes can be used to **regenerate process baths** and extend their useful life. Similarly, flushing water can be used several times over with **continuous dirt and oil separation** (recycling via ion exchangers, emulsion cracking and cascade flushing techniques). Resulting wastewater quantities and pollutant loads are reduced. One might also attempt to process and reuse solvents in a **closed solvent circuit**. This technique is rarely successful in the case of reprocessing **surfactants**, so the **improvement of their bio-**

---

[2]   The best known are CFCs (chlorofluorocarbons) used in other application e.g. as refrigerants. CFCs are partially responsible for the destruction of the vital ozone layer in the atmosphere. CFCs and carbon tetrachlorides and certain other chlorinated hydrocarbons are banned in Germany under the CFC halogen prohibition directive of 6 May 1991 and the chloro-aliphatic compounds directive.

degradability is an important factor. Management should optimize selection of solvents based on technical and environmental factors[3].

The following **precautions** should be taken where **degreasing** is carried out with **organic solvents:**

- do not use substances which are unknown;
- use enclosed equipment where possible;
- ensure effective ventilation and aeration of work rooms;
- ensure good extraction at the workplace;
- avoid skin contact;
- use protective equipment;
- as solvents are heavier than air, they force the breathing air out of trenches, cellars, containers and depressions in the ground; suffocation can be avoided by means of floor openings and ventilation;
- use only non-combustible washing vessels with self-closing covers for cleaning small parts with inflammable solvents;
- keep only quantities of flammable solvents at the workplace as are required for the work and store in suitable containers with effectively sealed covers;
- avoid electrostatic charges;

---

[3] Only wastewater experts can definitively optimise the choice of solvents. Information is available in: Dagmar Minkwitz, "Ersatzstoffe für Halogenkohlenwasserstoffe bei der Entfettung und Reinigung in industriellen Prozessen" (serial publication of the *Bundesanstalt für Arbeitsschutz* (German Federal Institute for Occupational Safety and Health) GA 38) Dortmund, Bremerhaven 1991 (Wirtschaftsverlag NW) ISBN-3-89429-086-2. See also "Zeitschrift Oberflächentechnik, Bezugsquellennachweis für die Oberflächentechnik mit Trendübersichten und Tabellen, Munich, 4th edition 1991) (Seibt Verlag), ISBN 3-922948-70-7.

- in operating manuals indicate the solvents used, limitations on use and safety precautions, instruct personnel;
- secure and lock installations when not in use;
- avoid hand-spraying of degreasing agents with spray guns;
- avoid blow-drying with compressed air of surfaces which have been treated with chlorinated solvents;
- with open degreasing equipment, note the quantities of solvent displaced on immersion of the workpiece and dimension the system accordingly;
- workpieces should leave the system free of solvents.

2.1.3 Painting

Most **spray paints** and **brush paints** contain considerable quantities of hydrocarbon and chlorinated hydrocarbon **solvents** (spray paints as much as 90%, normally 50 - 70%) which evaporate on spraying and drying. Paints also contain finely dispersed **pigments**. Some of these are **highly toxic**. Depending on the application, paints may have to satisfy a wide range of quality requirements. Available paint systems are accordingly diverse.

There are three possible ways of **avoiding solvent emissions** from painting installations; these can be used separately or in combination:

- use of **low-solvent** paints

    "High solids" paints, water soluble paints and dispersion paints have been developed for this purpose. A further alternative is solvent-free powder paint, for which new applications are constantly being found.

- use of **high-performance application methods**

    Solvent emissions depend not only on the paint formulation but also on the application method. An important evaluation criterion is the application efficiency factor, which is defined as the ratio of paint remaining on the product to the total quantity of paint used. Lower efficiency means higher paint consumption and thus higher solvent emission. Application efficiency is primarily determined by the process and by the form of the parts to be painted.

The following **application efficiency** values can be taken as guidelines for the painting of large surface areas by various methods:

-	compressed air spraying	65%
-	"airless" spraying	80%
-	powder painting	98%
	(with recycling of spraying loss)	
-	electrostatic spraying	95%
-	dipping, flooding	90%
-	rolling, pouring	approx. 100%
-	brush, roller application	98%

The **choice of application method** depends on certain **quality requirements**, e.g. coating thickness, surface roughness etc. and is hence closely related to the purpose of the painted object.

The various levels of **waste gas** resulting from the different methods can be greatly reduced by **enclosing** the application zone and additionally by **air circulation.** This will reduce the outlay on waste gas cleaning.

- **collection and cleaning of waste gas** (with solvent recycling).

2.1.4 Electroplating

To obtain different surface properties **(surface refining)**, workpieces are electroplated with chromium, zinc, tin, copper, cadmium, lead or brass. For this the selected **metallic coating** is deposited from an electrolyte solution in an **electrochemical process**. To enable the metal coatings to be applied, **workpieces must be cleaned and degreased.**

Where cold cleaning and degreasing are carried out, the **hazards from cold cleaners** must be taken into account (see 2.1.2). The **boil-off** technique is also used for rough cleaning. Strong alkalines such as sodium or potassium hydroxide solutions are used for this purpose. These alkalines can damage eyes, skin and respiratory tracts if splashed or given off as mist or dust. An electrolytic process is often used for subsequent **fine cleaning**. Electrolytes are often alkaline solutions (5% sodium hydroxide solution) or cyanidic salts. Apart from the dangers posed by the boil-off technique, **extraction ventilation** is necessary in view of the large quantities of **hydrogen** produced, so as to avoid reaching the **explosibility limit** of the air-hydrogen mixture. **Safety in the workplace** is increased by installing **gas warning devices.**

Pickling degreasers and pickles are used to remove oxidation layers and casting or rolling scale from metal surfaces. These are acids (sodium hydroxide solution for aluminium) such as sulphuric, hydrochloric, phosphoric, hydrofluoric or nitric acid which **attack and dissolve the workpiece surface.** The main health hazards are skin diseases; **dangerous vapours and gases** can be inhaled in the case of inadequate extraction. Especially dangerous are nitrous gases which can occur when using nitric acids, also fluorine compounds from hydrofluoric acid and hydrogen chloride from hydrochloric acid.

Cyanides are used for cleaning in salt baths (fluorides), pickling (removal of thin surface films), with chemical and electrolytic polishing or burnishing, and also with surface coating and thermo-chemical hardening processes. These can cause **hydrocyanic poisoning** as well as **skin diseases** when solutions containing cyanide come into contact with acids. Therefore baths containing acids and cyanide must be **covered** and separated by **partitions. Containers and equipment are to be clearly marked** to prevent carry-over of substances which can mutually react. In all cases one should check to determine whether cyanide can be **replaced** by substances less hazardous to health.

The actual electroplating of the workpiece can be done in **countless different process variants and stages.** Materials posing just about every conceivable danger can be used in electroplating. The **dangerous properties** result both from the main components of the bath fluid and from different additives such as emulsifiers, foam inhibitors and wetting agents.

Strong aerosols can occur during bath **filling** and further preparation. Dangerous substances may enter the breathing air due to the production of gas (hydrogen) during the electrolytic process.

The main hazards with **coatings** are **skin complaints** and in particular allergies due to nickel and chromates. If consumed, both nickel and chromates can be **carcinogenic.** Nickel in fluid particle form is subject to a **TRK value**[4] 4 of 0.05 mg/m^3 breathing air.

---

[4]   TRK value: German technical directive on concentration of carcinogenic substances

2.1.5   Welding

Welding is the **joining of materials** using heat and/or force, with or without the use of welding fillers (anti-oxidation substances).

The **individual processes** most commonly used are gas welding, arc welding and inert gas shielded welding.

Polluting factors in welding workshops are:

-       chemicals in the generated gases, vapours and dusts

-       high temperatures (approx. 3,200°C - 10,000°C)

-       radiation   (ultraviolet radiation): eye damage, severe inflammation of unprotected skin;

                        infrared radiation: can penetrate the vitreous body of the eye, reaching the retina and causing cataracts)

-       noise (up to 110 dB(A))

Diverse hazards occur depending on the fuels, inert gases, filler materials, work-piece coatings etc. in use. The following table summarizes the pollutants occurring with the different welding methods. The **carcinogenic** and **mutagenic** elements chrome and nickel are especially relevant. Certain hazardous elements are detectable in welding fumes in concentrations of over 1% and can lead to health damage. Clinical and epidemiological investigations indicate a frequent occurrence amongst welders of **chronic bronchitis** and increased **impairment** of the respiratory tracts.

Pollutants occurring in various welding processes include:

Pollutant		Causes	Welding process	MAK * mg/m^3
Lead	PbO	Welding of lead or lead-coated workpieces	all	0.1
Chromium	$Cr_{2/3}$	Welding with alloyed electrodes, Cr Ni steel	all	
Cadmium	CdO	Cadmium-coated workpieces	all	0.05
Carbon monoxide	CO	Welding with basic coated electrodes, gas flame	all	30
Carbon dioxide	CO_2	Gas welding with coated electrodes, inert gas	all	5000
Copper	CuO	Welding of copper, copper-coated workpieces	all	0.1
Manganese	Mn O	Welding of workpieces containing Mn, all electrodes	all	5
Nickel	NiO	Welding of Cr Ni steel, alloyed electrodes	all	
Nitrogen	NO_2	Welding in confined spaces, trenches, tanks	all	9
Zinc	ZnO	Welding of zinc, galvanized workpieces, zinc paint	all	5

*   MAK: maximum allowable concentration

Pollutant		Causes	Welding process	MAK mg/m^3
Aluminium	Al$_2$O$_3$	Welding of aluminium, almost all types of electrodes	Arc welding	-
Iron	Fe$_2$O$_3$	Welding of steels, all electrodes	Plasma arc welding	8
Fluorides	F	Welding with basic and alloyed electrodes	Arc welding	2.5
Calcium	CaO	Welding with coated electrodes	Arc welding	5
Sodium	Na$_2$OH	Welding with coated electrodes	Arc welding	2
Oxygen (ozone)	O$_3$	Strong UV radiation	Plasma arc welding	0.2
Titanium	TiO$_2$	Welding with coated electrodes	Arc welding	8
Vanadium	V$_2$O$_3$	Welding workpieces containing vanadium	Arc welding	0.5

The **welding of metals** with anti-corrosion coatings may also have adverse toxicological consequences. Pollutants may be released depending on the type of coating.

Alkyl resins:	acrolein, butyric acid
Phenolic resins:	phenols, formaldehyde
Polyurethane:	isocyanates, hydrogen cyanide
Epoxy resins:	phenols, formaldehyde, hydrogen cyanide.

Although the **inert gases** carbon dioxide, argon and helium are **not toxic**, in poorly ventilated rooms they can **displace the breathing air** and under extreme conditions cause **suffocation. Ozone** may be produced during arc welding. Even low concentrations (0.1 parts per million (ppm)) of ozone can cause irritation to the eyes and upper respiratory tracts and in the event of exposure to 5-10 ppm over several minutes, **pulmonary oedema.**

At high temperatures **nitrogen oxides** are formed and emitted from the nitrogen and oxygen in the air on the periphery of the welding flame. Nitrogen oxides are highly toxic and, after a relatively long asymptomatic period, can lead to **radical lung changes**, pulmonary oedema and death. If the workpiece has been degreased with solvents containing chlorine and not properly dried, **phosgene** may be produced during welding. Phosgene is **highly toxic** and can also cause **pulmonary oedema** after a long asymptomatic period.

Since the **welding of plastics** is not yet widespread in many countries, it is not dealt with here. It must be pointed out however that the **hazards** for man and the environment are also **considerable** with the welding of plastics. Protective measures and special disposal procedures are necessary to guard against the release of solvents and other waste gases containing pollutants.

2.1.6 Soldering

Soldering is the **thermal joining of two materials** using a material (solder) whose melting point is below that of the workpiece.

If the solder melts above 450°C the process is termed **"hard soldering or brazing"** and at lower temperatures **"soft soldering"**. Apart from additional hazards due to the base material binders, the **hazards** involved in soldering are mainly associated with the **flux** and the **solder.**

The **composition of a flux** depends on the base material, the solder and the intended use. More than 300 different types of flux are currently available, all of which contain **aggressive chemicals.** Soldering paste usually contains colophonium, talc and salmiac, while soldering fluid contains zinc chloride or tin chloride. Chlorine and chlorine compounds cause irritation of the respiratory tracts and the skin and, in high concentrations, lung damage. Fluxes also frequently contain fluorine compounds (irritation of the respiratory tracts, burns). Fluxes often contain substances responsible for allergies. These are mainly colophonium and hydrazine. **Hydrazine** is additionally classed as **carcinogenic.**

Tin-based solders containing lead are used for **soft soldering** and silver solder containing cadmium for hard soldering. **Flux vapours** carry metal particles which can be inhaled.

Environmental protection measures to combat the emission of liberated gases and component substances of solders and fluxes include the **installation of extractor systems** with downstream **separator filters** (cyclone method). This method may also be used to contain the environmental impact of the production stage discussed in the next section.

2.1.7   Grinding

Grinding is the **cutting** of a workpiece with a geometrically undefined cutting process.

Grinding processes are **characterised by high temperatures, workpiece removal** and **abrasive wear.** In addition to **noise**, the health hazards from grinding are principally **emissions from abraded dust or particles** from the abrasive tool, workpiece and any coating, and - in the case of **wet grinding** - from **coolants.** There is an attendant risk of **health disorders** especially affecting the skin and respiratory tracts. Additives in coolants and the metal dusts produced (e.g. from chromium, cobalt, nickel or beryllium) can result in **allergies.** These metals may also be **carcinogenic.** The following table shows the potential pollutant sources when grinding metals.

Potential pollutant sources in the grinding of metals

Material-dependent	**Process-dependent**
Grinding tool	Formation of superfine dust with
- abrasive material containing zircon	- profiling and dressing of grinding discs
- lead chloride, antimony sulphide in separating cutting in stationary operation	- tool grinding
	- fettling, due to adhering mould residues
- additives in grinding belts containing fluorine	- manual grinding, usually carried out without extraction ventilation
Coolant additives, with respect to toxicity, carcinogenicity and mutual reactivity	- coarse grinding
	- use of magnesite binders

Material containing	Combustion and pyrolysis products which can occur with the thermal decomposition of rubber or synthetic resin
- more than 80% %/wt. nickel (e.g. depositing materials)	
- less than 80% %/wt. nickel (e.g. high grade corrosion-resistant steel)	
- Lead (e.g. in automatic steel)	Accumulation of heavy metals and superfine particles in coolant due to inadequate filtration or overuse
- Cobalt (e.g. hard metal, co-alloys)	
- Beryllium (e.g. Ni Be alloys)	
	Atomization of coolant and thus also of additives, reaction products, dissolved heavy metals and superfine, non-separated particles.

Protective measures include **environment-oriented** selection of grinding tools, coolant and - where possible - materials, **extraction** of the abraded materials and **personal breathing and hearing protection.**

2.2   Mechanical engineering and operation of workshops and shipyards

Special environmental problems which are not found elsewhere arise in mechanical engineering, in workshops and in shipyards. This is because the work is not carried out in one location alone, and because **pollutants are diffused and vaporised** throughout the site. Estimation of their environmental relevance is difficult due to their often low concentration and they frequently seem harmless, so it is not easy to communicate the problem to workers and managers. Therefore **environmental training measures** should be taken into account as early as the **planning phase.** Much depends on the **attitude** in the workplace, the **choice** of working equipment and materials and the **observance** of worker protection measures. Planning must additionally include **early integration** of **technical environmental protection measures** (filter systems, wastewater collection installations, cleaning installations etc.).

2.2.1   Waste air

Environmentally relevant **waste air flows** can be released into the environment through **forced ventilation** (e.g. fan systems) and/or random emission[5] from diverse areas of the site.
These include **emissions from:**

-       production extractor systems
-       workplace extractor systems
-       room air extraction
-       production processes
-       mechanical cutting
-       thermal joining and parting (welding, cutting)
-       joining (e.g. bonding, soldering)
-       surface treatment (cleaning, coating, hardening and tempering)
-       drying

Emissions into the air can be **divided** into:

-       coarse and fine dust
-       aerosols
-       organic and inorganic gases and vapours

Harmful components of the waste air are essentially:

-       organic solvents and halogenated hydrocarbons from metal cutting
        (coolants), cleaning, degreasing, bonding and painting of workpieces in the
        form of gases, vapours and aerosols
-       dusts from the mechanical processing of materials

Whether or not cleaning of the waste air is an absolute necessity depends e.g. on the solvents in use, the presence of other contaminating operations, weather conditions etc., also therefore on **ambient factors.** Long-term **risks** for man and the environment may be posed even by relatively small workshops.

---

[5]   Emissions are defined as air impurities (gases, dusts), noise, radiation (heat,
        radioactivity etc.), vibration and similar phenomena given off to the environment by
        a (fixed or mobile) system.

In the interests of **worker protection,** room air pollutants occurring in the production process must not exceed certain **MAK values**[6]. Where necessary work should be carried out in **enclosed equipment. Efficient aeration and ventilation** must be guaranteed, or pollutants must be extracted at the point of origin. Extracted flows of (pollutant) substances are to be cleaned by suitable processes before being expelled to the environment.

Possible processes are:

- **Dust separation:**

Dust is a mixture of particles of different grain size, particle size depending very much on the process. **Various processes** are used for dust separation. These are classified as follows:

A: inertia separators (cyclone, "multiclone", mechanical separator)
B: wet type separators (scrubbers, wet separators)
C: electrical separators (dry and wet electrostatic filters)
D: filtering separators (fabric filters, cloth filters, bag filters, vibratory sheet filters and tubular filters).

- **Aerosol separation:**

Waste gases containing droplets are also termed aerosols and hence distinguished from dust-laden waste gases. Droplets can be separated using the same physical principles as for dust. The **greater adhesion** of the separated droplets compared with dust however rules out use of the principal dust separators such as electrostatic filters and filtering separators. Only **wet separators,** i.e. scrubbers and wet type electrostatic filters are suitable without modification for separating aerosols.

- **Separation of vaporous or gaseous substances:**

The principal methods for reducing emissions of gaseous inorganic and organic substances are absorption, adsorption and thermal processes. With **absorption** the gaseous air pollutant is absorbed by a washing fluid. Absorption is either **physical or chemical**, depending on whether the absorption is based exclusively on the solubility of the gas, or whether additional chemical reactions occur in the liquid phase. Absorption processes, and also thermal and catalytic processes, are used particularly for reducing levels of organic substances.

---

[6]   in Germany

Water-soluble organic substances, e.g. methanol, ethanol, isopropanol and acetone can be effectively separated from waste gases through absorption by means of **scrubbers**. The contaminated washing fluid can normally be regenerated by fractionation[7].

Separation of **large amounts of solvents** is done by the **condensation process**. Recently, **biological processes** such as biofilters or biowashers have also become popular for cleaning waste gases with highly **odorous components** and/or solvents.

Adsorption is the **attachment** or accumulation of foreign molecules on the surface of a solid (adsorbent). **Regeneration** of laden adsorbents is normally done by desorption of the adsorbed substances in the gas or liquid phase (so-called desorption phase), i.e. by reversal of the adsorption process. As the desorption phase (usually a gas) contains the substance removed from the waste gas in an **enriched concentration,** recycling or reprocessing is possible. **Solvent recycling** is an especially important area of application for the adsorption process. Adsorbents are mostly activated carbons.

The **residual substances** yielded by the separation of the solid and gaseous waste gas pollutants (filter dusts, scrubbing water residues etc.) are normally **hazardous materials** and must be disposed of as **special waste** (giving rise to waste problems). The price of solving emission problems is often soil and water contamination, and land may become so contaminated it will eventually have to be rehabilitated (see also the environmental brief Disposal of Hazardous Waste).

2.2.2 Wastewater

In mechanical engineering, the **recycling** of process materials from wastewater is frequently only possible with **disproportionate technical effort** or not at all, because of the low concentrations involved. Concentrated liquid and spent process and production materials can and must be collected and disposed of as **(hazardous) waste.**

---

[7]   Fractionation is the separation of fluid mixtures by repeated distillation.

Wastewater is returned to natural bodies of water (lakes, streams, rivers, sea) after **preliminary and final cleaning**. There, any **inorganic pollutants** will lead to poisoning and deposits. **Organic impurities** may also be toxic and/or non-degradable. Degradable, non-toxic waste substances damage the environment by initiating excessive growth (eutrophication) of bacteria and minute life forms (algae, fungi) due to the nutrient supply. In combination with cell metabolism, this results in high **oxygen consumption** and finally to phenomena such as the "overturning" of the water (anoxic waters).

Heavy metals mostly enter the wastewater as metal salts produced by chemical reaction of the metals with the acids. The acidic pH value which prevails in heat treating and pickling shops promotes the **solubility** of the heavy metals in wastewater and hinders their removal.

Heat treating and **pickling shops** rinse workpieces in fresh water prior to further processes. After use, pickling fluids also contain heavy metals. In **electroplating** operations, rinsing water contains cyanide and is polluted with the heavy metals which are used (depending on the type of surface finishing).

Halogenated hydrocarbons are **insoluble** in water. They mainly enter the wastewater via **rinsing water** after degreasing in surface treatment plants and when cleaning engines and other objects in motor vehicle and general workshops by the use of **cold cleaners** and **pickling agents.** Further emission sources are coolant carry-overs and losses, workpiece rinsing and workshop floor cleaning.

Organic solvents can enter the wastewater via absorption and spray cleaning processes. **Mineral oils** occur with the cleaning of workpieces and floors and degreasing, and through losses during processing. Emission sources are repair, motor vehicle, factory and maintenance workshops. In surface treatment workshops they occur in the form of the workpiece anti-corrosion and rust-protection oils used in preliminary cleaning.

Acids and alkalis enter the wastewater in pickling shops and heat treatment shops in connection with degreasing. Other wastewater burdens occur due to nitrogen (ammonium) and phosphor compounds (phosphates from pickling shops).

Wastewater can be cleaned by **chemical, physical** and biological processes or a combination of these. **Three-stage purification** of industrial wastewater is now generally considered the state of the art.

Only **organic** and **non-toxic wastewater impurities** can be removed **biologically.** Tests in the laboratory will determine whether or not the components contained in the wastewater inhibit biodegradability.

With **biological processes** a distinction is made between **aerobes** (with oxygen) and **anaerobes** (without oxygen). With high burdens (chemical oxygen demand (COD) in excess of 15,000 mg/l), anaerobic processes are used for preliminary cleaning before aerobes carry out the final cleaning, since otherwise the oxygen supply costs are excessive.

High-performance biological processes with high pollutant decomposition rates are now available to build small but nevertheless efficient systems. Newly developed processes are achieving success in the biological neutralisation of organic pollutants previously considered non-biodegradable, e.g. CHCs, by optimising the living conditions for special bacteria.

Flocculation/precipitation processes can be used to **remove heavy metals from wastewater,** also **sedimentation processes** in the case of undissolved wastewater. Chemical oxidation and precipitation processes can be used for the removal and de-toxification of cyanide.

Emulsions originating from the use of coolants can be separated by the **membrane filtration process** in conductive wastewater (approx. 90%) and concentrate.

Ultrafiltration is used with **electrostatic immersion painting** for the separation of solid paint residues. This is increasingly replacing simple sedimentation with the separation of undissolved pollutants in wastewater because it is more efficient, though more costly. **Wastewater containing acids and alkalines** must pass through **neutralization systems. Ion exchange systems** cannot selectively remove metals but are highly suitable for cleaning water carried in a circuit and **recycling raw materials.** For recycling pure raw materials, the different waters must be carried and used separately.

2.2.3 Waste matter

Waste matters generated by these plants can be divided into **three groups:**

A. **Residues of the used raw materials.** These include both ferrous and non-ferrous (NF) waste (scrap/chips and swarf) which may be highly contaminated with coolant, cutting oil and leaked lubricating oil.

B. Waste from process residues resulting from the processing of semifinished products and auxiliary materials. **Metalliferous residues** are e.g. burnt slag from torch cutting, metal sludges, used salt and acid baths from electroplating or pickling shops.

C. **Non-metalliferous** waste can be paint and adhesive residues, oil and oily waste, organic acids, alkalis and concentrates. Finally, waste may also be produced by wastewater and waste air **cleaning processes.** These include purification sludges from the works' own sewage treatment plant plus dusts and sludges from the cleaning of waste air and extraction flows in the form of filter residues.

Nearly all waste in the **second and third groups** can be regarded as **hazardous waste.** They **demand special monitoring** and special disposal methods. Waste from the **first group** should mostly be **recycled. Separate collection** of scrap types (structural steel, alloyed steel, NF metals) in different containers is important for simple and comprehensive recycling.

In order to **reduce scrap quantities** with torch cutting and punching, care must be taken to achieve a **systematic geometrical arrangement** of contours on the sheet metal. **Recycling** should be considered where there are high concentrations of costly raw materials in liquid or sludge waste. To **reduce waste** further, **fluids** should where possible be **cleaned** with filters or **baths regenerated.**

2.2.4 Soil

The **effects** on the soil can be **problematic** in terms of both **quality** (e.g. toxicity or persistence) and **quantity** (e.g. acidification or leaching). Airborne emissions are normally small in quantity, therefore the main causes of pollution are the discharging of residual and waste materials (filter dusts, washer and scrubber residues, purification sludges) and improper handling of auxiliary materials. Of the large number of **chemical substances** used in metal processing, only a **few substance groups** have to be regarded as representing a **soil hazard** and thus in general also **a groundwater hazard :**

-   anions (chlorides, sulphates, ammonium, nitrates, cyanides etc., produced e.g. in heat treatment and pickling shops)
-   heavy metals (lead, cadmium, chromium, copper, nickel, zinc, tin etc.).

- solvents (halogenated and pure hydrocarbons)
- other oil-containing substances

The areas in which contamination occur are:

- all production stages using the named substances
- storage of new and used chemicals
- transport, loading and unloading on the works site (containers, tanks, pipe-lines, extraction system)
- cleaning and repair processes.

To **protect against contamination** in these areas, the **ground** must be **"sealed"** (i.e. provided with a protective layer to prevent penetration of the materials into the ground, or with pollutant collection devices, e.g. containment basins). Insufficient attention is often paid to the **storage** of hazardous materials. This can result in **severe environmental pollution** with **long term consequences**, also and in particular for third parties (e.g. due to groundwater pollution). Containers and pipelines used for **transporting** the materials are to be regularly **checked for leaks**. Care must be taken to ensure an efficient flow of work and materials, with clear rules on the depositing and disposal of waste/residues (see environmental brief Disposal of Hazardous Waste, also the reference literature).

2.2.5 Noise

Deafness and **loss of productivity** result from **noise pollution** above a certain level. A noise level in the workplace of around 85 dB(A) or higher for the greater part of the working shift, over a number of years, is regarded as detrimental to hearing[8].

For **comparison:** Leaves blown by a light wind emit a noise level of 25 to 35 dB(A); normal conversation is between 40 and 60 dB(A). Note also that the medium and higher frequencies between 1,000 and 6,000 Hz are the most damaging.

---

[8]   It is as harmful to be exposed constantly to a uniformly low noise level as to a higher one for a short time.

When considering noise immissions[9], a distinction must be made between the direct **effect on workers at the workplace** and the indirect effect due to radiation and immission in the **environment**. In assessing noise therefore, **three aspects**, each requiring **different reduction measures** must be considered.

A.   Noise origination

B.   Noise propagation

- Noise transmission (propagation of sound waves in different media, e.g. transmission of machine vibrations to foundations);

- Noise radiation (stimulation of air vibrations by solid-body vibrations - loudspeaker diaphragm principle).

During operations in this sector, noise is generated by machinery, by hammering, nailing, or chipping, by internal transportation processes, impact upon depositing or lifting of semifinished products, air and gas movements, fan outlets, pneumatic components, cutting torches etc.

A **fan**, e.g. with 50 kW, 970 rpm and a diameter of 1,800 mm, without noise damping produces a noise level of 100 dB(A). A **compressed air jet** produces a noise level of 108 dB(A) with an air pressure of 5 atmospheres. Welding and cutting generate noise levels of up to 101 dB(A) and **pneumatic riveters and chippers** produce between 100 and 130 dB(A). **Manual grinders** develop up to 106 dB(A). **Metal band saws** develop up to 106 dB(A). **Turning** generates between 80 and 107 dB(A). **Screw presses** produce noise levels up to 103 dB(A).

Noise pollution in the neighbourhood of a factory is mainly caused by **radiation** through the walls of the production sheds and buildings and by outward blowing **fans**.

Structural measures to reduce noise should therefore be taken into account as early as the **planning phase** (noise-absorbing walls, choice of windows, type of building materials etc.). The expected noise conditions cannot be determined simply by adding together the known noise levels of the planned machines and processes. Due to interaction and the different damping and reflection circumstances, only **on-site measurements** can yield accurate data on noise conditions. The **maintenance of adequate distances** reduces the effect on the **neighbourhood**.

---

[9]   The term immission is the effect of air impurities, radiation (e.g. thermal radiation) on man, animals, plants and property.

With regard to **noise protection,** a distinction is made between **primary** and **secondary measures.** Active primary measures signify the use of **machines constructed according to low-noise principles.** For example, sheet metal forming can be made quieter by replacing impact methods with hydraulic pressing. Priority should be given to the implementation of active primary measures.

Active secondary measures are **sound insulation** (prevention of propagation by obstacles) and **sound absorption** (absorption of sound energy and its conversion to heat). A distinction is made between structure-borne and airborne noise:

-       **Insulation of airborne noise** is achieved by partition walls, full or partial enclosure, cladding or screening.

-       **Insulation of structure-borne noise** can be achieved by machine feet of elastic material which prevent the transfer of vibrations.

-       **Absorption of airborne noise** over large areas can be achieved with sound-absorbing cladding material of foam or glass fibre matting. Silencers should be fitted to reduce noise at gas and air outlets. Composite silencers combining an absorber and a resonator should be provided for dust-laden gases.

-       **Absorption of structure-borne noise damping** is achieved by means of soundproof coverings in the form of foam rubber mats on sheet metal or in sandwich form (metal - covering - metal).

Passive noise protection signifies all equipment and measures for preventing the immission of noise and vibration to the environment and the human ear. These include personal ear protection, noise protection for control rooms, noise-insulated cabins etc.

Workers must wear **ear protection** in the workplace where noise levels are higher than **90 dB(A).** Such workplaces must display suitable **warning signs;** the observance of protective measures must be **monitored.**

Proven methods of reducing noise immissions include the use of **soundproof walls or partitions** and **increasing the distance between industrial buildings** and residential areas. With uninhibited propagation the acoustic power level is reduced by 3 (house wall) or 6 (point source of noise) dB(A) by doubling the distance.

3. Notes on the Analysis and Evaluation of Environmental Impacts

This section describes underlying reference material which, unless otherwise indicated, refers to the situation prevailing in Germany. Obviously these rules cannot be applied wholesale to other countries without modification. The material is intended at least to serve as a **reference** where no national regulations are available. The **INFOTERRA National Focus Points of the UNEP** are a valuable source of information. These contain environmental information records for the member state in question. The reference service is free of charge. The **Environmental Guidelines of the World Bank** are an important source of **application-related information**, e.g. for dust emissions, waste matter and wastewater.

The Catalogue of Environmental Standards **(Vol. III of this Environmental Handbook)** also deserves special mention. This lists standards and limits for assessment purposes.

Regulations on the protection of persons from danger and injury in the workplace (worker safety, industrial medicine) are contained in the *Unfallverhütungsvorschriften der Berufsgenossenschaften* (**accident prevention regulations of the employers' liability insurance associations**) and in their other publications such as *"Sicherheitsregeln"* (**safety regulations**) and *"Richtlinien"* (**guidelines**). Also worthy of note are the **Occupational Health and Safety Guidelines** of the World Bank and the **Encyclopedia of Occupational Health and Safety** of the International Labour Organisation (ILO).

3.1 Air

TA-Luft (Technical Instructions on Air Quality Control) regulates the technical standards concerning pollutant emissions and immissions for installations subject to licensing.

The **guidelines on maximum immission concentrations (MIK)** published by the Association of German Engineers (VDI) lay down limits for certain air pollution levels. These are defined as the concentrations in the ground-level open-air atmosphere or in dust and the quantities precipitated on the land below which **man, animals, plants and property are guaranteed to be safe** according to the present level of scientific knowledge (see also reference to **MAK values**).

Also of importance are the **EC Directives on sulphur dioxide and suspended particulates, lead and nitrogen dioxide** (EC Directives 80/779/EWG, 82/884/EWG, 85/203/EWG), also the **WHO Air Quality Guidelines** for 28 chemicals on the basis of toxicological findings.

3.2 Wastewater

Discharge conditions for wastewater are laid down in the *Wasserhaushaltsgesetz* (**Federal Water Act**), the *Abwasserangabengesetz* (**Wastewater Charges Act**) and the associated *Verwaltungsvorschriften* (**administrative regulations**). These prescribe limits of individual pollutants for different sectors.

Annex 40 currently applies to metalworking and processing with direct discharge (discharge into bodies of water). This gives details of the maximum concentrations for COD[10], BOD[11], heavy metals, hydrocarbons, ammonium, phosphorous and halogenated hydrocarbons.

The limits for **indirect discharge** (discharge into waste-water purification plants) for COD and substances classified as non-hazardous are less strict. The limits for indirect discharge are detailed in the **ATV work sheet** *Arbeitsblatt A 115* according to sectors. This work sheet is currently being adapted to the new wastewater management regulations.

For their projects the World Bank stipulates that the **temperature** of **discharged wastewater** must be no more than 3°C higher than that of the receiving body of water. If the temperature of the receiving body is 28°C or less, the temperature of the discharged wastewater must be no more than 5°C above that of the receiving water.

3.3 Waste matter

The definition of waste types according to Section 2 of the *Abfallgesetz* (**Waste Avoidance and Waste Management Act**) for the metal working and processing industry is contained in the *Verordnung zur Bestimmung von Abfällen* (**Regulation on Waste Definition**). A further designation of waste types according to code numbers is contained in the **publication on waste types of the German state working group on waste** *Landesarbeitsgemeinschaft Abfall (LAGA)*. Of relevance here are waste groups 35 (metal waste), 51 (oxides, hydroxides, salts) with electrolytic sludges, 52 (acids, alkalis, concentrates) with waste from surface treatment, 54 (mineral oil products) and 55 (organic solvents, paints, lacquers, adhesives, cements and resins). For electroplating and painting, refer to the **regulation on waste identification** *Abfallnachweis-Verordnung*. For motor vehicle and general workshops, see the **regulation on waste oil** *Altölverordnung*.

---

[10]   COD: Chemical Oxygen Demand
[11]   BOD: Biological Oxygen Demand

For the use and handling of special waste, refer to the three-volume edition of **"The Safe Disposal of Hazardous Wastes"** and the manual **"Techniques for Assessing Industrial Hazards"**, both published by the World Bank; also the environmental brief **Disposal of Hazardous Waste.**

3.4   Noise

Accident prevention regulations are published by the employers' liability in-surance associations or *Berufsgenossenschaften*, and these deal with the protection of workers from noise in the workplace. The **noise protection and information sheets of the *Hauptverband der gewerblichen Berufsgenossenschaften*,** are important publications in this regard. The Association of German Engineers (VDI) has issued numerous regulations and directives concerning noise in the workplace, the impact of noise on the environment and noise protection measures. Guidelines on noise protection for installations subject to licensing (according to the implementing ordinance of the Federal Immission Control Act) and neighbourhood immission guidelines are contained in **TA-Lärm (Technical Instructions on Noise Abatement).**

4.   Interaction with Other sectors

Mechanical engineering and the production of semifinished products for machinery in other sectors represent a **highly diversified capital goods industry,** so there is often close interaction with other sectors. Interaction is not necessarily regional, on account of the high specific added value, therefore interacting environmental problems do not occur generally but rather in individual cases.

With mechanical engineering works, workshops and shipyards of a certain size, attention must be paid to the **effects on infrastructural sectors.** In this regard see the environmental briefs Spatial and Regional Planning, Planning of Locations for Trade and Industry, Overall Energy Planning, Water Framework Planning, Urban Water Supply, Rural Water Supply, Wastewater Disposal, Solid Waste Disposal, Transport and Traffic Planning, Road Building and Maintenance - Building of Rural Roads, Road Traffic, Railways and Railway Operation, Inland Ports, Shipping on Inland Waterways, Ports and Harbours - Harbour Works and Operations, Shipping, River and Canal Engineering.

There may also be an **interaction** with sectors covered in the following environ-mental briefs Surface Mining, Underground Mining, Minerals - Handling and Processing, Power Transmission and Distribution, Iron and Steel and Non-ferrous Metals.

5.   Summary Assessment of Environmental Relevance

The aim of this environmental brief has been to summarize the **environmental relevance** of mechanical engineering works, workshops and shipyards. **Detailed investigations** are to be made in each **specific case** as to the possible environmental hazards. Even **small environmental problems** which appear at the outset to be of marginal importance can, in certain contexts, result in **failed projects** or **serious damage. Countermeasures** must be integrated into the planning and execution at an early stage. From the point of view of environmental protection, **mechanical engineering** should involve a combination of **precautionary measures** and appropriate **management decisions.** Therefore **training** in environmental protection must be given high priority in every project. **Workers** should be **trained** in occupational safety and environmental protection. The **management** should be familiar with and apply **additional precautionary measures** (knowledge of suitable pollutant disposal methods or plant optimization with a view to environmental protection, e.g. choice of paints/solvents with low pollutant contents).

An important further precondition of applied environmental protection is the existence of **effective waste disposal facilities,** especially for problematic hazardous waste. Technical staff must also be on hand, e.g. to **maintain filtration and wastewater treatment plants,** the erecting and operation of which are described in the present brief.

6.   References

Abwassertechnische Vereinigung (ATV) (Ed.): Lehr- und Handbuch der Abwassertechnik, Bd. I - VI, Ernst Verlag, Berlin, various years.

Abwassertechnische Vereinigung (ATV) (Ed.) Arbeitsblatt A 115, Hinweise für das Einleiten von Abwasser in eine öffentliche Abwasseranlage, draft of 22.03.1990.

40. Anhang zur Allgemeinen Rahmen-Verwaltungsvorschrift über Mindestanforderungen an das Einleiten von Abwasser in Gewässer, GMBI. (joint ministerial circular) 1989, Nr. 25, p. 517 ff.

Batstone R. et al.: The Safe Disposal of Hazardous Wastes, The Special Needs and Problems of Developing Countries, Vol. I, II, III, World-Bank Technical Paper No. 93, Washington, 1989.

Brauer, H.: "Die Adsorptionstechnik - ein Gebiet mit Zukunft", Chem.-Ing.-Tech. 57, (1985), Nr. 8, p. 650 - 663.

Deutsche Forschungsgemeinschaft (DFG) (Ed.): Kühlschmierstoffe, Liste von Komponenten, in: Toxikologisch-Arbeitsmedizinische Begründung von MAK-Werten, Weinheim 1983.

DIN 45635: Geräuschmessungen an Maschinen.

EC Council Directive on sulphur dioxide and suspended particulates, lead and nitrogen dioxide (EC Directive 80/779/EWG, 82/884/EWG and 85/203/EWG).

Fischer, H. et al.: "Galvanotechnik", in: Ullmanns Enzyklopädie der technischen Chemie, Band 12, p. 137 - 203, 76th year.

Geretzki, P.: "Erkrankungen durch Kühlschmierstoffe in der Metallindustrie", in: Dermatosen 31, 1983, Nr. 1, p. 10 - 14.

Gesetz über Abgaben für das Einleiten von Abwasser in Gewässer (Abwasserabgabengesetz -AbwAG, BGBl. I (Federal Law Gazette I), p. 2432, 1990).

Gesetz zur Ordnung des Wasserhaushalts (Wasserhaushaltsgesetz - WHG, BGBl. I (Federal Law Gazette I), p. 205, 1990).

Gewerbliche Berufsgenossenschaften, Unfallverhütungsvorschriften: VBG 7, VBG 15, VBG 23, VBG 24, VBG 57, VBG 113.

Häusser, M., et al.: Kühlschmiermittelbestandteile und ihre gesundheitliche Wirkung", in: ZbC. Arbeitsmed., 35, 1985, Nr. 6, p. 176 - 181.

Hartinger, H.: Handbuch der Abwasser- und Recyclingtechnik für die metallver-arbeitende Industrie, Munich, Vienna, 2. Auflage 1991 (Carl Hanser Verlag) ISBN 3-446-15615-1.

Hauptverband der gewerbliche Berufsgenossenschaften e.V.:
    ZH 1/81   Merkblatt für gefährliche chemische Stoffe
    ZH 1/194 Merkblatt für Chlorkohlenwasserstoffe
    ZH 1/425 Kaltreiniger-Merkblatt
    ZH 1/562 Sicherheitsregeln für Anlagen zum Reinigen von Werkstücken mit Lösemitteln (Lösemittel-Reinigungsanlagen)
    ZH 1/566 Merkblatt für Explosionsschutzmaßnahmen an Lösemittel-reinigungsanlage
Other ZH 1 publications.

Hauptverband der gewerblichen Berufsgenossenschaften e.V.: Lärmschutz-Arbeits-blätter und Lärmschutz-Informationsblätter.

Koenigs, M.: "Schweißverfahren, Gefährdungen und Schutzmaßnahmen", BAD-intern 2/83.

König, W., et al.: Schadstoffe beim Schleifvorgang, Schriftenreihe des Bundes-anstalt für Arbeitsschutz, Forschungsbericht 427, Dortmund 1985.

Air Quality Guidelines for Europe, WHO regional publications European series: No. 23/1987.

Mahler, W., Zimmermann, K.F.: "Aktuelle Hinweise zur Einhaltung der verschär-ften Arbeitssicherheits-und Umweltschutzbestimmungen beim Verarbeiten cadmiumhaltiger Hartlote", in: Schweißen und Schneiden, 1986.

Mannheim: "Sicherheitsmaßnahmen bei der Verwendung von Halogen-Kohlen-wasserstoffen bei der Metallentfettung", in: sicher ist sicher 7/8, 1983, p. 333 - 338.

Maschinenbau- und Kleineisenindustrie-Berufsgenossenschaft, Kampf dem Arbeits-lärm 3, Lärmminderung für Betriebspraktiker, 1983.

Maschinenbau- und Kleineisenindustrie-Berufsgenossenschaft: Broschüre "Kühlschmierstoffe", January 1991.

Menig, H.: "Luftreinhaltung durch Adsorption, Absorption und Oxidation", Deutscher Fachschriften Verlag, Wiesbaden 1977.

Ministerium für Ernährung, Landwirtschaft, Umwelt und Forsten, Baden-Württemberg, Altlasten-Handbuch Teil 1, Stuttgart 1987.

Müller, R.: "Arbeitssituation und gesundheitliche Lage von Schweißern", Forschungsbericht Nr. 252 der Bundesanstalt für Arbeitsschutz und Unfallforschung, Dortmund.

Muster-Verordnung über Anlagen zum Umgang mit wassergefährdenden Stoffen und über Fachbetriebe der Länderarbeitsgemeinschaft Wasser (LAWA), draft of 31.08.1990.

Rosenkranz, D., Einsele, G., Harreß, H.M. (Ed.): in: Bodenschutz-Handbuch, Erich Schmidt Verlag, Berlin 1988.

Schütz, A.: "Öl-Aerosole an industriellen Arbeitsplätzen", in: Staub RL 44, 1984, Nr. 6, p. 268 - 272.

Seebohum, K.W.: "Beurteilung von Schweißarbeitsplätzen", in: sicher ist sicher - Zeitschrift für Arbeitsschutz, 9/85, p. 454.

Szedkowski, D.: "Gesundheitsgefahren durch Lösemittel", in: Württ. Bau BG, Mitteilungen 2/1985, p. 25 - 27.

TA-Luft, Technische Anleitung zur Reinhaltung der Luft of 27.02.1986, GMBl. (joint ministerial circular),p. 95 ber. p. 202.

TA-Lärm, Technische Anleitung zum Schutz gegen Lärm of 16.07.1968, annex to Federal Law Gazette (BAnz.) No. 137 of 26.07.1968.

Technica, Ltd.: Techniques for Assessing Industrial Hazards, A Manual, World Bank Technical Papers No. 55, Washington, 1988.

Umweltbundesamt (Federal Environmental Agency (Ed.)): Handbuch Abscheidung gasförmiger Luftverunreinigungen, Erich Schmidt Verlag, Berlin 1981.

Umweltbundesamt (Federal Environmental Agency (Ed.)): Branchentypische Inventarisierung von Bodenkontaminationen, Forschungsbericht 03001, Berlin 1986.

VDI (Verein Deutscher Ingenieure): VDI-Handbuch Reinhaltung der Luft, Beuth Verlag, Berlin and Cologne.

VDI (Verein Deutscher Ingenieure): Technische Sorptionsverfahren zur Reinhaltung der Luft, VDI-Bericht 253 (1975).

VDI (Verein Deutscher Ingenieure): Abgasreinigung durch Adsorption, Oberflächenreaktion und heterogene Katalyse, VDI Richtlinie 3674.

VDI (Verein Deutscher Ingenieure): VDI-Richtlinien zur Geräuschmessung, Schallschutz, Schwingungstechnik: 2560, 2564, 2567, 2570, 2571, 2711, 2714, 2720, 3727, 3749, 3731, 3742.

Verordnung über die Herkunftsbereiche von Abwasser of 03.07.1987, BGBl. I (Federal Law Gazette I), p. 1529.

Zschiesche, W., et al.: "Neue Erkenntnisse zur Berufspathologie der Schweißer", Arbeitsmed., Sozialmed. Präventivmed. 20 (1985), p. 140 ff.

Trade and Industry

Agro-industry

52. Agro-industry

Contents

1. Scope

The agro-industry is based on **agricultural and forestry production,** and its purpose is to **preserve** and **refine raw produce** and to **extract** and **concentrate** the valuable **constituents.** The **food industry** constitutes the most important sector of the agro-industry.

Many agro-industries have developed from skilled manual production processes and accordingly can be carried out at varying technical levels. The following information, however, applies to **small and medium-sized operations.** The definition of small and medium-sized operations varies from country to country but a maximum of **100 employees** can be taken as an upper limit. There are **environmental briefs** which focus **specifically** on a number of agro-industries, particularly **large plants.**

In no other area are development and environment so closely intertwined as in that of the agro-industry. Unforeseen implications can turn intended impacts on their head, and medium and long-term damage may prove to be of short-term benefit. Nowhere are effects on the biosphere -including human society - so all-embracing as in the agro-industry. And no other sector is so dominated by female employment; all the activities in this sector are of major importance to and have major effects on women. All agro-industry activities depend essentially on the limited time women have available, their extensive responsibility and on limited water and energy resources. This is why the socio-economic parameters and influences are priority issues in agro-industry projects.

A distinction can be made within the agro-industry between **primary, secondary** and even **tertiary processing. Primary processing is** basically **most suited to small industrial operations,** as technical input increases in line with processing complexity.

2.   Environmental Impacts and Protective Measures

2.1   The agro-industry generally

As the agro-industry will probably increase the demand for certain commodities, or alternatively push towards different forms of land use and farming, the following environmental impacts in the area of agricultural production should be mentioned:

Problems relating to the direct expansion and intensification of resource usage include impairment of soil fertility, problems of soil losses and sedimentation, problems of desertification and irrigation problems (soil and water salination, fluctuating water table and water pollution), which in turn reduce resource productivity. The problems of fertility losses, desertification, and salination are generally greatest in countries where the population pressure on land is greatest. Here, agriculture expands most markedly in peripheral areas and marginal resources are utilised intensively.

The most successful efforts lie in the promotion of soil-conservation measures: reducing the intensification of soil usage, and introducing programmes for minimum or soil-conservation farming (contour line farming, terrace farming, strip farming, extension of dry and green fallow land), programmes to control flooding and wind erosion and programmes for the improvement of crop rotation. What needs to be examined is the extent to which these measures should be implemented as an alternative or in addition to the establishment of agro-industrial production operations.

The economic and social parameters in place and those sought are decisive factors in the agro-industrial sector generally. The maintenance and promotion of subsistence production and agro-industrial activities without restricting subsistence are major axioms in this respect.

Commodity processing gives rise to environmental impacts on the atmosphere (odours and dust emissions), water (quantity and wastewater), primary energy sources (mainly timber) and the soil.

The following comments are confined to certain branches which have been in the greatest demand in recent years.

2.2 Selected branches

2.2.1 Mills handling cereal crops

Only **dry milling** is carried out in such plants, thus account must be taken of **noise and dust emissions** which affect not only the specific operational area but also the area surrounding the mill. Suitable countermeasures are **technical installations** (extraction, soundproofing) and **individual measures** (breathing apparatus, hearing protection), priority being given to the first group, since the use of individual safety equipment requires explanatory and supervisory measures.

Surface water quality is impaired in cases where streams and rivers are used for waste disposal, for example. Further usage or controlled dumping are suitable **countermeasures** (cf. environmental brief Mills Handling Cereal Crops).

2.2.2 Processing of starch sources and root crops

If the **biologically polluted wastewater** from washing and processing is discharged into surface water untreated, the result can be **overfertilization, reduction in the oxygen content** and therefore a general **impairment of water quality, changes in the micro flora and fauna** and, in the medium term, **disruption of water biotopes**.

Appropriate minimum measures are **mechanical separators and aeration ponds** in which the biological oxygen demand is reduced to an acceptable level. Since a reduction in the biological pollution of wastewater is associated with improved yield, **optimised process technology** can also be an economically beneficial environmental measure. Finally, highly polluted wastewater which can normally be avoided where a process is appropriately optimised, can be used as a **substrate for biogas production**.

2.2.3 Processing of oil-bearing seeds and fruits

In small and medium-sized works, only **pressing processes** are used for oil extraction, with solvent extraction reserved for large plants (see also the environmental brief Oils and Fats). Oil-bearing fruits are heated directly or with steam or hot water to improve yields. This produces **steam emissions** and oil-laden **wastewater**. **Wood** is often used for **energy production**, and this can lead to **over-use** of tree stocks.

Because **steam emissions** affect mainly operating personnel **extraction** should take place at the point of production. Once again, process optimisation, the use of better separators and treatment in aeration ponds should be used to **reduce waste-water pollution**. Consumption of wood or other commercial **fuels** can be reduced by incinerating the waste produced in the processing operations and also by opti-mising energy circuits and consumption in the processing plant.

2.2.4 Sugar beet and sugar cane processing

The essential environmentally relevant aspect of beet and cane processing is the **energy required** for the concentration of the sugar solution. While this require-ment can be met in cane processing by burning bagasse, energy consumption in sugar beet processing must be optimised and, if necessary, alternative energy sources must be identified.

Mention should also be made of **organically polluted wastewater** from purifi-cation and condensate.

There is an **environmental brief** specifically relating to Sugar.

2.2.5 Fruit and vegetable processing

Biologically polluted washing water and the **energy requirement** for thermal preservation processes are of environmental relevance in this area, and the same comments as in the previous sections apply. **Solar driers** can also be used, thereby reducing the energy required for the production of top quality dried products quite considerably.

2.2.6 Dairies

As milk and dairy products are ideal breeding grounds for microorganisms, **hy-giene requirements** are relatively stringent, a factor which prompts the use of **ag-gressive cleaning agents**. If they are discharged at certain concentrations, the quality of **surface water** is impaired and micro flora and fauna are affected.

Countermeasures are the sparing use of **biodegradable cleaning agents** and dilution in tanks.

Mention should also be made of **percolating milk** in rinsing and washing water as a source of organic pollution.

2.2.7 Processing of semi-luxury goods and spices

The operations having the greatest environmental relevance in the production of semi-luxury goods and spices are fermentation and waste disposal. **Fermentation is** generally carried out in fixed locations, and the **pollutants** thereby produced can **accumulate in the soil** over long periods, **damaging micro flora and fauna**. The washing operations sometimes carried out after fermentation (e.g. coffee) give rise to **biologically polluted wastewater** which, if discharged untreated, can impair **surface water quality**. The impacts of this are restricted to harvest time, and are then found over longer intervals.

Fermentation should be carried out in the immediate vicinity of an abundant supply of running water at appropriately prepared places (cement bases). The heavily polluted wastewater produced must either be suitably diluted before discharge or used for **biogas production**. As washing water is not generally so heavily polluted, special measures (aeration ponds) are only required in exceptional cases. Spices are often irradiated as a method of preservation, although the consequences of **irradiation** on human **health** are as yet unknown.

2.2.8 Plant fibre extraction

In many countries, **microbiological retting** is practically the only method of plant fibre production in use. It involves the degradation of non-fibrous components by a microbiological process and is carried out by immersing the raw material either in a slow flow of water or in specially prepared tanks, whereupon the retting is spontaneously initiated. Since this process and the subsequent fibre washing require **large quantities of water**, these installations are always built close to abundant supplies of running water. In these circumstances, the water exchange required once the retting process is complete is no problem (except perhaps for any dissolving pesticides used during farming).

The retting process is associated with a certain **odour nuisance** which cannot be avoided at reasonable cost. The only remedy is **not** to site these plants **close to residential areas** and to take account of prevailing wind directions.

Because fibre production is a low-input technology in every respect, negative environmental effects can only be avoided by selecting a suitable site and making use of what nature has provided.

2.2.9   Tanneries

Of all the agro-industries tanneries harbour the greatest **risk potential** for the environment.   This is due on the one hand to the considerable **odour nuisance** and on the other to the **dyes** and other **chemicals** (particularly chromium compounds) used in the tanning process which complicate the wastewater treatment operation.   And there is also biological pollution. Besides a substantial impairment of the quality of the nearby **surface waters**, an enrichment of the hazardous substances in the **soil**, and possibly also in the **groundwater** must also be expected.

The **elimination of odours** at source is only possible if the tanning is carried out in **enclosed rooms** and any air escaping is cleaned in technically sophisticated **filter systems**.   The nuisance can be limited indirectly by **concentrating** plants of this kind **on sites a suitable distance from residential areas**.   This would also create the conditions essential for the relatively complex process of **multistage wastewater treatment,** which is essential in this industry but which is really too costly for an individual small plant (see also the environmental guidelines of the World Bank).

2.3   Socio-economic impacts

The overwhelming majority of **jobs** in the agro-industry call for **little in the way of qualifications** and most workers are women.   However, as mechanisation and machine-based jobs increase, the proportion of male workers rises - as do monotony and isolation of the individual working processes, and the risk of accidents.   The extent to which the employment of women leads to changes in their own food production needs to be examined. The jobs are of poor quality in ergonomic terms, and nuisances in the form of **dust, damp, smells and noise** may attain levels which can affect the health of employees, constituting a considerable risk to women in particular. Because different types of jobs are done by the two sexes, **qualification and training programmes** must be established at an early stage, with the emphasis on female employment.   These programmes should take account of the overall form of production and lifestyle of female employees and their families.

3. Notes on the Analysis and Evaluation of Environmental Impacts

The environmental impacts in the agro-industry can be assessed in terms of space, time and in relation to various resources and employees.

In the "agro-industry" sector assessments are based directly or indirectly on the following **test criteria**:

- impacts on employees in the factory
- impacts on people living near the factory
- environmental changes due to the emissions from the factory
- environmental changes caused indirectly by the factory (e.g. change in the quantity of water or extra energy required).

Short-, medium- and long-term impacts and likewise direct and indirect impacts must be considered in the light of these test criteria.

The **evaluation** involves comparing the project with other possible projects, and also considers the economic, ecological and social costs involved.

The evaluation of effects on health faces the problem of the frequent lack of national limits or recommended values for individual substances, and this is further complicated where a number of substances are emitted at the same time, thereby increasing their impact due to synergistic effects. One initial approach to this problem area may be provided by publications of international organisations such as the World Health Organisation (WHO) (see in this regard Volume III, Compendium of Environmental Standards.

4. Interaction with Other Sectors

There are close links with the **plant and animal production sector** which supplies the raw materials, and with the **marketing** sector, not forgetting the **metal** and **mechanical engineering industries** which manufacture the processing equipment, and the packaging materials industry.

Other factors in the equation are **veterinary services, livestock farming, irrigation** and **health and nutrition**. Projects in the field of the **economy** and also **infrastructural measures**, particularly in the **rural hydraulic engineering sector**, are significant issues in the assessment of agro-industrial projects, while cross-sectoral concepts of **general resource management, location planning and regional planning** must not be forgotten.

5.   Summary Assessment of Environmental Relevance

Agro-industries often serve as **pilot projects** for more general industrialisation, and must therefore be examined very closely in terms of their direct and indirect impacts on the **food supply and economic prospects of the country concerned**, its general environmental conditions, and the lives of its female population in particular.

Agro-industrial projects are extremely important to a country's independent development and this is closely allied to general subsistence production.

Direct environmental pollution from small and medium-sized agro-industrial factories on an individual level is relatively slight in the short-term, but the **more general effects** can be quite considerable.

One **exception** to this is **tanneries** because of the chemicals used -which are problematic in environmental terms - and the odour nuisance.

All factories which use water as the extraction, cleaning and transport medium produce **wastewater** which is biologically polluted to varying degrees, and this generally requires treatment in aeration ponds or treatment plants. Noise and dust emissions are normally **restricted in terms of the area affected**, and therefore affect primarily the employees themselves.

6. References

Bundesimmissionsschutzgesetz BImSchG of 15.03.1974.

Environmental Guidelines, The World Bank, Environment Department.

TA-Luft 27.02.1986.

TA-Lärm 1968.

Verwaltungsvorschriften to § 7a WHG, Mindestanforderungen an das Einleiten von Schmutz- bzw. Abwasser in Gewässer.

Trade and Industry

Slaughterhouses and Meat Processing

53. Slaughterhouses and Meat Processing

Contents

1. Scope

This sector embraces **slaughterhouses, meat-processing plants** and **animal carcass disposal plants**.

To date no standard project types have gained prevalence for **slaughterhouses**, particularly in terms of size, as each project is dependent on a a number of factors, such as:

- regional population density;
- specific consumption (kg/person and year);
- animal stocks in the region, catchment area;
- distance from nearest slaughterhouse;
- export potential, restrictions;
- eating habits;
- religious constraints.

Nor is there a standard size of **meat-processing plant**, as their design is also influenced by these same factors.

Animal carcass disposal plants (ADP) process dead animals, confiscated carcasses (where the meat or organs of slaughtered animals is found to be unfit for human consumption), blood, bones etc., the **end products** of which are - depending on the raw material - technical fats and meat meal, bone meal, blood meal etc., used for fodder and in some cases as fertilizer. Project size is determined primarily by the capacity of the neighbouring slaughterhouse.

For reasons of hygiene, **cattle** are hung for slaughter. The slaughter line feed system is manual in small operations and mechanical in plants with a medium or large line capacity.

Different processes are used in the **bleeding stage**, for example, since animals must be hung for bleeding to comply with EC Guidelines, but laid flat ("bleeding with the neck pointing to Mecca") in accordance with the dictates of Islam. (Sheep and camel slaughter similar to cattle slaughter).

Pigs can be **slaughtered** either hung or lying. A number of processes have been developed for the scalding and skinning of pig carcasses (scalding tank and depilation machines, production line systems where the animals are suspended or laid flat) depending on line capacity. **Ritual slaughter** is carried out for export.

Sheep are hung for slaughter and a number of methods are used for bleeding.

Because of the large number of different meat and sausage products, a wide range of processing stages[1] are required. However, the following can be regarded as **basic operations** for all products:

carcass splitting - grinding of meat - seasoning - filling of natural or synthetic skins with sausage meat - heat treatment - cooling - dispatch - long-life meat products - tinned food.

The various processes used in the **manufacture of meat and sausage products** depend on the particular meat and sausage products in question, with processing carried out **within different temperature ranges**:

- uncooked sausage   process temperature approx. 14 -28°C
- cooked sausage   process temperature approx. 50 - 80°C
- tinned meat and sausage products process temperature approx. 80 -121°C

In **animal carcass disposal plants**, the material used and the waste material is largely processed by the **pressing process** following heating.

The **extraction process** is rarely used today because of the **residues** it leaves **in the meal**.

---

[1]   for the processing of raw products and by-products.

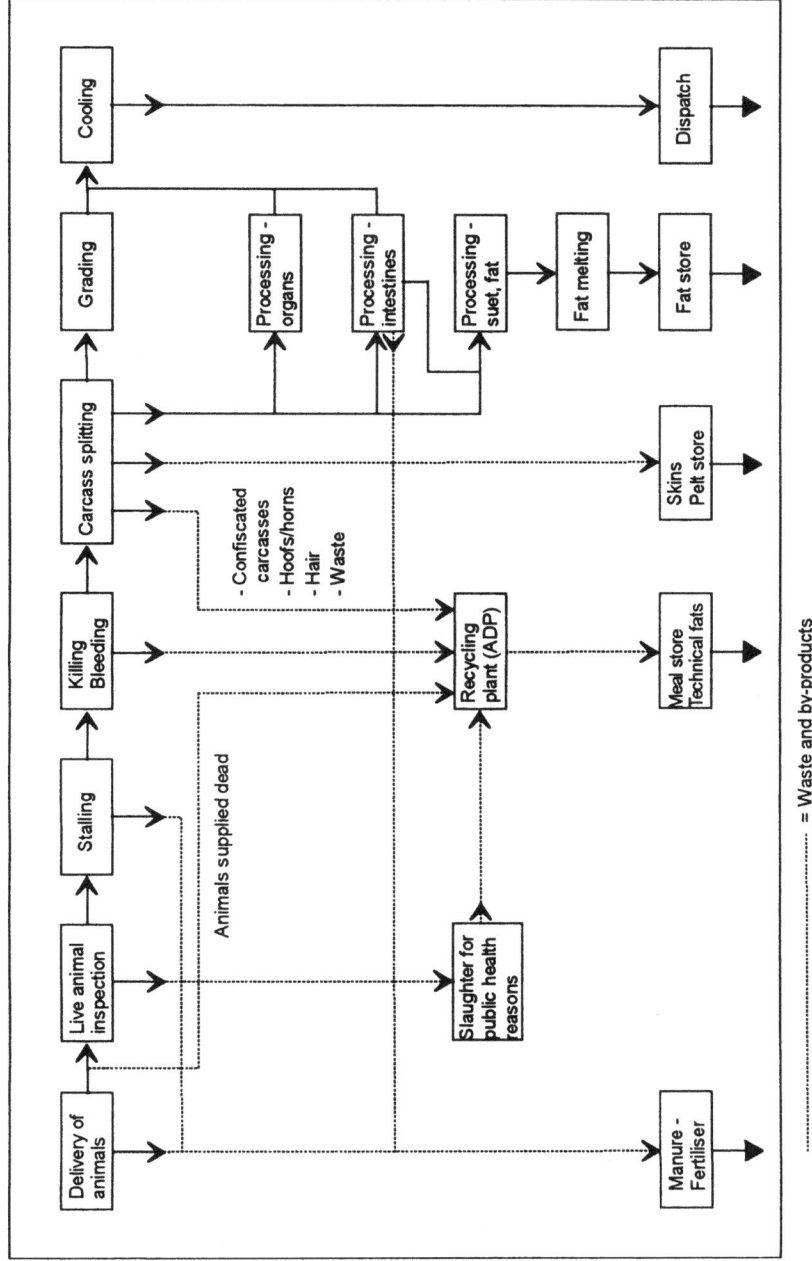

Fig. 1

Flow chart of slaughterhouse

Figure 2
Flow chart of slaughter procedure

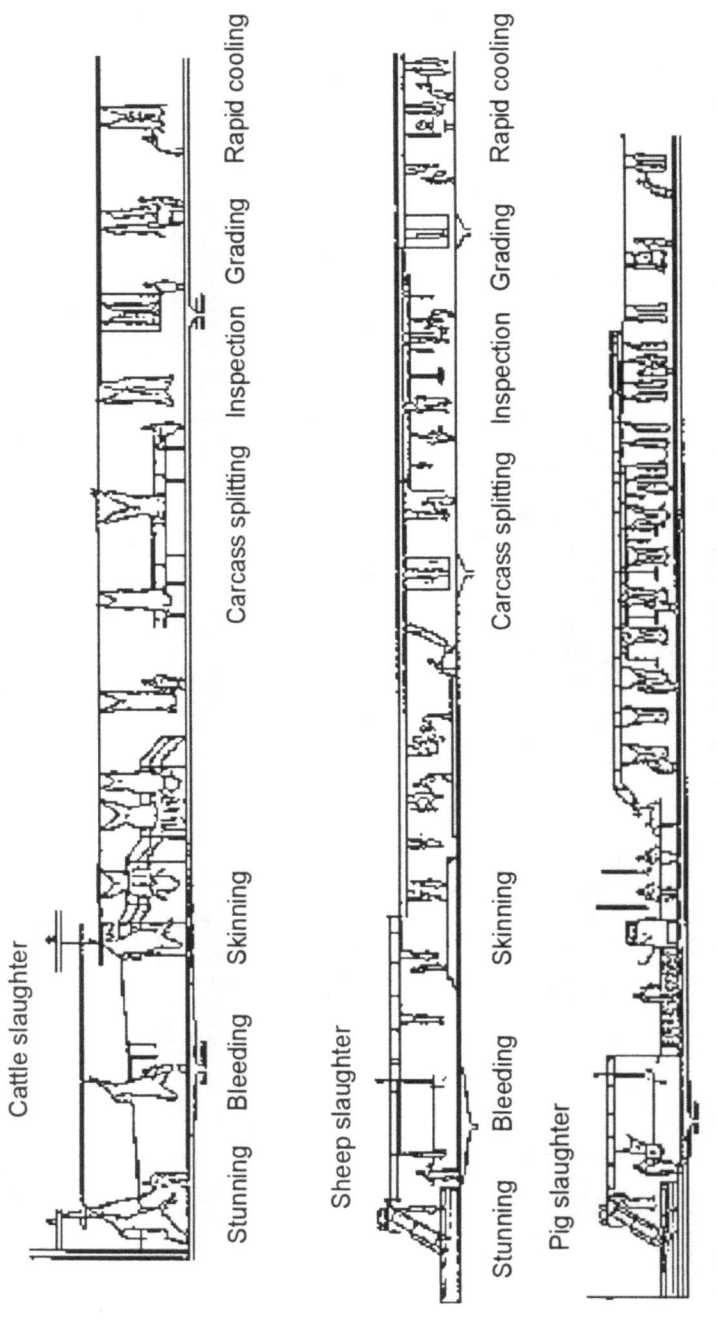

Fig. 3
Flow chart of a meat product factory

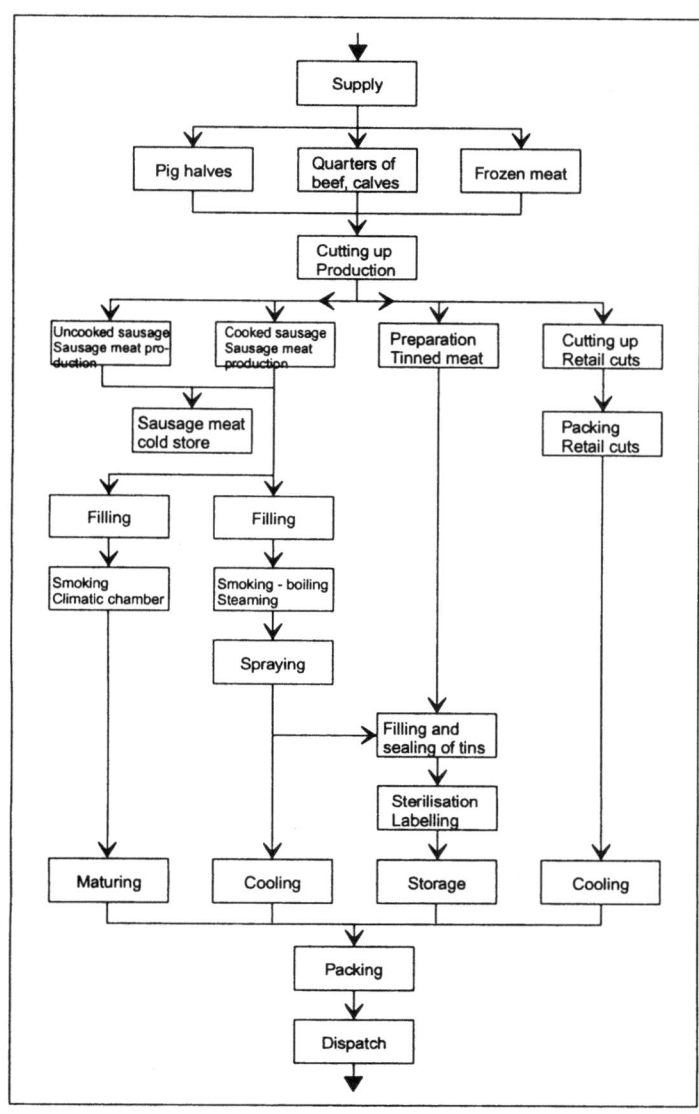

Fig. 4
Flow chart of ADP press installation

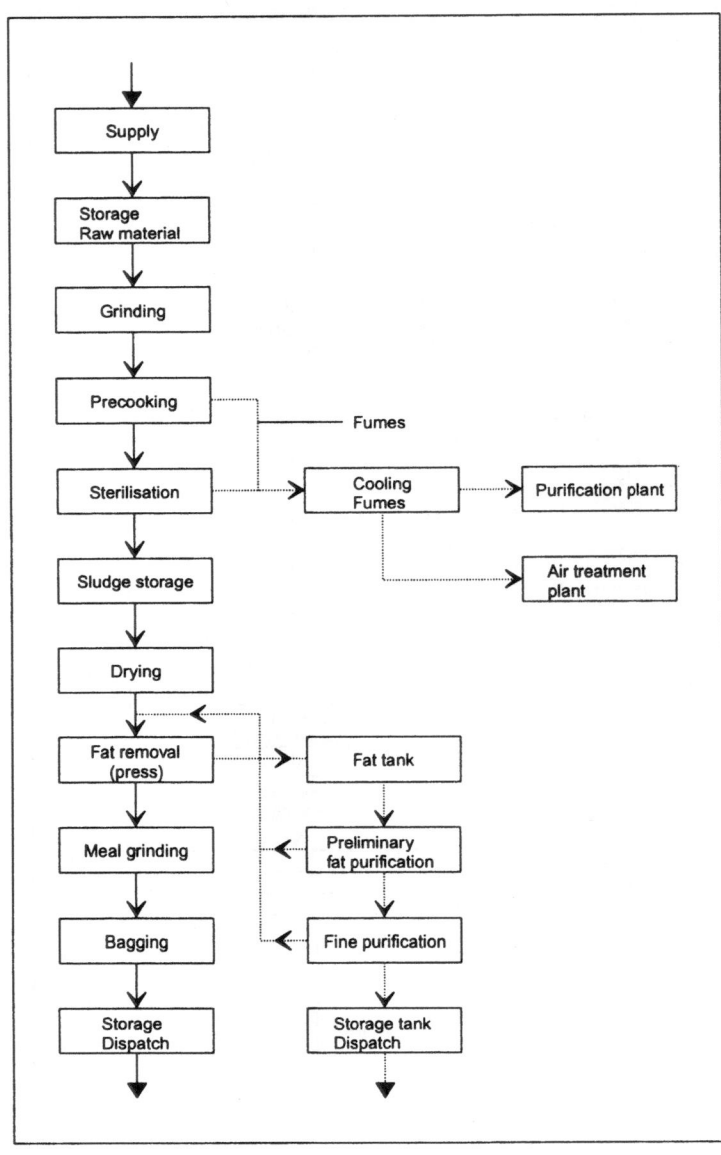

2. Environmental Impacts and Protective Measures

Meat industry facilities cause environmental impacts due to:

- wastewater;
- spent air/waste gases;
- noise;
- animal waste;
- waste heat;
- residues in the end product;
- waste.

German regulations relating to environmental pollution caused by meat processing are taken as the reference in the following as they have also gained acceptance as the international standard.

Table 1
Environmental impacts from meat industry plants

Type of plant	Waste-water	Odour	Waste gases	Noise	Waste	Waste heat
Fattening and breeding operations	X	X	X	X	X	
Slaughterhouses	X	X	X	X	X	X
Recycling plant	X	X	X	X		X
Meat-product factories	X	X	X	X	X	X

2.1 Water pollution

Water consumption and the **degree of contamination of the wastewater** arising from the process **depend on a number of factors**, and are determined principally by the following:

- species of animal;
- type and capacity of plant;
- intensity of cleaning of carcasses and
- working accommodation during the process.

The following values apply for **slaughterhouses** (average values):

-    cattle   600 - 800 l/animal
-    swine   300 - 500 l/animal
-    sheep   200 - 300 l/animal.

Water consumption in **meat-product factories** is largely **product dependent**. Wastewater pollution is higher, for example, in plants producing mainly cooked sausage and tinned products than in those which, for example, produce only uncooked sausage (salami). Consumption is around 10 - 15 m^3 per tonne of sausage and meat products.

Water consumption in ADPs is relatively low. The quantity of wastewater produced depends on the quantity processed, as some 65% of the raw material must be evaporated. On average, the wastewater level is approximately 1 m^3/t raw material.

The **degree of water pollution** in the meat-processing industry is **extremely high**, particularly in slaughterhouses and ADPs. In Germany, the following **minimum requirements** on the discharge of dirt or wastewater into watercourses must be observed by the meat industry to prevent water pollution.

Table 2
Degree of pollution of wastewater

Type	BOD5 value mg/l	Causes and factors of influence
Slaughterhouses	approx. 4,000	Some blood, content of stomach and intestinal tract, urine, liquid manure, animal waste etc.
Meat-product factories	approx. 10,000	Animal waste, type of processing (boiling and steaming of raw material and end product)
ADPs	approx. 10,000	Type and quality of the raw material

Table 3

Minimum requirements for wastewater disposal in water

Type	Matter which can be removed by settling[3]	BOD$_5$[1]	COD[2] [4]
Slaughterhouses and meat-processing plants	< 0.3 ml/l	< 35 ml/l	< 160 ml/l
ADPs	< 0.5 ml/l	< 40 ml/l	< 30 ml/l

Key:

[1] BOD$_5$ =   Biochemical oxygen demand over a 5-day period, with oxygen consumption determined in this period (g O_2/l wastewater at T = 20°C)

[2] COD =   Chemical oxygen demand in the reaction with $KMnO_4$ or $K_2Cr_2O_7$ as the oxidation agent (mg O_2/l wastewater)

[3]   Random sample

[4]   2hr mixed sample

Slaughter costs are **increased** because of increased investment costs and running costs for **wastewater treatment** in relatively **expensive treatment plants**. Consequently animals may be slaughtered outside instead of inside slaughterhouses and thus no comprehensive check on hygiene conditions can be guaranteed.

After **eliminating solids** by mechanical purification, **pond systems** or the **seepage** of wastewater into the ground can be considered a substitute for biological treatment systems, provided that this does not pollute the groundwater mains or groundwater collection installations used for the drinking water supply.

The following can help slaughterhouses and meat-product factories **reduce their wastewater pollution** and dispose of their effluent correctly:

-      better understanding of environmental issues by personnel;

-      installation of technical facilities for improved separation of blood from the wastewater system;

- removal of waste of coarser consistency from production area floors before wet cleaning;

- fitting of sludge buckets in floor drains;

- fitting of wastewater screens to separate solids from the wastewater (these solids have a high protein content and can be passed on to ADPs);

- installation of sludge trap and fat separator;

- flotation plants (mechanical flotation treatment);

- supplementary biological clarification as a second treatment stage following mechanical treatment in plants which discharge their wastewater directly into surface water.

Wastewater from **ADPs** has to be **sterilised.**

2.2   Air pollution

Emissions occur primarily in the form of **air discharged from the following areas**:

Table 4
Emissions from outgoing air

Type	Source area
Slaughterhouses	Stalling, possibly also storage, confiscated meat
Meat-product factories	Processing, smoke (cooking plant)
ADPs	Delivery, processing

To reduce the smell nuisance, **slaughterhouses** in Germany must, wherever possible, be sited at a **distance of at least** approx. 350 m from the nearest residential building.

Odours arise due to the **odour of the animals themselves** and **changes to organic materials**. As all smells arising in slaughterhouses are **biodegradable**, bioscrubbers and biofilters can be used to reduce smells, as can adsorption and absorption processes.

Table 5
Immission values (IVs) (*TA-Luft* [Technical Instructions on Air Quality Control])

Pollutant	IV 1 continuous operation	IV 2
Airborne dust (regardless of dust content)	0.15	0.30 mg/m^3
Lead and inorganic lead compounds as components of airborne dust -expressed as Pb -	2.0	- µg/m^3
Cadmium and inorganic cadmium compounds as components of airborne dust - expressed as Cd -	0.04	- µg/m^3
Chlorine	10.0	0.30 mg/m^3
Hydrogen chloride - expressed as Cl -	0.10	0.20 mg/m^3
Carbon monoxide	10	30.00 mg/m^3
Sulphur dioxide	0.14	0.40 mg/m^3
Nitrogen dioxide	0.08	0.20 mg/m^3

Waste gas from **meat-product factories** can be treated in a number of ways, including:

-       post-combustion;
-       condensation;
-       absorption - adsorption;
-       electrical separators for particulate substances in conjunction with the above processes.

The emission reference value is the **total carbon** in the organic compounds.

In new continuously operating plants, emission values can be contained with **technical installations** so that:

-       the established immission values (see Table 5) are not exceeded, and

-       as experience has shown, no odour nuisance occurs where the chimney is of the correct height for gas disposal.

The installation of ventilation and air-extraction systems, waste gas systems etc. incurs high investment costs and this in turn can lead to high slaughterhouse fees which users cannot afford.

The following values are recommended to reduce substances which cause odours in **ADPs**:

-       thermal post-combustion:
        20 mg/m^3 carbon in the combustible substances.

-       other post-treatment systems:
        The total frequency of odour assessments of emitted spent air, measured by the **olfactometry process** with 50% negative evaluations (ADP odour not perceivable) must produce a dilution factor of 100. A solids emission value of 75 mg/m^3 can be observed in the **air emitted from meal, conveyor and storage systems**. The air emitted from heating and air purification systems must be removed through a **chimney** of an appropriate height.

Odour emissions can be generally **reduced or prevented** by:

- designing enclosed working and production areas with windows which cannot be opened;
- closed process circuits;
- fitting of air locks;
- preventing any accumulation of materials which could result in the development of odours;
- spent air systems with appropriate air treatment, as shown in Table 6.

Table 6
Reduction of odour emission due to spent air treatment

Type	System
Slaughterhouses	Biofilters, waste gas scrubbing, active carbons
Meat-product factories (Smoking installations)	Post-combustion, condensation, absorption, adsorption
ADPs	Wet scrubbing (multi-stage), heat treatment, biological treatment, earth filters, biological scrubbers

2.3   Noise

Potential **sources of noise** in slaughterhouses and/or meat-product factories and ADPs are:

Table 7
Sources of noise

Source	Slaughterhouses	Meat-product factories	ADPs
Animal delivery	X		
Animal slaughter area	X		
Machine and process area	X	X	X
Spent air system recooling chamber	X	X	X

As the operations under discussion here are not noise-intensive, **technical measures** - such as the fitting of sound absorbers etc. - are usually sufficient to comply with local limits/guide values. The possibility of keeping **at an adequate distance** must be checked first.

It is possible to **avoid or reduce noise** by:

- installing sound dampers in ventilation systems;
- enclosing machines;
- using sound-barrier walls;
- making allowance for the main wind direction at the design stage in terms of sources of noise.

2.4   Waste and residues

There are two types of waste in the meat-processing industry:

- waste material which can be **reused** for the manufacture of by-products;
- **waste** to be destroyed or stored in dumps.

Odour emission in the **processing of waste** to **by-products** is **reduced** by:

- immediate processing of waste;
- cold storage of waste until reprocessing;
- use of closed containers;
- spent air treatment by appropriate installations.

If possible, a wet extraction process should be avoided in ADPs in view of the residues this leaves behind in the end product (animal meal);   the **pressing process** should be used instead.

Waste which goes for **further processing, destruction or storage in dumps**, should be collected **in separate containers** (metal, plastic, paper etc.).

Manure should be reprocessed as far as possible for **agricultural purposes**.

53

2.5   Waste heat

The operations considered here produce waste heat primarily from:

- boiler house installations;
- cooking and smoking installations;
- open-hearth furnaces (pig slaughter);
- extract cooling (ADP).

State-of-the-art **heat recycling plants** must be used in new plants, to ensure a lower consumption of primary energy (see also environmental brief Renewable Sources of Energy).

2.6   Industrial safety

The well-being of people employed in the meat-product processing industry is affected in relatively few areas. **Noisy machines** are used, for example, to saw carcasses into pieces (approx. 90 dB(A)) and to grind meat with cutter mixers (approx. 80-90 dB(A)), and so here appropriate **hearing protection** is to be worn.

ADP personnel is exposed to **odours** for a short time when the raw material is delivered, but suitable **ventilation and air-extraction systems** can reduce this problem, with protective mouth masks recommended in some cases.

2.7   Location planning

Modern slaughterhouse sites are divided by a fence into a clean and an unclean zone, each of which has its own entrances and exits.

The **unclean zone** houses all activities where cleanliness is not an issue, such as the cattle market, stabling, transport of waste, confiscated carcasses, preliminary clarification, middens etc.

The **clean zone** is for all areas where hygiene is a major consideration, such as the slaughter plant, cold chambers, cutting plants, dispatch etc.

When designing slaughterhouses, the siting of the clean zone must be analysed and specified appropriately for hygiene reasons in terms of wind direction and emissions from existing or planned works or factories.

3. Notes on the Analysis and Evaluation of Environmental Impacts

Limits and approximate values are specified for **wastewater and air pollution** and are laid down in Germany, for example, in the *Wasserhaushaltsgesetz* [Federal Water Act] and in *TA-Luft* [Technical Instructions on Air Quality Control] or in the guidelines *(Richtlinien)* of the Association of German Engineers VDI; they also describe the **correct analysis procedure**. **Constant** wastewater and outgoing air **control** is necessary to conform to these values, and this also involves checking that **technical laboratory conditions** are adequate. It must also be ensured that suitable numbers of **qualified personnel** are available for the analysis work.

The **adverse effect of noise on nearby utilities** can be reduced by **keeping appropriate distances**; in Germany, for example, a distance of 350 m from the nearest residential building must be observed. Within the plant, **hearing protection** must be provided for personnel at noise-intensive workplaces and the wearing thereof checked. Values for the maximum admissible noise nuisance at the workplace are included, in Germany for example, in the *Arbeitsstättenverordnung* [Ordinance on Workplaces].

Waste recycling produces, in the main, **odour emissions**, but this nuisance can be minimised in the neighbourhood by appropriately designed **operating procedures** (immediate processing, cold storage, closed containers) and adequate plant distances.

Appropriate dumping facilities must be guaranteed for waste.

The possibility of any **residues** remaining in the end product must be ruled out by an appropriate **choice of process**; end products must be controlled by **continuous analysis**. In new installations, **waste heat** is **returned** to the process.

If there are no national provisions, **analyses** should be carried out to define the preconditions for protecting the population from pollution, e.g. in the form of groundwater pollution, the storage of waste and the associated risk of disease. This applies correspondingly to industrial safety.

Factors **of a socio-economic nature** must also be analysed, with due consideration given to employment opportunity issues and working conditions, differentiated by sex, and an examination of sources of income for women etc.

4. Interaction with Other Sectors

Raw material procurement for the meat industry - in this case live animals - and the waste and by-products arising from animal slaughter and the meat-processing industry, give rise to a range of interactions within this sector of industry.

The following special **recycling facilities** are therefore provided for waste and by-products from slaughterhouses and meat-processing plants, as shown below.

Table 8
Waste recycling potential

By-products and waste product	Secondary industry	Product	Use
Blood	Blood reprocessing	Plasma	Food industry
Technical blood	ADP	Blood meal	Animal fodder
Hair	Brush processing	Brush/paint brush hair	General
Manure	--	Compost	Fertilizers
Content of intestines	--	Biogas	Energy
Skins, pelts	Tanneries Leather industry	Leather	Leather items
Bones (not fit for human consumption)	Fat production	Technical fats Bone meal	Soap industry Animal fodder
Bones (fit for human consumption)	Fat production	Fat gelatine	Food industry
Hoofs	ADP	Hoof meal Technical oils (acid free)	Fodder Lubricants
Suet	Fat production	Edible fat	Food industry

Since slaughterhouses are provided for the general **supply of meat** and to serve meat-processing factories for the production of meat and sausage goods, and since the **by-products and waste** constitute the **raw material base for these secondary processing plants,** there are close links between the businesses in this sector.

The following environmental briefs provide more detailed information about the surveying, evaluation and reduction of environmental impact caused by the meat-processing industry:

- Wastewater Disposal
- Solid Waste Disposal
- Livestock Farming
- Veterinary Services
- Planning of Locations for Trade and Industry

5.   Summary Assessment of Environmental Relevance

The main environmental impact from slaughterhouses and meat-processing plants derives from **wastewater**, as the pollutant waste load produced in the process is absolutely enormous. If this wastewater (effluent) is discharged into receiving bodies, **a fee** should be charged, based on the pollutant load.

In addition to the wastewater, serious environmental implications (e.g. **odour**) can be caused if the critical areas/plants are not maintained as required and waste storage or removal is not carried out with due care.

Unless they are subsidised, slaughterhouses are financed solely by **slaughter fees** paid by users. The fees are higher the greater the investment and maintenance costs.

The provision of state-of-the-art slaughterhouses can therefore lead to increased **meat prices**.

In the light of these factors, there is a risk that animals will not be slaughtered in municipal slaughterhouses under veterinary control, but **unregulated** outside the slaughterhouse (e.g. on the road side) to avoid slaughterhouse fees.

A further crucial point which must be considered when designing such plants, is the availability of technically **trained personnel**.

Correct plant operation with due allowance made for environmental imperatives can only be guaranteed if the technical installations are correctly designed and the **following conditions** are met:

- availability of adequately trained personnel;
- understanding of environmental protection constraints;
- implementation of preventive maintenance;
- adequate spare part supply.

6. References

ArbStättV § 15 - Schutz gegen Lärm

ATV-Arbeitsblatt A 107, Hinweise für das Ableiten von Schlachthofabwasser in ein öffentliches Kanalnetz

Bundes-Immissionsschutzgesetz BImSchG of 15.03.1974

Zweite Durchführungsverordnung zum Vieh- u. Fleischgesetz (VFIG), amended on 20.08.1979

Vierte Durchführungsverordnung zum Vieh- u. Fleischgesetz (VFIG), amended on 10.11.82

Sechste Durchführungsverordnung zum Vieh- u. Fleischgesetz (VFIG), newly issued on 16.12.1986

Siebente Durchführungsverordnung zum Vieh- u. Fleischgesetz (VFIG), in the version of 10.11.1982

EC Directive of 15 July 1980 relating to the quality of water intended for human consumption

Fleischhygienegesetz in der Fassung der Bekanntmachung of 24.02.1987 - BGBl. I (Federal Law Gazette I), p.549 - FIHG

Gesetz über den Verkehr mit Vieh und Fleisch (Vieh- und Fleischgesetz - VFIG) of 25.04.1951 - BGBl. I (Federal Law Gazette I), p.272, in the revised version of 21.03.1977 - BGBl. I, p.1477, most recently amended on 10.06.1985 - BGBl. I, p.953

Gesetz über die Beseitigung von Tierkörpern, Tierkörperteilen und tierischen Erzeugnissen (Tierkörperbeseitigungsgesetz - TierKBG) of 02.09.1975, BGBl. I (Federal Law Gazette I), p.2313 and 2610

Gesetz über die Neuorganisation der Marktordnungsstellen of 23.06.1976 - BGBl. I (Federal Law Gazette I), p.1608

Handelsklassengesetz of 05.12.1968 - BGBl. I, p.1303 in der Fassung der Wiederlautbarung of 23.11.1972 - BGBl. I (Federal Law Gazette I), p.2201

Council Directive No. 64/433/EEC of 26 June 1964 on health problems affecting Intra-Community trade in fresh meat, in the version of the Council amendment directive no. 83/90/EEC of February 07 1983 (Official Journal of the European Communities L 59 of March 05 1983, p.10), last amended by the Council Directive no. 88/288/EEC of May 03 1988 (Official Journal of the European Communities L 124, p.29)

TA-Luft of 27.02.1986

TA-Lärm - Genehmigungspflichtige Anlagen in accordance with § 16 of the Gewerbeordnung

VDI-Richtlinie 2590 Auswurfbegrenzung, Anlagen zur Tierkörperbeseitigung

VDI-Richtlinien der Luftreinhaltung, Nr. 2595, Blatt 1 Emissionsminderung bei Räucheranlagen

VDI-Richtlinien der Luftreinhaltung, Nr. 25965, Emissionsminderung bei Schlachthöfen

Verordnung über gesetzliche Handelsklassen für Schweinehälften, coordinated with Verordnung of 18.12.1986, valid from 01.04.1987

Verordnung über gesetzliche Handelsklassen für Schaffleisch of 27.01.1971, BGBl. I (Federal Law Gazette I), p.77 - in the version of the amendment of 11.11.1977 - BGBl. I, p. 2139

Verordnung über gesetzliche Handelsklassen für Rindfleisch, coordinated with Verordnung of 13.11.1982, valid from 01.01.1983

Verordnung über die hygienischen Anforderungen und amtlichen Untersuchungen beim Verkehr mit Fleisch (Fleischhygiene-Verordnung - FlHV of 30.10.1986, BGBl. I (Federal Law Gazette I), p. 1678)

Allgemeine Verwaltungsvorschrift über die Durchführung der amtlichen Untersuchungen nach dem Fleischhygienegesetz (VwVFlHG) of 11.12.1986 - B Anz. (Federal Gazette) Nr. 238a of 23.12.1986

Verwaltungsvorschrift on § 7a WHG, Mindestanforderungen an das Einleiten von Schmutz- bzw. Abwasser in Gewässer.

Trade and Industry

Mills Handling Cereal Crops

54. Mills Handling Cereal Crops

Contents

1. Scope

The sector embraces **mills handling cereal crops**, including **warehousing for raw materials** and **end products**, and also **animal feed production** and **seed dressing**, operations which are almost always linked to the cereal processing complex.

The only **milling industries** to be considered in the study are those involved in manufacturing **end products** for **human consumption** from raw materials imported or grown in rural areas, with **animal feed** simply a by-product.

Below, the environmentally relevant factors of **noise, dust, process water** and **pesticides** are considered.

The sector concerned can be divided essentially into **four parts**:

- storage, drying and seed dressing,
- flour mills,
- hulling mills,
- heat treatment.

Projects for the **drying** and **storage** of locally grown cereal and for seed conditioning have been given a boost recently and have been priority now that it is realised that **raw materials** for food need to be **protected** from spoilage and perishing due to climatic conditions, pests etc. and that better seed **increases production**.

Modern industrial mills have an **integrated silo and warehouse capacity** for the raw material to be processed and the end products and by-products manufactured. Depending on the site, ownership and the general purpose of the plants, **drying and seed cleaning plants** may be **incorporated**. **By-products** are often recycled as **animal feed components**.

2. Environmental Impacts and Protective Measures

Given the processing techniques employed today, it may generally be assumed that large volumes of **air** are required to **produce milled and hulled products** (flour, wholemeal products, flakes, grains etc.), in addition to **power** for cleaning, hulling, grinding (milling) and the transport of intermediate and end products.

This air is used mainly for vertical and horizontal transfer inside the milling or hulling system and for dust extraction from the processing units and the entire mill complex. Furthermore, under certain climatic conditions **cool air** is required to ventilate power plant and processing machinery as well as the entire building complex.

Industrial wastewater is produced only in the **cereal washing department** in the mill industry, and even then only where granular or wholemeal products are to be produced. The modern mill industry makes particular use of a **dry cleaning** process which separates out impurities by means of screens and weighing sorters. If the plant also produces **bulgur and parboiled rice, process water** with a low **starch content** is produced.

The **wastewater from waste-recycling** power generating plants, particularly that from **rice husk gasification** for the production of lean gas for gas-engine powered plants, has a **phenol content** of over 0.03 mg/l. When **husks are burnt** to produce steam, a residual quantity of 18% **ash** in relation to the quantity input must be disposed of. The same applies to gas plants.

It can therefore generally be stated that the **environmental impacts** of the mill operation lie in the **following areas**:

-       dust emission,
-       noise nuisance,
-       hazard of dust explosions and fires,
-       odour nuisance to a limited degree,
-       hazard of toxic gas,
-       recycling of residual substances and waste disposal,
-       process water.

2.1   Cereal storage and handling

2.1.1   Port and transshipment silos, mill silos

Storage installations of this kind are used for the **storage** and **transshipment** of cereal for import and export. They are found in all major ports where imported cereals (wheat, maize, rice, millet etc.) as well as raw products and semi-finished products for the food and animal feed industry are put into store for intermediate storage, and from which the domestic industry is supplied with raw materials or goods for export are shipped (maize, rice, millet, tapioca etc.).

The following table shows the **dust content of the service air** from the various mill sections and admissible **emission values** in Germany.

Table 1

Pollutants produced and admissible emission values in Germany

Type of mill industry	Dust content of service air	Permissible emission values
Silo installations	12 to 15 g/m^3	50 mg/m^3
Drying plants	15 to 18 g/m^3	50 mg/m^3
Mills handling cereal crops	approx.96 g/m^3	50 mg/m^3
Hulling mills	6 to 8 g/m^3	50 mg/m^3
Seed cleaning	8 to 10 g/m^3	50 mg/m^3

In **storage installations with preliminary cleaning plants** and in **mills, dust emissions** are **collected** in aspiration pipe systems during cleaning, and **separated** with the help of cyclones and filters. To achieve the best possible removal of dust from machines and buildings, all equipment handling materials and machinery should be **enclosed** and fitted with appropriate aspiration connections. The **extraction of dust** with so-called mass separators or filter separators is described and explained in the guidelines nos. 3676 and 3677 of the Association of German Engineers VDI. The safety measures in these guidelines should be observed.

With the high degree of mechanisation in modern mills, the only **workplaces** where **dust is a problem** are the **loading and packing operations**; here too, ex-**traction devices** must be used wherever possible.

All the dust from aspiration systems and cleaning in transshipment silo installations is collected and bagged.

The cleaning waste, which may contain **live pests**, is to be **destroyed** immediately.

In mill-cleaning plants, **dust waste** and **granular cleaning waste is treated** and added to mill afterproducts (bran) (feed ingredient).

Noise is another environmental problem. The increasing use of **high-speed technical equipment** and the intensive use of machines in the smallest possible space give rise to an increasing noise nuisance which is becoming a hazard to man.

Precautions must be taken to protect employees and local residents. Structural measures, such as the lining of ceilings and walls with **soundproofing materials**, must be taken, and **vibration isolation** materials must be used for machine foundations.

The **TA-Lärm** [Technical Instructions on Noise Abatement] in Germany lays down, for the various industrial and residential areas and mixed use areas, **safety guidelines** for **immissions** which must be observed in the planning and erection of industrial plants.

Personnel must be issued with **hearing protection** where they are constantly exposed to noise levels of over 70 dB.

Information and **training** must therefore be provided for **personnel**, and compliance with safety measures **monitored**.

People, buildings and the machine stock can be at risk from **dust explosions** and **fires**. Following any such explosion, there is a chemical conversion of a dust/air mixture, which accelerates as **heat is generated**, causing a sudden **pressure effect** from existing or newly formed gases. **Three components** constitute the basis for a dust explosion: dust, air (oxygen) and ignition energy; the latter can be in the form of heat or electricity (electrostatic charging).

Silo installations are particularly **at risk from dust explosions**. Mechanical sparks, pockets of glowing materials, mechanical heating, hot surfaces, welding work, electrostatic discharge sparks and the like are possible **ignition sources**. They must be **eliminated** as a **safety measure**, and the formation of **explosive dust concentrations** must be **prevented**, for example by enclosing machines. **Structural precautions** can also be taken, namely the creation of compression-proof rooms and pressure release and explosion suppressing systems. The following **organisational precautions** are also **effective** in terms of fire and explosion safety:

-       welding and cutting works only to be carried out during factory shutdowns;
-       regular cleaning with dust-explosion-proof equipment;
-       training of employees in the handling of fire-fighting equipment and
-       information to employees about the causes of dust fires and explosions.

Finally, in the **planning phase**, provision must be made for taking all the measures required to limit the risk of explosion (cf. in Germany, VDI guideline, 2263 on dust fires and dust explosions - *Staubbrände und Staubexplosionen*).

Gases are most commonly used to **protect stocks (pest control)** in the silo installations and warehouses, but under certain circumstances **sprays** and **vapours** are an option.

The types of **pest control agent** for cereals currently used and approved in **Germany** include gaseous insecticides:

- hydrogen phosphide,
- methyl bromide,
- hydrogen cyanide.

In addition to gases, fumigants and sprays can be used for the **disinfestation of silos and stores** - without any need to include stocks in the treatment.

The following are **approved in Germany**:

- lindane
- bromophos,
- malathion,
- dichlorvos,
- piperonyl butoxide,
- pyrethrum
- and combinations of these.

The incorrect use of **agents** for **pest control** for **stock protection** purposes can lead to **hazardous substances** seeping into adjacent production or residential buildings (e.g. hydrogen phosphide). Therefore, particular attention must be paid to the technique of pest control (e.g. silo fumigation using a circulation system).

Specific bans or restrictions on the use of these agents are recorded in the **plant pesticide register** of the country concerned or may be requested from the **registration office** for these substances. The **manufacturer's instructions** must be strictly **observed** and made available in the **local language**.

After treatment, **waiting periods** must be observed to ensure that plant products do not contain **higher residue** levels than are permissible where they are to be brought into circulation or eaten (cf. environmental brief, Analysis, Diagnosis, Testing and Volume III, Compendium of Environmental Standards (CES)).

Authorised contractors must be employed for the **application** of agents to protect commodities stored in silos and warehouses; their **personnel** must be appropriately **trained** and able to use **the special equipment** and **safety installations**.

2.1.2 Cooperative stores and warehouses

Simple storage installations (including raw material stores) are **warehouses** for bagged commodities or for **horizontal storage**. **Bagged commodities** or **loose grains** are cleaned, stored, ventilated and may also be treated as a pest control measure. Most maize, rice and sorghum harvests are still stored in this way in many countries, with possible **storage losses** of 15% or more.

Figure 1
Diagram of a port and transshipment silo installation

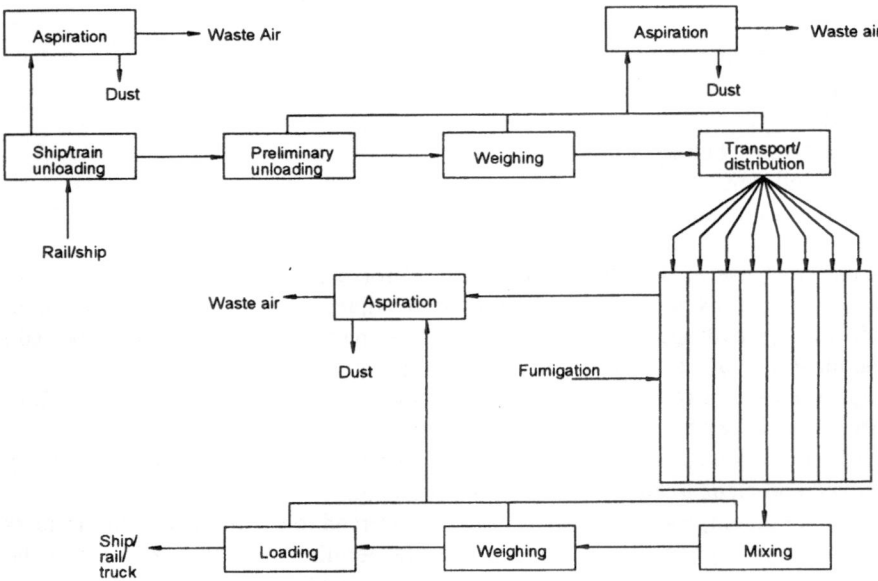

Standard warehouses should also have installations for cleaning, ventilation and fumigation.

The **risk of dust explosions** can be largely avoided in warehouses by a **light and open design**, although this does not protect against normal types of fires which can occur. Otherwise, the environmental implications are as described in 2.1.1. **Pest control** in warehouses may take the form of **sprays** although **fumigation** is also commonplace.

The safety measures for silo installations described in 2.1.1 are also applicable to warehouses, with the exception of measures against the risk of explosion.

Special precautions must be taken where gases are used for pest control purposes. As **bagged commodities** cannot be **fumigated** in airtight areas, **gastight fumigation tarpaulins**, sealed underneath with sand, are required if this type of pest control is to be used.

2.1.3 Seed cleaning installations

Seed dressing is not considered part of a mill's activities, but in many countries it is one of the services a cooperative storage facility may offer its members.

In using these facilities **seed** is produced **with a higher grade purity** thanks to air, screen and specific **weight classification**. The lower content of other types of grain and the improved growth conditions resulting from chemical treatment **improve quality** and thus yields per hectare.

The service air from seed cleaning plants contains **primary dust**. It and the cleaning waste produced (rejected grain, weed seeds etc.) can be used for **animal feed production**.

Treatment involves the wet or dry application of fungicides and insecticides which - as **pesticides** - protect the seed and are classified as seed treatment agents. All such treatments approved by the Biologische Bundesanstalt für Land- und Forstwirtschaft (Federal Biological Research Centre for Agriculture and Forestry) in Germany are listed in the pesticide register *Pflanzenschutzmittel-Verzeichnis* (1990).

These plant pesticides are used in **seed improvement operations** either alone or in combination depending on the purpose of the treatment.

Common pesticides (active ingredients) are:

- anthraquinone,
- bibertanol,
- bendiocarb,
- fuderidazol,
- bromophos,

- lindane,
- carboxin,
- fenfuram etc.

Environmental measures in seed cleaning operations are confined, in terms of aspiration and service air, to keeping the production rooms and the outgoing **waste air** clean. The filter installations listed in paragraph 2.1.1 and the emission values shown in table 1 are applicable here. ،

When **protecting seeds,** appropriate precautions must be taken to protect personnel and, subsequently, users.

The **approval regulations** of the individual countries must be obeyed as must the **manufacturer's recommendations for use** (see too Volume III, CES).

Figure 2
Diagram of a seed cleaning installation

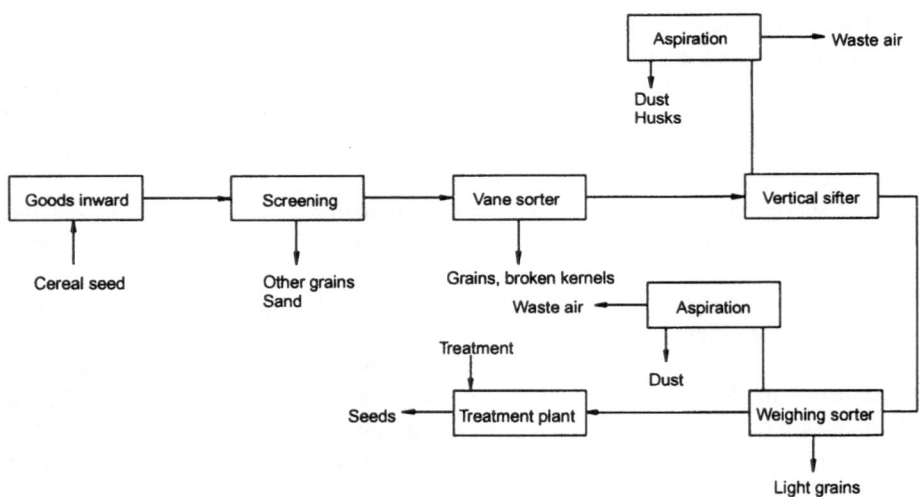

2.1.4 Drying installations

Grain drying is a **thermal process** in which water is **removed** from the pre-cleaned damp commodities (cereal, maize, rough rice (paddy), sorghum etc.) by evaporation. Obviously, an adequate **supply** of **heat** is essential. The drying of the moist harvested produce is usual in warehousing facilities and the agricultural trade (cooperatives), and mill and silo facilities often have drying installations too. Thus, wherever large quantities of damp grain (moisture content of over 15%) are supplied, rapid drying is called for. Drying installations are used where **natural sun drying** is **not practicable** in view of the weather conditions (rainy season). Only **dry commodities** can be safely stored for prolonged periods without any deterioration in quality.

The **service air** of drying installations and the preliminary cleaning machines contains **coarse to fine particles of dust** which need to be **separated** by means of the dust separators described in paragraph 2.1.1. **Drying installations** are only used at harvest time, and should preferably be sited close to the (sparsely populated) growing areas. **Noise** is another **problem**. Very often mill and silo facilities also have drying installations.

The safety measures listed in 2.1.1 to protect against dust and noise must be taken here too.

Figure 3
Diagram of a drying installation

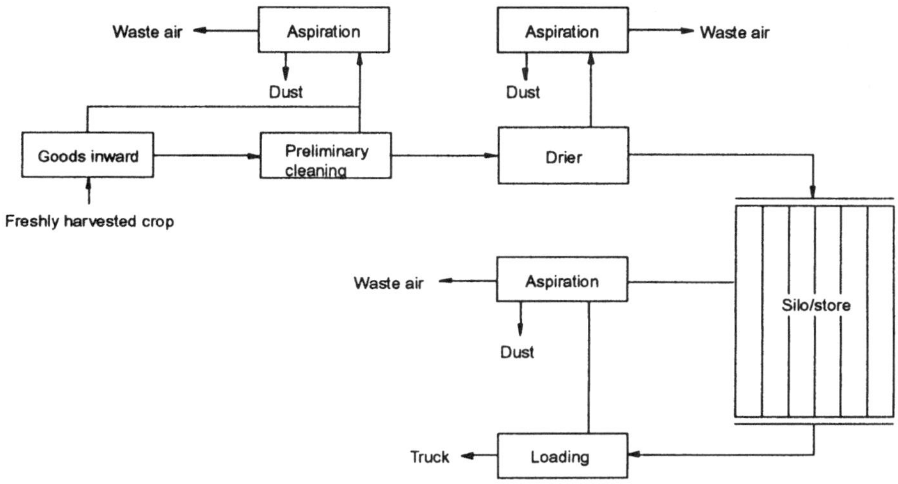

2.2   Flour mills (wheat mills)

The purpose of a **flour mill** is to obtain **large quantities of flour** which also meet flour product requirements in terms of quality. The **by-products and after-products** (bran, middling and cleaning waste) are recycled in agriculture or in the animal feed industry in the form of **feed components**. Mills also make **wholemeal products**.

In some cases, very **antiquated washing machines** are still used today for cereal cleaning in the wheat mill industry, requiring monitoring of wastewater quantities (up to 1000 l/t), thus safe distances from residential areas must be maintained. In the **modern cereal mill, water** is only used for conditioning (wetting) the cereal and it is **fully absorbed by the grain**. Today, the entire cleaning process is carried out by means of air, screening and weight classification. **Scouring machines** have largely replaced the washing systems, thus **no industrial wastewater** is now produced.

In conventional mills handling cereal crops, some 5 - 10 cubic metres of **air** are required per milled tonne. This quantity drops to just 15% where machines working on the **circulating** air principle are used for cleaning. All **service air** discharged into the atmosphere must be filtered.

There is also a **fire risk** due to dust explosions in mills, and mills handling cereal crops as a whole generate **noise emissions** which have environmental implications for humans.

All **safety measures** described for cereal storage are appropriate for mills handling cereal crops in all respects. If silo installations are structurally linked to the mill, not only must **automatic fire valves** be fitted in the interconnecting materials handling equipment, but also the **connecting walls** between the installations must constitute **fire barriers**[1]. A **settling tank** for organic substances (husks, pieces of stem, fines etc.) must be provided.

---

[1]   For this reason, safe distances from populated areas must be maintained.

Figure 4
Diagram of a wheat mill

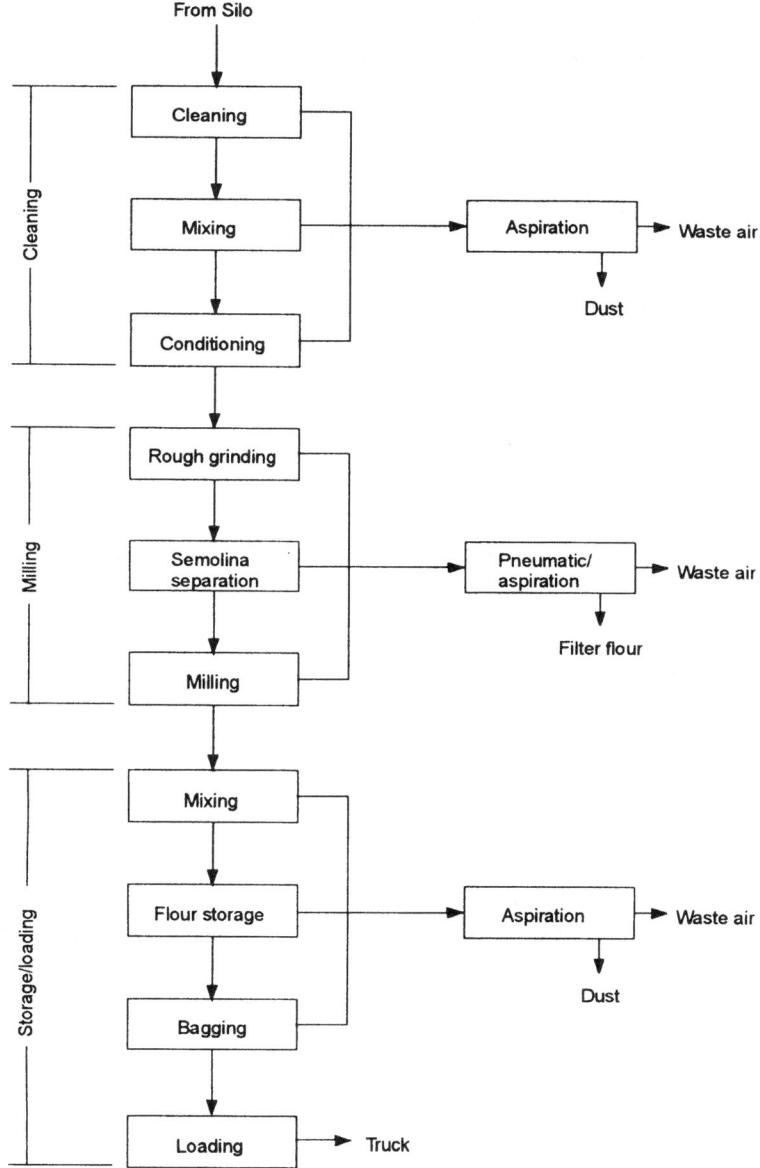

2.3   Hulling mills

Hulling mills handle the following cereal types: oats, barley, rice, sorghum and millet, as well as pulses.   While hulling mill technology is very different from that used in mills handling cereal crops, the **environmental pollution** and the resultant **safety measures** are largely similar in both types of mill.

2.3.1   Rice mills

The process from **rough rice** (paddy) through to ready-to-use **white rice** passes from **cleaning** with air, screening and weighing, **dehulling** and the **polishing process** (removal of the aleurone layer) through to **sizing**.   Some countries have their own production from low to medium capacity rice mill facilities (China, Taiwan, Malaysia, Thailand, India and some South American countries).

The **environmental pollution** from rice mills in these countries is considerable if there are no complete aspiration systems or such systems are not designed to appropriate technical standards.   Often, cyclones alone are used for dust separation although they only achieve a separation level of 90 - 95%, with dust emissions standing at 70 to 150 mg/m^3 air.   **Dust filters** must be **used**.

The major **disposal problem** for rice mills is presented by the **rice husks** from the production process (20%).   One possible means of **economically recycling** rice husks is that of pyrolysis to produce **energy** in steam power plants or **lean gas** for gas-engine plants (see environmental brief Plant Production).

Hot **industrial water** (approx. 65°C) and **saturated steam** are used to **make parboiled rice** (rice precooked in the husk).

Apart from the rice husks, all other **by-products** are either used locally as **animal feed** or exported (rice polish/emery flour).

A residue of about 18% **ash** is produced by rice hull pyrolysis.   Ash can be disposed off locally as a **soil structure improver** and more recently **rice hull ash** has been used in steel works **as an insulant**.

Where **parboiled rice is produced**, **organic substances** are found in the wastewater, but in such small quantities that their recovery is not economical.   Approx. 1 cubic metre of drinking quality water is required per tonne of rough rice (paddy), approximately 30% of which is absorbed by the grain.

Otherwise, the environmental impacts are as described in paragraphs 2.1.1 and 2.2.

The **environmental protection measures** to be observed for rice mills are listed here in order of priority:

-   **Dust emission values** as applicable in mills handling cereal crops must also be observed in rice mills, i.e. modern aspiration systems with separators and filter installations must be used.

-   **Noise emissions** are a nuisance to the population living in the surrounding area and so the details per sections 2.1.1 and 2.2 apply for this industry.

-   The use of biodegradation tanks is recommended for **wastewater disposal** from parboiling facilities because of the higher starch concentrations.

-   Measures must be taken to ensure proper **hull disposal**. In addition to pyrolysis and soil structure improvement, the hulls can be used in the brick industry, distilleries and possibly for furfurol production. Other uses are technically feasible and should be considered in the light of the particular site.

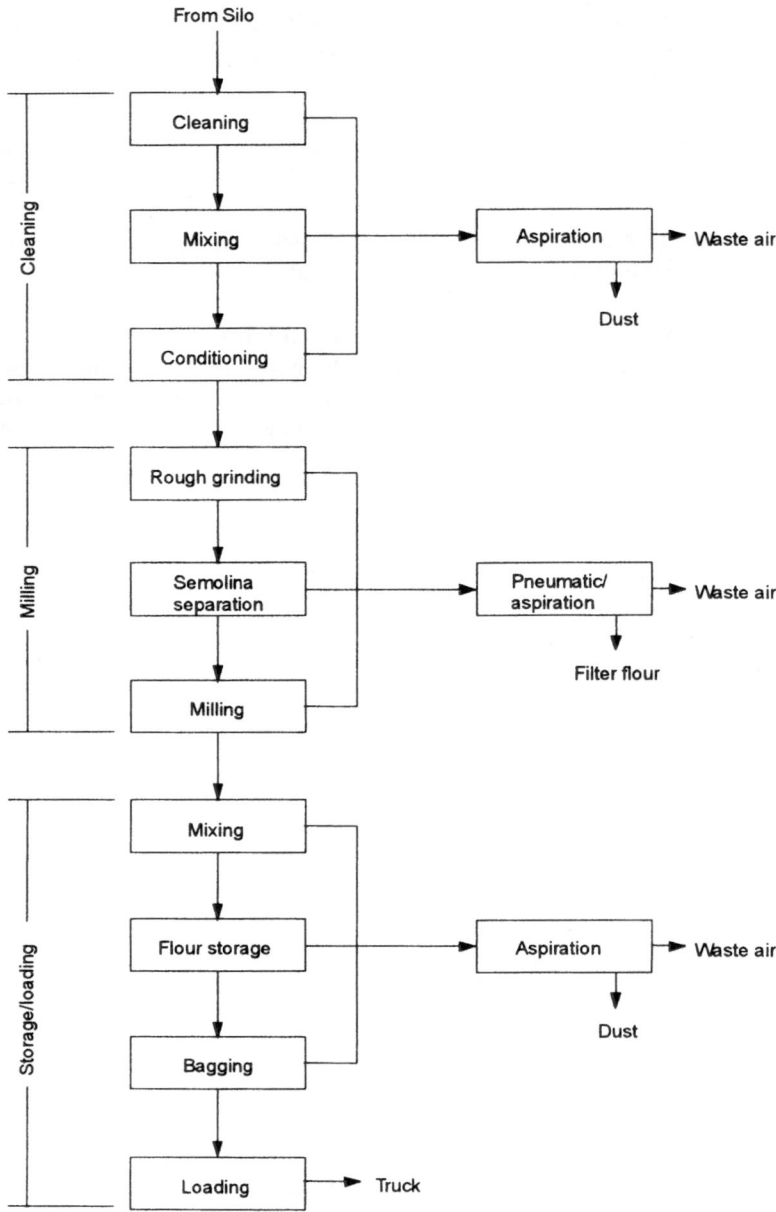

2.3.2 Hulling and processing of sorghum and millet

The industrial processing of sorghum and millet **produces flours with good keeping properties** and enables **quality to be controlled** in the light of the end products to be made. **Better quality flours** and **higher yields** are therefore achieved.

The upturn in the fortunes of this new branch of milling has been further enhanced by the possibility of **blending** this flour with wheat flour (composite flour). This has enabled a number of countries to use **local raw materials** by producing flours of this kind (blends of up to 20%).

The pollutants produced and safety measures to be taken are based on the data in paragraph 2.2.1.

2.3.3 Pulse hulling

The produce processed in hulling mills includes a range of **pulses** which are grown in temperate climates as well as the tropics. Pulses such as chickpeas, lentils and local bean varieties are brought onto the market in their **hulled/split form** or as **flour**.

The pollutants produced and safety measures to be taken are similar to the substances and measures described in paragraph 2.1.1.

2.4 <u>Location planning</u>

When **planning the location** of a **food industry** facility, it must be assumed that **medium-sized and large operations** will be accommodated. In such **mass-production plants** where food is processed, manufactured, transported, loaded and unloaded or stored, the following **environmental factors** must be taken into account, for which detailed information can be found in the relevant environmental briefs:

-   An organised **transport system** is necessary as facilities of this kind turn over substantial quantities of raw materials and finished products (environmental brief and Transport and Traffic Planning).

-   When planning larger facilities, a sea/land/sea **transshipment facility** should also be provided (environmental briefs Inland Ports and Ports and Harbours, Harbour Works and Operations).

-   As facilities of this kind run day and night, they must be sited at appropriate **distances from residential areas**. Dust, and above all noise nuisance, must be prevented (environmental brief Planning of Locations for Trade and Industry).

-   There should be a **reliable energy supply** on site to ensure safe operation of larger plants (environmental brief Overall Energy Planning).

-   Furthermore, for safety reasons, they must be sited **further from other industrial plants** so that in the event of incidents (fire, dust explosion) extensive damage can be avoided.

-   **A water supply** and **organised disposal facilities** are absolutely essential (environmental brief Water Framework Planning, Wastewater Disposal).

    The fundamental factors in site selection (e.g. avoidance of agricultural areas or rare/valuable areas of countryside) are likewise to be found in the environmental brief on Planning of Locations for Trade and Industry.

2.5   Energy from husk waste

The **energy requirement** of milling plants ranges from 30 to 70 kWh per tonne of end product, and that of rice mills 30 kWh. One economic and environmental objective should be the use of the **husks** from rice production (approx. 20%) as a **source of energy**.

Waste gas emissions from the chimneys of steam power plants are becoming an environmental problem due to their ash particle content. When incinerated, a **residual quantity** of about 18% **ash** remains.

Where gas is produced from husks, industrial water is used for gas scrubbing to separate tar and dust, and also as cooling water for the gas reactor. This wastewater contains up to 1.6 mg/l phenol. The ash produced by husk pyrolysis must also be disposed of.

All **residues** from husk burning in the ash disposal section of steam power plants should be **collected** in its dry form; after **cooling** and **intermediate storage**, the ash can be **reused by agriculture and industry**. Fly ash in the chimney must be separated by scrubbing with dust separators before it enters the waste gas chimney.

Wastewater from gas generators may only be released once it has been **neutralised** and **freed from solids**. **Venturi scrubbers** and **biological plant tanks** must be used for tar separation.

Even at the **planning** stage of these power plants, ash disposal, flue gas emission and wastewater disposal must be considered in the light of **parallel municipal development**.

2.6 Further processing of cleaning waste and mill afterproducts

Normally **waste** from mills handling cereal crops is immediately milled in hammer mills and then, together with mill afterproducts, supplied to the **animal feed industry**. Other **mill afterproducts** are bran and low-grade flour, and from the hulling mill, hull bran.

This industry, often regarded as a **secondary operation** in mills, produces **fodder concentrate** for livestock farming, which contains protein, carbohydrate, fats, mineral substances and vitamins as its main blending components.

2.7 Dust disposal

Dust requiring disposal is produced only in the goods inward sections of agricultural trade establishments and cooperatives. This dust actually consists of **sandy impurities** which are separated when the commodities enter the preliminary cleaning plant, and can, for example, be **returned to the supplier**.

3. Notes on the Analysis and Evaluation of Environmental Impacts

Large quantities of service air are used for transport, separating processes, heat discharge and aspiration in **flour and hulling mills** where the process technologies comprise milling and sifting techniques and abrasive and centrifugal hulling. This dust-laden air must be treated, making **dust removal from air** a priority concern.

Furthermore, the **noise produced** in such installations must of course also be given due consideration.

When planning mill projects, the various **national limits** must be taken into account at the design stage; if there are no adequate legal provisions, **international standards** must be applied. The provisions and limits applicable in Germany are quoted below by way of example.

The following **legal standards** and **technical regulations** are valid in the **Federal Republic of Germany** for the area under consideration:

- Bundes-Immissionsschutzgesetz, Neuauflage (New Edition) 1990
  TA-Luft, 1986
  TA-Lärm, 1986
- VDI-Handbuch Reinhaltung der Luft, Band 6;
  VDI 2264, VDI 2263
  VDI 3673
- Arbeitsblätter der Abwassertechnischen Vereinigung (ATV), November 1980 Issue
- Biologische Bundesanstalt für Land- und Forstwirtschaft Pflanzenschutzmittel-Verzeichnis 1990.

Almost all **afterproducts** from flour and hulling mills can be recycled in the **animal feed industry**, are potential **sources of energy** or can be used as raw products or auxiliary materials in **subsequent processing industries** (oil mills and breweries, steel industry and foundries).

Emission values for **local environmental protection** in Germany are those specified in the TA-Lärm (Technical Instructions on Noise Abatement). The guideline of the German Association of Engineers no. 2058 requires **noise protection at the workplace**, with the issue of personal hearing protection from 85 dB(A) upwards, and the wearing of this protective gear from 90 dB(A). These workplaces should be labelled and compliance with safety measures monitored.

Noise level and **air quality measurements** in the flour milling industry provide information about environmental impact and the safety measures required to be taken.

Table 2
Noise levels in mills handling cereal crops

Machine/part of building	Noise level dBA	Frequencies Hz
Separator floor	105	1000 to 2000
Sifter floor	100	800 to 1200
Roller floor	105	1500 to 1800
Hulling machines	108	1800
Compressors	95	2000
High-pressure ventilators	100	2500

This data shows that not only external noise emissions, but also workplace conditions inside the plant need to be dealt with by effective safety measures[2].

The **Compendium of Environmental Standards** provides information on assessment with regard to individual substances.

Where **combustible substances** are used for drying, only those with a maximum **sulphur content** of 1.0% are permissible.

54

4. Interaction with Other Sectors

Mills handling cereal crops perform **numerous additional** activities **upstream and downstream**, for example plant production, transport, and the handling and use of products obtained, e.g. as food. There are therefore a number of **connections with other sectors**; these are indicated in the text by reference to the relevant environmental briefs.

5. Summary Assessment of Environmental Relevance

Afterproducts resulting from the process or raw materials used, which are in almost all cases used as **fodder concentrate components**, are produced in flour and hulling mills and in the associated drying and seed cleaning plants which process cereal and tropical grains into food for human consumption. In contrast, rice mills give rise to **environmental pollution** from the recycling of the **husks** (steam production).

Dust, which is an ever-present combustible, is a health hazard, potential source of ignition and a **danger** in the storage and milling industry. Precautions are taken in the form of **constant maintenance and monitoring** of extractor plants, dust accumulation, with temperature and humidity adjustment as associated preventive measures; **personnel** must be suitably be **trained** for this.

The emission guide values applicable for local environmental protection with regard to **noise**, insofar as they concern the area of production, are binding on the industry and must be observed by measures such as **providing safe distances** or **soundproofing**. Inside stores and mill buildings, the **noise level** is a serious **problem for the employees**. In the light of all we know about the long-term effects on hearing, due account must be taken of noise protection requirements at the workplace. **Personal hearing protection** must be issued and its use monitored.

---

[2]   such as machine enclosures and the wearing of hearing protection.

The **use of insecticides and pesticides** for pest control purposes and seed disinfestation also presents problems since the substances used are **highly toxic** and **health problems** can be caused through their uncontrolled use and circulation. Only **specially trained personnel** using **the correct equipment** should be employed in these operations.

A certain amount of experience has been acquired in the field of the **degradation of organic substances** in process water with the help of fermenters, but a **degradation tank** is recommended where the wastewater is heavily polluted.

Modern mills do **not** give rise to **substantial emissions and residues** which pollute wastewater and dumps. Industrial water is only used for conditioning purposes in mills handling cereal crops and is absorbed by the grain. **Mill installations with washing plants** have to meet **minimum requirements** for the disposal of organically loaded wastewater. The same applies to industrial wastewater from bulgur or parboiled rice production.

6.   References

Abwassertechnische Vereinigung (ATV), Arbeitsblatt A 115, 1980; Hinweise für das Einleiten von Abwasser in eine öffentliche Abwasseranlage.

Ammermann, K: Ausrüstung von textilen Filtermedien zur Staubabscheidung.

Bartknecht, W: Staubexplosionen (1987), Springer-Verlag.

Berufsgenossenschaft Nahrungsmittel und Gaststätten: Staubexplosionen, Mack & Metz GmbH, 68 Mannheim.

Biologische Bundesanstalt für Land- und Forstwirtschaft (German Federal Biological Research Centre for Agriculture and Forestry): Pflanzenschutz-mittelverzeichnis 1990, Teil 1, 36. Auflage 1990; 6.1 Saatgutbehandlungs-mittel, Teil 5, 37. Auflage 1989/90; Vorratsschutz.

DSE (German Foundation for International Development): Möglichkeiten, Grenzen und Alternativen des Pflanzenschutzmitteleinsatzes in Entwicklungsländern (1987).

DSE (German Foundation for International Development); Zeitschrift: "entwicklung und ländlicher raum" (1988).

Erste Allgemeine Verwaltungsvorschrift zum Bundes-Immissionsschutzgesetz (Technische Anleitung zur Reinhaltung der Luft - TA-Luft), 1986.

FAO: Rice parboiling, Bulletin 56, 1984.

FAO: Rice-husk conversion to energy.

Gerecke, K-H: Vademecum, Teil I - IV (1986), Technische Werte der Getreide-verarbeitung und Futtermitteltechnik, Verlag Moritz Schäfer, Detmold.

GTZ: Aus Abfallbergen Strom für die Energieversorgung.

Heiss, Rudolf: Lebensmitteltechnologie (1990), Springer-Verlag.

Löffler, F: Staubabscheiden (1988), Georg Thieme Verlag, Stuttgart.

Luh, B S: Rice production and utilization (1980), AVI-Publishing Company, Inc., USA.

Mühlbauer, W: Verminderung des Energiebedarfs und Reduzierung der Staube-mission bei Trocknungsanlagen.

Pomeranz, Y: Modern Cereal Science and Technology (1987), VGH-Verlags-gesellschaft, Weinheim.

Rohner, A W: Maschinenkunde für Müller (1986), Versandbuchhandlung DIE MÜHLE, Detmold.

Schäfer, Flechsig: Das Getreide. 5. Auflage (1986), Verlag Moritz Schäfer, Detmold.

Technische Anleitung zum Schutz gegen Lärm - TA-Lärm, 1986.

VDI-Handbuch Reinhaltung der Luft, Band 6
    VDI-3676, Massenkraftabscheider
    VDI-3677, Filternde Abscheider
    VDI-3679, Naßarbeitende Abscheider
    VDI-2263, Staubbrände, Staubexplosionen
    VDI-3673, Druckentlastung von Staubexplosionen
    VDI-2057, Einwirkungen von mechanischen Schwingungen auf den Menschen
    VDI-2711, Schallschutz durch Kapselung.

Trade and Industry

Vegetable Oils and Fats

55. Vegetable Oils and Fats

Contents

1.   Scope

This environmental brief discusses the **extraction** and **processing of oils** and fats from **vegetable sources**.

Vegetable oils and fats are used principally for **human consumption**, but are also used in **animal feed**, for **medicinal purposes** and for certain **technical applications**. They are **extracted** from a range of different **fruits, seeds** and **nuts**. Unlike industrial oils and fats, which are mostly produced from petroleum, they are **generally non-toxic and biodegradable**, without requiring any further treatment. However, they pollute the environment as they degrade due to their **oxygen demand** and their capacity to break down into **water emulsions**. An overview of the main types used is shown in Table 1[1].

Table 1
Use of various fruits, seeds and nuts

Use*)	Seeds	Nuts	Fruit and fruit flesh
For human consumption or medicinal purposes and animal feed	Cotton seed Sunflower seed Soya beans Palm kernels Cocoa beans Sesame seed Corn (germ) Rapeseed Linseed	Coconut Hazelnut Walnut Peanut	Palm fruit Olives
For technical applications and fuel	Castor oil plant Linseed Perilla seed Oiticica seed	---	---

*)   The subdivision into use for human consumption and use for medicinal and technical applications is based on the principle application and may change. For example, rapeseed, palm kernels, soya beans, sunflower seeds and peanuts are potential raw materials for fuel production (Elsbett motor).

---

[1]   Table 1 shows only the most common types. In many countries, a range of other varieties is used in part on a small industrial scale, e.g. rice bran, cashew nut, safflower, mahua, neem, mustard, tobacco, rubber plant, khakhan, dhupa, kokum, thumba seed and others besides.

Production processes for vegetable oils and fats differ according to the required yield and raw material type. They can be categorised as follows:

- fruit processing
- processing of seeds and nuts by mechanical extraction (pressing)
- processing of seeds and nuts by solvent extraction.

Processing, in which the raw materials are separated into **oils** and **oil-bearing solid residues,** comprises the following **operations** after harvesting and any storage:

1. Preparation by raw material husking and cleaning, crushing and conditioning[2].

2. a)   Boiling of the fruit or
    b)   Pressing or pressing and/or
    c)   Solvent extraction of oil-seeds/nuts.

3. a)   Skimming of the liquid oil phase if boiling is carried out
    b)   Filtration of the pressed fat if pressing is applied
    c)   Separation of the crude oil while at the same time evaporating and recovering the solvent where solvent extraction is carried out.

4. Conditioning (drying) and reprocessing of residues.

5. Crude oil improvement by refining
    a)   Degumming
    b)   Neutralisation
    c)   Bleaching
    d)   Deodorisation.

6. Further processing of the refined crude oil.

---

[2]   Conditioning means treating the raw material so that it has certain chemical or physical and chemical conditions in order to obtain the highest possible oil yield from the subsequent pressing operation.

2. Environmental Impacts and Protective Measures

The **intensification of land use** in connection with projects for oil and fat production can have **negative environmental implications** (single-crop agriculture, erosion, water and soil contamination, loss of soil fertility, destruction of wildlife habitats). **Farming methods** and **harvesting practices** must be **controlled** and **optimised** from the outset.

2.1 Hazard potential of the different processing stages

The forms of environmental pollution shown in Table 2 below can arise during **intermediate storage** and the different stages of **processing**.

Table 2
Hazard potential during storage and processing

Type of pollution	Storage	Cleaning Crushing Conditioning	Pressing Boiling	Extraction	Refining Improvement	Packing
Dust		X		X		X
Noise		X		X	X	
Pollutants (including smell)	X	X	X	X	X	X
Wastewater	X		X	X	X	X
Flue gas			X*)			
Waste/special waste		X	X	X	X	

*) From the burning of palm fruit stems, which have a residual oil content of 0.38%, in charcoal kilns.

2.2 Processing of fruits (palm fruit, olives)

Fruits are processed in the producer countries in the tropics (palm fruit) or around the Mediterranean (olives) by **relatively small rural concerns** and by **medium-sized industrial companies**. Figure 1 gives an overview of the various **production processes**, and in the following we examine in detail palm fruit processing.

Fig. 1
Oil production from fruits

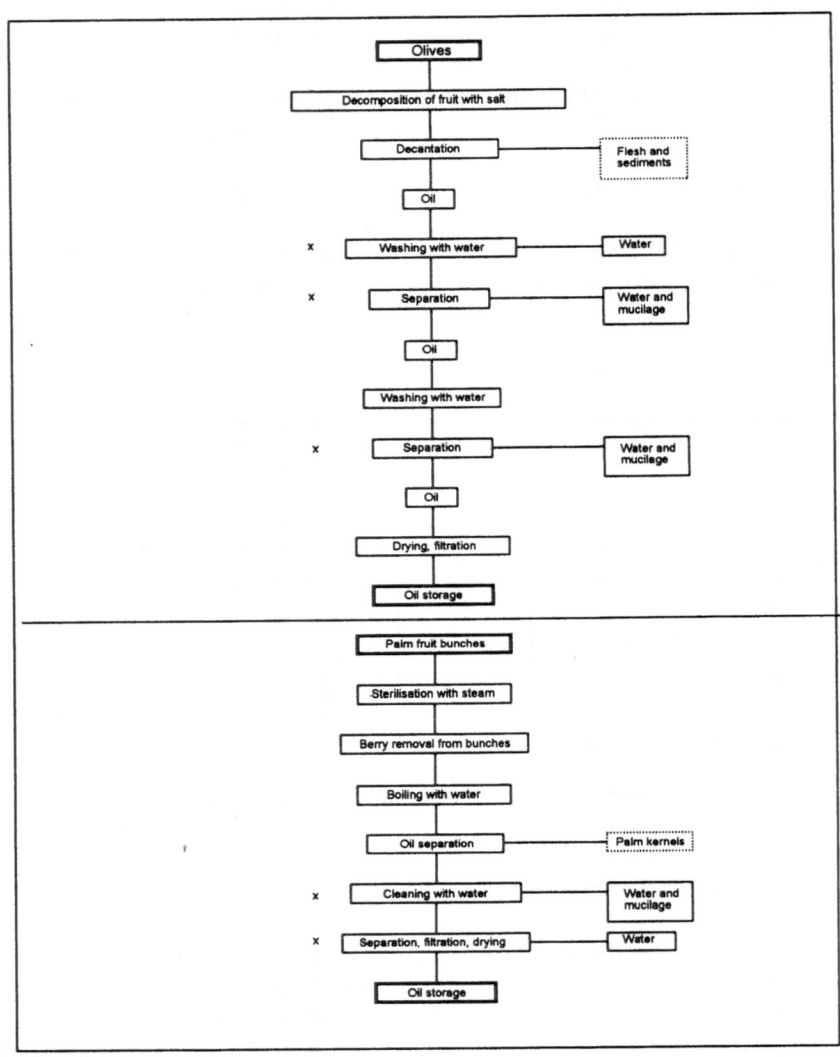

Key:

x   Points of wastewater pollution

           Initial and extraction products

           Intermediate products, process stages

           Waste products

With **palm fruit**, some **2 to 3 tonnes of wastewater** are produced **per tonne of crude oil**. Due to its **organic residues**, the wastewater has a particularly **high biological and chemical oxygen demand** for cleaning (water pollution). Moreover, **dissolved solids** (sludge particles), **oil and fat residues, organic nitrogen** and **ash residues** are the principle constituents of the wastewater.

The first operation in the **treatment and reprocessing of wastewater** is that of **separating settleable solids**. The **residual oil content** is collected in an **oil trap**. There are also **combined sludge and oil traps** which are oil traps with an integrated sludge chamber and are 92% **effective**. A **100% reduction in wastewater and pollutant discharge** into surface water can be achieved by any of the **following measures**:

- discharge by spraying
- discharge by other irrigation systems
- drainage into settling tanks
- drainage into municipal and urban sewage treatment systems.

No **soil conservation problems** due to wastewater penetration have **been reported to date**.

Additional **storage facilities** and areas should be kept in reserve in case of **leaks of solvents, lyes and acids in the event of accidents**, and **equipment** to deal with such accidents should be to hand at all times.

Figure 2 shows a **percentage analysis**, based on 100% palm fruit bunches, which can be used to estimate the potential **waste and wastewater volume**.

55

Fig. 2
Palm fruit processing with percentage analysis

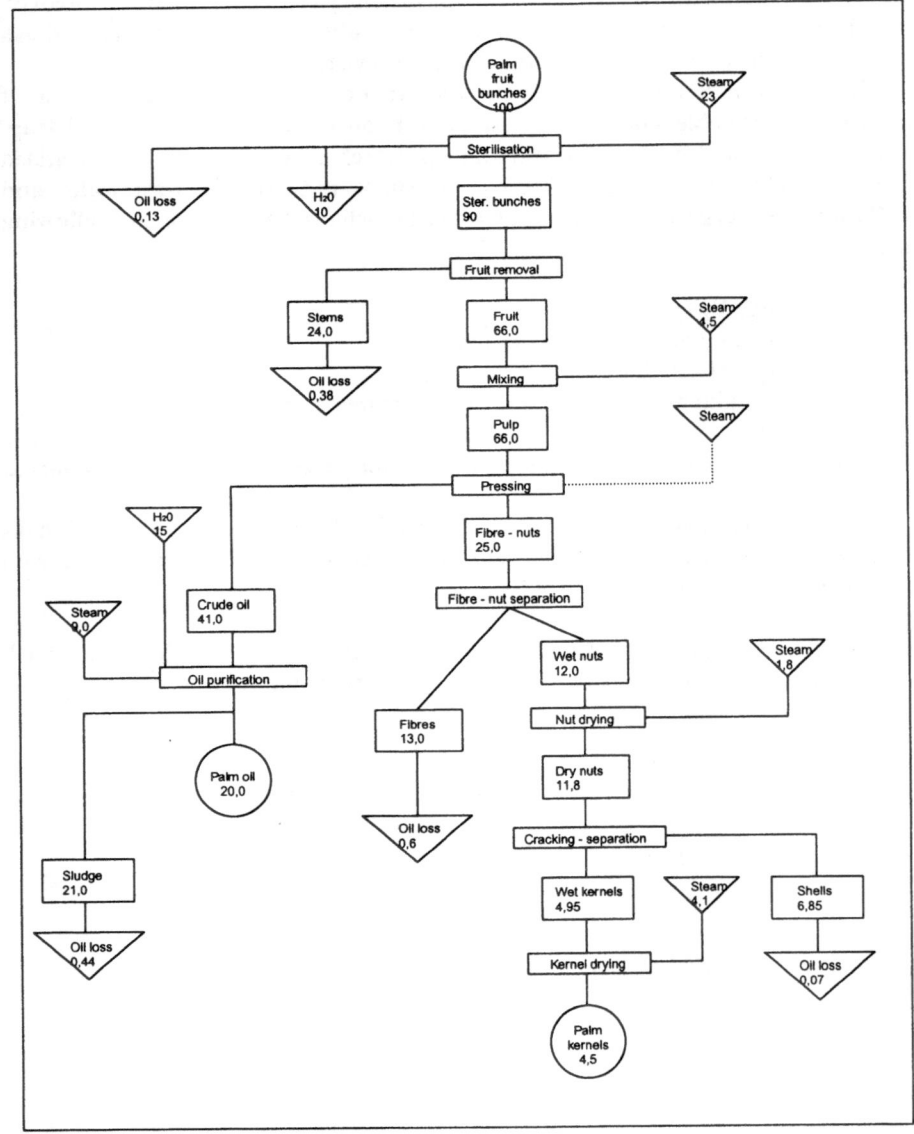

Minimum requirements for **wastewater drainage into watercourses in Germany** are laid down by the 4. Abwasser-Verwaltungsverordnung (4.AbwVwV) [4th Wastewater Administrative Regulation] of February 1987, some details of which are shown in extract form in Table 3 below as a guide.

Table 3
Minimum requirements (from 4.AbwVwV)

	Quantity of waste-water in m³/t initial product	Settleable solids ml/l	Chemical oxygen demand (COD) mg/l		Extractable substances mg/l	
		Random sample	Mixed sample*)		Mixed sample*)	
			2 h	24 h	2 h	24 h
Seed dressing	10	0.3	200	170	30	20
Edible fat and oil	10	0.3	250	230	50	40
refining	10-25	0.3	200	170	30	20

*) Within 2 to 24 hours

An alternative, more environmentally friendly method than draining wastewater into surface water consists of **recycling** the **wastewater** as **process and boiler feed water (circuit system).** The World Bank "Environmental Guidelines" (see item 6 in References) gives a technical description of biological wastewater treatment methods for palm oil extraction plants as practised in Malaysia.

Considerable quantities (per tonne of raw material, approx. 0.7 to 0.8 tonnes) of **waste of vegetable origin** (cellulose, husks, stems, pressing residues) arise during production, and the disposal of them must be taken into account when such facilities are planned. Due to their **content of oil-bearing, organic components,** the stripped bunches pose a **major odour problem,** as do pressing or extraction residues. Transport and dumping should be organised on this basis (e.g. dumping far from populated areas). The remaining solid residues are often **incinerated** to **produce process steam,** although this is **not an ideal form of recycling** as the waste contains **silicates** which vaporise when burnt and form a glassy **coating in the furnace.** It should be ensured that the **incineration process** is **controlled** and **waste air is not** used to **separate the husks from the kernels** (contamination with silicates) as is frequently observed. **Heat exchangers with integrated self-cleaning** systems are one possible solution. The **incorporation of organic waste** (mulch) in farmed **arable soils raises a number of problems** as the **soil cover** could, under certain circumstances, be **destroyed** (erosion risk) if waste were **ploughed into** it. On the other hand, prior **mechanical comminution of the waste** - which would facilitate its application to arable soils - could **nullify its cost effectiveness,** although under certain circumstances it would make a **practical contribution to soil structure improvement.**

2.3   Processing of oil-seeds and nuts

Three different processes may be used to **extract the oil from oil-seeds and nuts**:

-      pressing
-      solvent extraction
-      a combination of pressing and solvent extraction.

Processing produces **waste, dust and odorous substances as well as wastewater** in a quantity of some 10 m^3/tonne seed. Cylinder mills, fans and pneumatic conveyors are also **sources of noise**.

Figure 3 provides an overview of the processes used.

The environmental implications arising and the environmental protection measures which can be taken are described below in the sequence of the individual processing stages.

Fig. 3
Oil production from oil-seeds and nuts

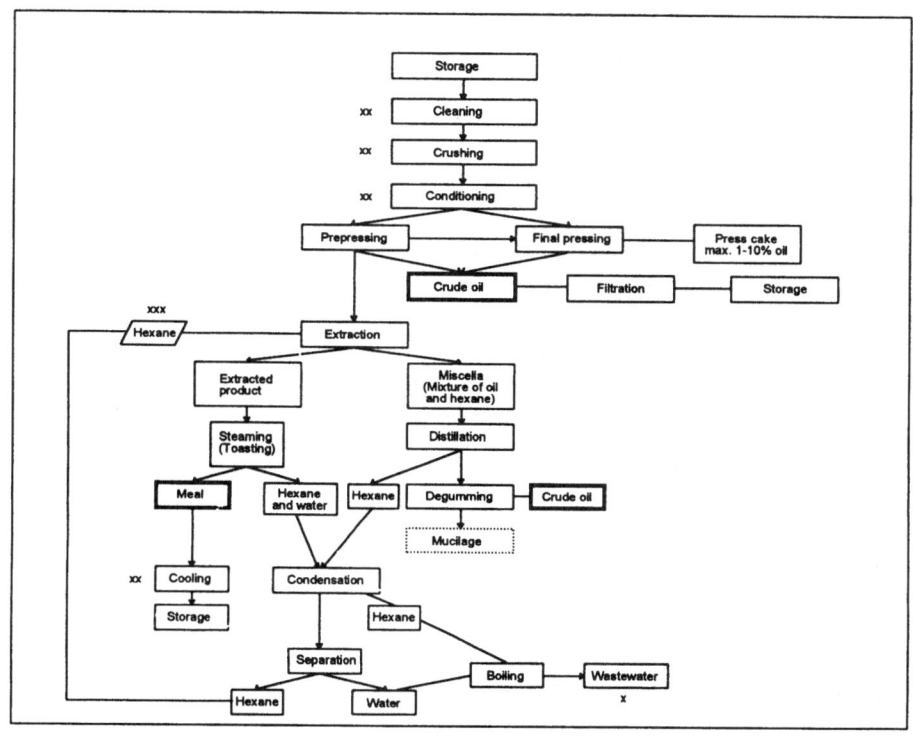

Key:

- ⬭ Additives
- ☐ Intermediate products, process stages
- ◼ Extraction products
- ⸬ Waste

Points of:

- x   wastewater pollution
- xx   Waste air pollution with dust
- xxx waste air pollution with hexane

55

2.3.1   Storage

There are **three methods** of **storage**:

- bagged under cover
- loose in a warehouse
- loose in a silo.

Dust is produced during the **filling** operation in the latter two cases, in variable quantities depending on the equipment used. The dust is of **organic origin and relatively harmless** (direct contact is unpleasant and can cause skin irritation, visual and respiratory difficulties). If only because of **dust explosion risk**, aspiration **(extraction)** is essential for the mechanical processes described below (cleaning, crushing, conditioning). Thus instead of quantities of dust being released during the cleaning, screening or crushing operations, the **dust-laden air is extracted**, collected and cleared of solids via a **central dust-removal installation**, normally **cyclones** (maximum separation efficiency of 95%) or, better still, via **filters** (separation efficiency of up to 99%).

If **mould** should be found and if the presence of **aflatoxins is suspected (in peanuts)**, there is **no risk of contamination of the soil or groundwater under the stores**, as the metabolism of the particular mould fungi limits the presence of aflatoxins to the food product only (peanut kernels). **Preventive measures** (air humidity control and monitoring) and the **regular checking and sorting of stocks** are essential here. Any possibility of the **fungal spore dissemination** must be **eliminated** (prevention of strong air currents, stores to be protected from the wind), otherwise **peanuts not yet affected** can be **infested**, causing **health risks to employees** as the spores can enter the lungs and, once established there, can multiply.

2.3.2   Cleaning and crushing

The **mechanical cleaning and crushing** of oil-seeds and nuts generate **noise and dust**, which can be controlled by aspiration and **dust-extractor installations** (collecting filters, electrostatic precipitators/cyclones) - thereby also **preventing dust explosions**.

2.3.3   Raw material conditioning

Raw materials are generally **conditioned** by the **addition of steam** (heating), an operation which enables the degree of wetting of the product to be controlled. The so-called vapours, the **odorous substances**, are released as **condensate. Gaseous** emissions and emissions of **odorous substances** can be limited by **cleaning the**

outside of machines and pipes with **alkalis** (caustic soda, caustic potash). The **sulphur content** can be **determined** by the **analysis of the local raw material** to be processed, and on the basis of this appropriate **emission monitoring equipment** can be **developed**.

2.3.4 Pressing process

No environmentally relevant substances other than the vapours are produced in the **preliminary and final pressing of oil-seeds**. However, during the **washing** (usually with steam jets) of the fat-sprayed machines, **oily water** is drained into the **wastewater system**. Here too **oil traps** are required. The **heat from the vapours** can be recovered in **heat exchangers** as an **energy-saving measure** and to **reduce odours**.

2.3.5 Solvent extraction

In the **fluid extraction process**, the oil in the unpressed or prepressed products is **chemically dissolved with solvents** and discharged in the form of miscella (oil-solvent mixture) (see figure 4).

The solvent most commonly used is **hexane (C_6H_{14})[3]** which is to be regarded **as both a nerve and an environmental poison**. **Hexane-contaminated production residues** must therefore be **treated or disposed of**. The following can be contaminated with hexane: the air, extracted product, miscella (residual oil-solvent mixture) and water.

---

[3] Hexane is a hydrocarbon of the paraffin group. It constitutes a fire hazard and must be regarded as a nerve poison. At high concentrations, hexane is narcotic and states of intoxication may be observed, although these are overcome quickly and without any consequences for health where oxygen or fresh air is provided. In the case of prolonged exposure, paralysis together with cardiac and respiratory problems arise. Severe poisoning can result in death, in some cases weeks later. Constant exposure causes death by suffocation. Some cases of skin irritations through to necroses (tissue destruction) have been observed as a result of hexane and employees must therefore be given training in the handling of hexane. Surplus quantities, which cannot be released into the environment under the terms of discharge regulations (e.g. 4. AbwVwV in Germany) must be disposed of as special waste. In storage, the general regulations applicable to the handling of chemical products should be observed. Hexane can be stored in drums under stands fitted with extractor systems and collector sumps. Another solvent which is sometimes used is benzol, but it is not recommended in view of its high level of toxicity and other problems.

2.3.5.1 Air polluted with hexane

- is formed due to **leaks** in the **plant** and the **conveying pipes**.

Hazards: Air-hexane mixture is explosive once the explosion threshold of 1 to 7% is reached.

Remedy: The concentration is measured with probes at suitable points (conductivity meters) and an alarm triggered if the threshold is exceeded. Particular care must be exercised when entering tanks and in all cases fumes must first be removed.

- is formed during the **extraction process** in the extractor and during the subsequent **steam treatment** of the extracted product in the toaster.

  The **waste air** can be treated by **absorption plants**, in which the air is fed through a mineral oil bath and the hexane transfers from the air into the mineral oil. The **hexane pollution** in the **waste air** released into the atmosphere should not exceed 150 mg hexane per m^3 air at a mass flow of 3 kg/h. The explosion safety threshold is 42 g/m^3 air.

2.3.5.2 Extracted product polluted with hexane and residual hexane-oil mixture (miscella)

The **solid raw material residues** and the **miscella** are largely **stripped of hexane** by **steam** distillation, in which **meal** (animal feed) and a **water-hexane mixture** are produced, or where **hexane and crude oil** are separated out from the **miscella**. The **hexane can be collected and reused** (hexane recycling).

The **hexane content of the meal** must not exceed 0.03% for transport safety reasons. As hexane is heavier than air, there is a risk with lengthy transport times that the hexane could sink and concentrate, thereby exceeding the explosion safety threshold (42 g/m^3). As hexane **vaporises relatively quickly**, no **consequences** have yet been **observed** with regard to the **health of cattle fed on the meal**.

2.3.5.3 Hexane-water mixture

If **hexane-contaminated wastewater** is to be disposed of, 50 parts per million (ppm) hexane, for a total wastewater quantity of 3 - 5 m^3/t feedstock, should not be exceeded.

Hexane-water mixtures are **separated** by the density difference and the (theoretical) insolubility of the two media in each other, in order to **condition (produce) disposable wastewater**. They are separated by the drawing off of the two fractions in a settling tank at 40°C. Water, as the heavier fraction, is drawn off at the bottom, while the lighter hexane, which floats, is pumped off from the top. Cooling to 40°C is essential so that the separation operation is carried out well below the boiling point of hexane (68°C). The **residual hexane content in the wa-**

ter is **reduced** by **evaporation** in a boiler (90°C, to stay below the boiling point of water).

2.3.5.4 Wastewater polluted with hexane

The total quantity of water supplied in the form of steam which is added is 12%, related to the quantity of raw material used in the steam treatments (see 2.3.3). 50% of this remains in the meal, the other half being converted into the liquid state by condensation. Thus **some 0.06 m³ wastewater per tonne of feedstock is contaminated with hexane.** It is not possible to give more precise details about **potential risks to the environment** arising in tropical areas due to non-compliance with this limit (long-term consequences of possible damage to the ecotope) as **research** in this area is woefully **inadequate**.

2.3.6 Refining

The **oils** produced by extraction must - for reasons of durability, taste, appearance and consistency - be cleared of **impurities** such as free fatty acids, particles of dirt and seed, lecithin, carbohydrates, fats, gummy or mucilaginous substances, pigments, waxes and oxidation products. **The purpose of refining** is to **remove undesirable by-products whilst retaining desirable ones,** e.g. vitamins, antioxidants (tocopherols) or certain technical properties. Refining comprises basically the **degumming, neutralisation, bleaching and deodorisation** of the crude oil, and it is in these processes that most of the **wastewater** and **unpleasant odours** are produced. The lyes and acids used in the process bring with them a potential **risk of injury to personnel** (safety measures and training necessary). Figure 4 illustrates the refining process schematically.

Either a **chemical or physical method** can be used for **oil neutralisation** (removal of free fatty acids). The **chemical process** involves the neutralisation of acid using **caustic soda**, whilst the **physical process** neutralises by **steam distillation.** Physical neutralisation is the norm for palm, coconut and palm nut oil, whereas cottonseed and sunflower oil are generally also neutralised chemically as steam distillation is inadequate in view of the high lecithin content.

55

Since the **treatment of the wastewaters formed is easier, and the quantity of wastewater lower during the physical process,** efforts are being made the world over to develop processes which separate off the lecithin in the said oils so that they can be neutralised physically.

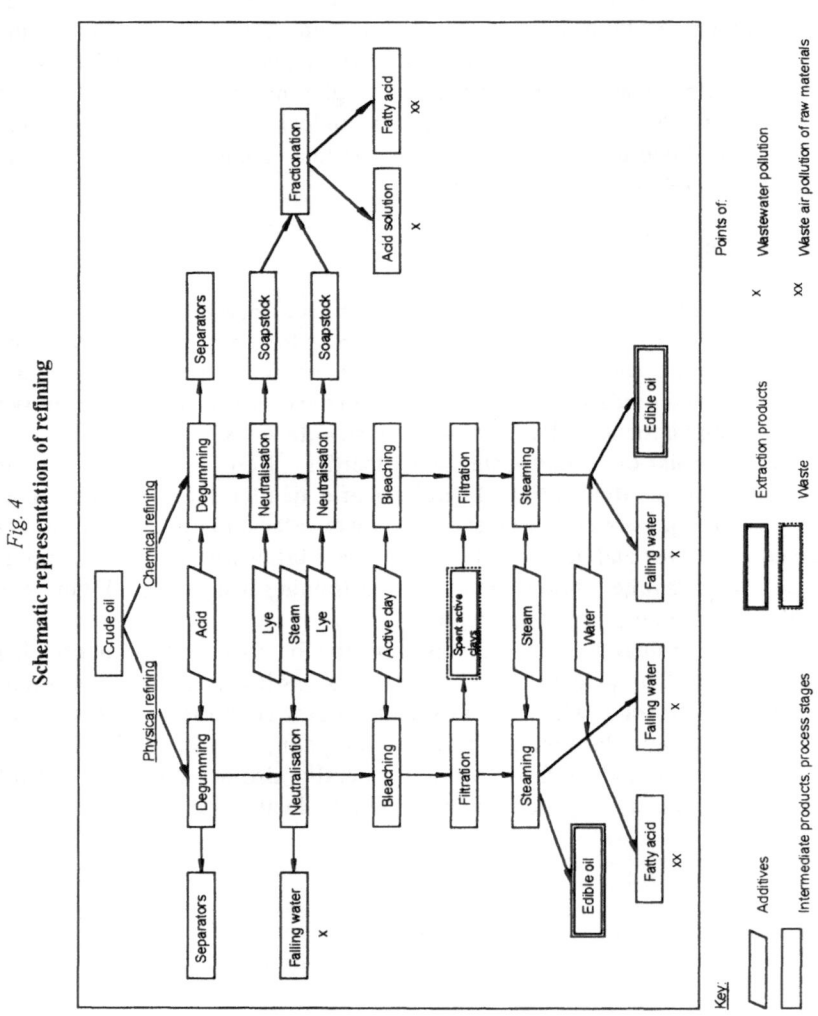

Fig. 4

Schematic representation of refining

2.3.6.1 Physical refining

During the physical process, the **preliminary stage** involves **degumming** the oil, normally with **phosphoric acid**, which coagulates and precipitates proteins which are then removed in separators. The **separated solid matter** is added to the meal, from which **animal feed** is made. To prevent phosphate discharge into the refinery wastewater, phosphoric acid is now being replaced by citric acid, which does not degrade into pollutants because of its organic origin, among other things.

The **degummed crude oil** is then **bleached** with active clay (clay with a high silicate content)[4], since the natural pigments of the crude oil are adsorbed into the active clay and absorbed into the active clay bed. One of two possible processes is used to **recover the residual oil** which the **spent active clay** contains. In smaller plants, a **steam treatment** is used to recover at least some of the oil, but **wastewater** is also produced. In large-scale plants, all the oil is removed from the active clay in **special extraction installations**. The **oil recovered** in this way is of an **inferior quality**. The process itself produces **wastewater** and **waste air** which contain **solvent residues**, and must be **clarified** or **purified** (separators, filter installations).

Extracted active (bleaching) clay can be **dumped** without **harming the environment**, and provision must be made for dumps at the planning stage. **Non-extracted active clay** can also be dumped without any direct environmental hazard, although there is the problem of **odour** as the oils contained degrade enzymatically thereby producing, amongst other things, sensorily active fatty acids which give off a rancid odour. The proportion of active clay used is around 3 - 5 percent by mass in relation to the crude oil used.

During the subsequent **steaming** process, **odorous substances (aromatics) and flavourings** and approx. 20 - 100 kg of **fatty acids** per tonne of oil are **stripped** (at 180 to 270°C under a light vacuum of 4 to 10 mbars) by steam distillation. The steaming vapour is first fed through separator devices, such as hydrocyclones (centrifugal separators) to remove the oil droplets entrained with it and the fatty acids, and then condensed by direct contact with cooling water and recirculated. Using this method, only **small quantities of wastewater** are produced and they can **be treated biologically** with a maximum fat quantity of 20 - 25 mg/l wastewater. The oil-contaminated **fatty acids** can in turn be processed further in soap factories for **soap** production or in the **chemical industry** to manufacture other products.

---

[4] In some countries, charcoal is used for bleaching, but should be avoided in view of the shortage of resources.

2.3.6.2 Chemical refining

In the chemical process, the **crude oil** is first **degummed** and then immediately **neutralised** in one process stage. First, **phosphoric acid** (or more recently **citric acid**) is added to **degum** the crude oil **by precipitating the protein**. Then, in contrast to the physical separation, the **acidic crude oil** - acidic due to the free fatty acids it contains (2 - 10%, depending on the oil-seed and storage conditions) and the citric or phosphoric acid added - is **neutralised** by the addition of lyes, usually soda lye. This yields a mixture of neutralised oil, mucilages and soap-stock.

After separation, the **crude oil** obtained is **bleached** and **steamed** as in physical refining. The same by-products are also produced, although the **active clay consumption** is considerably **lower**. Moreover, the steaming operation yields only about one tenth of the oil droplets and fatty acids obtained in physical refining.

2.3.6.3 Processing of soaps and mucilage

Disposal problems are associated with the **processing of soapstock and mucilage**. The **soap** is first **boiled** and **separated** with sulphuric acid (to break up the emulsion). This produces fatty acids which can be separated from the acid solution in settling tanks. The acid solution is then neutralised and cooled with slaked lime. **Organic substances** should be separated by **mechanical or biological processing**, and the remaining **wastewater** must comply with the following conditions for drainage (standard German values as a guide):

- maximum temperature 35°C
- max. sulphate content due to addition of sulphuric acid 600 mg/l.

The **quantity of wastewater** from chemical wet neutralisation and the subsequent soapstock fractionation is around **0.5 m³/t** of initial product under modern production conditions. This is only equivalent to about 5% of the total wastewater from a refinery, but because of the **high organic content** and consequently the much **higher Chemical Oxygen Demand** (COD), this alone amounts to 50 - 60 % of the admissible total COD load of a refinery in Germany. The **discharge of wastewater** must therefore be **inspected** to ensure compliance with the relevant limit values.

2.3.6.4 Comparison of physical and chemical refining based on environmental factors

Wastewater quantities from neutralisation, particularly where there is a preliminary condensation vapour stage, can be considerably **reduced** by the use of the **physical** distillation process. However, this process, compared with the chemical refining process, consumes a far higher quantity of active (bleaching) clay. For reasons of economy therefore, **chemical refining** is popular although - as described above - it is characterised by the generation of large quantities of **heavily contaminated wastewater** which requires checking at the point of discharge into sewers and/or natural bodies of water to ensure that limit values are observed. **Physical refining** is **preferable to chemical refining** as active clay has a lower environmental impact.

3. Notes on the Analysis and Evaluation of Environmental Impacts

3.1 Air

Limit values for air pollution are laid down in Germany by the guidelines of the *Technische Anleitung zur Reinhaltung der Luft* (*TA-Luft* - Technical Instructions on Air Quality Control), and comply with the *Bundes-Immissionsschutzgesetz* [*BImSchG* - Federal Immission Control Act]. Further reference material is to be found in the ***Richtlinien des Vereins Deutscher Ingenieure*** [Guidelines of the Association of German Engineers VDI] for **maximum immission concentrations** [MIK] which are concerned with the establishment of limit values for certain air contaminants.

Under the terms of TA-Luft, the **dust emission** for organic substances in industrial plants may not exceed 50 mg/m³ air at a mass flow of 0.5 kg/h. The **waste air from extraction** today may not contain more than **150 mg hexane/m³**.

3.2 Noise

Where **noise levels** are over 70 dB(A), **noise-reducing measures** such as hearing protection (ear muffs etc.) or silencing devices on machines must be provided. For comparison: leaves rustling in the wind produce a noise level of 25 - 35 dB(A) and normal conversation ranges from 40 - 60 dB(A). Years of exposure to around 85 dB(A) or more during most of a person's time in work is deemed **damaging to the hearing** at the **workplace**. It is, in fact, just as harmful to be exposed constantly to a uniformly low noise level as to a correspondingly higher one for a short time.

In Germany, the *Technische Anleitung zur Lärmbekämpfung* [*TA-Lärm* - Technical Instructions on Noise Abatement] establishes principles for noise protection and **approximate emission values**. The following approximate atmospheric immission values[5] apply to:

- areas containing mainly commercial premises: daytime 65 dB(A), night 50 dB(A),

- areas containing mainly residential accommodation: daytime 55 dB(A), night 40 dB(A).

It should be borne in mind that noise immissions inside buildings can have more damaging consequences because of building materials and methods used, in which case the values should be assumed to be correspondingly lower.

3.3   Wastewater

In **crude oil refining**, a **wastewater quantity** of **10 to 25** m³/t initial product must be assumed (see Table 3, details from Germany's 4. AbwVwV). The following are the principal **constituents of the wastewater**:

- sodium sulphate or sodium chloride
- calcium phosphate
- fatty acids (in part as calcium soap)
- mono-, di- and triglycerides
- glycerin
- protein
- lecithin
- aldehyde
- ketones
- lactones
- sterines.

A **refinery's wastewater output** can be **reduced by up to 90%** if the **vapour cooling water** is managed in a **circuit** - a system however, which results in **higher COD concentrations in the circuit water**. The minimum requirements for the final discharge of refinery wastewater must take account of this circumstance. However, despite the higher COD concentration where the cooling water is managed in a circuit, there is an **overall general reduction in pollutant load**. Biological wastewater treatment cannot yet be described as the most modern state-of-the-art

---

[5]   Immissions are the effect of air-borne contaminants, vibrations, radiation and noise on humans, animals, plants and property (e.g. buildings).

process in view of the land required, the higher energy consumption and the problem of sludge disposal.

In Germany, the *4. Allgemeine Verwaltungsvorschrift [4. AbwVwV* - 4th General Administrative Regulation] establishing **minimum requirements** for the discharge of wastewater from oil-seed processing and fat and edible oil refining applies with regard to **wastewater control.**

Table 4 indicates the minimum requirements imposed on the wastewater. In works with **biological wastewater treatment**, a **5-day biological oxygen demand** (BOD$_5$) of 25 mg/l and a **chemical oxygen demand** (COD) of 100 mg/l is prescribed[6].

55

Table 4
Minimum requirements of the 4.AbwVwV

Parameters	Quantity of contaminated water	Settleable solids	COD		Extractable substances	
Dimension	m^3/t (1)	mg/l	mg/l*)		mg/l*)	
Type of sample		random sample	2h	24 h	2 h	24 h
Oil-seed processing	10	0.3	200	170	20	20
Crude oil refining for edible oil production	10	0.3	250	230	50	40
	10 - 25	0.3	200	170	30	20
Analysis process		DEV H 2.2 (2)	Appendix to 2. AbwVwV of 10.01.80 (3)		DEV H 17/18-1	
Analysis of measured values		Single value or mean value (4)	Single value or mean value			

*) Within 2 or 24 hours

---

[6] BOD$_5$ stands for the biological oxygen demand which is required by microorganisms in a five day period for the processing of organic substances in industrial water.
   In the case of COD, the quantity of oxygen produced by an oxidant to oxidize organic substances in wastewater is calculated.

(1)   Initial product
      In crude oil refining for the production of edible oil and edible fat, the fol-
      lowing initial products are used:
      -       crude oil, produced in the oil extraction process
      -       reject and reprocess batches passing through the refining process once
              again.
(2)   German standard process for water testing.
(3)   If the value specified for settleable solids are exceeded in a single sample, 0.3
      ml/l can be used to obtain the arithmetical mean if the dry mass of the filter-
      able substances (DEV H 2.1) does not exceed 30 mg/l.
(4)   Analysis of the precipitated sample.

The values given in Table 5 should apply for the **discharge of acid solutions** from
soap fractionation.

Table 5
Limit values for the discharge of acid solutions from soap fractionation

Quantity	0.3 m^3/t oil
Maximum temperature	3°C
pH value	6.0 - 9.0
Settleable solids which can pre-cipitate in 30 mins.	10 mg/l
Fat	250 mg/l
SO$_4$	600 mg/l

Generally speaking, **oil processing** is **linked to a laboratory** in which checks are
constantly carried out using a standardised measuring procedure. **COD, BOD, special waste** requiring disposal, **dissolved solids** and the **oils** and **fats** should be constantly tested for content. Regular **temperature checks** should also be carried out
in situ.

In addition, the World Bank gives the following **information** relating to the **wastewater** in question here:

-       In principle cooling water should not be discharged; if it cannot be recycled
        in the circuit, it should only be discharged if the temperature of the water into
        which it is released does not rise by more than 3°C,

- The pH of the wastewater and liquid waste should be kept constant between 6.0 and 9.0,

- The BOD value of the wastewater should be less than 100 mg/l,

- The COD value of the wastewater should be less than 1000 mg/l,

- The dissolved solid content of the water should be less than 500 mg/l,

- Additional preservation and storage facilities and areas should be kept available in case of accidents resulting in leaks of solvents, lyes and acids. Equipment should also be kept to hand to deal with any such accident situations.

3.4   Waste

Types of waste defined in § 2 of the German *Abfallgesetz* [Waste Avoidance and Waste Management Act] are determined for oil- and fat- producing facilities in the ordinance on the determination of waste and residues *Verordnung zur Bestimmung von Abfällen und Reststoffen*. Waste types are further qualified by waste codes in accordance with the information bulletin entitled "Abfallarten der Landesarbeitsgemeinschaft Abfall" [LAGA - Waste types of the state working group on waste]. Waste groups 52 (acids, lyes, concentrates) and 55 (organic solvents, dyes, varnishes, adhesives, fillers and resins) are relevant.

3.5   Soil

Soil contamination problems in the production of vegetable oils and fats only occur in connection with **improper disposal** of **waste** and **wastewater** (see also sections 3.3 and 3.4).

3.6   Choice of site

The following is to be borne in mind when **choosing a site**:

- The plant or plant complex should not be sited near ecologically sensitive habitats (marshlands, protected areas, national parks etc.).

- Local resource protection agencies or authorities should be involved at an early stage in the selection of the site or alternatives.

- Because of the odour nuisance, the plant should not be sited in the immediate vicinity of residential areas. Generally speaking, plants should be sited on high ground above the local topography; the sites should not constitute areas through which air passes and the prevailing wind currents must not affect populated areas. Local climatic and meteorological studies can be used to obtain useful information.

- The plant or plant complex should be built close to surface water (preferably flowing water), and this water must have the maximum dissolving and absorption capacity for wastewater.

- At the site it should be possible to recycle wastewater - following minimal treatment - for agricultural and industrial purposes.

- The plant should only be built in a municipality if the production wastewater can be treated in the municipal sewage system.

Processing facilities for fruit are sited **in the actual growing areas** as the crop has to be processed immediately after harvesting. Economic processing capacities in industrial countries start at 15 to 20 t feedstock per day.

Oil-seeds and **nuts** are transported in some cases over **long distances to the processing industries**. Processing capacities for **pressing plants** start at around 200 t/day, and those of solvent extraction plants at around 100 t/day. In highly industrialised countries, however, capacities of 1000 to 2000 t/day are commonplace. **Refineries** can operate economically from 50 t/day but plants in industrialised countries have processing capacities of 100 to 300 t/day. The question to be answered when deciding on the capacities to be installed is whether it would be preferable (for reasons of environmental protection and employment) to have **small, decentralised plants rather than one large plant**. Wastewater and waste air treatment systems and likewise waste disposal can also be organised decentrally as can the operation of a test laboratory.

3.7 Transport

Decentralised processing can obviate the need - as sometimes happens in view of the amount of **traffic to and from large plants - to rethink local transport routes and traffic plans,** and can avoid the associated **noise, air pollution** and **traffic jams,** as well as the **risks to pedestrians** from heavy goods traffic transporting raw materials or products to or from the plant. A transport sector and traffic study should be produced for route selection and for the analysis of problems and possible remedies.

55

4. Interaction with Other Sectors

Oil-cake or **meal** are **by-products** of **crude oil production** and are frequently processed further in the same plant to make **animal feed** (see environmental brief Livestock Farming).

As **soap and fatty acids** are produced in the refining process, a **soap factory** can be built alongside the plant. This eliminates problems with the sale of fatty acids or the fractionation of the soapstock (acid solutions). Likewise, the production of edible oil or edible fat in the refinery can be linked to the production of baking or cooking fat, shortenings or margarine.

Filling installations are often linked to refineries as edible oil and edible fat are now sold almost exclusively in a packaged form. The linking of filling installations to refineries is advantageous as the oils and fats are packed immediately and so have no chance of becoming rancid, and **plant wastewater** produced during the filling process can be **treated** and **disposed of** with the wastewater from the refinery (see environmental brief Wastewater Disposal).

Steam is required to produce and refine vegetable oils in small and large plants, thus oil mills or refineries often operate their own **steam production plant.** National provisions for **large furnace installations** must be observed in this regard (in Germany: the TA-Luft).

5.   Summary Assessment of Environmental Relevance

5.1   Crude oil extraction

In the **extraction of vegetable oils** from fruit and seeds, the cleaning, crushing and conditioning operations generate **dust** which it must be possible to remove with **centrifugal cyclones**. Dust is also produced when meal and press cake are made and it must be possible to remove it in the same way.

As this **dust** is **of vegetable origin**, it can be **biologically degraded** without the need to take any technical environmental protection measures (small plants), or it can be used as **fertiliser** (castor-oil seed meal). Production plants should not, however, be sited close to populated areas. The same is true of the larger quantities produced by large-scale plants in which the dust, after extraction, must be collected and dumped in a well-organised manner.

When oil-bearing fruits are **extracted** and **boiled**, large quantities of **wastewater** are produced which can be degraded **biologically**, albeit at the expense of **a high oxygen consumption**. For this reason, **mechanical pre-cleaning** is necessary - an operation in which sediments are precipitated and removed from time to time.

All oil mill **wastewater** should be fed through **oil separators** as larger quantities of wastewater containing vegetable oil cause the formation of a thin film of oil on bodies of water which interferes with the oxygen supply. Wastewater with an excessively high oil content must also pass through a **biological treatment plant** in which organic substances can be degraded by constant aeration (oxygen supply).

Extraction processes produce **wastewater** which can **contain solvents**. Measures must therefore be taken to ensure that the maximum solvent discharge into the environment is not exceeded.

5.2   Crude oil refining

Large quantities of **falling water** are produced in the **refining** of crude vegetable oils and fats and these are passed through oil separators for disposal. The **water** is **fed back** into the refinery after passing through a recooler. Excess falling water can be discharged into the surrounding water once it has passed through a **biological treatment plant** in which organic substances are degraded.

When **soapstock is separated into fatty acids, acid solutions** are produced which can no longer be used in the circuit. **Before their discharge** into treatment plants or water, they must be **specially treated (neutralised)** as they are acidic and contain - in addition to fat and mucilaginous substances - **sulphate ions**, which must not exceed certain values as they lead to **salination of the wastewater** and **can destroy concrete drainage pipes**.

A **100% reduction in wastewater and pollutant discharges into surface water** is considered to be feasible if **one of the following measures** is taken:

- discharge by spraying
- discharge by other irrigation systems
- drainage into settling tanks
- drainage into municipal and urban sewage treatment systems.

In oil and fat production, the **environmental pollutants formed** and released in waste air or wastewater depend largely on **compliance with technical tests and countermeasures for environmental safety purposes**. **Constant supervision** of normal plant operation is essential to ensure that limit values are observed in terms of pollutant removal. **Personnel** is to be appropriately **instructed in the framework of continuous training and upgrading**. Training and preparation campaigns aimed at women are recommended for environmental protection jobs.

In view of the potentially serious negative implications for the environment and health in large plants in particular, the appointment of safety, environmental protection and works officers should also be demanded and the women concerned should be involved in the selection process.

55

6.   References

Abwassertechnische Vereinigung (ATV) (Ed.): Lehr- und Handbuch der Abwassertechnik, Bd. I - VI, Ernst Verlag, Berlin, various years.

Abwassertechnische Vereinigung (ATV) (Ed.): Arbeitsblatt A 115, Hinweise für das Einleiten von Abwasser in eine öffentliche Abwasseranlage, draft of 22.03.1990.

Adam, W.A.: Waschmittel und Gewässerschutz, FSA 77, 1975.

40. Anhang zur Allgemeinen Rahmen-Verwaltungsvorschrift über Mindestanforderungen an das Einleiten von Abwasser in Gewässer, GMBI. (joint ministerial circular) 1989, Nr.25, page 517 ff.

Baily's Industrial Oil and Fat Products, Vol.2, 1982.

Brammer, H.: Industrielle Verarbeiter von Speisefetten im Lichte von Umweltfragen, FSA 75, 1973.

Brauch, V.: Einsatz von physikalisch-chemischen Reinigungsmitteln in der Fettindustrie, FSA 84, 1982.

Brunner K.H.: Kontinuierliche Alkaliraffination und on-line Verlustanzeige, FSA 88, 1986.

Conze, E.: Abwasserschlämme in der Speisefettraffination und ihre Entstehung, Möglichkeiten zur umweltfreundlichen Beseitigung, FSA 84, 1982.

Deutsches Einheitsverfahren zur Wasseruntersuchung: Verlagsgesellschaft Deutscher Chemiker, Weinheim a.d. Bergstraße, loose-leaf collection, last issue on 26.03.92.

Dieckelmann, A., Hirsch A. et al.: Abluftverbrennung und Abluftnutzung aus ölchemischer Produktion, FSA 85, 1983.

Jennewein, H.: Über Wasserreinigung in Ölmühlen, FSA 75, 1973.

Jongenelen C. H. and Veldhoen: Fermentation von Abwässern in einem Säulenfermentator, FSA 82, 1982.

Kaufmann H.P. et al.: Technologie der Fette, Verlag Aschendorff, Münster, 1968.

Krause, A.: Abluftprobleme in der Fettindustrie und in verwandten Gebieten, FSA 80, 1978.

Krause, A.: Pflanzliche und tierische Fette und ihre Wirkung auf Mikroorganismen in biologischen Kläranlagen, FSA 85, 1982.

Lehr- und Handbuch der Abwassertechnik, 3. Auflage (Ernst u. Sohn), Bd. 5, 1985.

Liebe, H.G., Münch, E.W.: Neues Verfahren zur Reduzierung geruchsintensiver Emissionen, FSA 88, 1986.

Air Quality Guidelines for Europe, WHO regional publications European series: No.23/1987).

Mahatta, T.L.: Technology and Refining of Oils and Fats (Production and Processing of Oils and Fats). Delhi: Small Business Publ. without year of publication, 360 pages (SBP Chemical Engineering Series. 49).

Morger, M.: Abwasseraufbereitung in Betrieben der Speisen-, Fett- und Molkereiprodukteindustrie in werkseigenen Kläranlagen, FSA 88, 1986.

Niemitz, W.: Abwasser und Abfall. Schwerpunkte der Umweltprobleme industrieller Produktionen, FSA 75, 1973.

Nösler, H.G.: Umweltschutz zwischen Wunsch und Wirklichkeit, FSA 86, 1984.

Organisch verschmutzte Industrieabwässer in Nahrungsmittel-, Genußmittel und Getränkeindustrie, 1984.

Pardun, H.: 50 Jahre Technologie pflanzlicher Öle und Fette, FSA 85, 1983.

Schmidt-Holthausen, H.J.: Verfahren zur Abwasser-Aufbereitung in der Speisefett- und Fettverarbeitenden Industrie, FSA 81, 1979.

Segers, J.C.: Möglichkeiten und Beschränkungen bei Verringerung der Umweltbelastung infolge der Raffination von Ölen und Fetten, FSA 87, 1985.

Segers, J.C.: Superdegamming. A new degamming process and its effect on the effluent problems of edible oil refining, FSA 84, 1982.

TA-Luft, Technische Anleitung zur Reinhaltung der Luft of 27.02.1986, GMBI. (joint ministerial circular) p.95, rep. p.202.

55

TA-Lärm, Technische Anleitung zum Schutz gegen Lärm of 16.07.1968, appendix to volume no.137 of 26.07.1968.

Umweltbundesamt - German Federal Environmental Agency (Ed.): Handbuch Abscheidung gasförmiger Luftverunreinigungen, Erich Schmidt Verlag, Berlin 1981.

VDI (Verein Deutscher Ingenieure): VDI-Handbuch Reinhaltung der Luft, Beuth Verlag, Berlin and Cologne, Feb. 1992.

VDI (Verein Deutscher Ingenieure): VDI-Richtlinien zur Geräuschmessung, Schallschutz, Schwingungstechnik: 2560, 2564, 2567, 2570, 2571, 2711, 2714, 2720, 3727, 3749, 3731, 3742.

Verordnung über die Herkunftsbereiche von Abwasser of 03.07.1987, BGBl. (Federal Law Gazette) I, p.1529.

World Bank: Environmental Assessment Sourcebook, Volume II Sectoral Guidelines, Environment Department, Technical Paper Number 140, Washington D.C., 1991.

World Bank: Environmental Guidelines, Environmental Department, Washington D.C., August 1988.

Zockoll, C.: Sicherheitstechnische Fragen beim Silobetrieb, FSA 81, 1979.

Trade and Industry

Sugar

56. Sugar

Contents

1.   Scope

Sugar is the only food extracted from **two different plants** - the **sugar beet** (beta vulgaris) and the **sugar cane** (saccharum officinarum) - which grow in different regions. **Competition** between these two international cash crops only arises in **small border areas** where both are considerably below their physiological optimum, usually latitude 25 to 38 degrees north. The main **sugar beet** growing areas are located in the **temperate climate** of Europe and North America in regions with average **mid-summer temperatures** of 16° to 25°C and an annual rainfall of at least 600 mm, but sugar beet is also **grown** in the sub-tropics in the **winter months**. Irrigation is essential where the rainfall is less than 500 mm. Beet thrives best on deep **loamy soils** with a neutral to weakly alkaline reaction, and in intensive farming requires adequate mineral **compound fertilisation**. Since beet can only be grown **in the same field every fourth year** to ensure a healthy crop (avoiding, for example, beet nematode, the main cause of the disease known as beet sickness), the catchment area of a sugar beet factory is very large. The **vegetation period** is generally five to six months, with **yields** in a temperate climate ranging from 40 to 60 t/ha, and in the sub-tropics averaging 30 to 40 t/ha. The **sugar content** ranges from 16% to 18%. **Sugar cane** grows in **tropical lowland climates**, and is farmed almost exclusively between latitude 30° south and 30° north, and particularly between the north and south 20°C isotherms. Besides intensive sunlight, a **rainfall** of at least **1,650 mm** or irrigation is essential. **Heavy, nutrient-rich soils** with a high water capacity are preferred; pH values in the weakly acidic to neutral range are best. **Nutrient requirements are high** due to the huge mass production. Pest and disease attack have been reduced by resistance breeding and plant development, with **biological pest control** playing an increasingly important role. **Sugar cane** is **suitable** for monoculture and is indeed mainly grown as such. **Plant cane** is generally **harvested** after 14 to 18 months, and the new growth (ratoon) after 12 to 14 months. **Yields** are from 60 to 120 t/ha; the **sugar content** is on average 12.5%. Harvest quantity and sugar content decline as stocks age, with the result that the total useful life does not normally exceed 4 to 5 harvests.

 Sugar factories are agro-industrial centres which contract out the cultivation of their **raw material** or, alternatively, grow it themselves, have their own **energy and water supplies** and large, varied **workshop complexes**. The plant installed is designed to handle a single natural raw material. Where used for direct processing of the harvest, the seasonal processing period coincides with the period of use of the sugar production plant. New plants process between 5,000 and 20,000 t daily (24 hr), although in order to handle 10,000 t/day, sugar production plants must have an appropriate infrastructure. The **production plant** should be situated wherever possible **in the centre of the cane or beet growing area**, it should be close to water and should be connected to the public railway and road networks. The **by-products** arising during sugar manufacture - molasses, sliced beet and cane bagasse - are

56

either used or **processed further** in the plant, or alternatively form the raw materials for other industries.

●   **Harvesting, storage and cleaning of the raw material**

Sugar beet is harvested almost exclusively **mechanically** while **sugar cane** in contrast is harvested largely **manually** (cutting of stalks). The raw material is then transported to the factory by rail or road. Exceptionally, sugar cane is transported by water. **Sugar beet can be stored** for one to three days, depending on the temperature and method of storage, whereas **sugar cane cannot be stored** and must be processed immediately after harvesting or in any case no more than 12 hours later; **sugar losses** of up to 2%/24 hr are possible. Sugar beet is always **washed** before processing, but sugar cane is usually only washed if it has been harvested by machine.

●   **Cutting, crushing and extract purification**

Sugar beet is **chopped** in slicing machines, and **sugar is extracted** from the slices in the countercurrent with water at 60 - 70°C in an extraction plant; the water is then removed mechanically and before drying the extracted beet is usually mixed with up to 30% molasses, normally made into pellets and used as **animal feed**. Because of their residual sugar content (approx. 0.8%), the slices - after drying - can be used as silage (preserved by fermentation) and as an agricultural feedstuff.

Sugar cane is **prepared** by revolving knives, crushers and/or shredders and then the juice is extracted in four to seven rollers in line or is **extracted** like sugar beet in a diffuser. A fibrous residue (bagasse) with a low sucrose content is produced at a rate of 25 to 30 kg/100 kg cane. The fibre content is approx. 50%/bagasse.

- **Extract purification**

Beet and cane are processed in a very similar way after the raw juice has been extracted. The raw juice is **purified mechanically** and **chemically**. First fibre and cell particles are removed mechanically, then the **juice is purified chemically** by **precipitation** of some of the nonsucrose substances dissolved or dispersed in the juice, and the **precipitate is then filtered off**. In the **beet sugar industry**, repeated **precipitation with calcium carbonate** has proved successful, an operation in juice purification where **lime and carbon dioxide** are introduced into the juice at the same time. Synthetic flocculants, in particular polyelectrolytes, improve particle agglomeration and reduce sedimentation times in the decanter from the normal 40 - 60 minutes to 15 - 20 minutes. In the **cane sugar industry, simple liming** (defecation) is usually employed as the extract purification process, lime/sulphur dioxide treatment (sulpho-defecation) being less common and lime/carbon dioxide treatment rare. The **decantate** is then finely **filtered for a second time** and goes directly to the **evaporation station**. The sediment or **sludge concentrate** (approx. 25 to 30 kg/100 kg raw material) is usually separated in rotary vacuum filters into **filtrate** and **filter sludge/cake** (approx. 3 to 6 kg/100 kg raw material), the **filtrate returned to the process** and the **filter sludge separated.**

- **Evaporation and crystallisation**

The **clear juice** (from 12 to 15% dry matter/dry sugar content) is continuously **concentrated** by multiple stage evaporation until it has a dry content of 60 to 70%, each stage of this process being heated with the steam (steam-saturated air released when the clear juice is concentrated) from the previous stage. In the **boiling process,** more **water is removed** from the concentrated juice (**syrup**) in boilers operating at an approx. 80% vacuum. The juice is boiled at a lower temperature than normal because of the low pressure in the equipment, thus preventing any discoloration due to caramelisation. When a certain ratio of sugar to water (supersaturation) is reached, **crystals** form. By adding more syrup and evaporating more water, crystallisation continues under controlled conditions until the required crystal size and quantity are obtained. The boiling process is then complete and the resulting boiled mass, now called **massecuite**, is drained from the vacuum pans into crystallisers. As the massecuite is **constantly cooled**, the supersaturation changes, causing the sugar crystals to grow once again. The massecuite is then transferred from the crystallisers into **centrifuges**, in which the crystallisate is separated from the syrup, leaving behind the yellowy brown **raw sugar**. The centrifuged **syrup** is boiled to form a **massecuite** once again and the crystallisate obtained from it is centrifuged. The syrup produced from the centrifuging is called **molasses**. If, when the massecuite is centrifuged, the crystallisate is cleaned of the residual syrup still attached to it by a water and/or steam jet (**affination**), **white sugar** is extracted in just one process from beet or cane. In **refining** (recrystallisation), a plant-intensive technology, raw sugar and poor quality white sugar are dissolved, and then

decoloured and **filtered** by the **addition of activated carbon** or **bone char,** or ion exchange resins. **Refined sugar,** which meets the **most exacting requirements** in the sugar processing industry, is extracted from the subsequent crystallisation process. The quantity of molasses produced ranges from 3 to 6% of the raw material fed in, depending on the technological quality of the raw material and the end product. The sugar content of the **blackstrap molasses** is around 50%.

- **Storage**

The sugar extracted is cooled and dried before storage or packaging. It can be stored **loose, packed** (1 kg) or **bagged** (50 or 100 kg). The essential factor for **proper storage** is a **relative humidity** of around 65% in the store. This is approximately the point of equilibrium between the absorption and the release of moisture from the sugar crystals.

2. Environmental Impacts and Protective Measures

Typical **environmental impacts** caused by sugar manufacture are due to:

- **wastewater** from beet and cane washing, the boiler house (boiler blow-down water) and extract purification in the evaporation and boiling station (excess condensate and purification water), refining (regeneration water from the ion exchange resins), the manufacture of alcohol, yeast, paper or chipboard (where molasses and bagasse are further processed in the plant), site cleaning and rainfall;

- **emissions** into the air from the boiler plant (flue gases from processes in which solid, liquid and gaseous combustibles are burnt), airborne substances (soot and flue ash), raw material processing, extraction, juice purification and juice concentration (ammonia) and from biochemical reactions of organic wastewater components in lagoons (ammonia and hydrogen sulphide);

- **solid waste** from raw material treatment (earth, plant remains), the steam generator (ash) and extract purification (filter sludge).

2.1 Cultivation, harvesting, storage and cleaning of the raw material

Because of the demands placed on it, the **soil is heavily polluted** due, in particular, to many years of single-crop agriculture (sugar cane), the **main pollutants** being:

- the fertilizer and pesticide feed,
- adverse effects on the natural cycle due to soil compaction and salination, dehydration and microorganism decimation.

Precautions in agriculture:

- growing not to be allowed in marginal soils
- examination of soils for chemical and physical properties, water-retention capacity, drainage properties and workability (important where crops need to be irrigated),
- fertilisation to be in line with crop requirements in terms of time and quantity,
- checking of pesticides for suitability for targeted pest control and establishing of a precise dosing concentration and quantity,
- observation springs to be made for the constant monitoring of groundwater and any changes to it.

The environmental effects caused by the harvesting and transport of the raw material are basically air pollution from the **burning of sugar cane fields (flue ash)** and **contaminated access routes. Clarification agents containing lead** (lead acetate solution) should no longer be used for the polarimetric **sugar analysis** of beet and cane extracts; instead the environmentally friendly reagents **aluminium chloride or sulphate** alone should be used.

 Odours are **rarely a nuisance** in beet **storage**, occurring **more frequently** with **sugar cane** especially where it is stored for more than one day. **Sugar beet** is normally supplied with 10 to 20% **moist dirt** attached. In dry seasons, this percentage can be under 5% but can be over 60% in times of frequent and prolonged rainfall. Accordingly, the quantity of suspended solids in a **flotation and washing operation**, where some 750% water on one beet can be assumed, can range from around 7 to 80 kg per m^3 water. The water pollution, after one operation, stands at around 200 to 300 mg BOD_5/l, although this value can rise to over 1,000 mg/l if larger quantities of sugar and other beet components should transfer into the flotation water. Washing and flotation water are today kept in a **closed circuit**, continuously purified in basins with mechanical sludge removal and cleared of plant fragments using narrow-mesh sieves. The recycling of the flotation and washing water **reduces** the wastewater quantity to around 30 to 50% in the case of beet. This reduction in quantity means that **wastewater treatment** is at last a feasible option. The earthy sludge concentrate produced is deposited, wherever possible, in dips, boggy soils and lowlands which can then be put to agricultural use. The water is kept at a pH of about 11 using waste lime from lime kilns so as to prevent any **odour nuisance** due to microbial activity in the flotation water.

The **burning of sugar cane** before harvesting is still common practice, its only **advantage** being that it **facilitates** manual harvesting, as all the dry **parts of the plants** are removed by burning and the **harvest volume** is thereby considerably **reduced**. Cutting speed and thus earnings per worker are higher as the piece work pay is not based on harvested weight but unit of length/row. The **drawbacks** are the **adverse effect on cane quality** due to damage to the cell tissue and thus increased risk of infections at points of damage, **destruction of organic matter, damage to the soil structure** due to increased drying, **increased soil erosion** particularly on hilly sites, and finally **air pollution** in the form of fumes and flue ash emissions. Sugar cane field burning would therefore seem to be contraindicated for biological and ecological reasons.

The **level of contamination of the harvested crop** depends directly on the harvesting technique and soil and weather conditions during harvesting. Hand-cut cane can contain between 7 and 20% foreign matter, while the percentage weight when mechanically harvested is 3 to 5%.

If the **cane is washed**, one can estimate 3 m³ to 10 m³ washing water/t (fresh water, excess condensate, hot well water and recycled washing water).

If cooled excess condensate and hot well water is not fed into the circuit, it can all be used for washing instead of fresh water.

In this way, both the factory's water consumption and the pollutant freight of its wastewater can be reduced. Depending on the washing system, the BOD_5 value can be anything from 200 to 900 mg/l. Foreign matter could be removed by **pneumatic dry separation** to obviate the need for cane washing. The **sludge concentrate** is **treated** and the **odour nuisance** prevented in the same way as in the beet sugar industry.

2.2   Raw material cutting, crushing and extraction

Noise nuisance is produced by the high-speed **slicing equipment** for sugar beet and in the whole area of **mill extraction** (sugar cane). Individual ear muffs are required. **Dust** is generated with particular intensity in the **area of sugar cane intake and transfer** to the mill tandem. No direct hazard to personnel is caused where these operations are automated.

The intermediate products of the sugar industry are ideal nutrient media for a large number of microorganisms. The whole range of activities, ranging from the preparation of the raw material through to crystallisation, also provides a promising culture medium. The risk of **microbial contamination** is particularly high in extraction, where not even the most stringent technical hygiene measures and optimum process management can obviate the need to **use disinfectants**. Repeated high-pressure steam disinfection at the points in the mill tandem most at risk (links in the chain and connecting elements) is only about 60% as effective as biocides. A **chemical treatment** can also be used during mill operation, while steam treatment can be effectively applied when the mill is not working. Major disinfection operations can lead to heavy sugar losses and are therefore not economically viable. The substance most frequently used to **disinfect extraction plants** is still **formalin** (approx. 35% aqueous solution of formaldehyde). It is added in batches at a concentration of around 0.02 to 0.04% in relation to the quantity of raw material processed. The **formaldehyde concentration** in the juice decreases constantly throughout the subsequent process stages. In the clear juice, levels are less than 1 mg/kg while only traces can be found in the **syrup** and concentrations of around 0.10 mg/kg in **white sugar**. It is nonetheless clear that, however it is used, formaldehyde is removed from the sugar to the extent that the technically unavoidable **residues** are rendered **harmless**. Traces of formaldehyde are also found in the condensates which are produced from evaporation and which are returned to the factory's water circuit. **Formalin is controversial** in the light of the **carcinogenic effect** attributed to it, but is still the preferred disinfectant in extraction. **Alternative substances**, e.g. thiocarbamate, quaternary ammonia compounds, cresol derivatives, hydrogen peroxide and the like, have all been tested over recent years. Their **effectiveness as disinfectants** when used in extraction plants is **comparable** to that of formalin. Thiocarbamate, cresol and hydrogen peroxide - like formalin - are also removed by the extraction water during the process, thus only traces can be found in the extracted slices. **Quaternary ammonia compounds**, by contrast, are **irreversibly adsorbed** or **precipitated** along with other organic substances during extract purification.

2.3   Extract purification

The **filter sludge** produced in sugar factories has a **dry content** of 50 to 60%, up to three quarters of which is in the form of **calcium carbonate**, depending on the juice purification process, the rest consisting for the most part of **organic substances**. In **beet sugar factories**, it is usually pressed to a dry content of at least 70%. Because of its **phosphate and nitrogen content**, it is used mainly as **fertilizer** and **instead of lime for soil neutralisation**. In **cane sugar factories**, the sludge is either **passed on to farmers** or spread directly **on to the factory's own fields**. The **high protein content** in the dry cane filter sludge (14 to 18%) means that it can be used as **supplementary fodder** in cattle farming. The solid content is separated in almost all cases by continuous filtration through rotary filters and secondary filtration using precoated filters. The **quantity of wash water** produced is so **low** that it can be disposed of with the filtrate.

The **auxiliary substance** most frequently used in **extract purification** is **lime** (CaO). Depending on the purification process used, consumption ranges from around 0.75 kg CaO (defecation) to 20 kg per tonne of raw material (two-stage calcium carbonate precipitation). Lime and carbon dioxide is **obtained** from limestone in the **coke-fired shaft furnaces** of beet sugar factories. It is not economical for cane sugar factories to produce their own burnt lime because they need such small quantities of it. The CO_2-**rich gas** produced in lime burning **consists of around** 35 to 40% CO_2, the rest being N_2. If the oxygen supply is inadequate, carbon monoxide (CO) may be produced. Since the combustion temperature is below 1,200°C, no nitrogen oxides (NO_x) are formed. The washing water (8 to 10 kg/kg limestone) produced during gas purification is not organically loaded. The **crushing of burnt lime** is associated with the **generation of large quantities of dust**, necessitating the use of dust separators. **Breathing apparatus** must also be worn by personnel working in the immediate vicinity of the lime kiln, particularly personnel involved in cleaning work.

2.4   Evaporation, crystallisation and sugar drying

Some 4 - 6 m^3 **cooling water**/t raw material - depending on whether single or central condensation is used - is required to condense the steam from the last vessel of the evaporation plant and the evaporation crystallisers. In the cooling water circuit, the mixed condensate (hot well water) produced from the condensers (steam condensation) at 40 to 50°C must be recooled to 20°C maximum in cooling towers, **grading houses** or evaporation lagoons (**mist formation**) so that it can be reused. The level of pollution of the condensate produced is determined by the technical conditions in the evaporation and boiling station and the condensation plant. **Organic pollution** and sugar losses in the mixed condensate can be kept **very low** where correctly designed **juice separators** are fitted, with values of around 30 to 150 mg/l BOD_5 in raw sugar factories (cane processing) and 5 to 15 mg/l in the beet sugar industry of today. **Incrustations** are **removed** from evaporator pipes and other hot surfaces by cleaning with an approx. 5% soda solution followed by a 2 to 5% hydrochloric acid solution. The **acids and lyes** used for cleaning can be **neutralised and fed** into the water circuit.

Sugar dust from sugar driers or defective dust-extraction systems can **give rise to severe air pollution**. This is not only a **health hazard** but, at a grain size of < 0.03 mm, also highly **explosive** if the dust/air mixture concentration is within the explosion limit (approx. 20 to 300 g/m^3). A low dust level is 2 g/kg sugar. **Dust is separated** in dry electro-filters or in wet dust-extractors (scrubbers). If no dust separators are installed (e.g. older white sugar factories), **breathing masks** must be provided. The concentration must be kept low by means of adequate **ventilation** and precautions must be taken to **prevent ignition** of the explosive atmosphere (smoking ban, no repair work which produces frictional heat or sparks, installation of explosion-proof electrical systems) in order to **minimise the risk of explosion**.

2.5 By-product manufacture

Dry slices: the water obtained from the mechanical drying of the extracted beet slices is recycled in the extraction process. The slices are mostly dried in drum type driers (700 to 900°C) to a dry content of 25 to 90%. The mixture of combustion gas from natural gas or oil plus flue gas from steam production is used as the **drying gas**. A waste gas quantity of 1.5 to 4.5 m^3/kg extracted slices - depending on the drying system used - must be expected. The **dust content of waste gases from the drier** depends, inter alia, on the quantity of molasses added and process management (temperature, holding time). The **dust concentration** in the untreated gas ranges from 2 to 4 g/m^3. Similar dust concentrations may also arise after the cooling drum, the pelleting station, the bagging plant and the pneumatic conveyors. The **total carbon concentration** ranges from 300 to 1,200 mg/m^3, depending on process management. The SO_2 **concentration** after the slice drying plant depends, inter alia, on the fuel used, the drying process and slice composition. SO_2 concentrations of up to 1,000 mg/m^3 in the waste gas have been encountered.

Sugar extraction from molasses: more **sugar can be extracted from** the **molasses** if **additional expenditure** is allowed for operating costs and equipment. The **regeneration- and wash water** then produced is **heavily contaminated** and must be **handled separately** if the **chloride concentration** in the total wastewater exceeds 2,000 mg/l.

56

Biotechnical processing of molasses : the biotechnical recycling of beet molasses almost always takes place in different plants; in contrast, it often forms part of a sugar factory and also gives rise to high wastewater loads. **Yeast, ethanol, citric acid** and much more can be produced from molasses by fermentation. The **waste load** remaining from **baker's yeast manufacture** is around 156 kg COD/t molasses or 187 kg/t baker's yeast. The distillation **residue** (vinasse) which remains is highly diluted (aqueous) (up to 96% water) and can be used as **cattle feed** (approx. 50 l/day and animal). As these difficultly biodegradable substances cannot be removed by economically feasible treatment processes, the **process wastewater from yeast manufacture** has to be thoroughly **biologically treated,** and then **recycled** for **agricultural use** where possible. However, even after biological treatment, the process wastewater **must not be discharged** directly into the drains. Since the climatic conditions of sugar-cane-growing countries favour **eutrophication processes, stringent requirements** must be laid down with regard to **wastewater purification.**

2.6   Energy supply

Some 300 to 400 kg **steam** and 35 to 40 kWh **electrical energy** is required to process 1 t **beet** to white sugar. Every factory must be equipped with steam and **electrical energy** generation plants (heat/power linkage). Oil, gas and coal are used as **primary energy sources** and waste gas purification is essential where emission limits are exceeded.

The **steam and electrical energy requirement** for **sugar cane** processing (to drive the roller mill with steam turbines) stands at around 550 kg steam/t cane (raw sugar production) or 615 kg steam/t cane (white sugar production) and 35 to 40 kWh electrical energy/t cane. The mean **calorific value of bagasse** (approx. 50% moisture) is around 8,400 kJ/kg (mean calorific value of oil around 42,000 kJ/kg). The **bagasse produced** is sufficient to **cover** the factory's **energy requirements.** Incomplete burning of bagasse (water content > 50%) increases the emission of flue ash and carbon particles.

To **start up the factory** (start of campaign), **other energy sources** have to be used. If a refinery is also operated, it may also be necessary to back up the bagasse with other fuels. Maintenance firing is also essential where the plant is shut down for a prolonged period. A **complete conversion** to other energy carriers is essential where **bagasse** is used **as a raw material for paper or chipboard manufacture.**

2.7   Water management

There is no stage in sugar production where water in some quantity is not required. The technical **water requirement** of the sugar extraction process in a **beet factory** is around 20 m³/t beet. The amount of process water introduced can be reduced to 0.5 m³/t by the **introduction of water circuits**. For practical reasons, **highly contaminated** (transport, purification, regeneration, hot well water) and **slightly contaminated** (turbine cooling, pump cooling, seal and gas scrubbing water, condensate excess) water circuits should be kept **separate**, as water with a low level of pollution (in Germany < 60 mg/l COD or 30 mg/l BOD_5) can be drained into receiving streams. In a well-managed factory, the quantity of **highly contaminated wastewater** produced can be reduced to 0.2 m³/t beet; it should not exceed 0.5 m³/t, beyond which the cost of wastewater treatment would then become uneconomic. Pollution rises in the course of the campaign, reaching COD values of 6,500 mg/l or BOD_5 values of 4,000 mg/l and more. In **sugar cane** processing, large quantities of **cane washing water** (up to 10 m³/t) and mixed condensate are produced during steam condensation and raw sugar refining, which must be **managed in a circuit system** (large land areas required for evaporation lagoons, high investment costs for cooling towers). Cane washing water (260 to 700 mg/l BOD_5), **filter residue** (2,500 to 10,000 mg/l BOD_5) and **washing water from decolorising carbon and ion exchange resins** in refineries (750 to 1,200 mg/l BOD_5) are all heavily contaminated. The purification water also includes wastewater required for **cleaning the production areas and plant** during and after the campaign, and for **cleaning sugar transport vehicles**. There are also **juice and water overflows** at plant breakdowns (clear juice, for example, has a BOD_5 of about 80,000 mg/l) so that values of up to 18,000 mg BOD_5/l can occur. **Negligence** is the **main cause** of excessive wastewater contamination. If a plant is carefully managed, this wastewater does not exceed values of 5,000 mg BOD_5/l. **Low organic pollution and sugar losses** in the mixed condensate (30 to 150 mg/l) can only be achieved by the installation of **separators in the steam pipes**.

The aim of **establishing water management in a sugar factory** must be to eject or treat as low a quantity of polluted water as possible. **Water recycling** heads the list of measures to be taken inside the factory. Water management must be such that, once closed circuits are established, unpolluted or only slightly polluted water requiring no further treatment is discharged into the drains.

56

The **treatment processes** for wastewater which can be carried out in sugar factories are largely **determined by local factors**. The management of the wastewater and circuit conditions inside the plant have a major effect on plant size and the level of degradation which can be achieved.

Wastewater treatment begins with the **mechanical removal** of suspended particles followed by aerobic treatment. The **simplest** and by far the most desirable treatment method for the processing of organically polluted, concentrated sugar factory wastewater is its collection in a **series of lagoons** using an overflow system. The wastewater then **purifies** itself. The **time** required for adequate degradation of the wastewater in the lagoons is **determined by the following factors**:

-      height of water level in the lagoon
-      lagoon area
-      subsoil below the lagoon/adequate sealing against the subsoil
-      weather conditions
-      external flows of water.

The **lower the water level** and the **warmer the weather** during the degradation processes, the **faster** the water in the lagoon is **treated**. Lagoon depth should not exceed 1.20 m in temperate climates, while 1.50 m is acceptable for subtropical/tropical zones (sugar cane growing). If there is a high level of **evaporation**, the content of the wastewater is **concentrated**; if there are **external flows of water** and **rainfall** is high, the lagoon wastewater is **diluted**. The treatment of wastewater with a concentration of 5,000 mg BOD_5/l, as is usual in the domestic sugar industry, requires a constant degradation efficiency of 99% or more to reach a BOD_5 value of 30 mg/l. At a lagoon depth of 1 m and a degradation period of 6 to 8 months, values of 100 mg BOD_5/l can be achieved, which is equivalent to partially biologically treated water. In **sugar cane growing areas, complete biological purification** - a BOD_5 of less than 30 mg/l - can be achieved within five or six months with good lagoon management. This long-term process of degradation in lagoons could solve the wastewater problem for the sugar industry if sufficient **lagoon area** were available. The sugar cane processing industry, in fact, generally has sufficient land area available and so uses the lagoon process almost without exception.

Organic substances are degraded by both aerobic and anaerobic processes. In the anaerobic stages, fermentation and putrefaction occur, thus the possibility of **odour nuisances**, primarily due to the **formation of hydrogen sulphide and butyric acid**, cannot be ruled out. This drawback can be overcome, however, by selecting **suitable locations** and by adequate additional **aeration**. During the activated sludge process, oxygen is introduced into the water in the form of air via an aeration system.

Small-scale continuous systems operate with a substantially higher micro-organism density and a higher oxygen supply, and achieve a degradation level of around 90%. Atmospheric pollution is clearly higher at 2 to 7 kg $COD/(m^3/day)$ and the energy required for the air supply stands at around 3.5 $kWh/(m^3/day)$.

Anaerobic wastewater purification plants consist of large tanks (around 3,000 to 7,000 m^3) in which anaerobic bacteria degrade the organic pollutants to form **biogas** (approx. 75 to 85% methane). This is particularly effective where wastewater is heavily contaminated. The organic content is 80 - 85% decomposed, with the remaining degradation taking place aerobically in the aeration system. The **benefits** of this process are that the **methane gas** can be used directly **as an energy source** to heat the tanks and that the **problems of odour** can be **overcome**; an additional factor is that less space is required than in the case of lagoon systems.

Compared to the large **quantities of sludge** produced in the flotation and washing water circuit, and in some cases occurring in the form of lime sludge from juice purification, the quantities produced in **wastewater conditioning** are very low. Recycling is as for filter sludge (see 2.3). The form of "large-scale wastewater processing" most frequently used is overhead irrigation or, rarely, basin irrigation. The **preconditions** for this are level undrained areas, deep soils with no tendency to silting and a low water table (> 1.30 m). During the passage through the soil, the **following processes** take place:

- mechanical filtration on the surface
- absorption of the dissolved substances by bacteria in the soil
- biological oxidation of filtered and adsorbed material by bacteria in the soil in the intervals between the individual doses of wastewater.

Basin-irrigation is carried out mostly on small **parcels of land** surrounded by earth walls (so-called retaining filters). **Mechanisation** is severely restricted by parcel size and the earth walls. Only crops which grow vertically and which are not sensitive to overdamming are suitable, e.g. tree crops and meadow areas.

Sprinkler irrigation is the **most expensive biological treatment process**. All settleable solids must be removed to minimise malfunctions in the sprinkler irrigation installations. The load should be intermittent and the quantity rained onto each area must remain small (< 500 mm/vegetation period - individual doses not exceeding 80 mm). If the wastewater is first treated in lagoons to at least 180 mg BOD_5/l, drained areas can also be irrigated if the water table is suitably low. In addition to wastewater treatment in the soil, the **wastewater recycled** for irrigation also **acts as a fertilizer**.

3.   Notes on the Analysis and Evaluation of Environmental Impacts

3.1   Emission-limiting requirements

Two types of requirement are imposed on sugar factories: general and special. The provisions regarding **general regulations to limit emissions** contain:

-       emission values, which current technology can keep to admissible levels,
-       emission-limiting requirements conforming to the state of the art,
-       other requirements to protect against harmful environmental impacts by air pollution and
-       processes for determining the emissions.

The following **general requirements** are imposed:

-       reduction in the quantity of waste gas by enclosing plant components,

-       registration of waste gas flows,

-       circulating air management and process optimisation through more efficient use of waste heat,

-       waste gases must be released so that they are freely discharged without obstruction in the general airstream,

-       chimneys should be at least 10 m above the ground and project 3 m above the roof ridge, but should be no more than twice the height of the building,

-       in the field of wastewater treatment, including lagoons, anaerobic degradation shall be eliminated by technical or structural measures as far as possible.

The **special requirements** (e.g. in Germany in accordance with the TA-Luft [Technical Instructions on Air Quality Control]):

- the drum inlet temperature in sugar beet slice drying plants must not exceed 750°C or equivalent measures to reduce odours must be applied,
- **dust emissions** in the damp waste gas must not exceed 75 mg/m³(f),
- where **solid** or **liquid fuels** are used, the sulphur content by weight must not exceed 1 % in the case of solid fuels related to a net calorific value of 29.3 MJ/kg, or equivalent waste gas purification must be carried out.

A crucial factor in all emission considerations is the **emission load** resulting from the quantity of waste gas discharged from the chimney, multiplied by the pollutant concentration. This concerns primarily the load from **sulphur, nitrogen oxide, carbon monoxide and dust**.

56

The following **emission limits** apply to furnace installations with a furnace heat output of < 50 MW:

Emissions	Unit	solid	liquid fuels	gaseous
dust	mg/m³	50	80	5
CO	mg/m³	250	175	100
NO_x	mg/m³	400	300	200
SO_2	mg/m³	2000	1700	35

Source: TA-Luft [Technical Instructions on Air Quality Control]

The emission values relate to an oxygen content by volume in the waste gas of 3% for liquid and gaseous fuels. In the case of solid fuel, 7% applies where coal is used and 11% applies where wood is used.

Flue ash and **soot** are the main **air pollutants** where **bagasse** is used as **fuel**, but flue gases from bagasse do not contain any toxic substances. Where fuel oil is used in the cane sugar industry, a sulphur content of 0.5 to 1.0 % by weight in the fuel oil is permissible.

The **main parameter** of any biological treatment and in any watercourse is the **biochemical oxygen demand (BOD)**. This is the quantity of oxygen in mg/l which is consumed by microorganisms at 20°C within a degradation time of five days. The **chemical oxygen demand (COD)**, on the other hand, is the standard for the content of oxidisable substances found in water, i.e. the method covers not only biologically active substances but also inert organic compounds. It is essential to use the COD method (evidence provided using potassium permanganate or potassium bichromate) as a fast method of determining the level of water pollution.

In its **guidelines for the cane sugar industry**, the **World Bank** takes the view that three parameters are of fundamental importance when it comes to assessing sugar factory wastewater pollution with biodegradable substances and their impact on the environment:

- BOD_5 for determining the oxygen-consuming organic material;
- TSS (total suspended solids mg/l) for establishing the total quantity of suspended matter (primarily inorganic substances from cane and beet washing water);
- pH as extreme pH changes are harmful to water fauna.

The **minimum requirements** regarding the pollution levels in **wastewater** to be **released** into bodies of water are based on the treatment processes normally applied in the various industries and must conform to the state of the art.

For sugar production and associated industries (including alcohol and yeast production from molasses) the following minimum requirements are specified in **Germany** (source: (1)):

	A cm^3/l random sample	COD mg/l mixed sample		BOD5 mg/l mixed sample		(T_F) random sample
Seal and cond. water	0.3	60	--	30	--	--
Other water	0.5	500	450	50	40	4

A   =   volume of suspended solids
T_F =   toxicity to fish, expressed as the minimum dilution factor of the wastewater at which all test fish survive under standardised conditions within 48 hours.
             These values apply to random samples in the case of lagoons.

Due to **local circumstances** it may be necessary to limit **other parameters** for discharging into watercourses, e.g. temperature, pH, ammonia, chloride.

In the **USA**, the Environmental Protection Agency (EPA) has imposed **limit values for cane sugar factories** (raw sugar factories and refineries).

General limit values regarded as "best available technology economically achievable" (BATEA or BAT) are:

	BOD$_5$	A	
Raw sugar factory (kg/t cane)			
max.daily value	0.10	0.24	
30-day			pH 6.0-6.9
mean	0.05	0.08	
White sugar factory (only mixed condensate)	(kg/t raw syrup)		
max.daily value	0.18	0.11	
30-day			pH 6.0-6.9
mean	0.09	0.035	
Liquid sugar factor (only mixed condensate)	(kg/t raw syrup)		
max.daily value	0.30	0.09	
30-day			pH 6.0 - 6.9
mean	0.15	0.03	

With regard to sugar factories, the **noise immission guide values** are (in Germany) 60 dB(A) in the daytime and 45 dB(A) at night. **Cane sugar factories** are generally located **in the centre of the growing area**, very rarely in the vicinity of sizeable residential areas. The **design** of factories is **light and open** (due to the climate); cane is received and conveyed to the mill in the open air (large quantities of **dust generated**).

Noise emissions can be restricted by **structural and acoustic measures, with housings** provided around sources of noise and **soundproofing**.

Where the noise from certain tasks or areas of the factory cannot be restricted or insulated, **personnel** shall be issued with appropriate individual **protective gear**.

56

These include, in particular, **tipping and bagging plants, cane handling and roller extraction**, washing plants for the raw material and the **centrifuge station**. In the workshop area, they include mainly **work at rotary machines** with a diameter > 500 mm, **sheet metal processing machines and drilling and punching machines**. The acoustic power level in these areas ranges from between 80 and 130 dB(A). At values of > 85 dB(A), individual protective gear (ear plugs, ear muffs) must be worn. With acoustic pressure levels of > 115 dB(A), the combined use of both items is recommended.

3.2   Technology for the reduction of emissions and emission monitoring

Measures to **prevent damage due to atmospheric sulphur dioxide immissions** in flue gases comprise the retention of SO_2 in desulphurisation plants (e.g. absorption in lime milk) and the use of low-sulphur fuels. The installation of a scrubber before the chimney inlet has proved successful in reducing the emission load in waste gases. As well as its action on dust, this scrubbing operation achieves an SO_2 separation of some 30%. If **sludge from calcium carbonate precipitation** is used as the washing liquid, **pure gas dust concentrations** of less than 75 mg/m^3(f) are obtained. At the same time the SO_2 emission is reduced by 60 to 70 %. "Calcium carbonate precipitation scrubbing" is therefore a **particularly good dust and SO$_2$ separation** method as it does not pose any additional wastewater or residue problems.

Dust emissions occurring inside the sugar factory are reduced with **scrubbers** or **fabric filters** and the pure gas concentration is less than 20 mg/m^3. Dust levels are kept low in the same way during further processing stages.

In the **cane sugar industry**, the generally **high proportion of flue ash** necessitates flue gas purification measures. **Older furnace plants** can be **easily fitted** with wet or dry separators (cyclones: approx. 96% effective, more investment- and maintenance-intensive than wet separators). The **water requirement** figures for wet separation are approx. 0.025 m^3 water/25 m^3 gas.

Emissions and the temperature in the waste gas from steam generation and slice drying are **measured and monitored** by **integrated**, continuously operating **measuring instruments**. In the **cane sugar industry, portable, manually operated equipment** (e.g. Orsat apparatus) is used mainly to determine, for example, oxygen, carbon dioxide and monoxide. If state-of-the-art waste gas purification systems are installed in new plants, and if dust emissions are below 75 mg/m^3, **daily measurements** with a portable unit are adequate.

Any **odour nuisance** due to ammonia emissions is largely suppressed in advance by closed-**circuit systems**.

In principle **lagoons** should be fitted with additional aeration equipment, and aeration rollers have proved extremely successful here. They should not be located in the immediate vicinity of a factory or residential area (**upwind**).

A number of processes can be used to **measure rate of discharge**, e.g. measurement of flow rate with an impeller device and integration via the discharge cross section, or direct determination with a measuring weir.

The mixed samples taken **for wastewater assessment** are analysed for BOD_5 according to DEV (source: (5)) and for sludge deposits, COD and toxicity to fish according to DIN. The EPA has specified the analysis methods for the cane sugar industry in "Methods of Chemical Analysis of Water and Wastes". In the case of **lagoons**, random samples are adequate in view of the low fluctuations over time of wastewater composition and the long retention times.

Control services and control mechanisms should be put in place to **check** that environmental provisions are observed, e.g. **environmental protection consultants**. Their task would also involve checking that environmental protection installations are **in good working order** and are regularly **maintained,** and they would also be responsible for **personnel training and making personnel aware** of environmental issues. **Medical care** should be provided inside the works and for the local population.

3.3 Limit values issued to protect health

Substances for which **maximum workplace concentrations (MAK values)** or **technical approximate concentrations (TRK)** apply in Germany:

	mg/m^3		Application/source
Ammonia	35	-	Raw material treatment, extraction, juice purification, juice concentration, lagoons;
Asbestos dust	0.025	-	Heat insulation, filter aids (diatomaceous earth);
Lead	0.1	-	Laboratory: lead acetate solution for the clarification of juice samples for polarisation analysis;
Calcium oxide	5	-	Milk of lime manufacture: juice purification, juice neutralisation, wastewater treatment; lime burning;
Hydrogen chloride	7	-	Evaporator station: cleaning with dilute hydrochloric acid to remove incrustations (calcium carbonate);
Formaldehyde	1.2	-	Disinfectants: at places in the production area at risk from microorganisms, mainly extraction;

56

Hydrazine 0.13	- Anticorrosives for boiler feed water (chemical oxygen bonding with hydrazine hydrate);
Carbon dioxide 9000	- Juice purification (calcium carbonate precipitation); lime burning;
Sulphur dioxide 5	- Made from sulphur in sulphur furnaces, juice purification (calcium sulphite precipitation), acidification of the extraction water, waste gases where fossil fuels are used;
Hydrogen sulphide 15	- Raw material treatment, lagoons;
Dust (generally)................ 6	- Cane receipt and crushing, slice and sugar drying, sugar bagging; storage of excess bagasse.

Synthetic flocculants do not create dust or irritate the skin when handled, and do not constitute any **toxicological hazard**. **Carcinogenic substances** and substances which are suspected of having **carcinogenic potential** are: asbestos dust, alkali chromates and lead chromate (laboratory reagents), formaldehyde, hydrazine, fumes from VA welding.

The lethal dose (LD_{50}) of a 39% formaldehyde solution is 800 mg/kg body weight (oral: rat); according to the working materials ordinance *Arbeitsstoff-Verordnung* classified as "low toxicity" and labelled with the hazard symbol R22 ("harmful if swallowed!") (necroses of the mouth, oesophagus, stomach).

Measures: in principle toxic chemicals must always to be **kept** sealed; the **wearing of** rubber gloves is recommended for analysis work; vessels and instruments must be thoroughly cleaned; installation of effective extraction and ventilation systems.

4.   Interaction with Other Sectors

Sugar is produced jointly by agriculture (crop growing) and **industry** (processing technology), and there are close links in the ecological and technical fields. The **use of modern agricultural knowledge and methods** in the growing of the raw material, particularly with regard to fertilizer and pesticide application, largely determines the technological value of beet and cane (all physical, mechanical, chemical and biological properties of the raw material). **High-quality raw material** facilitates the tasks of extraction and juice purification and this in turn is reflected in improved technological - and hence in the final analysis economic - performance of the factory (**higher sugar yields**).

Excess bagasse can be used for the additional generation of electricity for the national grid (**power stations sector**) or for briquette production (domestic fuel supply). Bagasse is also a raw material for the manufacture of hardboard, cardboard or paper (**wood and paper sector**). Molasses, as well as extracts from cane and beet, are used as the raw material for fermentation processes (**fermentation technology and biotechnology sector**). Sugar is processed in numerous branches of the **food industry**. Refined sugar can be used in drug manufacture (**pharmaceuticals sector**).

All sugar beet and some sugar cane processing factories are equipped with lime kilns for the production of calcium oxide and carbon dioxide, and there are thus parallels with the **cement/lime sector**.

There are also links with the **water supply, wastewater treatment** and **solid waste disposal sectors** generally.

56

5. Summary Assessment of Environmental Relevance

The impact on the environment from the **sugar extraction process** and the processing of the **by-products** from it are manifold, but can be **kept** to a reasonable, and in part legally prescribed, minimum level by means of **established methods and processes**. In new beet sugar factories the **proportion of costs** required for installations to protect the environment stands at some 15 to 20 % of total investment costs, while the figure for cane sugar factories is 10 to 15 %.

The **wastewater produced** can be minimised by optimum design of internal water circuits and the use of established purification processes (lagoon degradation/biological treatment plants). The rational control of the process must **prevent any sugar solutions entering water circuits**. This not only reduces pollution but increases profitability. **Dumps for filter residues** and **earth** can be used for **soil conditioning** once the load has been degraded. The **production** of fuel in the form **of biogas** should be taken into account when planning new factories.

Emissions from power stations and drying plants can be contained with the treatment technologies which have now been developed. A **large quantity of soot and ash** must be expected in the waste gas, particularly where **bagasse is used** as fuel, and consequently installations to optimise the combustion process and waste gas purification must be provided.

The **open design** of factories in warmer climates appears to obstruct possible noise prevention measures, thus noise nuisance would seem to be avoidable only by siting factories **at an appropriate distance from residential areas**.

In principle the environmental impact caused by sugar factories can be minimised by current technology. In the preparatory phase, it must be ensured that the plant installed will **continue** to be both **fully operational** and **fully used** for many years. This calls for the **training** of specialists at a technical level who realise the need for regular maintenance work. **Training projects** for sugar technicians and workmen can be appropriately integrated in sugar factories.

Sugar factories contribute to the general **economic development** of a country, including the intensification of agriculture, infrastructural improvement, start of the general industrialisation of rural areas and job creation in agriculture and manufacture, all of which attracts the potential workforce in the surrounding area. This generally leads to the **uncontrolled growth** of local communities and over-burdening of the infrastructure and public services. **Settlements** in the immediate vicinity of the site must therefore be **prevented** from the outset. To minimise detrimental effects at the earliest possible stage, **close cooperation** must be sought at planning stage with the relevant authorities on the **regional development plan**. Likewise, **the affected population groups** - and this includes women - should be **involved** in the decision-making process at all planning phases so as to **resolve environmental problems** which may arise, e.g. land-use conflicts.

6. References

(1) Achtzehnte Allgemeine Verwaltungsvorschrift über Mindestanforderungen an das Einleiten von Abwasser in Gewässer (Zuckerherstellung), January 1982.

(2) Autorenkollektiv, Die Zuckerherstellung, Fachbuchverlag Leipzig, 1984.

(3) Bronn W.K.: Untersuchung der technologischen und wirtschaftlichen Möglichkeiten einer Abfallminderung in Hefefabriken durch Einsatz von anderen Rohstoffen anstelle von Melasse, Forschungsbericht, 1985.

(4) Davids, P. und Lange, M.: Die TA-Luft, Technischer Kommentar, Herstellung oder Raffination von Zucker, 672 - 678, Verlag des Vereins Deutscher Ingenieure, 1986.

(5) Deutsche Einheitsverfahren zur Wasser-, Abwasser- und Schlammuntersuchung, Fachgruppe Wasserchemie, 1979.

(6) Großfeueranlagen-Verordnung, Dreizehnte Verordnung zur Durchführung des Bundes-Immissionsschutzgesetzes, 1983.

(7) Hugot: Handbook of Cane Sugar Engineering, Elsevier Scientific Publishing Company, 1972.

(8) International Commission for Uniform Methods of Sugar Analysis, Report on the Proceedings of the 20th Session, 1990.

(9) Korn, K.: Harmonisierung von Umweltschutz und Kostenbelastung an Beispielen der deutschen Zuckerindustrie, Zuckerindustrie 12, 1987.

(10) Meade, G. P., Chen J. C. P.: Cane Sugar Handbook, John Wiley Sons, N.Y. 1985.

(11) National Institute of Occupational Safety and Health, Registry of Toxic Effects of Chemical Substances, 1984.

(12) Persönliche Mitteilungen des Instituts für Landwirtschaftliche Technologie und Zuckerindustrie zu Fragen über alternative Chemikalien zur Desinfektion und Reinigung von Säften in der Zuckerindustrie, 1991.

(13) Reichel, H. U.: Auswirkungen der TA-Luft und der Großfeueranlagen-Verordnung auf die Zuckerindustrie, 1985.

56

(14)   TA-Luft: Erste Allgemeine Verwaltungsvorschrift zum Bundes-Immis-
        sionsschutzgesetz, 1986.

(15)   Technische Regeln für gefährliche Arbeitsstoffe, Bundesarbeitsblatt, 1985.

(16)   UNEP - Industry & Environment Overview Series, Environmental Aspects of
        the Sugar Industry, 1982.

(17)   Untersuchungen über Desinfektionsmittel für deren Einsatz in Extraktions-
        anlagen, Nickisch/Hartfiel/Maud, Zuckerindustrie 108, 1983.

(18)   Zucker-Berufsgenossenschaft, Lärmbereiche in der Zuckerindustrie, 1978.

Appendices:

56

Flow chart - raw sugar manufacture from sugar cane

Key to symbols:
→ Process flow
⤑ Gaseous, solid and liquid emissions

Flow chart - white sugar manufacture from beet

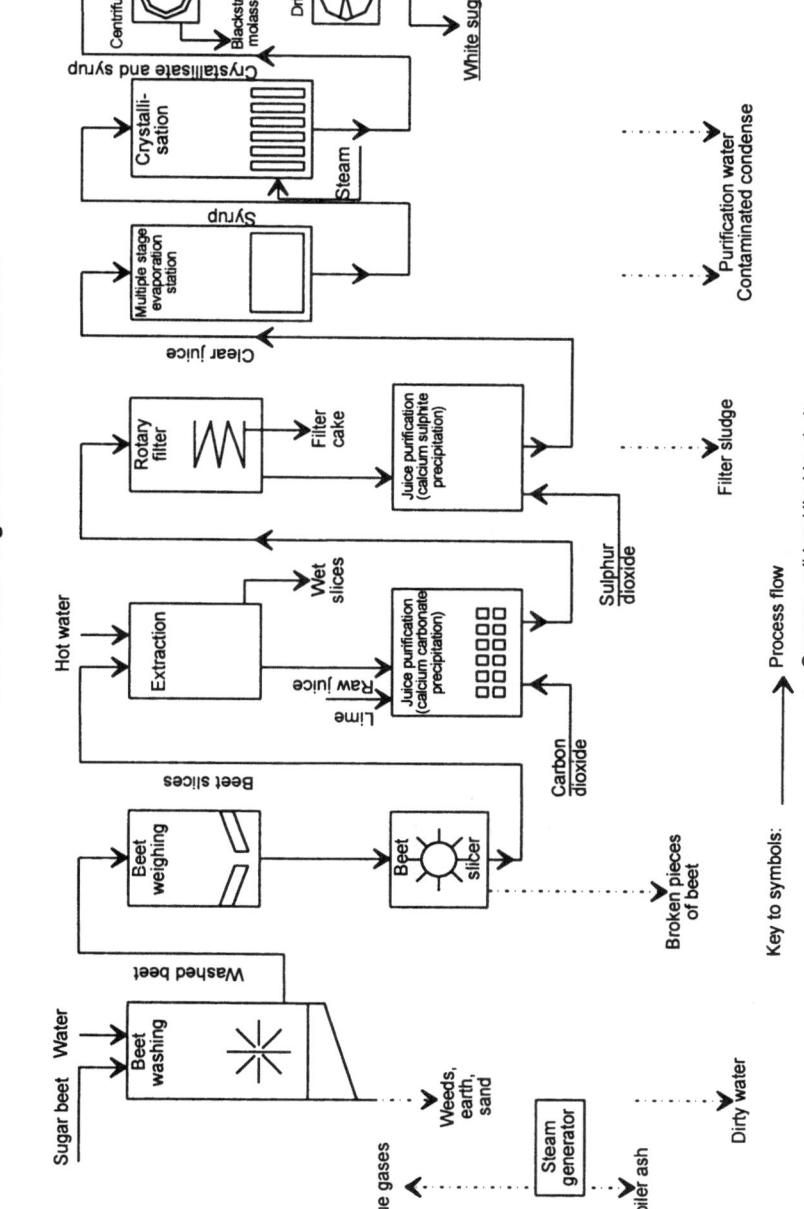

Flow chart - refining raw cane sugar

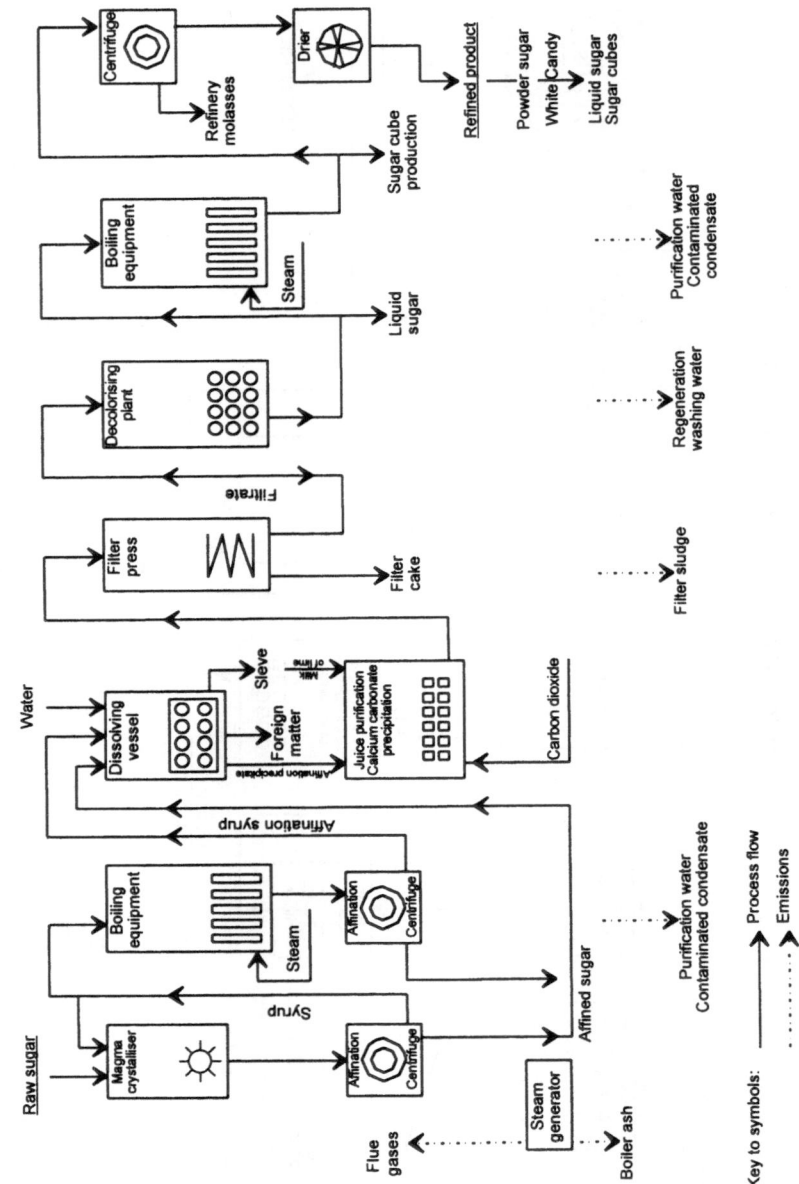

Water management in a beet sugar factory
- All figures in kg/100 kg beet -

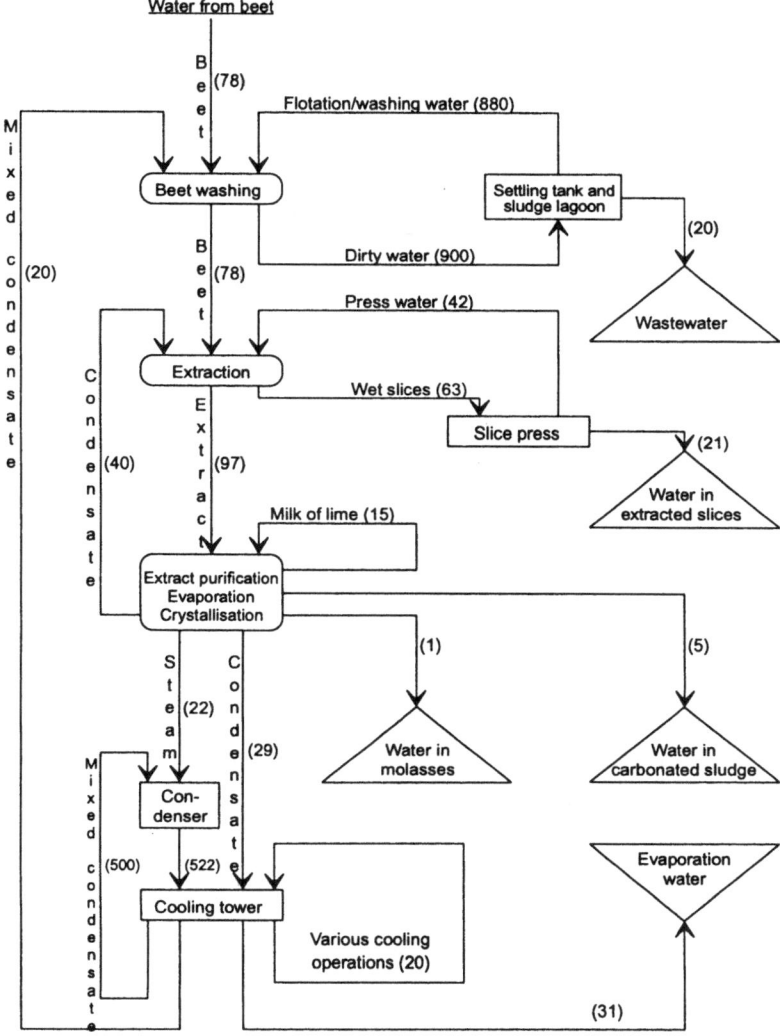

56

Wastewater control in a cane sugar factory

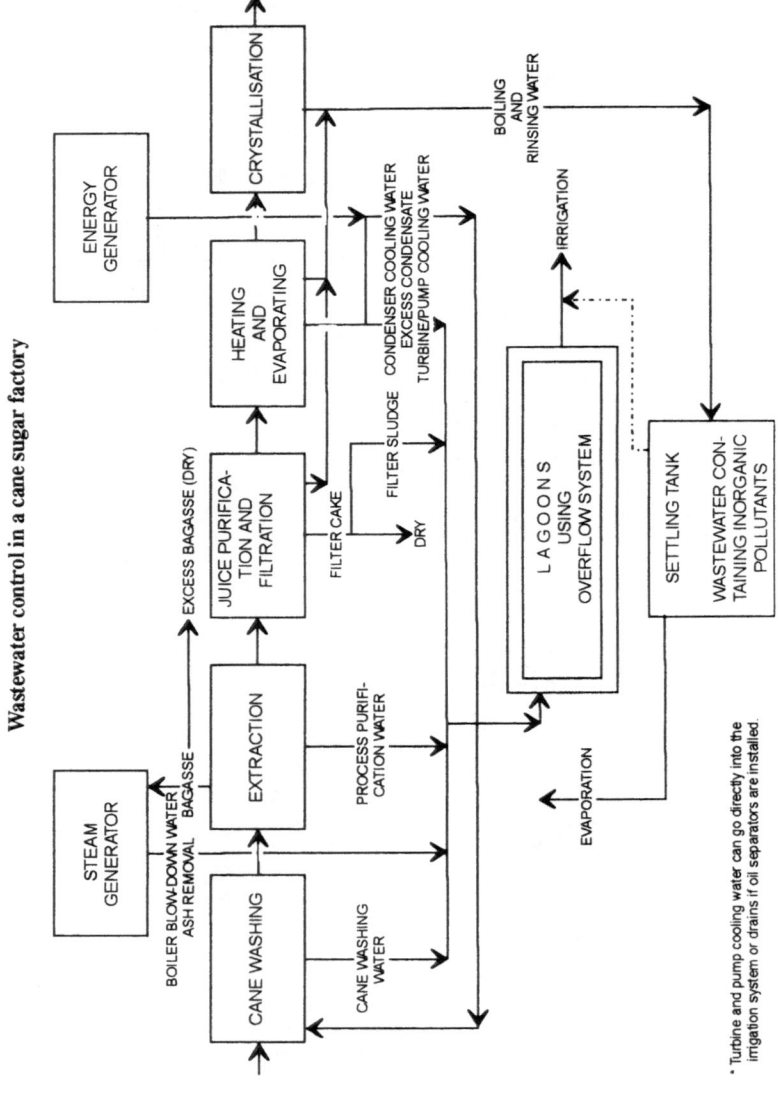

* Turbine and pump cooling water can go directly into the irrigation system or drains if oil separators are installed.

Trade and Industry

Timber, Sawmills, Wood Processing and Wood Products

57. Timber, Sawmills, Wood Processing and Wood Products

Contents

1. Scope

Wood is man's oldest **material** and **energy source**; it is particularly important as it is a **renewable resource**. Despite the availability of metal, synthetic (plastic/chemical) and mineral materials it is still important as a raw material. Because of their technological properties, tropical woods have been accepted - particularly in the last thirty years - as valuable functional and decorative materials. In most tropical and subtropical countries, wood still plays a vital role as an energy source.

The major timber sub-sectors are as follows:

- production (timber industry, incl. reforestation), felling and transport
- mechanical woodworking (sawing, shaping, milling, sanding)
- manufacture of wood board materials (plywood, chipboard and fibreboard)
- transformation into other products with extensive chemical modification of timber
- combustion.

This article focuses mainly on the primary, i.e. **mechanical processing** of wood, the manufacture of wood products and charcoal production, with only a brief look at combustion.

The **manufacture of paper and pulps** from wood is regarded as a separate, secondary processing sector and is not covered by this brief but by the environmental brief Pulp and Paper.

The environmental impacts of woodworking and wood processing operations, in the form of dust, noise and odours, can be countered to a large extent by appropriate **siting**, namely downwind from residential areas (see also the environmental brief Planning of Locations for Trade and Industry). The problem of **wastewater**, on the other hand, calls for closer attention. Direct effects on personnel can be at least reduced by the wearing of suitable **hearing protection and breathing equipment**.

In terms of the **scale of environmental impacts** it should be borne in mind that advancing **slash and burn (shifting cultivation)** can be the most dangerous environmental impact of timber felling, and is frequently the most significant factor in forest destruction.

The major effect on employment to be considered is that the workforce in the timber industry is almost exclusively male.

2.   Environmental Impacts and Protective Measures

2.1   Mechanical woodworking

Wood is a raw material which regrows and is obtained mainly from **natural woodland**, with **plantations** still playing only a minor role in many countries.

The dual system of national forestry authorities and private timber
concessionaries often results in a clash of business management and forestry policy
interests - being founded largely on conflicting principles.

Woodworking *per se* begins in the sawmill with debarking, unless this has already
been done in the forest, followed by the cutting into lengths and cutting to size of
timber supplied from the forest. The cut timber is either used **directly as building
lumber** or is **upgraded** by shaping, milling, sanding and painting or impregnation.

Sawmills are workshops in which **round wood** is processed **into sawn lumber**
(primary processing). Mechanical woodworking goes hand in hand with noise and
dust, and is often carried out in the same mills which carry out **surface treatments**
with paint, stains etc., processes which result in the formation of gaseous and substances with strong odours.

*   **Noise**

The mechanically driven transport, cutting, milling, shaping and dust extractor installations in the timber industry generate noise, a problem which is exacerbated
where sawmills are built to an open design in warmer climates.

However, because most **sites** are selected because they are **close to raw
material sources**, they tend to be far from residential areas. Thus it is only mill
personnel who are mainly affected.   For this reason, the wearing of hearing
protection should be compulsory and, where new plant or new equipment is to be
installed, emphasis should be placed on providing tooling which is enclosed and
designed to reduce noise.

Further negative effects on machine operators exist in the form of **vibrations**.
The reduction of vibrations is an important factor when laying the foundations and
erecting operating and control stations.

*   **Dust emissions**

Alongside noise, dust is emitted from mechanical woodworking processes. In
sawmills, wood cutting produces wood shavings, but because the wood is in most
cases supplied fresh from the forest or is fibre-saturated, dust emissions do not
present **a major problem in relative terms**, and fabric filters or wet extractors are
not generally required. However, where wood shavings are stored in the open air,
measures must be taken to protect against airborne dust.

Far more significant is the dust generated by mechanical woodworking in **joinery works, cabinet-making and similar businesses**, where both dust quantity and qualitative dust composition differ from that in a sawmill. The crucial factor is the **fineness** of the dust, expressed by its grain size and grain size distribution. Fine dust is naturally more difficult to remove than coarse dust and constitutes a **greater health hazard** to man, particularly where the particles are small enough to reach the lungs. The fine dust content is particularly high in sanding operations, and not so high in operations which produce shavings.

The **inhalation** of wood dust, particularly hardwood dust, can result in the absorption of harmful substances found in wood, which in turn can lead to **serious illnesses**. Thus, before any wood processing is undertaken, the health risks arising from working with wood must be thoroughly investigated and adequate precautions taken.

To reduce the quantity of dust generated at workplaces machines must be fitted with extractor systems, a measure which is justified as both a **health precaution** for employees and a **fire and explosion** prevention measure. Machines must be enclosed whenever possible and the extraction and transportation installations must be designed to handle the quantities of dust produced. If the extractor unit is likely to generate a high partial vacuum in the workroom, a pressure compensation system must be provided, but this must not cause any draughts in the workplace. Even where the industrial building is of an open design, every effort should be made to prevent draughts.

If **harmful substances** are released during the woodworking operations, the **exhaust air** cannot be **returned** to the work areas. Furthermore, where exhaust air is returned, the dust load at the workplace must not exceed permitted levels. The extracted dust must be discharged through non-flammable, fracture and wear-resistant extraction pipes, which must be designed and their rate of extraction dimensioned so that no undesirable deposits are able to form in the system.

Before the exhaust air is discharged into the environment the **dust** it contains must be **separated off**, for which purpose **centrifugal separators or fabric filters** are used. More costly and more effective fabric filters are required where extracted air contains sanding dust. Due to the risks of fire and explosion the extractor installations must be fitted with **preventive safety devices**, such as pressure relief valves, bursting discs, spark detection installations, smouldering fire alarms and fire extinguishing equipment.

57

- **Gaseous emissions**

When wood is dried, **volatile constituents of wood** in the exhaust air generate **odours,** and this exhaust air must therefore be released so as to avoid any odour nuisance.

Since **wood processing mills** are often sited in **isolated locations,** as already mentioned, the employees are those most subject to gaseous emissions.

This problem can be minimised by an appropriate **choice of site** (in terms of distance, allowance for the prevailing wind direction).

Otherwise, **gaseous emissions** are only of **minor significance in sawmills.**

- **Analysis and evaluation of environmental impacts**

In Germany timber mills are governed by the *Technische Anleitung zur Reinhaltung der Luft TA-Luft* [Technical Instructions on Air Quality Control] and the *Technische Anleitung zum Schutz gegen Lärm TA-Lärm* [Technical Instructions on Noise Abatement]. Accordingly the TA-Luft of 1986 restricts the **mass concentration of wood dust** in inhalable form to 20 mg/m^3 at a mass flow of 0.5 kg/h. Lower limit values apply correspondingly to various dusts from woods treated with certain wood preservatives.

For most of the **organic substances** involved in wood processing, the upper limit is 150 mg/m^3 at 3 kg/h. For airborne dust, which is a health hazard, concentration values of 0.45 mg/m^3 and 0.30 mg/m^3 are specified.

The acoustic pressure level is taken as the basic unit to describe the **noise situation.** Where measured and assessed values are indicated, three fundamentally different frequency weighting curves are used: single measured value, effective level and evaluation level (German DIN standards, guidelines of the Association of German Engineers VDI). In Germany the permissible **limit values** are 35 and 70 dB(A), depending on the preconditions for assessment.

In the case of **wood preservatives** their composition must be carefully examined (preservatives containing PCB's are banned in Germany). They must be kept sealed and accident-proof during storage. No wood preservatives dripping from treated timber are allowed to seep away in an uncontrolled fashion. Appropriate fire and accident prevention measures must be taken and **waste** must be disposed of correctly.

Where sawmills or mechanical woodworking mills are to be built or re-equipped, these statutory provisions must be applied as guidelines where there are no national regulations.

- **Interaction with other sectors**

The sawmill industry generally obtains its raw products from nearby forests. It must be ensured that the timber comes only from a **properly managed forest** (management strategy, coordination of individual usage plans, yield regulation, forestry and wood crop techniques) operating on the principle of **sustainability**.

Sawmills supply their products primarily to the **wood-processing trade and industry** (building, furniture and packaging sectors) and also for export. On the other hand the **waste material** produced contributes to **supplying** the derived wood product industry, particularly the chipboard industry, with **raw materials**.

The **burning of waste wood** concerns all aspects of timber usage and is therefore considered in a separate section.

Mechanical woodworking is primarily associated with **the generation of noise and dust**, with **gaseous emissions** and **odours** occurring only to a limited extent during artificial drying or treatment, and merely constituting a nuisance. Generally speaking, the sawmill industry does not damage or endanger the environment, except where **wood preservatives** are used, but even this problem can be avoided by careful **siting** of mills relative to residential areas.

2.2 Derived wood product manufacture

The term "derived wood products" covers **chipboard, fibreboard and plywood**. In addition to wood, these products - with the exception of a few types of fibreboard - contain an organic or inorganic **bonding agent** and in some cases **additives**.

The **bonding agents** used are mainly amino and phenolic resins, condensation products from an amino compound (urea, melamine) or a phenolic substance (phenol, resorcin, cresol or formaldehyde). Chipboard bonding agents on a diisocyanate adhesive base are a relatively new development. Polyvinyl acetate adhesives are used for **wood core plywood**.

- **Chipboard manufacture**

Practically all varieties of wood, wood waste and in some cases fibrous plant substances, bark and biomasses can be used as the feedstock for board. The first stage in the process is the **machining of the raw material**. Long cut or round wood is either cut into chips with drum chopping machines or processed directly into shavings with cutters. The next stage in shaving processing is **drying**, following by **sizing, intermediate storage, bonding and hot pressing** of the cut material. The initial production stages take place in enclosed installations without any significant **emissions**, these occurring only at the stage of hot bonding in the chipboard press at temperatures of 160 to 220°C. The final production stages comprise trimming, sanding and formatting of the board.

The main bonding agents used are amino and phenolic resins, condensation products from an amino compound (urea, melamine) or a phenolic substance (phenol, resorcin, cresol or formaldehyde). Chipboard bonding agents on a diisocyanate adhesive base are a relatively new development. Polyvinyl acetate adhesives are used for **wood core plywood**.

- **Plywood manufacture**

The term "plywood" covers **veneer plywood** and **wood core plywood**, the latter containing a central layer of rods, while veneer plywood is made by bonding together a number of individual sheets of veneer.

Suitable untreated wood is cut into veneer by sawing, cutting or scraping, dried and then bonded and pressed. The final production stage comprises trimming, sanding and formatting.

- **Fibreboard production**

A distinction is made between soft fibreboard, medium-density fibreboard (MDF) and hardboard.

Soft fibreboard contains no bonding agents. **Hardboard** likewise contains no adhesive, or at the most very small quantities of a phenol-formaldehyde bonding resin. **MDF**, like chipboard, contains 7 to 9% bonding agent.

The first stage of fibreboard production involves the production of fibres from wood, a process carried out by **heat or chemical treatment**.

It is then pressed by a number of different processes.

- **Manufacture of mineral-bonded derived wood products**

These products are made from **wood chips, wood shavings or wood fibres** and a **mineral bonding agent** such as cement, gypsum or magnesite. Wood is the main component, accounting for at least 85% of the dry weight. Manufacture is similar to that for chipboard, except that drying and hot pressing are not required.

3. Notes on the Analysis and Evaluation of Environmental Impacts

Noise emissions are produced in wood transport, cutting and preparation in all four manufacturing processes. **Dust emissions** may arise in stores. As in sawmills the resultant nuisance can be reduced by a sensible choice of site and by providing suitable precautions for employees (enclosed workstations, personal hearing protection).

Extremely **fine dust** is produced in the final stages of chipboard, plywood and fibreboard manufacture and this must be removed by means of centrifugal separators or fabric filters, as it is a health hazard to employees.

Gaseous emissions are produced only in the drying of wood shavings and the pressing of shavings and veneers.

In chipboard and plywood pressing, where amino bonding resins are used, **formaldehyde** is the main substance yielded in terms of the mol ratio of the bonding resin. Where phenol-formaldehyde bonding resins are used, only traces of **phenol** are found, and smaller quantities of formaldehyde are produced than is the case with amino bonding resins. **Phenol** and **formaldehyde** are both **potential health hazards**. In Germany formaldehyde emissions at the workplace must not exceed 0.6 mg/m^3 and the finished board formaldehyde content must not exceed 10mg/100g board weight, according to EC Directives. After **fitting the** boards formaldehyde concentration must not exceed 0.1 ppm in the ambient room atmosphere.

The Gefahrstoffverordnung [Ordinance on Hazardous Substances] of 1986 specifies the formaldehyde emission values for all derived wood products in Germany. These gaseous immissions do not occur during the manufacture of mineral- bonded derived wood products.

Wastewater problems occur during the cleaning of the bonding machines and presses. In fibreboard manufacture, wastewater is produced during the wet process and contains **wood particles, wood substances, bonding agents** and other **treatment agents** which can be cleaned using physical processes (sedimentation, flotation or filtration) and/or biological processes. Semi-dry and dry processes do not produce any wastewater.

Residues in the form of wood particles can be returned to the production process in chipboard manufacture, but are otherwise burnt.

In addition to the specific statements in the text, the information relating to mechanical woodworking is also applicable to the **analysis and evaluation of environmental impacts**.

4.   Interaction with Other Sectors

The wood product industry is reliant upon the **forest** as its raw material supplier, except where wood waste can be used, as is the case in the chipboard and fibreboard industry. The basic tenet here is that of the principle of sustained yield. The environmental brief **Forestry** contains detailed information on this subject.

Round wood can be fully utilised by linking sawmills wherever possible to fibreboard and chipboard manufacture.

Derived wood product factories are major **power consumers** which today generate their power with wood very rarely nowadays. The environmental briefs **Overall Energy Planning, Thermal Power Stations and Renewable Sources of Energy** should be consulted in this regard.

Wastewater management issues are also addressed in a separate environmental brief.

- **Charcoal production**

Charcoal is produced by the **thermal decomposition** of wood carried out **without air** (wood pyrolysis). This process also yields **gaseous and liquid reaction products** such as wood gas, wood vinegar, wood spirit and wood tar.

Charcoal is produced at temperatures of between 400 and 600°C, and is used as a **fuel**, a **reducing agent** in metallurgy and as a **raw material** for the chemical and pharmaceutical industries. Wood tar and the other liquid organic substances can be processed or alternatively burnt as an energy source.

Charcoal production is the only process still used today on an industrial scale in which wood is **chemically modified** to a substantial degree. Charcoal production is not therefore classified as part of the timber industry, but constitutes a separate **sector of the chemical industry**.

In many countries charcoal is an important **source of energy** for cooking and heating. The favourable ratio of weight to calorific value means that it can also be **transported considerable distances** from the place of production to its market. This fuel is in particular demand in cities because it produces heat without a great deal of smoke.

Charcoal is often produced in **small businesses** (with the exception of the East Amazon, Caracas), which either fell the raw material themselves or have charcoal production plants for **wood waste** in the immediate vicinity of sawmills. The latter arrangement is particularly useful where there is no derived wood product plant downline from the sawmill.

Gaseous emissions from charcoal production, in the form of smoke and strong odours do not merely constitute a nuisance: where the process is inefficiently managed pyrolysis derivatives, such as benzpyrene may constitute a **health hazard** to employees, and at high concentrations also to the general population (**cancer risk**). The comments made on the siting of sawmills apply here too.

The charcoal production process yields considerable quantities of **pyrolysis water** - up to 15% of the feedstock; this wastewater contains, *inter alia*, pyrolysis tar and water-soluble organic substances. Whereas liquid pyrolysis products must be **conditioned** according to the regulations for chemical industry installations in **charcoal production** on an industrial scale, **no such solution yet** exists **for the small business**.

If large-scale charcoal production from wood waste is sited close to wood processing mills appropriate measures must be taken to prevent pollutants from penetrating the water or soil.

- **Burning of wood waste**

The quantity of residual material (sawdust, splinters, bark, pieces) varies from process to process and product to product; tropical hardwood cutting produces extremely large quantities of waste (up to 60%). **Waste disposal** may take the form of combustion **for energy production**, as it is not possible to market the wood waste elsewhere because of the siting of the mill close to the saw of supply of the raw material. The presence of a **downstream pulp or paper mill** is a rare exception.

Complete incineration produces carbon monoxide, organic hydrocarbons, tar and soot. It is **almost impossible to influence** the **nitric oxide emissions** from wood furnaces.

5. Summary Assessment of Environmental Relevance

While plywood mills process high-quality round timber, chipboard and fibreboard are the result of the value-adding utilisation of different wood varieties, some of which are low grade.

Gaseous emissions represent **further** harmful environmental impacts of chipboard and fibreboard plants, the main principal hazardous substance being **formaldehyde**. By contract, bonding with phenolic resins and diisocyanates help reduce emission values. One exception, in terms of emissions, is the manufacture of adhesive-free fibreboard.

Gaseous emissions from **wood shaving driers** have few environmentally harmful properties, especially in the case of hardwood, although the **intensity of the odours** produced constitutes a nuisance. The same **siting** criteria apply as for sawmills.

6. References

Anonymous 1976: Planung von Absaug- und Entstaubungsanlagen; Internationaler Holzmarkt 67, Heft 24, p.1...6.

ASP; without year of publication: Arbeitsmedizin - Sozialmedizin - Präventivmedizin; ASP-Hefte 4, 7, 10; Gentner Verlag Stuttgart.

Baldwin, S.; Geller, H.; Dutt, G.; Ravindranath; N.H.; 1985: Improved Woodburning cookstoves: signs of success; Ambio; 14; 4/5; 280-287; 1985, 47 ref.

Baller, Gerd; 1987: Lärmschutz im Tischlerhandwerk; dds; No. 5; p.115; No. 6, p.67.

Baller, Gerd; 1987: Staub- und Späneabsaugung im Tischlerhandwerk; No. 8, p.47; No. 9, p.65.

Baums, M.; Brötzmann, U.; 1986: Die Gefahrstoffverordnung ist in Kraft getreten; Kommentar und Hinweise für die holz- und kunstoffverarbeitende Industrie; Holz-Zentralblatt 112 (1986), p.1833...1841.

Bernert, J.; 1976: Emissionen von Holzspanplattenwerken, Wasser, Luft, Betrieb 20, p.27...34.

Birjukov, V.; Oskov, N.; Zamaraev, M.; Sokolov, V.: Vervollkommnung der Verfahren zur Beseitigung schädlicher Emissionen holzverarbeitender Betriebe; Holztechnologie 18, p.235...238.

Blanchet, G.; 1984: Beehive charcoal kiln in Zaire; Bioenergy 84 Proceedings of conference 15-21 June 1984, Gothenburg, Sweden, Volume III, Biomass conversion (edited by Egneus, H; Ellegard, A) 160-162; 1984.

Bringezu, St.; 1988: Zur Prüfung und Bewertung der Umweltverträglichkeit von Holzschutzmitteln. Holzschutz und Umweltschutz haben gemeinsame Ziele; als Roh- und Werkstoff, 1989; p.421 ff.

Brocksiepe, G.; 1971: Holzverkohlung; in Chemische Technologie Band 3; p.417...492; Carl Hanser Verlag, Munich.

Brocksiepe, G.; 1976: Holzverkohlung; in Ullmanns Enzyklopädie der Technischen Chemie Band 12, p.703...708; Verlag Chemie Weinheim.

Bundesgesundheitsamt [German Federal Health Office], without year of publication: Vom Umgang mit Holzschutzmitteln; eine Informationsschrift vom BGA; own publication Berlin.

Busch, B.; Energiegewinnung aus Rinde; Holz-Zentralblatt 107, p.351...352.

Buslei, Wilfried; 1989: Holzstaub, Verdacht auf Krebs; dds; No. 9, p.44; No. 10 p.70; No. 11 p.54; No. 12 p.96.

CRIR; 1985: Forest Product Research International - Achievements and the Future; Carbon from Biomass - Woodgas in Practice; Proceedings Volume Five; own publication Pretoria/RSA.

Deppe, H.-J., Ernst, K.; 1990: Taschenbuch der Spanplattentechnik, 3. Auflage edition, 1990; DRW-Verlag Stuttgart.

DRW 1970: Maschinen und Maschinenstraßen in der Holzindustrie; DRW-Verlag Stuttgart.

Ernst, K.; 1987: Umweltfreundliche Holzwerkstoffe; Holz als Roh- und Werkstoff, 1987; p.411.

Ernst, K.; Schwab, E. Wilke, K-D.: Holzwerkstoffe im Bauwesen Teil 1: Materialkunde; EGH Entwicklungsgemeinschaft Holzbau; own publication, Munich.

Ferreira, F.A.; Alfenas, A.C.; 1985: Injurias em folhas de Eucalyptus spp. causados por condensados pirolenhosos originiarios de fornos de carvoejamento; Foliar injury in Eucalyptus spp. caused by condensed pyrolignins from charcoal kilns; Revista arvore; 9; 2; 186-190; 1985; 7 ref.

Graf, E. 1989: Ökologische Aspekte zur chemischen Holzbekämpfung; Holz als Roh- und Werkstoff, 1989; p.383 ff.

Hartmann, E.; Havla, R., 1984: Technologie und Technik der energetischen Nutzung von Holzresten unter besonderer Berücksichtigung der Wärmegewinnung durch Verbrennung; VEB Wissenschaftlich-Technisches Zentrum der holzverarbeitenden Industrie; own publication: Dresden.

HII, G.S.C.; Tay S.S.; 1980: An assessment of sawmill pollution in Sarawak; Malaysian Forester; 43; 2; 238-243; 1980; 5 ref.

Kauppinen, T.; Lindroos, L.; Mäkinen, R., 1984: Holzstaub in der Luft von Sägewerken und Sperrholzfabriken; Staub-Reinhaltung der Luft 44, p.322...324.

57

Knigge, W.; Schulz, H., 1966: Grundriss der Forstbenutzung; Entstehung der Eigen-
    schaften, Verwertung und Verwendung des Holzes und anderer Forstpro-
    dukte; Verlag Paul Parey: Hamburg/Berlin.

Koch, D.; Funke, T.; Grosse Wiesmann, G.; Wiemer, H-J.; Wüllenweber, H-J.;
    1985: Werkstoffe und Gefährdungen im Tischlerhandwerk; Schriftenreihe
    der Bundesanstalt für Arbeitsschutz; Forschungsbericht Nr. 441; Wirt-
    schaftsverlag NW: Bremerhaven.

König, E.; 1972: Holz-Lexikon, Bd. 1 und Bd. 2; DRW-Verlag Stuttgart.

Kollmann F. 1951: Technologie des Holzes und der Holzwerkstoffe Bd. 1 und Bd.
    2; Springer-Verlag.

Lemann, M.; 1981: Abgasreinigung mit Wärmerückgewinnung; Holz-Zentralblatt
    107, p.41...42.

Lingelbach, K.; 1982: Maßnahmen zur Senkung von staub- und gasförmigen Luft-
    verunreinigungen an einem Spanplattenwerk; Bericht 50 441-3/8 des TÜV
    Kassel; Umweltbundesamt [German Federal Environmental Agency]: Berlin.

Lorenz, W.; 1982: Heizen mit Holz; Technik am Bau 2/82, p.117...121.

Maier, G.; 1988: Späneabsaugung an Maschinen. Überlegungen zu Strömungs-
    technik und Konstruktion; Holz als Roh- und Werkstoff 1988; p.311.

Mamit, J.D.; Wee, H.B.; Lai, C.J.; 1985: The survey of the disposal of woodwaste
    by sawmills in Sarawak; Technical Report (Timber Research and Technical
    Training Centre); Sarawak; No. TR/4; 15pp.; 1985; 3 ref.

Marutzky, R. 1977: Untersuchungen zum Terpengehalt der Trocknungsgase von
    Holzspantrocknern; Holz als Roh- und Werkstoff 36, p.407...411.

Marutzky, R.; Mehlhorn, L.; May, H.-A.; 1980: Formaldehydemissionen beim Her-
    stellungsprozeß von Holzspanplatten; Holz als Roh- und Werkstoff 38,
    p.329...335.

Marutzky, R.; 1981: Emissionstechnische Erfassung von luftverunreinigenden Stof-
    fen aus Anlagen zur Herstellung von Holzspan- und Holzfaserplatten;
    Gesundheitsingenieur - Haustechnik - Bauphysik - Umwelttechnik - 102,
    p.300-335.

Marutzky, R.; 1981: Möglichkeiten zur Verkohlung und Vergasung von Holz und anderen pflanzlichen Reststoffen; Holz-Zentralblatt 107, p.315...317.

Marutzky, R.; 1984: Holzreststoffverbrennung - Techniken, Umweltschutzmaßnahmen, Wirtschaftlichkeit; Holz-Zentralblatt 110, p.1693...1694 and 1713...1714.

Marutzky, R.; 1987: Grenzen der Emissionsminderung bei Holzspänetrocknern unter Berücksichtigung der neuen TA-Luft; Holz als Roh- und Werkstoff 1987; p.421.

Marutzky, R.; Flentge, A.; Mehlhorn, L; 1987: Zur Messung der Formaldehydabgabe von Holzwerkstoffen, Baustoffen und Möbeln mittels der 1m³ Methode-Kammer Methode; Holz als Roh- und Werkstoff, 1987; p.339.

May, H.-A.; Mehlhorn, L.; Marutzky, R.; 1981: Gefahrlose Spänetrocknung bei der Spanplattenfertigung; Schriftenreihe "Humanisierung des Arbeitslebens" Bd.21; VDI Verlag Düsseldorf.

Mayrhofer; W; Pimminger, M; Gritzner, G; 1987: Untersuchungen zur Abgasreinigung von Spänetrocknern; Holz als Roh- und Werkstoff, 1987; p.379 ff.

Nantke, H-J.; 1986: TA-Luft - Was ist bei der Spanplattenherstellung zu beachten? Holz-Zentralblatt 112, p.2183.

Nimz, H.H.; 1988: Probleme, Kentnisse und Hoffnungen zum Thema "Holzstaub"; Holz als Roh- und Werkstoff 1988; p.117 ff.

Patao, D.N.; 1987: Sample Survey of charcoal and fuelwood consumption in Region 1; FPRDI, Vol. 16, No. 1, Jan-June 1987; p.58 ff.

Peters, F.; 1983: Energieerzeugung aus Holzreststoffen; Industriefeuerung 25, p.49...58.

Philipp, W.; 1980: Verbrennung von Rinde und deren Abwärmenutzung; Holz-Zentralblatt 106, p.1457...1458.

Robertson, D. (Ed); 1983: The Sixth International FPRS Industrial Wood Energy Forum 82, Volume II: Power and Heat Plants; Forest Products Research Society: Madison, WI/USA.

57

Robertson, D. (Ed.); 1983: The Sixth International FPRS Industrial Wood Energy Forum 82, Volume II: Gas and Charcoal Production from Wood/Biomass Fuels; Forest Products Research Society: Madison, WI/USA.

Rong, M; 1981: Herstellung von Holzspan- und Holzfaserplatten -Verfahrenstechnik und Emissionen luftfremder Stoffe; Gesundheits-Ingenieur - Haustechnik - Bauphysik - Umwelttechnik 102, p.287...295.

Saeman, J. (Ed.); 1976: Wood Residue as an Energy Source; Forest Products Research Society: Madison, WI/USA.

Salje, E. (Ed.); 1975: Umweltschutz bei der Holzbearbeitung; Tagesbericht; Deutsche Messe- und Ausstellungs-AG: Hannover.

Salje, E.; Geerken, J. 1988: Verringerung der Staubemissionen beim Fräsen; Holz als Roh- und Werkstoff, 1988; p.340 ff.

Schlotterhausen, R; 1986: Holzwerkstoffe - Merkmale und Verarbeitung; Wedra Verlagsgesellschaft: Stuttgart.

Wolf, Dr. J; without year of publication: Sicherheitswissenschaftliche Monographien Gesellschaft für Sicherheitswissenschaft; Bergische Universität Wuppertal, Wirtschaftsverlag NW GmbH.

Soine, H.; 1982: Energiegewinnung aus Holzabfällen; Holz als Roh- und Werkstoff 40, p.217...222, 263...268 and 281...286.

Smith, K.R.; 1986: Biomass combustion and indoor air pollution: the bright and dark sides of small is beautiful; Environmental Management; 10; 1; 61 to 74; 1986; 52 ref. BLL.

Strehler, A.: Wärmegewinnung aus Hackschnitzeln und Scheitholz; Holz-Zentralblatt 110, p.292...294.

Tsai, C.M.; 1983: Study on the quality improvement of urea-formaldehyde resin bonded plywood; Memoirs of the College of Agriculture, National Taiwan University; 23; 2; 80-82; 1983.

Tsai, C.M.; 1984: Effect of adding urea or melamine to urea-formaldehyde resin on the elimination of formaldehyde release from plywood; Technical Bulletin, Experimental Forest, National Taiwan University; No. 155; 14pp.; 1984; 65 ref.

VDMA 1986: Holzfeuerungsanlagen - Emissionsvorschriften, Holzbrennstoffgruppen; Einheitsblatt 24178 Teil 1; Verband Deutscher Maschinen- und Anlagen- bau e.V./Frankfurt.

VKE 1985: PVC - Ursache für Dioxin-Bildung? Informationsschrift des Verbands Kunstofferzeugende Industrie/Frankfurt.

United Nations Industrial Development Organisation; 1983: wood processing industry; Sectoral Studies Series, Division for Industrial Studies, United Nations Industrial Development Organisation; No. 4; 83 pp.; 1983; 59 ref. UNIDO/IS. 394, Limited distribution.

Vorreiter, L. 1958: Holztechnologisches Handbuch, Bd. I und II.; Verlag Georg Fromme: Vienna/Munich.

57

Trade and Industry

Pulp and Paper

58. Pulp and Paper

Contents

Appendix A: **Tables**

Appendix B: **Glossary**

58

1.   Scope

1.1   Introduction/general information/terminology

● **Pulp**

-   **Pulp** (in the context of this brief) is **the common** name for **vegetable cellulose** in the form of fibres or bundles of fibres, more or less free of residual plant parts.
    Cellulose is the principal **supporting component** of all **plants** and thus in theory, pulp can be obtained from any type of plant. However, as the various **fibre properties** and also the **fibre content** can **differ widely**, in practice only **relatively few types of plant** are used for **pulp production**.

-   Wood is the chief **raw material** used in pulp production. Generally speaking, softwood has longer fibres and hardwood shorter ones.

-   In addition to wood, **annual plants** are used for pulp production, mainly in countries with scarce wood resources (China, India etc.).

-   Pulp is produced for **mechanical processing into board, paper etc.** and for **chemical processing to films, synthetic fibres etc.**

-   Pulp is produced from the stated raw materials by means of **chemical, chemical-mechanical or mechanical processes**.

-   Pulp can also be obtained from **waste paper** (recycling) although in this case it can only be used for paper and board production.

-   The **auxiliary materials** used are water, steam, mechanical and electric energy and chemicals.

-   **By-products and waste** are produced which can cause direct or indirect atmospheric and water pollution, but which can be **reduced by measures within the production plant** (internal means) and by installations downstream from them (external means).

- **Paper and board**

- In the context of this brief, **paper** is a **thin fibrous mat made** principally from **pulp**, with or without surface treatment, manufactured from the above-mentioned types of cellulose.

- **Board** is thick (stiff) paper.

- **Cardboard** is strong board (thickness), made by a special process.

- Depending on the **type of pulp** or **waste paper** and how they are **pretreated**, the properties of the paper, board or cardboard can be adapted for their intended uses, thus the range of **different paper and board types** is manifold.

- The **wastewater** produced during paper and board manufacture contains **pollutants** which can be removed by appropriate **cleaning processes**.

58

1.2   Pulp production

Table 1.2 summarises the **basic data on pulp**, such as yield, specific energy consumption, relative consumption of chemicals, relative pollutant quantities and relative environmental pollution. Some **basic terms** are also defined below:

1.2.1   Raw materials

- **Wood**

a)   Softwood: Mainly pine, fir and spruce varieties, for pulp with long fibres (high strength).

b)   Hardwood: Mainly beech, birch, eucalyptus, poplar, other types and mixtures (medium strength) too, depending on the location.

- **Annual plants**

a)   Agricultural by-products: Various types of straw (wheat, rice etc.), ba-
                                 gasse, i.e. sugar cane following extraction (low
                                 strength). For special papers: linters, i.e. by-
                                 product from the cotton oil industry (high purity
                                 and strength).

b)   Others: Reed, bamboo, jute, kenaf etc. (less commonly
                                 used).

- **Waste paper**

Different grades, sorted into unprinted clippings from paper-processing (e.g. printing) through to mixes from domestic collections.

1.2.2 Products and processes

Apart from the hygiene sector (e.g. for nappies) **pulp is not an end product** but an
intermediate product for paper manufacture, and as chemical conversion pulp, **is
the raw material for the chemical industry** (fibres, films, plastics).

Pulp is available in a number of **different** forms, the most important of which are:

- **Groundwood (Mechanical wood pulp)**

Groundwood fibers produced mainly from softwood with a grinding stone and
containing practically the same constituents as the original wood, except for some
extractives. It provides the highest yield and generally is not bleached or only to a
low to medium brightness.

Applications: For mass-produced papers at the lower end of the quality scale.
Typically: newsprint, writing and printing paper containing wood, duplex board.
Characteristics: Products are of low strength, they yellow in daylight and are not
very resistant to ageing.
The chemicals most frequently used (for bleaching): sodium didithionite, peroxides
- peroxide being the least polluting bleach.
Plant sizes: 50 - 600 t/day.

- **TMP - thermomechanical pulp**

Comparable with groundwood, but defiberation between rotating discs. Slightly lower yield but better strength properties. Is bleached like groundwood.

Applications: as for groundwood.
Characteristics: as for groundwood.
Plant sizes: 300 - 600 t/day.

- **CTMP - chemi-thermomechanical pulp (includes APMP)**

In contrast to TMP, pulping is facilitated by a chemical pretreatment. The yield is slightly lower, but despite this the mechanical properties of the fibres are improved. Is usually bleached to medium or high brightness.

Applications: As absorbent in the field of hygiene products (nappies etc.), for mass produced printing and writing paper midway along the quality scale.
Characteristics: Depending on the raw material, products are medium to low strength; they yellow quite readily and are not particularly age-resistant.

- **SC (or NSSC) - semichemical pulp**

Still contains considerable quantities of non-cellulose substances. Wood chips or other fibrous raw materials are pretreated with chemicals in pressure vessels and with steam under pressure. Pulping is then carried out in refiners with a relatively low power consumption. Is not usually bleached.

Applications: Packaging papers, particularly corrugated medium in corrugated board.
Characteristics: Produces fairly stiff paper and board, depending on the raw material used.
Chemicals most often used: sodium sulphite, sodium hydroxide and/or sodium carbonate. Recycling or disposal within the mill is necessary.
Plant sizes: 50 - 500 t/day.

- ## Chemically produced pulp

Contains only low to very low quantities of non-cellulosic substances, low yields due to the removal non-cellulosic materials. Wood chips or other fibrous raw materials are pulped with chemicals and steam under pressure. This is usually followed by bleaching and then either drying and pressing into bales as commercial pulp or, in an integrated plant, further processing into paper products.

Applications:
- Unbleached: mostly as packaging paper, also added to lower strength pulps (reinforcing).
- Bleached: mostly for writing and printing papers, also as an additive to lower strength pulps, cellulose for chemical feedstock (dissolving pulp), mostly produced from hardwood.

Characteristics: High strength in the case of softwood products. Bleached substances yellow only slightly and are highly age resistant. High purity for chemical raw materials.

Pulping chemicals: Sodium hydroxide, Na_2S (alkaline processes: "soda", "sulphate") and Ca, Mg, Na and NH_4 bisulphite (acid processes: "sulphite"). Recovery and regeneration of chemicals is a precondition for economic and non-polluting operation. Some of the spent liquor from the sulphite process can be processed by fermentation to form yeast and alcohol or, in its dried form, can be sold as a binding agent.

Bleaching chemicals: Chlorine (the use of which is on the decline), sodium hypochlorite, chlorine dioxide, oxygen, sodium and hydrogen peroxide.

Plant sizes: softwood as raw material: 500 - 1300 t/day; annual plants as raw material: 50 - 250 nnt/day.

- ## Waste paper pulp

Waste paper contains a mixture of pulps from various sources, depending on the composition and sorting of waste paper, and is a substitute for fresh pulp (cheaper, energy-saving). It is pulped mechanically. May be de-inked and bleached following removal of non-paper fibres and other impurities.

Applications: In principle, for all paper and board types, with or without the addition of fresh pulp.

Characteristics: Quality slightly or moderately inferior to fresh pulp, depending on the quality, sorting, cleanliness etc. **Chemicals** for de-inking and bleaching are detergents, fatty acids, dispersants, dithionite, peroxide.

Plant sizes: 50 - 400 t/day.

1.3 Paper and board production

1.3.1 Basic fibrous material, pulp (raw materials for paper and card production)

All the products listed under 1.2.2 are basic materials for the paper industry. In most cases a **mixture** of two or more of them are used to give the paper the **required characteristics** or for **reasons of economy**.

1.3.2 Products and processes

Paper and board types are usually classified into the following **main groups** on the basis of intended use:

-      printing and writing (graphic) papers
-      industrial papers
-      special papers.

Practically all papers and boards are produced on continuously (or in the case of boards sometimes semi-continuously) operating machines, the principle of which is the **dewatering of the aqueous fiber suspension** on a wire to form a fibrous mat which is then **pressed and dried**. The sheet of paper thus produced is packed in the form of **rolls** or **packs of sheets**. The pulp fibres are **pretreated in "beating" machines (refiners)** to give them the **properties required** for the individual type of paper, and **additives** are used to give properties such as ink absorbance, water-resistance, stiffness, colour. **Fillers** such as kaolin (alumina), and more recently calcium carbonate and sulphate improve the **paper surface** for certain printing processes.

The **product groups** are characterised as follows:

•   **Printing and writing (graphic) papers**

These **writing and printing papers** are basically subdivided into those containing wood[1] (coated and uncoated)[2] and those not containing wood pulp (likewise coated and uncoated), the former mainly as mass-produced papers, the latter for **high quality and special applications**. Today, both kinds contain **increasing quantities (in some cases up to 100%) of waste paper**.

---

[1]   "wood containing":   containing not only pure, chemically produced pulp but also groundwood, CMP etc.

          "wood free":   containing only chemically produced material.

[2]   coated: surface treated with dominatingly inorganic pastes.

- **Industrial papers**

These include mainly **packaging papers** and **boards**, comprising many brands of grey common wrapping papers (made from recycled paper) through to high-quality packaging materials for food and luxury goods, in some cases surface treated, multi-layer or coated for costly print processes. **Corrugated board**, made from fresh unbleached pulp or recycled corrugated board (a rising trend) depending on quality, accounts for a **portion** of industrial papers in quantity terms.

- **Special papers**

These cover a wide range of paper types which **cannot** be specifically **allocated** to the two product groups described above, e.g.:

- papers for hygiene applications (tissues, kitchen rolls, toilet paper)
- filter papers for use in industry, the home, the laboratory etc.
- transparent papers for drawing
- photographic papers
- base paper for parchment, vulcanized fibre
- cigarette paper
- capacitor paper etc.

1.4   Secondary and auxiliary installations

- **Energy supply**

Energy is required in the form of **mechanical energy** (electricity) and **heat** (steam). Where no **hydraulic power** is available, electrical energy is obtained either from the **national grid** or generated by a **power plant inside the mill** (steam or gas turbines). **Fossil fuels** (heating oil, natural gas, coal), and also **wood and wood waste** (bark) or other **waste substances** are used for steam production.

> The **spent liquor from chemical pulp production** is an important "waste" product in terms of energy. It is burnt in special boilers ("recovery boilers")to produce steam to cover process energy needs.

- **Water**

The **availability of fresh water** is a **basic requirement** for pulp and paper production. The water demand may exceed **150 m³/t of product**, but in **very modern mills** it may be no more than **2 m³/t**, although this also depends on the quality of the process management.

- **Wastewater treatment**

Mechanical, biological and/or **chemical wastewater treatment** are now standard in any pulp and/or paper mill.

2. Environmental Impacts and Protective Measures

2.1 Area: Raw and auxiliary materials

2.1.1 Fibrous raw material

- **Wood**

The **afforestation and reforestation** of suitable areas for the raw materials supply of paper and pulp mills are **advantages in terms of climate, water resources and the labour market**.

The use of timber must be planned with a view to maintaining a balance between the **cutting and growth rates**.

Vegetable fiber resources are renewable - in the case of wood by reforestation. **Special measures** are essential for such **single-crop agriculture**; in-depth studies of **cultivation measures** as well as **socio-economic aspects** (e.g. competition for land usage) are essential.

- **Annual plants**

Agricultural products used as raw materials should **not automatically** be regarded **as environmentally advantageous**. For example, if **straw** is not ploughed back into the soil, increased fertiliser use is necessary, whilst the **humus** content in the soil will **drop**. The widespread **burning** of straw is also **undesirable** and the **collection** of straw is relatively **energy intensive** (pressing in bales for transport, yet still bulky, truck capacities by no means fully utilised). Furthermore, the large **stocks** which have to be **held** because of the relatively short harvest period create a **fire risk**.

In the case of **bagasse** (waste from cane sugar production) used as a fibrous raw material for paper, conditions are more favourable in that it does not have to be **collected separately**, but nonetheless **large stocks need to be held** to cover periods when the sugar factory is closed. The **competition between raw material for paper and fuel in the sugar factory** is described in the environmental brief Sugar.

In short, the **use of annual plants** is only **environmentally** positive under **certain conditions**. It is generally **insignificant** and only **relevant in special cases**.

•   **Waste paper**

This raw material enables **significant savings** to be made on **energy** compared to fresh pulp, with the exception of fully chemical pulp as modern pulp mills are energy selfsufficent. However, paper **cannot** be recirculated **ad infinitum**. For every circuit there is a **sacrifice in quality** due to fibre damage. However, the **use of waste paper** must be **regarded positively** from the environmental viewpoint, in most of today's cases of application.

2.1.2 Water

Production water (river and well water) is required in relatively **large quantities** (see 1.4 above) and must meet **certain minimum purity criteria**. It must be **processed**, but can be **recycled** in internal circuits a number of times. In less favourable instances the use of wells can result in a **long-term change in the groundwater table**. As far as water requirements are concerned **detailed analyses** with a view to **meeting the needs of competing usages** are essential at the **paper mill** design stage.

2.1.3 Energy

The **environmental impacts** of electricity generation and the use of **fossil fuels**, also used in pulp and paper mills, are known and can be found in the environmental briefs on Thermal Power Stations and Power Transmission and Distribution.

Sector-specific fuels arising in mills during pulp production or in the wood-processing industry are

-      spent liquor from the cooking and impregnation process
-      bark, sawdust, splinters.

Concentrated spent liquors are burned in recovery boilers specially designed for this purpose, thereby releasing the **pulping chemicals** in the form of molten ash for **regeneration**. Spent liquors replaces some, and in modern chemical pulp mills all, of the fossil fuels.

Wood waste is likewise burned in special boilers and **thereby replaces fossil fuels**.

(For relevance of emissions, see 2.2). Typical **specific energy consumptions** are listed in table 1.2.

2.1.4 Chemicals, auxiliary materials

Although some of the **chemicals to be added**, particularly bleaching agents, such as chlorine, sodium chlorate, caustic soda and peroxides, are bought in by pulp and paper mills, their **production** requires **considerable quantities of energy**. A reduction in bleach consumption requires a generally **greater acceptance of less bright paper** on the part of the consumer, but would be a **major environmental protection measure**.

The production of other **auxiliary materials**, such as dyes, starch, clay and resin, is likewise heavy on energy, but is less significant because of the relatively **small quantities used**.

2.2 Emissions from pulp and paper mills

2.2.1 Aqueous emissions

Table 2.2.1A gives a detailed survey of sources, substances emitted, impacts, and reduction measures and the degree of reduction of aqueous emissions, while table 2.2.1B provides information on typical emission limits in terms of quantity.

Considered first are:

- emissions
- their impacts and
- reducing and protection measures

before the downstream treatment plant (wastewater treatment plant). This is **followed by** the effect of these reducing and protection measures.

A: Quantity

The **quantity of wastewater** is approximately the same as the **quantity of fresh water used**. Thus **a reduction in fresh water consumption** by creating internal circuits results in **a reduction in wastewater quantities**, which is also a **major cost factor** when designing wastewater treatment plants.

B: Quality

Quality factors in aqueous emissions are

-       content of undissolved substances (settleable/filterable)
-       content of dissolved substances, comprising
    •   reaction products from pulping and chemical recovery
    •   reaction products from pulp bleaching
    •   concentrated condensates from chemical recovery
    •   chemical residues and soluble content of waste paper cleaning
    •   dissolved substances from paper manufacture and coating
    •   dissolved substances from secondary installation wastewater.

In terms of **impact** they are all capable of:

> changing pH, consuming O_2, causing discoloration or turbidity, they may be toxic, either individually or in combination.

Primary reduction measures are **internal recirculation** before the remaining wastewater and pollutants are transferred to the

C: Wastewater treatment plant

(secondary treatment, downstream plant) where they are purified to such a degree that they can be **discharged into public sewage systems**.

2.2.2   Emissions into the air

There is an wide variety of **major sources** in **pulp mills** and they are affected in some cases by highly complex technical factors. They range from dust produced in raw material crushing, through **vapours and gases escaping from reaction vessels and liquor tanks to flue gases from recovery, bark, sludge and oil/coal boilers, the waste gases from lime burning and degassing systems of bleach tanks and bleaching towers**.

In **paper mills**, the situation is less complex, with fewer factors involved, the **main source** being **waste air from dryers**.

Table 2.2.2A gives a detailed **overview** of source, substances emitted, impacts, reduction and protection measures which can be taken inside the mill and the degree of reduction in the most important departments, while **table 2.2.2B** provides **quantity data** relating to typical emissions in the sector, values currently achievable and limit values.

External plants are **avoided wherever possible** in the case of waste air cleaning plants; they are **incorporated** in the various process stages so that the media extracted can be **recirculated**.

The main **emission components** are carbon dioxide and monoxide, dust (wood and mineral), steam, sulphur dioxide, reduced sulphur compounds (mercaptans and the like), nitric oxides and hydrocarbon compounds.

Their most significant **impacts** are:

- they range from being hazardous to health to toxic, they present a fire risk, they give rise to odours and smog, they may be a contributory factor to acid rain and they reinforce the greenhouse effect.

Reduction or protection measures range from internal collection, recirculation, combustion or other chemical conversion processes to downstream (external) gas scrubbers, filters and absorbers.

2.2.3 Solid waste

Table 2.2.3 describes the **sources, materials, impacts** and **possible counter-measures** specific to solid waste.

The **main sources** are as **varied** as for gaseous emissions. They comprise for the most part **wood waste** such as sawdust, bark, fiber bundles, and also **mineral waste** such as lime sludge, sand and **spent auxiliary materials** such as screens, felts, plastic film, wire etc. The main **impact** is the dumping space requirement.

Reduction and protection measures involve essentially a reduction in volume by incineration and the return of recyclable materials to the manufacturer (metal parts, for example).

58

2.2.4 Noise

The principle **sources of noise** are:

> raw material preparation, such as wood debarking and chopping, transport equipment, refiners, vacuum pumps, finishing machines, steam blow-off in the boiler house, drive units.

The **impacts** may range from nuisance and disturbance to nearby residential areas at night to physical health problems and hearing impairment.

Possible **reduction measures** are:

> wood debarking and chopping and heavy goods transport in daytime only, since intermittent operation is possible. Also enclosure of machines, and if applicable sound-proofing materials, steam blow-off with silencers only, new plant sited a suitable distance from residential areas. (With few exceptions, machines are already being designed with noise reduction in mind).
> internal measures: prescription of hearing protection to be in the relevant departments.

3. Notes on the Analysis and Evaluation of Environmental Impacts

3.1 Aqueous emissions

The **monitoring** of these emissions demands continuous or semi-continuous **sampling** and appropriate equipment, both for the individual wastewater flows and for the combination of them.

Routine analyses can be confined to temperature, pH, settleable or filterable solids, biochemical oxygen demand (BOD_5), chemical oxygen demand (COD, measured as potassium chromate consumption), fish toxicity, adsorbable organic halogen compounds (AOX) in relevant cases (i.e. where chlorine or bleaching agents containing chlorine are used).

Special analyses include, *inter alia*, the determination of turbidity, colour, odour, conductivity, colloids, oils and fats.

Analysis methods for routine and special tests are listed in **table 3.1.1**.

Minimum requirements for the discharge of wastewater into public sewage systems have been established and have come into effect in a number of countries in order to **assess the environmental impact of aqueous emissions**.

For the **German pulp and paper industry,** the *Wasserhaushaltsgesetz* [*WHG* - Federal Water Act], the *Abwasserabgabengesetz* [*AbwAG* - Wastewater Charges Act] and the *Bundesimmissionsschutzgesetz* [*BImSchG* - Federal Immission Control Act] are of particular relevance. These laws and their enforcement ordinances lay down minimum requirements which must be observed following application of emission-reducing purification or cleaning processes (see in this regard tables 3.1.2A and 3.1.2B).

In **Switzerland,** assessment criteria are established in the ordinance on water discharge "Verordnung über Abwassereinleitung", in Austria the standards ÖNorm cover this area, and in the **USA,** the "Effluent Limitations Guidelines and New Source Performance Standard for the Bleached Kraft, Groundwood, Sulfite, Soda, Deink and Non-integrated Paper Mills Segment of the Pulp, Paper and Paperboard Mills" in the EPA (Environmental Protection Agency) programme apply.

In **some other countries,** although similar values exist, they are all too frequently in the form of guidelines, and **compliance** with them is **seldom monitored.**

Likewise, **pollutant concentration data** is **often** given **as an assessment criterion,** though it would be more correct to restrict the absolute quantity of the emission. The concentration assessment may well lead to a dilution of emissions to the limit values if adequate fresh water is available, but "dilution is no solution to pollution".

To obtain a **reliable assessment** of the expected emission situation for an extension or new-build project, consideration must be given not only to **emission quantity and quality** but also to the **drainage situation.** The essential **determining factors** here are:

- flow rates with seasonal minima and maxima
- initial loading of watercourses/rivers
- use of water downstream from the discharge point (drinking water, irrigation, fishing, industry).

3.2   Emissions into the air

As the **major part of emissions** into the air and the main emission components (dust, CO_2, CO, NO_x) stem from the **combustion plants** for steam generation, the **relevant environmental brief Thermal Power Stations** should be consulted.

58

Typical sectoral emissions (particularly pulp mills) are:

> sulphur dioxide (SO_2), reduced organic sulphur compounds (TRS), chlorine/chlorine dioxide gas (Cl_2, ClO_2), certain hydrocarbons (HC).

The **routine monitoring** of these emissions is in some cases carried out by **continuous display and recording equipment**, which must be **inspected** and **calibrated** by supervisory bodies at prescribed intervals.

Non-continuous inspections (**special inspections**) are carried out on intermittently collected samples by laboratory staff. The **measuring methods** for this are prescribed in Germany by the *TA-Luft* [Technical Instructions on Air Quality Control]. As with wastewater (see above), other countries have their own specific measuring methods which are defined in the relevant clean air regulations.

The substance-specific **emission limits** currently applicable in Germany, in accordance with the TA-Luft, are listed in **table 2.2.2B**. It should be noted in this regard that the TA-Luft does not contain a limit value for **TRS compounds** which are responsible for the odour nuisance generated by sulphate factories. For this reason, the **limit values** currently **contained in the USE-EPA** - which are largely in line with the state-of-the-art - are recommended as guidelines.

The limit values given in table 2.2.2B can be used as guidelines for extension and new-build projects in countries where provisions in this area are inadequate or do not yet exist yet.

3.3   Solid waste

There are only a limited number of **analysis methods** for the **environmental impacts** of solid waste; for example, there are none for bark, sawdust-type wood waste, bale wire, plastic bags and felt. Other solid waste (**constituents of waste sludge, lime sludge, waste gas dust** etc.) is **routinely examined**.

As solid waste cannot be described as altogether typical of the sector, the analysis methods described in the **relevant environmental briefs** (Solid Waste Disposal, Timber) should be consulted.

Limit values to assess the environmental impacts of the aforesaid substances in the form of ordinances are **rarely** prescribed. In Germany they exist in respect of the suitability of **substances for disposal** (see in this regard the environmental briefs Solid Waste Disposal and the Disposal of Hazardous Waste). The Compendium of Environmental Standards also contains information about **substances which** can be **critical in waste sludge** (heavy metals from printing inks, toxic compounds etc.).

3.4 Noise

Noise is **measured and assessed as noise immission**. The unit of measure used in Germany is the dB(A) to DIN standard 45 633.

Immission limits vary according to the type of area, ranging from 70 dB(A) for purely industrial zones to 35 dB(A) for health resort and residential areas as a night limit.

No special measures are required to comply with a noise immission level of around 50 dB(A) in the immediate surroundings of modern pulp and paper mills, provided that the mills are housed **inside buildings** and are fitted with state-of-the-art sound-proofing.

In Germany, for example, this means that a pulp and paper mill can only be built in industrial or principally industrial areas.

If the relatively large amount of land required to build a paper mill is available, the noise restrictions do not in many countries represent a major barrier to such projects. In fact, the noise immission values may be conformed to even outdoor design of mills, as is frequently the case in tropical countries, as long as it is at an **adequate distance** from neighbouring areas used for residential or other purposes requiring protection.

4. Interaction with Other Sectors

4.1 Areas typical of the sector

4.1.1 Raw materials

Pulp and paper mills are extremely **capital intensive** and have a **very long service life** (some are over 100 years old). For this reason, the **long-term reliability of raw material supplies is of fundamental importance.**

* **Wood**

Planned and organised cooperation with the **forestry** sector is essential. This collaboration can be structured in a number of ways: it ranges from the simple **purchase** of trunk timber, thinning timber or wood chips and sawmill waste through to the **management** of a mill's own forests. In view of the large time span between the planting and the felling of trees, **long-term planning** is required for new-build projects. Proper **afforestation** work is essential, along with the appropriate capital investment and the necessary organisation.

One important aspect of interaction with other sectors is that of **competition** in the standing timber and sawmill waste market (sawmills, plywood manufacture, the wood processing and wood product sector).

Interaction with the **agricultural technology** sector exists in the context of so-called "agro-forestry", in which trees are used to provide shade and as windbreaks. This could be interesting if the correct types of tree for paper production were chosen, and it could also be an additional source of income for farmers.

Other **competition** with pulp wood derives from use of wood as fuel (see the environmental brief Renewable Sources of Energies) or wood for charcoal or building.

- **Annual plants**

With few exceptions, the **annual plants** which can be used for pulp and paper production constitute **waste or by-products** from agriculturally based industries (e.g. sugar production).

In global terms this group is **not significant** as far as paper manufacture is concerned, but may be **important locally** if no wood is available. The raw material potential for pulp and paper production therefore depends on the market for the products of these other industries. Relatively short-term changes in farming pro-grammes can greatly reduce raw material supplies or lead to arable land being withdrawn from production for reasons of pricing policy.

- **Examples of annual plants**

Straw:
Large quantities produced worldwide, but with **low potential for use** due to the costs incurred in its collection, transport and storage; significant emission problems in its processing to pulp. Change of crop programme, e.g. due to introduction of short-stemmed cereal varieties, could jeopardise projects. **Competitive situations** could arise with **a demand for straw as bedding for cattle** or as **fuel for heating and cooking**.

Bagasse:
The sugar cane residues following sugar extraction are **traditionally used in the sugar factory itself as a fuel** (energy self-sufficiency). It is therefore in **compe-tition with fossil fuels** if it is to be used for paper manufacture, hence the interaction with the agro-industry.

Of minor importance in terms of quantity, but interesting in terms of fibre properties are raw materials such as:

Jute:
In the form of jute sticks, the waste from the ailing **jute (textile) industry**. Competition with fuel. Jute cuttings: competition with the textile industry.

Flax:
Flax straw as **waste from the linseed oil industry**. Transport and cleaning very costly. Competition with textile manufacture.

Sisal:
Since sisal is little used now for ship hawsers, efforts are being made to cultivate its use as **a raw material for special papers**. Very high transport and preliminary cleaning costs. Competition with sack and bag production.

Abaca:
Plays a (minor) role in **special paper** production (Philippines) alongside (minimal) use for textile purposes.

Linters:
Waste product from cottonseed oil factories. Is a raw material for special, chemically pure cellulose for the chemical and pharmaceutical industries, also for special papers and filter material. Advantage: produced centrally - at the oil mill. Competition on the product side from wood dissolving pulp.

Bamboo:
Important building material in all countries where it grows (in practice only a natural crop, cultivation difficult), therefore **only limited quantities available** for pulp and paper. Competition exists for use as a vegetable (bamboo shoots).

Esparto or alfa grass:
Like bamboo, it is not cultivated and is collected only for **insignificant quantities of paper production** (special papers) (North Africa and Spain). Competition exists with use as a braiding material.

58

- **Waste paper**

The **potential for waste paper supply** is directly **dependent** on **paper consumption** in the region or catchment area and on **market price** (determined by the economic climate). In countries with large stocks, lower qualities are most vulnerable to the economic climate.

Waste paper can be an **important raw material** in countries where a paper industry is to be established for a relatively low initial investment.

There are **links** here with the **paper-processing industry**.

In many countries, "competition" exists in the form of the **recycling of paper** which in the end is too dirty for reprocessing (e.g. newsprint which is used first as wrapping matter and finally as toilet paper).

4.1.2 Auxiliary materials and additives

- **Water:**

Since the pulp and paper industry requires large quantities of **water**, there is competition with other sectors, e.g.:

- water for domestic and industrial use,
- agriculture (irrigation),
- other industries which consume water.

This can have a **direct** effect (surface water) or an **indirect and delayed** effect (well water). The competitive situation vis-a-vis agricultural and other businesses can be **mitigated by appropriate wastewater treatment** so that the water can be reused to a greater or lesser extent. However, all aspects of the possible **salination of the soil** must be taken into account here.

The **availability of water** is one of the most important factors in **selecting a suitable site** for a new mill.

The environmental impacts of other **auxiliary materials and additives** are not typical of the sector (chemicals, energy).

4.2 Areas not typical of the sector

Areas which are not typical of the sector but which are **essential** for the operation of a pulp and paper mill and concern mainly the **infrastructure** are simply listed here as an addition to the information given in the various texts (although this list is not exhaustive):

- water supply
- chemical industry, for alkaline chemicals (caustic soda, soda, aluminium sulphate, sulphuric acid, chlorine, sodium and hydrogen peroxide, sulphurous acids etc.) (in this connection refer to the Compendium of Environmental Standards)
- mineral oil industry, for fuel oils, lubricating oils, natural gas, LPG
- mining, for coal and possibly clay, limestone
- power stations, electricity transmission
- transport, roads, railway connections, waterways

and:

- workshops for repair work and maintenance of mechanical and electrical machines and instruments
- general and technical colleges for the basic education of personnel
- hospitals, clinics, for medical care
- social areas.

There is therefore an interaction with many wider **areas** such as regional development, planning of locations, general energy planning, schools, health services, water supply and distribution planning, transport and traffic planning etc.

5. Summary Assessment of Environmental Relevance

Given the state-of-the-art in the pulp and paper industry, and the technology developed or adapted for it in the field of **recirculation, reduction or prevention of emissions which pollute the environment** with **proper operation monitoring**, the following points can be made:

- With regard to wood as a renewable raw material, this industry is environmentally sustaining as long as the **quantity of wood used is less, or possibly equal to the quantity growing to replace it.** A further environmental protection feature arises with the processing of wood residues and waste (sawmill and brushwood) in pulp mills.

- In the case of **annual plants** the environmental impact is considerably less positive - the alternative use (fuel in the case of bagasse) involves **replacement in the form of fossil fuels** and therefore has negative impacts for the global CO_2 balance, among other things.

- The still increasing **use of waste paper** as a raw material has generally **positive environmental impacts**: the processing of waste paper consumes less primary energy than fresh pulp and, overall, reduces the wood consumption per tonne of paper.

- **Chemically produced pulp** merits particular mention: in terms of both raw materials and energy, a modern production plant uses renewable feedstock (wood) only and therefore has no impact on the global CO_2 balance.

- **Aqueous emissions from pulp production** are minimal in the case of unbleached pulp (as long as the mill is equipped with a recovery system), and the increasingly common **replacement of chlorine and chlorine compounds as bleaching agents** with chlorine-free media (oxygen and peroxide) already enables a large number of bleaching installations to comply with the relevant limit values.

- **Aqueous emissions from paper manufacture** can also be kept below the relevant limit values without difficulty by the **use of water recirculation measures** and highly **efficient water treatment plants**.

- **Gaseous emissions from power station and recovery installations** can be kept below limit values with the cleaning/scrubbing techniques developed. **Odour emissions** (mercaptans) are still a **problem**, particularly in the case of sulphate pulp factories. However, systematic collection and control measures in modern European plants are also achieving acceptable reductions below the (EPA) limit values in densely populated areas.

- Only a small quantity of **solid waste** is produced, and a large proportion of it can be **used for energy** (bark, wood waste). In the area of waste sludge disposal (incineration, dumping), attention must be paid to the **problems of heavy metals from printing inks** where mills use waste paper.

- The **outdoor design** of mills, normally applied in warmer climates, makes **noise prevention measures more costly** than in enclosed plants, thus noise nuisance can only be prevented by siting such mills **further** from residential areas.

In temperate and/or cold climates, the question of noise can easily be resolved by building design, insulation and process management (avoiding operating noise-emitting departments at night).

The **early involvement of the population groups affected,** particularly women, in the planning and decision-making process, means that **their interests** can be **taken into account** and helps **reduce environmental problems** (e.g. competition for the use of water, wood and land).

The **implementation and monitoring of emission limit values** and a generally **environmentally oriented operation** are only possible if the necessary **control bodies** are **institutionalised** and **operate effectively.** One option is to **appoint industrial environmental protection officers,** who should also be responsible for the **training and further education** of personnel and **increasing the awareness of personnel** for environmental matters.

58

6.   References

W Brecht and H. L. Dalpke: "Wasser, Abwasser, Abwasserreinigung in der Papierindustrie".

Deutsches Wasserhaushaltsgesetz (WHG), Abwasserabgabengesetz (AbwAG) and the Bundesimmissionsschutzgesetz (BImSchG).

EPA (Environmental Protection Agency) "Effluent Limitations Guidelines and New Source Performance Standards for the Bleached Kraft, Groundwood, Sulfite, Soda, De-ink and Non-integrated Paper Mills Segment of the Pulp, Paper and Paperboard Mills".

NCASI, USA (National Council of the Paper Industry for Air and Stream Improvement), Bulletins (div.).

ÖNorm M 94 64.

Allan M. Springer: "Industrial Environmental Control, Pulp and Paper Industry".

SSVL Sweden (Sriftelsen Skogsindustriernas Vatten och Luftvardsforskning), The SSVL Environmental Care Project.

Table 1. 2.   Basic Data Relating to Pulp

Abbrev.	Type	Yield as a % of raw material		Spec. energy consumption (kWh/t pulp)		Rel. chemical use in relation to raw material		Rel. pollutant output after process, untreated		Rel. environmental impact after treatment (a)	
		Wood	Annual plants	Wood	Annual plants	Wood	Annual plants	Wood	Annual plants	Wood	Annual plants
GW	"mechanical" stock (wood pulp)	99-100	--	600-2000	--	very low	--	low	--	low	--
TMP	Thermo-mech. pulp	97-98	--	800-2400	--	very low	--	low	--	low	--
CTMP	Chem. thermo-mech. pulp	91-97	--			low	--	medium	--	low	--
CMP	Chem. mech. pulp	82-96	(70-80)	min. 930	min. 600	medium	(medium)	medium	medium	low	low
SCP	Semi-chemical pulp	62-82	50-60	800-900	500-600	high	medium	high	high	low	low/ very high (c)
CP	Chemical pulp	40-50	30-40			very high	high	very high	very high	low/ medium (b)	low/ very high (c)
AP	Waste paper stock	80-95		200-400		low		low/medium		low	

(a):   state-of-the-art emission treatment
(b):   noxiousness dependent on the type of bleaching chemicals: with or without chlorine
(c):   dependent on the technical or economic feasibility of recovery or destruction of spent pulping chemicals

Table 2.2.1A	Aqueous Emissions Pulp and Paper Mills			Page 1
Sources/causes typical in sector	Substances emitted	Impacts	Reduction measures (state-of-the-art) in mill	Degree of reduction (%)
Open water circuits	Large quantity of wastewater	Large treatment plant, high energy and chemical consumption	Closing of internal circuits	To approx. 80%
Undissolved sub-stances from various sources/ careless process control	Organic fibre com-ponents and inor-ganic components (dirt), filler remains	Turbidity, color, oxygen con-sumption, smell	Division of process water flows, closing of circuits in mill, improved filtration	Varies depending o production type
Dissolved sub-stances from pulp production and recovery	Ligno-sulphonates, other lignin decom-position products, crude tall oil etc., organic sulphur compounds, Na salts	Marked brown coloration, oxygen consumption (in part difficultly degradable), odour nuisance	Optimisation of process stages, leak prevention, recycling of leaked liquid	Up to 90%
Pulp bleaching	Decomposition products of lignin and hemicellulose, chlorinated organic compounds, Na and Cl salts	Oxygen con-sumption (in part difficultly de-gradable), discol-oration, toxicity	Recycling of filtrates in plant, leak prevention, conver-sion to chlorine-free/ low-chlorine bleaching	Chlorinated com-pounds up to 100%, others only slightly
Condensates	Organic compounds (methanol, ethanol, uncondensed gases)	High oxygen consumption , color, odour nui-sance	Liquor stripping in/before conden-sation, combustion or separate proc-essing of stripper gases	Up to over 90%

Table 2.2.1A	Aqueous Emissions Pulp and Paper Mills			Page 2
Sources/causes typical in sector	Substances emitted	Impacts	Reduction measures (state-of-the-art) in mill	Degree of reduction (%)
Chemicals from waste paper processing	Printing ink components (in part containing heavy metals), dyes, processing chemicals, complex salts	Turbidity, oxygen consumption, toxicity (with heavy metals)	Closed circuit management (restricted), toxicity can be reduced indirectly by using printing inks without heavy metal components	Oxygen consumption low, toxicity high
Paper manufacture	Remains of chemical additives (dyes, brighteners, anti-foaming agents, retention and cleaning agents, fillers)	Turbidity, oxygen consumption (toxicity, if additive toxic)	As for waste paper	As for waste paper
Coating plant	Coating materials (latex, clay, emulsifiers, starch etc.)	Turbidity, oxygen consumption	Careful process management to prevent losses	--
Wastewater from secondary installations	Chemicals from water softening/ demineralisation, clarification salts etc.	Salt content	--	--

58

Table 2.2.1A Aqueous Emissions Pulp and Paper Mills				Page 3
Sources/causes typical in sector	Substances emitted	Impacts	Reduction measures (state-of-the-art) in mill	Degree of reduction (%)
Wastewater treatment plant	A) In the wastewater: oxygen-consuming substances (lignin and cellulose decomposition), dyes	Turbidity, discoloration, oxygen consumption	Mechanical (sedimentation, filtration, flotation), biological (aerobic, anaerobic) and possibly chemical (precipitation, adsorption with active carbon etc.) wastewater treatment	Colour: 95%, oxygen: up to around 60%, (pulp) and up to 95% (paper), colouring: up to 100%
	B) In treated sludge: organic and inorganic solids (incl. toxic components), products of biodegradation	--	Sludge incineration (possibly with flue gas scrubbing)	Over 90%

Table 2.2.1 B — Examples of Quantitative Emission Values

Aqueous emissions, pulp production, untreated

	Wastewater quantity m³/t	BOD kg/t	COD kg/t	ss kg/t	AOx kg/t	TOX (TEF)
	1) 2)	1) 2)	1) 2)	1) 2)	3) 4)	
Wood pulp						
TMP	8 / 20	10-30				
CTMP	50	15-28				
SC	50	3 [1) 5)]				
C sulphate	225	10-20 / 40-60 [5)]	3xBOD		1-2 [5)]	
C sulphite	450	250-500* / 60-200**	3xBOD		0-0.2 [5)]	

Aqueous emissions, paper production, untreated

	Wastewater quantity m³/t	BOD kg/t	COD kg/t	ss kg/t	AOx kg/t	TOX (TEF)
Graphic papers						
Newsprint	25 / 80	1-2		10 / 40		
MF writing and printing papers	70 / 180			30 / 80		
Industrial papers						
Common wrapping papers	0 / 50	3	0-3	0 / 10-30	--	6)

*:   no chemical recovery
**:   with chemical recovery
1)   with water circuits largely closed in the mill
2)   with water circuits largely open
3)   with chlorinated bleaches largely avoided
4)   with chlorinated bleach
5)   value range SSVL (Sweden)
6)   printing inks containing heavy metals can yield toxic sludges
TEF   Toxicity Emission Factor

Table 2.2.2 A

Emissions into the Air
Pulp and Paper Mills

Page 1

Sources/causes typical in the sector	Substances emitted	Impacts	Reduction measures (state-of-the-art) in the mill	Degree of reduction (%)
Raw material crushing and cleaning (chopping of wood, straw etc.)	Organic dust	Fire risk constituting health hazard	Extracting of air and cleaning in cyclones (and/or filtering, recycling, burning or dumping of dust)	up to 100%
Waste gases from digesters, steaming out of equipment and vessels	Steam, turpentine, other HC compounds, SO_2	Fire risk, odour nuisance, health hazard, acid rain	Condense steam and turpentine, recycle turpentine, burn residue, recycle SO_2 in the process, scrub residual gases	99 +
	TRS	Odour nuisance	Collect and burn TRS (cannot be condensed)	99 +
Fumes of spent liquor condensation plant	Steam, terpenses, methanol, TRS	Odour nuisance	Collect and burn gases	95 +
	Steam	--	--	--
	SO_2	Acid rain	Absorption in alkaline gas scrubbers, recycling in process	99 +
	NO_2	Ozone formation	In development: noncatalytic conversion	0
	TRS	Odour nuisance process	State-of-the-art process	99 +
Recovery boiler (waste gases)	CO	Health hazard	Minimise by process	0
	CO_2	Greenhouse effect	Unavoidable, does not pollute global balance	0
	Dust	Health hazard	Electro-filters, recycling in process	99 +

Table 2.2.2 A	Emissions into the Air Pulp and Paper Mills			Page 2
Sources/causes typical in the sector	Substances emitted	Impacts	Reduction measures (state-of-the-art) in mill	Degree of reduction (%)
	Steam	--	--	--
Lime kiln (waste gases)	SO_2	Acid rain	Use of S-free fuel oil or natural gas; in development: wood and bark gas	95 +
	CO	Health hazard	Minimise by process management	0
	NO_X	Ozone formation	Reduction not yet state-of-the-art (cf. cement sector)	0
	TRS	Odour nuisance	Minimisation possible by good process management	99 +
	Dust	Health hazard	Electric filters and recycling in process	99 +
Steam boiler fired by bark or waste wood (waste gases)	Steam	--	--	--
	CO_2 and CO	Greenhouse effect, health hazard	Unavoidable, but does not affect the global balance, minimisation by process management	as above
	Hydrocarbons	Greenhouse effect, health hazard	Minimisation by process management	
	NO_X	Ozone formation	In development: conversion from non-catalytic to catalytic	
Furnaces to destroy sludges and residues	Steam	--	--	--
	CO_2 CO	Greenhouse effect, health hazard	As above , minimisation by process management	as above
	NO_X	Ozone formation	Currently not state-of-the-art	
	Dust	Health hazard	Scrubbers, cyclones, dump	

58

Table 2.2.2A	Emissions into the Air Pulp and Paper Mills			Page 3
Sources/causes typical in sector	Substances emitted	Impacts	Reduction measures (state-of-the-art) in mill	Degree of reduction (%)
Fumes of bleaching towers, bleach preparation, chlorine transport	Chlorine Chlorine dioxide SO_2	Health hazard " "	Extract fumes and wash in scrubbers, return to process	Up to 100
Waste air from transport equipment for raw materials and products	Motor exhaust NO_x, CO, HC, CO_2	Health hazard, atmospheric effects	Catalysts, diesel operation with soot filters, use of electric vehicles where possible	Up to 90
Paper dryer, paper machine (ditto coating and laminating machines)	Steam	--	--	--
	Organic solvents	Health hazard	Gas scrubbing, carbon filters with recovery, also use of water-soluble auxiliaries	Up to 95
Processing of additives, waste air from vacuum pumps	Steam	--	--	--
Waste air from transport equipment for raw materials and products	Motor exhaust, NO_x, CO, HC, CO_2	Health hazard	Catalysts, diesel with soot filters, use of electric vehicles where possible	Up to 90

Table 2.2.2B	Emissions into the Air Typical for the Sector, State-of-the-Art, Limit Values		
Emission	Source	State-of-the-art mg/Nm^3	Typical limit values mg/Nm^3
Dust	- Power boiler - Absorption plant, Mg, Ca bisulphite and magnefite process - Lime-burning kiln - Smelt-dissolving tank	less than 50 less than 50 less than 50 less than 50	50 (ÖNorm) 50 (ÖNorm) 50 (ÖNorm)
SO$_2$	- Power boiler - Absorption plant, Mg, Ca bisulphite process - Ditto magnefite process - Lime-burning kiln with TRS burning	less than 50 less than 250 less than 250 less than 400	400 (Önorm) 700 (ÖNorm) 300 (ÖNorm) 400 (ÖNorm)
CO	- Power boiler - Lime kiln	less than 100 less than 250	cf. TA-Luft generally: oil-fired: 170 solid fuel: 250
Organic C	- Lime kiln	less than 50	150 mg/m^3 (TA-Luft)
NO$_x$	- Power boiler - Lime burning kiln	less than 200 less than 900	400 mg/m^3 HMW (LRV-K, 1989) (1,500 TA-Luft, rotary kiln for lime)
TRS	- Power boiler - Lime-burning kiln - Smelt-dissolving tank	less than 5 ppm V less than 8 ppm V 8.4 g/t BLS	5 ppm V (EPA) 8 ppm V (EPA) 8.4 g/t BLS (EPA)
Inorganic Chlorine/ Chlorides	- Bleaching plant - Chemical processing	Cl^2 and CL: less than 10 mg/m^3	Cl^2: 5 mg/m^3 (TA-Luft Cl: 30 mg/m^3 as HCl (TA-Luft)
HMW: Mean hourly value LRV: Luftreinhalteverordnung (ordinance on clean air)			

58

Table 2.2.3	Solid Waste Pulp and Paper Mills			Page 1
Sources/causes typical in sector	Substances emitted	Impacts	Reduction measures (state-of-the-art) in mill	Degree of reduction (%)
Raw material transport and preparation: Wood	Bark Wood shavings	Space required	Burning for energy generation	> 95
Straw	Binding wire	"	Collection, compacting, scrap trade	--
Pulp cleaning	Knots, bundles of fibre, sand	" "	Incineration for energy generation, dumping	> 95 0
Quality Control	Rejected product	"	Return to process	> 85
Chemical recovery, removal of foreign ions	Lime sludge* or lime	Ground water pollution	Recycling in lime-processing industries, dumping	0 - 80
	Sulphate soap**	Process problems	Recycling as raw material for chemical works	Up to 100
Waste paper treatment	Iron wire, plastic film, string	Space required	Dumping	--
Waste paper de-inking	Printing ink sludge (may contain heavy metals)	Ground water contamination	Incineration or special dump	Up to 85
* in sulphate and soda pulp mills ** ditto for softwood				

Table 2.2.3	Solid Waste Pulp and Paper Mills			Page 2
Sources/causes typical in sector	Substances emitted	Impacts	Reduction measures (state-of-the-art) in mill	Degree of reduction (%)
Water and waste-water treatment	Fibre sludge, inorganic sludge, biological sludge	Space required	Recycle or burn fibre sludge	Up to 85
			Dump inorganic and biological sludge, under certain conditions also use for soil improvement	--
Wear of consumables	Metal, plastic screens, synthetic textiles (felts), lubricants, cleaning agents	Space required	Return to manufacturer for recycling, dump, burn	--
Mill maintenance	Defective machine parts Packaging material		Return to manufacturer for recycling, (scrap), burn or dump	--

58

Table 3.1.1			Page 1
Methods of Wastewater Analysis and Possibilities for Reducing Environmental Impact			
	Pollutant/ properties	Unit	Analysis method	Methods for elimination or reduction

	Pollutant/ properties	Unit	Analysis method	Methods for elimination or reduction
1	Undissolved substances	mg/l	DEV H_2, SM 148, 224	Mechanical treatment, flocculation, biol. treatment
2	Substances which can be precipitated	mg/l	DEV H_2, SM 224, z x/1/76	Mechanical clarification, flocculation, biol. treatment
3	Suspended matter	mg/l	Difference from 1 and 2	Flocculation (filtration), biol. treatment
4	Turbidity	cm visibility	DIN standard 38 404-C_2, SM 163, 232	Flocculation (filtration), biol. treatment
5	Colours		DIN standard 38 404-C_1, SM 118, 206	Flocculation, coagulation, flotation
6	Temperature	°C	DIN standard 38 404-C_4, SM 162	Cooling (towers, lagoons, trickling filters)
7	Odour		DEV $B_1/_2$, SM 136, 217	
8	pH		DEV C_5/S_5, SM 144, 221	Neutralisation
9	Conductivity	µS/cm	DEV C_8, SM 154, 226	

DEV: Deutsches Einheitsverfahren [German standardisation procedure]
SM: Standard methods (APHA)
Z: Zellcheming code of practice

Table 3.1.1			Page 2
Methods of Wastewater Analysis, and Possibilities for Reducing Environmental Impact			
	Pollutant/ properties	Unit	Analysis method	Methods for elimination or reduction

	Pollutant/ properties	Unit	Analysis method	Methods for elimination or reduction
10	Total, evaporation, and incineration residue	mg/l	DEV H1/S3, SM 148, 224	
11	BOD	mgO_2/l	DEV H_5, SM 141, 219 DIN 38409-H_{91}	Biodegradation, aerobic, anaerobic
12	COD	mgO_2/l	SM 142, 220 Z x/2/76 DIN 38409-H_{41} and -A_{30} ARAVwV[1] no.303	Biodegradation, aerobic, anaerobic
13	Total organic carbon TOC	mgC/l	SM 138	
14	Oxygen	mgO_2/l	DEV G_2/J_8, SM 140, 218	
15	Total nitrogen, organic	mg/l	DEV H_{11}, H_{12}	Biodegradation, aerobic, anaerobic
16	Colloids	mg/l	DEV H_3	
17	Oils, fats	mg/l	DEV H_{17}, H_{18} SM 137, 209	Separators
18	Lignin, tannin	mg/l	SM 160	
19	Hydrocarbons	µg/l	DEV H_{15}	
20	Organic poisons	µg/l	SM 139	Biodegradation

1) ARAVwV: Appendix to the Rahmen-Abwasser-Verwaltungsvorschrift of 08.09.1989
[General Administrative Framework Regulation]

58

Table 3.1.1   Page 3
Methods of Waste Water Analysis, and
Possibilities for Reducing Environmental Impact

	Pollutant/ properties	Unit	Analysis method	Methods for elimination or reduction
21	Phosphorus	mg/l	ARAVwV no.108	
22	Nitrogen	mg/l	ARAVwV no.106/107, 202	
23	AOX	g/l	DIN standard 38409-H_{14}, ARAVwV no.302	
24	Chloride etc.	mg/l	DEV D_{5-7}, D_{15}, J_7, SM 156 - 158, 228	Ion exchange, ultra-filtration, rev. osmosis
25	Nitrate, nitrite	mg/l	DEV D_{9-10}, E_5 SM 131-135, 212-21	Biological decomposition
26	Heavy metals	mg/l	SM 211, ARAVwV no.207 (Ca) no.209 (chromium)/214 (Ni)/206 (Pb)/213 (Cu)/215 (Hg)	Flocculation
27	Na^+ etc.	mg/l	DEV H_{13-15}, SM 126, 147, 153	Ion exchange, ultra-filtration, rev. osmosis
28	Toxicity and bio-degradability		DEV L_{2-3}	
29	Population equivalent		DEV L_1	
30	Toxicity to fish	TF**, %, TEF*	DEV L_{15}, SM231 DIN standard 3842-L_{20}, ARAVwV 401	
31	Biological/ ecological water in-spection (water quality classes)		DEV M_{1-7}, SM 601-606	

*   TEF: toxicity emission factor   **   TF: toxity to fish

Table 3.1.2A

(Minimum) Wastewater Requirements (specific)
As at January 1990 in Germany

Type of pulp or paper	COD kg/t max. [2]	BOD kg/t max.	BOD mg /l	AOX kg/t max.	Toxicity to fish T_F max.	Substances which can be precipitated ml/l max.
Pulp (generally)	70	5		1**	2	
Paper:						0.5
writing and printing papers, depending on type	5 - 7	0.7 -6	25		0.5	0.5
Based on waste paper	6	1.2				0.5
Parchment	12	6			0.5	0.5

[2] Tonne, air dry = 0.9 t absolutely dry

**) Not applicable to dissolving pulp until 31.12.1992

58

Table 3.1.2b

Pollutant Units for the Measurement of Wastewater Discharges in Germany

No.	Pollutants and pollutant groups assessed	The following units of measurement are equivalent to one pollution unit in each case	Threshold values by concentration and annual quantity	
1	Oxidizable substances in chemical oxygen demand (COD)	50 kilograms oxygen	20 milligrams per litre and 250 kilograms annual quantity	
2	Phosphorus	3 kilograms	0.1 milligrams per litre and 15 kilograms annual quantity	
3	Nitrogen	25 kilograms	5 milligrams per litre and 125 kilograms annual quantity	
4	Organic halogen compounds as adsorbable organically bound halogens (AOX)	2 kilograms halogen, calculated as organically bound chlorine	100 micrograms per litre and 10 kilograms annual quantity	
5	Metals and their compounds:			and
5.1	Mercury	20 grams	1 microgram	100 grams
5.2	Cadmium	100 grams	5 micrograms	500 grams
5.3	Chromium	500 grams	50 micrograms	2.5 kilograms
5.4	Nickel	500 grams	50 micrograms	2.5 kilograms
5.5	Lead	500 grams	50 micrograms	2.5 kilograms
5.6	Copper	1,000 grams of metal	100 micrograms per litre	5 kilograms annual quantity
6	Toxicity to fish	3,000 cubic metres of wastewater divided by T_F	$T_F = 2$	

T_F is the dilution factor at which the wastewater is no longer toxic in the fish test.

Appendix B: Glossary

Absolutely dry	also called "oven dry" (= b.d. bone dry)
AP	Waste paper
APHA	American Public Health Association
APMP	Alkaline peroxide mechanical pulp (special form of CTMP).
BLS	Black liquor solids.
Brightness	Measure of the "whiteness" of the paper, expressed as a percentage of "absolutely" white = 100% light reflection under blue light of a certain wavelength. Very bright papers have a brightness of 85 - 90%, newsprint around 60 - 65%.
Corrugating medium	Relatively strong paper for the corrugated central layer of corrugated board.
De-inking	Removal of printing inks from printed waste paper.
Dissolving pulp	High quality chemical pulp for further chemical processing (films, fibers, etc.)
EPA	The Environmental Protection Agency, USA.
"Fresh" pulp	"Virgin" pulp obtained from the various raw (wood, etc.) fibre materials (as opposed to waste paper pulp).
Lime burning kiln	Installation for chemical recovery in sulphate and soda pulp mills for the regeneration of burnt lime for the reprocessing of dissolving chemicals.
Mg and Ca bisulphite magnefite process	Chemical pulping process.

58

Refiner

Grinding machine for the pulping and fibrillation of fibrous raw materials. Mostly disc refiners with toothed or grooved counter-revolving discs at low adjustable distances.

Rigidity

Resistance to bending and folding of paper and board. Important for pressure loads where used for packaging purposes.

Shelf life

Generic term for paper and board which concerns resistance to light, moisture and resistance to heat.

Smelt-dissolving tank

Dissolving tank for ash from black liquor burning (sulfate process).

Strength

For paper and board, a non-scientific generic term for tensile strength (breaking length), bursting strength, folding coefficient, edge tearing resistance.

TRS

Total reduced sulphur, generic term for H_2S and organic compounds with bivalent sulphur, e.g. mercaptans. Usually have an unpleasant odour at the tiniest concentrations.

Yield

Ratio of quantity of product (absolutely dry) to quantity of raw material (absolutely dry) expressed as a percentage. Because in the trade, pulp is calculated at 90% dry matter content (10% moisture, called "air dry"), the yield can also be given on this basis. Vegetable raw materials contain from around 30% (for straw) to around 50% (for wood) pulp; yields are limited by this.

Trade and Industry

Textile Processing

59.   Textile Processing

<u>Contents</u>

1. Scope

1.1 Terminology

A **"textile mill"** is generally defined as an **industrial production plant** which processes **materials which can be spun**, such as fibres, threads, yarns, twines, fabrics, knitted fabrics, fleeces, felts, synthetic skins and such like.

The **"clothing"** industry further **processes** the majority of **products from the textile industry**, but this environmental brief only considers the "textile industry".

1.2 Raw material

The **textile industry** originally processed exclusively **natural**, and for the most part **indigenous raw materials obtained from both plants** and **animals** (plants: cotton, flax, sisal, ramie, jute; animals: wool, silk, hair). However, the **proportion of synthetic fibres** (regenerated cellulose fibres such as synthetic silk, viscose staple fibre from wood and cotton waste, and later fully synthetic fibres such as polyamide, polyacrylic and polyester fibres, the **raw material** for which is **petroleum**) of the total fibre demand is **on the increase** throughout the world. In **1990, the chemical fibre industry** covered some **45% of the worldwide demand for textile fibres**, standing at 42.9 million tonnes.

Chemical fibre manufacture and its attendant **environmental problems** are not the subject of this environmental brief as it is in fact **part of the chemical industry**.

1.3 Production stages

1.3.1 Fibre conditioning

All **natural fibres** are polluted with **extraneous matter** and **substances**, and first have to be rendered "spinnable" by **conditioning processes**, some of which are expensive. By far the **majority of natural fibres** are produced in **tropical** or **subtropical countries**, not the industrialised world. **This** is where **primary conditioning** is also carried out, involving, for example:

- cotton ginning
- degumming of sisal, hemp and flax
- unreeling and degumming of silk and
- washing of wool.

Natural fibres are now being produced on a large scale, i.e. in **"agrarian facto-ries"**, characterised by **single-crop agriculture**. The **problems** which inevitably ensue (uprooting of new farmland, terracing, prevention of reforestation by over-grazing etc., social problems) must be taken into account when **siting and planning such facilities**.

1.3.2 Spinning and yarn production

Yarns/threads/twines are manufactured in specialised **spinning mills** according to the raw material and intended future use: cotton or 3-cylinder spinning mills, worsted, woollen and bast fibre spinning mills etc., the first of these being the most commonly used type. Indeed, over the last two decades, it has become firmly en-trenched alongside the traditional OE rotor spinning process (OE = open end) as it produces considerably more cheaply, especially in the case of coarser yarns.

All spinning mills operate more or less on the basis of the **same process**: the **fibrous material** is (where necessary) further **cleaned, aligned** and, while being **stretched** and **rotated about its axis**, spun to a **thread**. Some of the yarn produced in this way is then twisted, i.e. two or more threads are combined by rotation to form a "twisted" yarn.

The **finished yarns** are today **supplied** to the subsequent processing stages in the form of so-called **cheeses**, i.e. bobbins weighing between 0.8 and 3.5 kg.

1.3.3 Weaving and knitting mills

Of these **textile production** techniques, **weaving** is by far the most important. It involves the production of a fabric from a set of threads aligned in one direction, called the "warp", by interlacing "weft" threads at right angles to it. **Considerable technical improvements** have been made to the looms used for this purpose in the last twenty years, resulting in a marked **increase in productivity**.

One **particular feature** of the **production of woven fabrics** is **sizing**. For cer-tain articles one of the two thread systems, the warp, must be protected by a kind of **glue coating**, which involves "sizing" the warp threads with a protective coating - e.g. **modified starch** or a **synthetic polymer** - by **immersion**.

Unlike woven goods, **knitted products** have only one thread system, i.e. the threads which are made into a mesh run diagonally or longitudinally. The items are produced on linear or circular knitting or hosiery machines or warp knitting ma-chines.

1.3.4 Textile finishing

The term **textile finishing** covers the **bleaching, dyeing, printing** and **stiffening of textile products** in the various processing stages (fibre, yarn, fabric, knits, finished items). **The purpose of finishing** is in every instance the **improvement of** the **serviceability** and **adaptation** of the products to meet the **ever-changing demands of fashion and function.**

Finishing processes can be categorised into **purely mechanical** and **wet processes.** The **liquid phase** for the latter type is primarily **water,** and - to a lesser extent - **solvents** and **liquefied ammonia gas.** Another important medium is **steam.** To **achieve the desired effects, a range of chemicals, dyes** and **chemical auxiliaries** are used.

Compared with textile production, **textile finishing mills** are generally **smaller,** offer a more **diverse** range of services and are **less automated.**

1.4 Mill sizes

1.4.1 Spinning mills

59

Spinning mill capacities are expressed in terms of **"spindles"** or **"rotors",** i.e. the **number** of these **units** installed. For purposes of comparison, **one rotor** equals **approximately 3 - 4 spindles.**

In the **Federal Republic of Germany, cotton spinning mills** have an average of **13,000 spindles** and **1,100 rotors** with **150 employees,** while the figures for **worsted/woollen mills** are some **9,500 spindles** and **120 employees.**

In **raw-material-producer countries mill sizes** generally **far exceed** the average values stated above (**up to 120,000 spindles/mill**).

1.4.2 Weaving and knitting mills

Weaving mill capacities are normally expressed in terms of the **number of looms.**

In the **Federal Republic of Germany** the average **mill size** is **46 looms** and **106 employees.** Weaving mill projects in **raw material producer countries** are mainly for **far larger units** with up to 2,000 looms/mill.

In **knitting mills**, the **number** of installed production units gives **no** direct **clue** to **production capacity** because, in addition to **piece goods sold by the metre**, a wide **range of finished goods** (outer wear and underwear, hosiery etc.) is manufactured.

Capacities are therefore expressed in a number of ways according to product type: in **tonnes** (yard ware), **1,000 units** (outer wear and underwear), **1,000 pairs** (hosiery) or **1,000 m²** (curtains). In 1991, the average mill size in the Federal Republic of Germany was 91 employees.

1.4.3 Textile finishing

Depending on the type of products finished, **finishing capacities** are expressed in **t/year** (fibre, yarn, knitted goods), **million m²/year** (fabric) or **1,000 items/year** (ready to wear articles).

In the **Federal Republic of Germany**, the **average production output** for the total of 320 mills in the sector was calculated to be **2,500 t finished goods/mill and year** in 1990 with an average of **116 employees** per workshop.

Large mills in the USA and in Asia can have **capacities of up to 50,000 t/year**.

1.5 Site issues

In textile industry projects particular attention must be paid in all cases to the **large quantities of water** required for the finishing operations, which is why appropriate **disposal facilities** must naturally be provided for the **process water**.

With increased automation the **space required** by textile mills has **declined**. **New mills** have a **more compact structure** for **shortening transport distances**, among other things. The **land requirement** for a medium-sized vertical mill, including yard and access route areas and infrastructure systems with expansion options, is **some 60,000 m²**.

2. Environmental Impacts and Protective Measures

2.1 Fibre conditioning

Before they are processed, **all natural fibres** must be **conditioned**. This involves the **removal** of all manner of **extraneous matter**.

The **by-products of conditioning** (cottonseed and linseed, wool fat, silk gum) are **in some cases commercially viable substances,** and **dry waste** can in theory be returned in the soil as **fertiliser**, although another **method of treating** it is to **compost** it. A large proportion of the "waste" produced when native fibres are cleaned may therefore be regarded as **commercial commodities**, although all this requires **additional treatment**.

With regard to **emissions**, only **cotton ginning** and **raw wool washing** are of **special significance**.

Only about **one third of the weight** of a **cotton boll** is in the form of **spinnable fibres**. When the cotton is **ginned** (cleaned) - an operation normally carried out in situ - **cotton seeds** are obtained as a **by-product**, from which **cotton seed oil** and **meal** are produced. Another **by-product** is known as **linters,** which are used amongst other things in the **manufacture of synthetic silk and viscose (spun rayon)**. The actual **husks** and **waste** constitute some **15% of the weight** of the cotton boll and can be **ploughed back into the soil**.

Ginning is a **dry, mechanical process** which **generates considerable noise and quantities of dust**. While the latter of these two problems can be greatly relieved in modern mills by the **use of extractor and filter systems, the wearing of ear protection** is absolutely **essential** to combat the former.

Emissions from **raw wool washing** are **far more problematic**. In contrast to cotton ginning, **raw wool is washed centrally in large industrial facilities** remote from the place where the wool is obtained. The operation yields between **300 and 600 g attendant material** per kg of washed wool. In addition to the **wool fat which is a saleable commodity** per se, being used for technical and cosmetic purposes, **biocides** and the like from sheep's wool are present in the washing water. **Raw wool washing** may therefore be regarded as one of the **major wastewater pollution problems in the textile industry**.

Today, **wool fat** is generally **extracted before the wastewater is discharged**.

A further factor to be considered is that **complex purification plants** are used to **condition** the still **highly polluted wastewater** (COD around 15,000) **so that it can be discharged into the drains** (on the subject of wastewater purification, see also the environmental brief Mechanical Engineering, Workshops, Shipyards and Wastewater Disposal).

2.2   Spinning and yarn production

Because of the **high spindle speeds** reached on new machines (ring spindles up to 20,000 rpm, rotor up to 110,000 rpm), **spinning mills** can generally be assumed to **generate a great deal of noise**.
Noise levels of 70 to 100 dB(A) are commonly recorded in **work rooms**.

As the **spinning process** calls for a specific **room climate**, i.e. **temperature** and **relative humidity,** which must be as **constant** as possible and not affected by time of day or night, or season, practically **all spinning mills** today are fitted with **powerful air-conditioning systems**. To **limit the cost** of **air-conditioning**, the **production plant** must be **well insulated against** external **temperature changes**. In the **sixties and seventies** this led to a **windowless architecture** for textile mills with **extremely good values for insulation** against **temperature variations** and against **noise**.

Since the **eighties, windowless designs have been frowned upon** by the building authorities in the light of the **physiological and psychological problems caused to employees**. **To improve workplace design fields of vision** of around 2 to 5% of the floor area of the workrooms must be provided, and **special glazing** is needed for this.

The **high mechanical stress on fibres** during the **spinning process** results in the production of **considerable quantities of dust**, which must be carefully **extracted** for **industrial safety reasons** and in order to **keep the product clean**. **Emissions** can be **prevented** and **dust extracted** by means of special **machine enclosures** and **extraction systems** and via the **air-conditioning system** which keeps the **air** in the rooms **circulating**. **Air is not reintroduced** until it has been passed through **automatic filter installations**. The **filter dust** is **not dangerous** and its **disposal** is therefore **not a problem**.

2.3   Weaving and knitting

Although considerable progress has been made in the **weaving sector** over the last twenty years, the **whole area of noise nuisance** and, closely associated with it, **vibrations** coming from looms, cause **major problems**.

Noise levels of **85 to 107 dB(A)** must be expected in **weaving rooms**, according to the design, type, fitting, erection and number of looms used, fabric structure, building type and size etc. The **vibrations** transmitted from the running looms to the building can, under certain circumstances, cause a **nuisance to the local population** and **damage to nearby buildings,** and to **avoid** this special **vibration absorbers** are now provided. Generally speaking, the **comments** made about **noise** and **dust emissions** with regard to **spinning mills** apply here too.

The **sizing of warp threads** in weaving mills gives rise to **emission problems**.

Natural substances such as starch and cellulose products and **synthetic products** such as polyvinyl alcohol, acrylates, PVC, oils and fats are used as **sizing material**.

The **evaporation fumes** produced during the **drying stage** consist mainly of **steam**. The small amounts of **sizing agents** contained are **not regarded as environmental pollutants**.

On the other hand the **sizing liquor** may **no longer** be **discharged into wastewater** in **Germany** as it is a "concentrate" according to *Anhang 38* [Appendix 38] to the *Rahmen-Abwasserverwaltungsvorschrift* [General Administrative Framework Regulation on Wastewater].

However, the **sizing coating** itself has **much more serious repercussions** in the subsequent **textile finishing** stage where the size first has to be completely removed. In **finishing works** which handle primarily woven goods, up to **half** the **wastewater pollution** may derive from **dissolved sizes** during washing.

There is currently a trend in the **USA** towards the use of **sizing agents which can be recycled**; this aids the work of the **finisher** in terms of **disposal**, and also saves up to 90% of sizing agent due to **recycling**. In **Europe**, where the textile industry, in contrast to the USA, is largely horizontally structured, this trend **has yet become established**. From the point of view of recycling, which is attracting increasing economic interest, this could provide a certain **environmental benefit** in the woven products sector.

Emissions in the **knitting industry** are substantially **lower than in the weaving sector**, with **noise levels** of around **77 to 90 dB(A)**, and **dust** and **vibration emissions** are a **rarity**. **More problematic** are the **slip agents** applied to the yarn which, as in the case of the warp sizes, only appear at the **textile finishing** stage, where they **pollute** either the **wastewater** (from washing) or the **waste air** (in thermofixing).

2.4   Textile finishing

In contrast to textile production, **noise** plays only a very **minor role in textile finishing**. On the other hand **odour emissions** are generated **from drying and thermofixing processes**, and particularly **wastewater pollutants**, which transfer into the water during cleaning and the various textile finishing processes. The **textile finishing industry** both **consumes relatively large quantities of water** and **generates large quantities of wastewater**.

Only **general comments** can be made about **water consumption** in the **finishing stage** (see table 1), as this is determined not only by the type of fibres processed, but also by the article, type and extent of finishing, as well as the technology applied (continuous/discontinuous process) and batch size structure. For the **textile finishing industry** in the **Federal Republic of Germany** the latest figures of the TVI-Verband for 1988 set the specific **water consumption at 120 l/kg goods** (3).

59

Table 1

**Water consumption in the textile industry
(amended per (1))**

	Fibre type/make-up	Mean water consumption in l/kg material
a)	by fibre type	
	cotton	50 - 120
	wool	75 - 250
	synthetic fibres	10 - 100
b)	by make-up	
	flock/yarn	100 - 200
	knit	80 - 120
	printing	0 - 400

2.4.1 Wastewater contamination

If **textile finishing mills** are sited in the catchment area of efficient **municipal sewage works**, textile wastewater should preferably undergo **mechanical biological treatment** before being fed to these works and discharged into the drains (indirect discharge). If this option is not provided, and the **wastewater** therefore has to be discharged **directly** into the **drains**, it must first be **treated** in **the mill's own treatment plant** to meet legal requirements (direct discharge). (For more detailed information on the requirements applicable in Germany, see 3.3).

Although **most** of the substances in the wastewater are **biodegradable**, discharges into open drains can in some circumstances, during the biological decomposition stage, **reduce** the **oxygen content** of the drain water to below the level required for a healthy water quality and lead to **fouling of the water**.

The **textile finishing** industry also uses a range of **compounds** which are **not biodegradable** per se.

Below, we look briefly at the following **types of wastewater pollution**:

For **dyes** and many **surfactants which do not degrade readily** (but which are increasingly being replaced by more easily biodegradable ones), the said **mechanical biological treatment is inadequate if there is insufficient biomass to absorb the dyes** (the bonding of excess sludge to bacterial protein is the principle method by which water-soluble dyes are eliminated, see also 3.3). Past experience has shown that a **combination of physico-chemical** and **biological wastewater treatment** is required in most cases to achieve a **satisfactory treatment** level, with a facility for **treating heavily polluted partial flows** (e.g. dye liquor) **separately**.

- **Settleable solids**

The values normally found in the wastewater relating to **settleable solids** are subject to **enormous fluctuations**; they are dependent on a **number of factors** such as finishing process, fibre type, fibre make-up and whether batch or continuous treatments are used. **Sometimes the undissolved substances** remain **in suspension** and cannot simply be filtered off. Most **values** should be **below 50 ml/l**.

- **Heavy metals**

The **pollution of textile wastewater** with **heavy metals is limited by the current state-of-the-art**. **Cadmium** is hardly **ever** present and **mercury** is present only insofar as it is introduced via soda lye and hydrochloric acid (manufactured with **mercury electrodes**). **Chromium, cobalt and copper** may penetrate the **wastewater** from a number of **dyeing processes**, and **zinc** from **more sophisticated finishing processes** (wash and wear cotton articles) (zinc in the lower milligram range, the others mostly under 1 mg/l).

- **Hydrocarbons**

Hydrocarbons are **more serious pollutants**. They derive mainly from **yarns** which are **coated with oil** to give them the right **slip properties**, and to a lesser extent from the **residues of sizing agents** (impregnation).

59

- ### Organic halogen compounds

Other **serious pollutants** are **chloro-organic compounds**. The **AOX total parameter** (adsorbable organic halogen compounds) introduced for wastewater covers a **spectrum of substances** which are **hardly comparable** in terms of their **ecological** and **toxicological properties** (highly volatile chlorinated hydrocarbons, PVC, non-toxic green pigments, toxic chlorophenols etc.).

Sources for the **AOX total parameter** are primarily **chlorinated bleaches, anti-felting finishing** of wool, **dye accelerators (carriers)** used to **dye synthetic fibres, chlorinated reactive dyes** and **solvent soaps** with solvents from the chlorinated hydrocarbon range (per), which are incidentally also used as the sole "dry cleaning" agent for **degreasing polyester articles**.

The **number** of **compounds used which are regarded as particularly environmentally pollutant** has already **fallen considerably**. Whereas up to 5% **carrier** (colour accelerators for polyester fibres), based on product weight, was previously used, dyeing is now carried out mainly in **high-pressure installations** where only around 0.5% carrier is needed, basically as a precautionary measure, and even then only **aromatic ester-based** substances are used.

Pentachlorophenol (PCP), formerly used occasionally as a **preservative for heavy fabrics**, has been banned since 1986. However, products of a similar composition are used with **native fibres**, frequently from the range of **pesticides** manufactured; examples of such products are chlorinated phenoxyacetic acids, hexachlorocyclohexane, DDT and allied substances.

- ### Surfactants/detergents

An equally serious pollution problem is posed by **surface-active agents**, called **surfactants** or **detergents**, which are used as **washing, emulsification** and **wetting agents**, as **adjustment agents for dyeing processes**, as **auxiliaries to improve smoothness and softness** and for a range of other purposes, and to a far larger extent as dyes, and which in some cases are **not fully biodegradable** either. However, in the Federal Republic of Germany and other central European countries, a **minimum 80% degradability** under the conditions prevailing in treatment plants is required for products used specifically as **washing agents**. Since this ban has been imposed there has been a definite **improvement** in **water quality**.

Surfactant water pollution is due not only to its organic load, but also its **surface-active action**, which on the one hand **hampers the self-purifying capacity of rivers** and on the other **causes problems for the micro flora and fauna, and fish**.

- **Colour**

Water-soluble dyes are a further **environmental pollutant specific to the textile finishing** industry. If **heavily coloured** - something which cannot generally be determined in textile effluent after biological treatment - the **light reaching plants is reduced.** If wastewater from **dyeing plants** constitutes **20% or less of the municipal wastewater** to which it is **added**, a mechanical biological treatment plant is generally able to bind this **dye content** to the excess sludge (by a sort of dyeing process) and then **degrade** it in the digestion tower.

Where this **is not the case**, substantial quantities of dye may pass through the treatment plant into the outlet and cause a perceptible **coloration** of the **drain water.**

Now that certain **limit values** for the **density of colour** of **textile wastewater** have been specified in *Anhang 38* (appendix 38) of the *Rahmen-Abwasser-VwV* [General Administrative Framework Regulation on Wastewater], this wastewater now **requires** additional **decolouring** in some cases.

A partially anaerobic treatment stage, comprising the **addition** of ferrous (II) salts in conjunction with lime, **active carbon biology** and, more recently, processes using **membrane technology** (ultra-filtration, reverse osmosis) have proved to be **effective processes.** In special cases, **partial dye recovery** and **process water recycling** are possible in conjunction with **membrane technology.**

59

- **Water temperature**

Wastewater temperature is another **major** form of **pollution.** In dyeing processes, so much hot water is discharged that, in the absence of any counter-measures, **total wastewater temperatures** may exceed **40°C**, even though **35°C** is the maximum **permissible** temperature. In many cases this heat can be **recovered in heat exchangers**, then **returned to the process.**

- **pH**

A **pH of between 6 and 9** is prescribed for wastewater discharged from treatment plants in the Federal Republic of Germany, as in most other European countries.

Because **partially acidic** and **partially alkaline wastewater** is produced in the mill, according to the treatment stage and process, a **balancing tank** to hold around 50% of the daily wastewater quantity is normally required by the approving authorities. Following a **partial mutual neutralisation** of the various flows, the wastewater is **routed** from here to the **wastewater plant** at a **fairly balanced pH** and at a **constant quantity.**

While wastewater from **wool processing plants** generally has **excess acidity**, which requires to be buffered with alkali, the **pH** of **wastewater from cotton processing plants** is usually in the **alkaline range**. In this case the simplest, and at the same time, most environmentally friendly method of **neutralising** the wastewater involves the use of **flue gas**.

- **Major incidents**

In principle **major incidents** may only be expected as a result of **negligence**, and a useful preventive measure is to appoint an **"industrial water pollution control officer"**.

In some of the **German Federal states**, *Abwasser-Eigenüberwachungsverordnungen* [*AbwEV* - wastewater self-monitoring ordinances] have been issued which oblige businesses in the **textile finishing industry** to **fit control and measuring installations** to **keep an internal record** of certain wastewater parameters and to **report major incidents** (27).

2.4.2   Gas and steam emissions

Gas and steam emissions are generated by **textile finishing processes** when **fumes** penetrate the exhaust air during the **dyeing and drying** operations, although these more general operations do not present **any real environmental pollution hazard**. **Table 2** provides an **overview** of the main sources of **exhaust air emissions in textile mills**.

Exhaust air from the **thermofixing** of synthetic fibre articles is not only **more unpleasant** but also **more noxious**, for it entrains **oligomers of fibres** and **fragments** of **smoothing agents** (including ethylene oxide) which may constitute up to 0.2% of the weight of the goods. **Heat recovery installations** - which are a must in all cases for energy reasons - **arrest** a **considerable proportion** in the form of a **fatty condensate**, but these substances nonetheless penetrate the **wastewater** during the **cleaning operation** (high-pressure cleaners).

Formaldehyde, which makes the eyes water and irritates the skin, may be produced in connection with the **high-grade finishing of cotton articles**, but **formaldehyde pollution** has been **dramatically reduced** with the introduction of modern, low-formaldehyde, etherified products, which were also required due to the effects on pregnant women.

In **Europe**, an increasing number of plants are resorting to **thermal** and/or **catalytic afterburning** to **treat exhaust air from tenters**, and all organic substances are therefore combusted to form CO_2, CO and NO_x.

A further **source of gas and fume emissions** are **coating installations**, which yield **solvents**. There is a simple **remedy** - provided that **chlorohydrocarbons** are **not** used - that of **elimination** via the **combustion air** in the **boiler plant**.

In Germany, the requirements of the *2. BImSchV* [Second Ordinance on the Implementation of the Federal Immission Control Act] apply to **all the above-mentioned emissions**. The **plants** concerned must be **monitored** on the basis of **emission measurements**.

2.4.3 Noise emissions

No significant noise is emitted in the **textile finishing industry** other than from the ventilation units commonly found elsewhere.

2.5   General impacts on the environment

In addition to **textile-specific emissions** which are the real object of this brief, forms of **environmental pollution** which are also found in many other branches of industry can occur.

- **Furnace installations**

Thermal energy consumption is relatively **high,** particularly in the **finishing stage** (~ 13 kWh/kg goods). The **firing power** from the boiler plant required for process heat (steam, hot water) and for space heating usually lies within the range of **6 - 10 MW (9 - 15 t steam/h)**. Moreover, because of the high energy efficiency in plants which **need power and heat energy simultaneously, power/heat couplings** are used. The environmental brief Thermal Power Stations contains further environmental information about these installations.

- **Water treatment plants**

A certain **quality of process water** (i.e. it must not contain iron or manganese, it must be not very hard but must be clear) is required for **textile finishing** (washing and dyeing processes) - a quality which is **not often attained in surface, spring or tap water**. The **rinsing wastewater** deriving from **regeneration in treatment plants** generally has a **high salt content** and must be fed to **partial or complete treatment plants with the process wastewater**.

- **Traffic**

Goods traffic: The **large material turnover** results in a **constant stream of goods traffic** for **supplying raw materials, taking away finished goods** and for **transport within the plants** from one processing stage to another.

Passenger traffic: Textile factories often operate a **two to three shift** system, and times of **shift changes** can result in **traffic jams and other problems**.

Reference is made to the environmental briefs Planning of Locations for Trade and Industry, and Transport and Traffic Planning, for information on environmental impacts and environmental protection measures.

- **Socio-economic and socio-cultural factors**

These days, **plants for cloth manufacture**, i.e. the textile production stages of spinning, weaving, knitting and finishing, are **capital intensive operations**. (In contrast to subsequent processing in the garment industry, where the wage bill accounts for a high proportion of costs.)

The **high capital input** means that the **machines** must **run**, especially in the spinning and weaving sector, in **3-shift** and sometimes even **4-shift operation** around the clock, and in many areas at the **weekend** too. In an **average vertical operation**, i.e. with a spinning mill, weaving mill and finishing section, and with a **production of about 6 million running m/year**, about **300 people** are now employed in **three shifts** in **industrialised countries**.

The **proportion** of **women** employed has **declined sharply**, but **traditional family structures** are nonetheless **greatly affected** by **multiple shift operation**. Furthermore, the **personnel structure** has been moving in the direction of trained industrial workers, and **constant further training** is necessary even for **supervisory personnel**.

The **legislative framework** and **options for the enforcement of provisions** in individual countries have a **substantial influence** on the **impacts a textile mill** has on the **environment**.

On the one hand there are **statutory regulations** and their **enforcement** for **clean water, air** and **soil** and for the **rational use of energy**, and on the other there are regulations to meet the requirements of the employees in terms of **working conditions**. Due mention should also be made of environmental and industrial safety provisions which are inadequate or simply do not exist, excessive working hours, low pay levels and child labour. These **factors** all have an influence on the **quality of life** of those directly affected, and the economic situation of **the textile industry** generally.

3.   Notes on the Analysis and Evaluation of Environmental Impacts

3.1   Air pollution control

Over a long period the **fine dust in cotton mills** can trigger **damage to the respiratory tract** and the **lungs,** resulting in a disease called byssinosis. For this reason **industrial countries** prescribe **maximum dust concentrations,** but these vary from country to country and are also measured and assessed by different methods (maximum occupational limits - MAK values) in the FRG and the OSHA rules in the USA), e.g.:

-     Germany, Switzerland   :   1.5 mg per m^3 total dust content

-     USA, Australia   :   0.5 mg per m^3 fine dust
      Netherlands   :   15 μm in the preliminary works
      0.2 mg per m^3 fine dust in the spinning mill

-     Great Britain   :   0.5 mg per m^3 total dust
      Sweden   :   without fibres.

In **Germany,** the TA-Luft [Technical Instructions on Air Quality Control] (2) applies to the **operation of furnace plants,** and more recently also to drying plants (tenters), i.e. there is an **approval obligation.** Similar regulations apply in some other European countries too.

    For other mostly **gaseous substances** which constitute **a health hazard,** the **occupational limits** of the Deutsche Forschungsgemeinschaft (German Research Foundation) are applicable.

59

Table 2

Main sources of waste air emissions in textile finishing operations

Process/make-up Substrate	Substances emitted	Comment/ counter-measure
Drying and thermofixing of textured synthetic products in their washed state	Mineral oil components from vaporising flushing oils	For washed goods, approx. 0.3% of the weight of the goods, due to the residual fat content of the products. Discharge purification for high loads from: cooling and aerosol separation, air scrubbing or thermal afterburning.
Disperse dyeing at atmospheric pressure	Carriers (aromatic halogen compounds)	Move to closed HT plant and thermosol process (without carrier) usually possible. Can be removed by powerful scrubbers or thermal afterburning.
Drying and fixing after printing	Heavy benzene components from emulsion concentrations of the pigment print process	Largely forced out by benzene-free swelling agents. Removal by activated carbon filters or thermal afterburning.
Drying and fixing after dyeing and after a water-repellent finish	Fumes and smells (cationic softeners, some dyes etc.), paraffins	Particularly where high temperatures are used (thermosol process), organic substances may evaporate or sublimate. Can be partially removed by air cleaning with chemical additives which have an absorptive action.
Drying after a synthetic resin finish (melamine, amino-formaldehyde preliminary condensates)	Formaldehyde	Physiologically harmful, but hardly a problem these days thanks to low-formaldehyde wetting agents. Only used for a few articles.
Drying after solvent treatment	Solvents included	In pre-cleaning, secondary cleaning and print post-treatment processes. To be removed by condensation and activated carbon filters.

3.2 Noise protection

Noise is a serious problem in spinning and weaving mills. As mentioned earlier in section 2.2, sound levels of 70 to 110 dB(A) are commonplace.

In Germany, guideline no. 2572 of the German Association of Engineers VDI (Geräusche von Textilmaschinen), must generally be complied with (25).

Today, to **prevent any adverse effects on health, individual hearing protection** (ear plugs, ear muffs) **must be provided from 85 dB(A) and are essential from 90 dB(A)**, according to the ordinance on workplaces *Arbeitsstätten-Richtlinien* and VDI guideline 2058, sheet 2.

Because of **building insulation,** which is standard and indeed compulsory (VDI 2571) today, **external sound propagation** is relatively **low. Where other facilities are close by, DIN standard 18005** applies in Germany (23). The **minimum distance** of textile manufacturing plants from **residential accommodation** is established in North Rhine Westphalia in the so-called *Abstandserlaß* [distance decree] of the Minister for Employment, Health and Social Affairs (Minister für Arbeit, Gesundheit und Soziales - 26).

3.3 Water pollution control

- **Wastewater pollution parameters**

The **substances found in wastewater** are classified by the **proportions which can be precipitated** and those which **can be oxidised biologically or chemically. Chloro-organic compounds and toxic substances** are also taken into account to a certain degree, e.g. some **heavy metals** used in textile finishing.

Anhang 38 [Appendix 38] of the *Allgemeinen Rahmen-Abwasser-VwV* [General Administrative Framework Regulation on Wastewater] for textile production and finishing (5) applies to the **textile processing industry** in the **German Federal Republic.** It contains a range of **wastewater requirements** throughout the treatment plant which may **not** be achieved by **dilution or mixing.** The main **contamination parameters** are described briefly below.

For **indirect discharges, state-of-the-art requirements** apply, and for **direct discharges** the **requirements of the generally accepted codes of practice** also apply.

The draft of the said Anhang 38 of the Allgemeinen Rahmen-Abwasser-VwV specifies, for example, (6) the following **requirements for wastewater discharged from textile mills** (with the exception of water from raw wool washing, cooling systems, stencil plate production and chemical cleaning (dry cleaning) which are covered by special provisions).

- requirements according to the **generally accepted codes of practice** (additional for direct discharges)

COD	160 mg/l	ammonium nitrogen	10 mg/l
BOD$_5$	25 mg/l	aluminium	3 mg/l
iron	3 mg/l	total phosphorus	2 mg/l

- requirements per the **state-of-the-art** (for direct and indirect discharges)

Cu	0.5 mg/l	AOX	0.5 mg/l
Cr VI	0.5 mg/l	HHHC	0.1 mg/l
Ni	0.5 mg/l	free chlorine	0.3 mg/l
Pb	0.5 mg/l	HC	15.0 mg/l
Zn	2.0 mg/l	sulphide	1.0 mg/l
Sn	2.0 mg/l	sulphite	1.0 mg/l
TF	dilution factor of 2		

Colour:	yellow	436 nm	7m - 1
	red	525 nm	5m - 1
	blue	620 nm	3m - 1

- **General requirements**

temperature	max. 35°C at the point of discharge
pH	6 - 9 at the point of discharge
settleable solids	1.0 ml/l after 0.5 h settling time
odour	no unpleasant odours
colour	no visible discoloration of the wastewater

Abbreviations:

- COD   chemical oxygen demand
- BOD$_5$ biochemical oxygen demand in 5 days
- TF   toxicity to fish
- AOX   adsorbable organically bound halogen
- HHHC   highly volatile halogenated hydrocarbons
- HC   hydrocarbons
- EDTA   ethylenediaminetetraacetic acid
- NTA   nitrilotriacetic acid
- PVP   polyvinylpyrrolidon

The **following substances,** mixtures etc. may **never be discharged** as **wastewater** or **with wastewater**:

- chromium VI compounds from the oxidation of sulphide dyestuffs
- chloro-organic carriers
- halogen-organic solvents
- arsenic and mercury and their compounds from use as preservatives
- pollutant concentrates, such as residual sizes, residual high-grade finishing agents, print pastes, dye preparations, residues of chemicals used, textile auxiliaries and dyes from drums
- surfactants which do not meet the requirements of the WRMG [law relating to the environmental compatibility of washing and cleaning agents] (22).

All these **substances** must be **collected, recycled** where technically feasible or **disposed of correctly**.

For **partial wastewater flows,** particularly from the following departments: desizing, bleaching, printing, dyeing, finishing, coating and backing, together with central barrel and drum cleaning, **threshold values** apply above which **treatment is required**. If various processes are carried out consecutively in a single mechanical unit, the wastewater from each must be dealt with as a partial flow.

Threshold values for partial wastewater flows beyond which **treatment is required**:

AOX	3.0 mg/l	Cr	2.0 mg/l
HHHC	1.0 mg/l	Cr VI	0.5 mg/l
HC	50.0 mg/l	Zn	10.0 mg/l
Cu	2.0 mg/l	Sn	10.0 mg/l
Ni	2.0 mg/l		

The **AOX threshold value** for **bleaching with chlorine** to obtain a particular shade of white and the non-felting finishing of wool is 8 mg/l (up to 31.12.1996 max.).

All **analyses** and **measuring procedures** required to determine the said pollutants have now been **standardised** in Germany to DIN.

Table 3

**Measuring procedures for wastewater parameters
in the wastewater treatment plant**

Parameter	Method
- Filterable substances	DIN[1] 38409-H2-2/3(July 80 edition)
- Settleable substances	DIN 38409-H9-2(July 80 edition)
- Chemical oxygen demand - COD - of the precipitated sample	DIN 38409-H41(December 80 edition)
- Biochemical oxygen demand - BOD$_5$ of the precipitated sample	DEV[2] H5A2 (4th issue 1966) with additional restriction on nitrification at 0.5 mg/l
- Toxicity to fish as TF dilution factor of the unprecipitated sample	DIN 38412-L20(December 80 edition)
- Zinc, copper, chromium	DIN 38406-E21(September 80 edition)
- Nitrogen from ammonium compounds from the unprecipitated homogenised sample	DEV E5.2(7th issue 75)
- Effective chlorine from the filtered sample	DEV G4 1.b (7th issue 75, glassfibre filters)
- Sulphide, total, from the unprecipitated sample	DEV D 7b(7th issue 75)
- Sulphite, total, from the unprecipitate sample	DEV D 6.2(1st issue 60)
- Hydrocarbons from the unprecipitated homogenised sample	DIN 38409-H18(February 81 edition)

[1] DIN = German Standard
[2] DEV = Deutsches Einheitsverfahren (German standardisation procedure)

4.   Interaction with Other Sectors

While, on the **raw materials side**, the **textile industry** has close connections with **plant production** (natural fibres), the **fibre industry** (synthetic fibres) and the **chemical industry** (chemical, dyes, auxiliaries), its activity on the sales side is characterised by its interaction with the **clothing industry downline**.

Further references to relevant project areas are given in the text.

5. Summary Assessment of Environmental Relevance

In investment projects in the **textile industry** sector a range of **environmentally relevant criteria** must be taken into account at the **location planning stage**. In **raw material producer countries** in particular **special consideration** must be given to the **effects of material production**. The early and full **involvement** of the **population groups affected**, particularly **women** in some cases, can help **resolve any problems which may arise**.

Special **attention** must also be paid to the **environmental impacts** of **raw wool washing** and **textile finishing** plants. While in the former the problem is posed by the **considerable degree of wastewater contamination**, in textile finishing mill projects due account must be taken of the **high water and energy consumption**, the **wide use of chemicals**, the **process-specific pollution of wastewater and exhaust air** and the **disposal of waste**. In this regard, special **industrial environmental protection officers** must be appointed.

The **current state-of-the-art** in the processes, process installations and supply and disposal plants, together with **relevant laws** and their **enforcement**, combine to ensure that **thoroughly environmentally sound textile production** is possible at all stages of manufacture.

With regard to the **socio-economic environmental impacts** of textile projects mention should be made of the much **more stringent requirements** relating to **personnel qualification**. The **capital-intensive nature** of modern, extensively automated spinning, weaving and knitting mills is leading to the **maximisation of machine operating times**, thus multi-shift operation, usually 7 days per week, is the norm.

The whole area of socio-economic and socio-cultural aspects of this type of operation and its legislative framework must be looked at carefully.

6.   References

1.   Düring, G.: VDI-Berichte, Nr.310, 1978.

2.   Technische Anleitung zur Reinhaltung der Luft - TA-Luft of 27.02.86;
       Gemeinsames Ministerialblatt GMBI 1986.

3.   Year books for 1988 and 1990 of the Gesamtverband der deutschen
       Textilveredelungsindustrie, TVI-Verband EV.

4.   83. Allgemeine Verwaltungsvorschrift über Mindestanforderungen an das
       Einleiten von Abwasser in Gewässer (Textilherstellung) - 38.
       AbwasserVwV - of 5 September 1984, GMBI (joint ministerial circular)
       1984.

5.   Anhang 38 to the Allgemeinen Rahmenabwasserverwaltungsvorschrift für
       die Textilherstellung - Draft of 20 December 1990.

6.   Dr. Heimann, S.: Textilhilfsmittel und Umweltschutz; Melliand Textil-
       berichte, 7/1991.

7.   Natke, H.G., Thiede, R., Elmer, K.: Curt-Risch-Institut für Dynamik,
       Schall- und Meßtechnik, Universität Hannover: Untersuchungen der von
       Webereien ausgehenden Schwingungsemissionen und Hinweise zur
       Websaal-Bauplanung.   Verband der Nord-Westdeutschen Textilindustrie
       Münster; Zeitschriftenreihe, Heft 66, 1985.

8.   Trauter. R.: Rückgewinnung und Wiedereinsatz von Webschichten mittels
       dynamisch geformter Membranen; Chemiefasern/Textilindustrie 37/89,
       1987.

9.   DIN 38409-H14-H14: Deutsche Einheitsverfahren zur Wasser-, Abwasser-
       und Schlamm-Untersuchung: Bestimmung der adsorbierbaren organisch
       gebundenen Halogene (AOX).

10.   2. Verordnung zur Durchführung des Bundes-Immissionsschutzgesetzes
       (Verordnung zur Emissionsbegrenzung von leichtflüchtigen Halogenkohlen-
       wasserstoffen - 2 BImSchV of 21 April 1986); BGBI (Federal Law Gazette)
       1986, Part 1.

11.   Kolb, M., Funke, B.: Die Entfärbung von textilem Abwasser mit Fe(II) +
       $Ca(OH)_2$; Vom Wasser, Bd. 65, 1985.

12.   Wysocki, G.; Höke, B.: Chemie-Technik, 1974.

13. Oehme, Ch.: Trägerbiologien in der Abwassertechnik; Chem.-Ing.-Tech. 56, 1984.

14. Croissant, B.: Efferenn K.; Frahne, D.: Reaktivfarbstoffe im Abwasser - sind sie durch ein bakterielles Symbiosesystem abbaubar? Melliand Textilberichte, 1983.

15. Erlaß über Richtlinien für die Anforderungen an Abwasser bei Einleiten in öffentliche Abwasseranlagen of 28 June 1978.

16. Wiesner, J.; Jochen, E.: Energieverbrauch und Möglichkeiten rationeller Energienutzung in der Verarbeitenden Industrie in Baden-Württemberg: Textilindustrie, Informationen zur Energiepolitik, Heft 11c, Wirtschaftsministerium von Bad.-Württ., Stuttgart, 1978.

17. Reetz, H.: Europäische Abwasserregelungen im Vergleich; Melliand Textilberichte, 11/1991.

18. Christ, M.: Wärmerückgewinnung und Abluftreinigung bei der Textilveredlung; Textilpraxis International, 46/1991.

19. Änderung der 4. BImSchV Nr. 5.3 - Genehmigungspflicht für Anlagen der Textilveredlung und von Feuerungsanlagen.

20. "Wastewater purification" working party J. Janitza, S. Koscielski, M. Schnabel of the ITV (Institut für Textil- und Verfahrenstechnik Denkendorf) Behandlung von Textilabwässern im Betrieb; Textilpraxis International, November 1991.

21. Gesetz zur Ordnung des Wasserhaushaltes (Wasserhaushaltsgesetz - WHG) of 23.09.1986.

22. Gesetz über die Umweltverträglichkeit von Wasch- und Reinigungsmitteln of March 05, 1987.

23. DIN 18 005; Schallschutz im Städtebau, Planungsrichtlinien.

24. VDI 2058; Beurteilung von Arbeitslärm am Arbeitsplatz hinsichtlich Gehörschäden.

25. VDI 2571 + 2572; Schallabstrahlung von Industriebauten.

59

26. Abstandserlaß (RdErl. d. Ministers für Arbeit, Gesundheit und Soziales NW of March 09, 1982).

27. Verordnung zur Eigenüberwachung von Abwasser, Bayern (AbwEV) of December 09, 1990.

28. ITMF; International Textile Manufacturers Federation - International Textile Shipment Statistics, Vol. 14/1991.

29. World Bank; Environment Guidelines; Washington 1988, p.451 ff.